Graduate Texts in Physics

Graduate Texts in Physics publishes core learning/teaching material for graduate- and advanced-level undergraduate courses on topics of current and emerging fields within physics, both pure and applied. These textbooks serve students at the MS- or PhD-level and their instructors as comprehensive sources of principles, definitions, derivations, experiments and applications (as relevant) for their mastery and teaching, respectively. International in scope and relevance, the textbooks correspond to course syllabi sufficiently to serve as required reading. Their didactic style, comprehensiveness and coverage of fundamental material also make them suitable as introductions or references for scientists entering, or requiring timely knowledge of, a research field.

Gordon Walter Semenoff

Quantum Field Theory

An Introduction

 Springer

Gordon Walter Semenoff
Department of Physics and Astronomy
University of British Columbia
Vancouver, BC, Canada

ISSN 1868-4513   ISSN 1868-4521   (electronic)
Graduate Texts in Physics
ISBN 978-981-99-5409-4   ISBN 978-981-99-5410-0   (eBook)
https://doi.org/10.1007/978-981-99-5410-0

This Springer imprint is published by the registered company Springer Nature Singapore Pte Ltd.
The registered company address is: 152 Beach Road, #21-01/04 Gateway East, Singapore 189721,
Singapore

Paper in this product is recyclable.

I want to express my profound thanks to all those who gave me the time and space to write this book, particularly Eileen, Leonarda, Amanda, Victor and many thanks to Eric Hiob for help with the diagrams.

Contents

1	**Prologue**	1
2	**Many Particle Physics as a Quantum Field Theory**	5
	2.1 Introduction	5
	2.2 Non-relativistic Particles	6
	2.2.1 Identical and Indistinguishable Particles	7
	2.2.2 The Example of Weakly Interacting Particles	9
	2.2.3 Hamiltonian and Stationary States	11
	2.2.4 Particles with Spin	13
	2.3 Second Quantization in the Schrödinger Picture	15
	2.4 Second Quantization in the Heisenberg Picture	20
3	**Degenerate Fermi and Bose Gases**	23
	3.1 The Limit of Weakly Interacting Particles	23
	3.2 Degenerate Fermi Gas	28
	3.2.1 The Ground State $|\mathcal{O}>$	28
	3.2.2 Particles and Holes	30
	3.2.3 The Grand Canonical Free Energy	33
	3.3 Degenerate Bose Gas	35
	3.3.1 Landau's Criterion for Superfluidity	38
	3.3.2 Vacuum Expectation Value	40
	3.4 Spontaneous Symmetry Breaking	46
4	**The Action Principle and Noether's Theorem**	49
	4.1 The Action	51
	4.1.1 The Euler–Lagrange Equations	52
	4.2 Canonical Momenta, Poisson Brackets and Commutation Relations	55
	4.3 Noether's Theorem	57
	4.3.1 Conservation Laws and Continuity Equations	57
	4.3.2 Definition of Symmetry	58
	4.3.3 Examples of Symmetries	59
	4.3.4 Proof of Noether's Theorem	61
	4.4 Phase Symmetry and the Conservation of Particle Number	63

5   Non-relativistic Space–Time Symmetries 65
      5.1   Translation Invariance and the Stress Tensor 65
      5.2   Galilean Symmetry 68
      5.3   Scale Invariance 73
             5.3.1   Improving the Stress Tensor 76
             5.3.2   The Consequences of Scale Invariance 77
      5.4   Special Schrödinger Symmetry 79
      5.5   Summary 80

6   Space–Time Symmetry and Relativistic Field Theory 85
      6.1   Quantum Mechanics and Special Relativity 85
      6.2   Coordinates 92
      6.3   Scalars, Vectors, Tensors 95
      6.4   The Metric 96
      6.5   Symmetry of Space–Time 98
      6.6   The Symmetries of Minkowski Space 99
      6.7   Natural Units 102
      6.8   Relativistic Fields 103

7   The Real Scalar Quantum Field Theory 105
      7.1   Constructing a Relativistic Lagrangian Density 105
      7.2   Field Equation and Commutation Relations 107
      7.3   Noether's Theorem and Poincare Symmetry 108
      7.4   Correlation Functions of the Real Scalar Field 111
      7.5   The Free Scalar Field 112
      7.6   Consequences of Spacetime Symmetry 117
      7.7   Spectral Theorem 120
      7.8   Normalization of the Spectral Function 123
      7.9   Analyticity 123
             7.9.1   The Reeh–Schlieder Theorem 126
      7.10   Conformal Symmetry 130

8   Emergent Relativistic Symmetry 137
      8.1   Phonons 138
      8.2   The Debye Theory of Solids 143
      8.3   Relativistic Fermions in Graphene 147

9   The Dirac Field Theory 157
      9.1   The Dirac Equation 157
      9.2   Solving the Dirac Equation 161
      9.3   Lorentz Invariance of the Dirac Equation 165
      9.4   Spin of the Dirac Field 167
      9.5   Phase Symmetry and the Conservation of Charge 169
             9.5.1   Conserved Number Current 170
             9.5.2   Relativistic Noether's Theorem for the Dirac
                      Equation 171
             9.5.3   Alternative Proof of Noether's Theorem 172

9.6   Spacetime Symmetry 174
    9.6.1   Translation Invariance and the Stress Tensor 174
    9.6.2   Lorentz Transformations 176
    9.6.3   Stress Tensor and Killing Vectors 179

10   Photons 181
10.1   Relativistic Classical Electrodynamics 182
10.2   Quantization 187
    10.2.1   Negative Normed States 190
    10.2.2   Physical State Condition 190
    10.2.3   Null States and the Equivalence Relation 194
10.3   Space–Time Symmetries of the Photon 201
10.4   Massive Photon 203
10.5   Quantum Electrodynamics 207
    10.5.1   C, P and T 209

11   Functional Methods 211
11.1   Functional Derivative 212
11.2   Functional Integral 214
11.3   Generating Functional for Free Scalar Fields 218
    11.3.1   Wick's Theorem for Scalar Fields 222
    11.3.2   Generating Functional as a Functional Integral 222
11.4   The Interacting Real Scalar Field 225

12   More Functional Integrals 229
12.1   Functional Integrals for the Photon Field 230
12.2   Functional Methods for Fermions 236
12.3   Generating Functionals for Non-relativistic Fermions 242
    12.3.1   Interacting Non-relativistic Fermions 246
12.4   The Dirac Field 247
    12.4.1   2 Point Function for the Dirac Field 248
    12.4.2   Generating Functional for the Dirac Field 250
    12.4.3   Functional Integral for the Dirac Field 252
12.5   Functional Quantum Electrodynamics 253

13   The Weakly Coupled Real Scalar Field 257
13.1   Counterterms 263
13.2   Computation of the 2 Point Function 267
13.3   Feynman Diagrams 270
13.4   Simplifications of Feynman Diagrams 273
13.5   Computation of a One-Loop Feynman Integral 280
    13.5.1   Dimensional Regularization 281
    13.5.2   Wick Rotation 281
    13.5.3   Feynman Parameters 282
    13.5.4   Integration in 2ω-Dimensions 283
    13.5.5   Asymptotic Expansion at $2\omega \sim 4$ 284
    13.5.6   Inverse Wick Rotation 284

        13.5.7   The Mass Tadpole 285
        13.5.8   Euclidean Quantum Field Theory 286
        13.5.9   The 2 Point and 4 Point Functions 287
    13.6   Subtraction Schemes 288
    13.7   Renormalization Group 292
    13.8   Appendix: Integration Formulae 301
        13.8.1   Euler's Gamma Function 301
        13.8.2   Feynman Parameter Formula 303
        13.8.3   Dimensional Regularization Integral 305

14 **More Theory of the Real Scalar Field** 307
    14.1   The S Matrix 307
        14.1.1   The T Matrix 310
    14.2   The LSZ Formula 311
    14.3   Elastic Two-Particle Scattering 315
    14.4   Connected and Irreducible Generating Functionals 319
        14.4.1   Connected Correlation Functions
                  and the Linked Cluster Theorem 320
        14.4.2   Connected Correlation Functions 321
        14.4.3   Cancelation of Vacuum Diagrams 325
        14.4.4   Irreducible Correlation Function 325
    14.5   Derivation of the LSZ Formula 335

15 **Perturbative Quantum Electrodynamics** 339
    15.1   Counterterms 340
    15.2   The Generating Functional in Perturbation Theory 343
        15.2.1   Wick's Theorem for Photons and Electrons 344
    15.3   Feynman Diagrams 346
    15.4   Feynman Rules 350
    15.5   The Electron 2 Point Function 351
    15.6   Feynman Rules in Momentum Space 353
    15.7   The Photon 2 Point Function 357
    15.8   Quantum Corrections of the Coulomb Potential 365
    15.9   The Electron 2 Point Function 370
    15.10   Radiative Correction of the Vertex 373
        15.10.1   Electromagnetic form Factors 377
        15.10.2   Anomalous Magnetic Moment 384
    15.11   Photon Production, the Soft Photon Theorem 388
    15.12   Furry's Theorem 393
    15.13   The Ward–Takahashi Identities 395

16 **Epilogue** 399

Index 401

Prologue

1

It is often said that quantum field theory is the natural framework in which to describe a physical system which has a very large number of degrees of freedom and whose behaviour is governed by quantum mechanics. In this textbook, we will begin by taking this point of view. We will study some examples of quantum mechanical systems with many degrees of freedom and we will learn how those systems can be described by a quantum field theory. In the course of doing this, we will learn some things about the structure of a quantum field theory itself. We will learn that it is described by field equations and commutation relations for the fields. We will learn about how to exploit symmetries of the system that we describe in order to simplify our work. Also, having gone through the trouble of reformulating a quantum system in the language of quantum field theory, we will learn how to extract some information about the physical systems that we are studying from the quantum field theory itself.

It is also often said that quantum field theory is the natural marriage of quantum mechanics and the special theory of relativity. This has the implication that, once we assume that a system obeys the laws of quantum mechanics, and we further assume that it is governed by Einstein's special theory of relativity, it must be very natural to treat it as a quantum field theory. Some intuition for why this might be so comes from Einstein's famous formula

$$E = mc^2$$

and the notion that, in some sense, energy and mass are indistinguishable and one can be converted to the other. Indeed this sort of conversion is done all of the time in phenomena that are studied in nuclear and particle physics. For example, when two protons collide with sufficient energy, as well as the protons, a number of pions could also appear in the final state. The pions have been created by the dynamical process. Or, for example, when an electron and a positron collide they can annihilate, leaving behind energy in the form of a high energy photon, which can then re-convert

G. W. Semenoff, *Quantum Field Theory*, Graduate Texts in Physics,
https://doi.org/10.1007/978-981-99-5410-0_1

to matter by creating some other pair of particles, like a quark and an anti-quark. This tells us that, when we fashion a model to describe such a relativistic quantum mechanical system, we must be flexible enough to allow for the possibility that the numbers of various kinds of particles will change when particles interact with each other. The quantum mechanical system containing only one, or only a fixed number of particles is simply not good enough.

Particle production aside, there is a deeper need for quantum field theory in relativistic quantum mechanics. This need is a little more abstract and it will be explained in detail in Sect. 6.1. Put briefly, in the quantum mechanics of a single particle, many events which could never happen in the classical world become possible. Quantum tunnelling through barriers is an example.

It turns out that, even in a theory whose classical version obeys all of the tenets of the special theory of relativity where the speed of light is an absolute speed limit, when the single particle theory is quantized, there is a small probability of propagation of the particle at speeds greater than that of light. This would be fine if nature behaved that way. However, faster than light particles are not observed in nature and we need to model them with theories which do not have faster than light propagation. As we shall see, relativistic quantum field theory gives us a way to do this.

In the spirit of the paragraphs above, we shall take the simplest possible path to an introduction to quantum field theory. We will begin with the second quantization of a system of many identical non-relativistic particles and we will cast the quantum mechanical many-particle system as a non-relativistic quantum field theory. In spite of its simplicity, it will have practically all of the features of a generic quantum field theory. We will exploit this fact to gain some intuition for how a quantum field theory works, particularly what it means to "solve the theory" and how to use the solution to extract some of the physical properties of the system that it describes. The examples that we focus on are the degenerate Fermi and Bose gases.

Then, once we are armed with our few concrete examples of quantum field theories we will develop some formalism which will help us later on when we encounter more abstract theories. This includes the introduction of classical field theory as an intermediate step in the logic of formulating the quantum field theory. The equations of motion of the classical field theory can be derived from an action and a Lagrangian density by using a variational principle. The existence of a Lagrangian formulation also gives us a way to formulate the concept of symmetry, to study the consequences of symmetry and to understand how symmetries are related to conservation laws. To the extent that the equations of motion, symmetries and conservation laws can be pulled back to the quantum field theory, this approach gives us some nontrivial information about it. Moreover, the classical action that we use for this turns out to be an important ingredient in the functional integral formulation of the quantum field theory. Later on, we will reformulate the description quantum field theory in terms of functional integrals and the action of the classical field theory plays an important role there.

After obtaining this solid footing with our non-relativistic quantum field theory, we will begin a discussion of relativistic field theories We will take the point of view that special relativity is simply a consequence of symmetry, specifically, the symmetry of spacetime. We will go through a detailed discussion of what we mean by this. Then we will develop some examples of quantum field theory by simply trying to formulate dynamical systems where the degrees of freedom are fields and where the dynamics respect the symmetries of the spacetime in which they are embedded.

Both historically, and in our context too, the logically clearest step from non-relativistic to relativistic field theory is in the context of a Fermi gas. The field equations of the non-relativistic field theory that describes a many-fermion system contains the Schroedinger wave operator

$$i\hbar\frac{\partial}{\partial t} + \frac{\hbar^2}{2m}\vec{\nabla}^2 + \mu$$

We can lift our non-relativistic quantum field theory to the relativistic setting by removing this Schrödinger wave operator from the field equation and replacing it with the Dirac wave operator

$$i\hbar\frac{\partial}{\partial t} + \frac{\hbar^2}{2m}\vec{\nabla}^2 + \mu \quad \rightarrow \quad i\hbar\frac{\partial}{\partial t} - i\hbar c\gamma^0\vec{\gamma}\cdot\vec{\nabla} - i\gamma^0 mc^2$$

The result is the electron theory which describes relativistic fermions and anti-fermions. We will study its properties in some detail. Then we will study a few other simple examples of relativistic fields, a scalar field and a massless vector field, which we will use to model the photon.

We will then spend some time on unraveling the complicated but very important question of interactions in quantum field theory, both in the non-relativistic and in the relativistic context. The only tool that we will develop in any detail is renormalized perturbation theory. Of course, this entails a discussion of renormalizability and renormalization. We will not prove renormalizability. We leave that to more advanced textbooks, of which there are many. But we will use renormalized perturbation theory to do some simple computations. We will also discuss some of the physical implications of the results of those computations.

We will end up with a rather elementary study of some of the properties of quantum electrodynamics. The absolute beauty of quantum electrodynamics as a physical theory of electrons and photons and their interactions cannot be overemphasized. We will write down a simple Lagrangian density,

$$\mathcal{L}(x) = -\frac{1}{4}F_{\mu\nu}F^{\mu\nu} - i\bar{\psi}(\slashed{\partial} - ie\slashed{A} + m)\psi$$

The terms that we write in this Lagrangian density are the only ones which are compatible with the symmetries of the theory and the criterion of renormalizability. This means that, when we use it to model the interactions of electrons, positrons and photons, for example, there are only a few parameters—e and m which appear explicitly—and a few possible re-scalings of the fields. Once those few parameters are fit to experimental data, everything else that we would ever use this theory to compute is a prediction of the behaviour of electrons and photons. Such predictions have been made to astounding accuracy and extremely precise experiments have been done and the two agree with each other to the limit of experimental accuracy, which is currently about eight decimal places. This is, at time of writing, the most precise agreement of theory and experiment in science. In the last part of this book we will learn how this all follows from the simple Lagrangian density written above.

There are many, many things about quantum field theory that we will not get to in this book. Our purpose here is to be introductory, not comprehensive. An attempt has also been made to be logical rather than historical. The guideline that has been used for the scope of the book is that the material should be presentable at a uniform level of pedagogy in a one-semester graduate level university class.

Many Particle Physics as a Quantum Field Theory

2

2.1 Introduction

In this chapter we will attempt to develop intuition aimed at answering the question "What is a quantum field theory?". We will do this by studying a system with many identical particles. For now, we will assume that the particles are non-relativistic in that their velocities are much less than the speed of light. The generalization to relativistic particles and relativistic quantum field theory will be discussed in later chapters. We will assume that the problem in front of us is quantum mechanical, that is, that we want to find a solution of the Schrödinger equation for the system as a whole and then use that solution, the wave function, to answer questions about the physical state of the system.

In order to describe the quantum mechanical problem for a large number of particles in an elegant way, we will develop a procedure which is called "second quantization". Second quantization presents the problem of analyzing such a system as a quantum field theory. This will be our first example of a quantum field theory. We will devote significant space in subsequent chapters to the analysis of this non-relativistic quantum field theory. Our goal is to set up our intuition for the structure of quantum field theory so that we can easily make the jump to relativistic field theories in later chapters.

Later, in subsequent chapters, we will generalize our quantum field theoretical description of the second quantized system to make it relativistic. In this generalization, second quantization will be essential. Relativistic quantum mechanics is necessarily a many-particle theory and, in some sense, the number of particles is always infinite, so there is no convenient description of it using the analog of a many-particle Schrödinger equation.

2.2   Non-relativistic Particles

Particles are non-relativistic if their velocities are much smaller than the speed of light. We will begin by focusing on this case. Particles are identical if all of their physical properties, such as their mass, electric charge, spin, et cetera, are identical. The framework of quantum mechanics gives us a sense in which such identical particles can also be indistinguishable. We will examine what it means for identical particles to be distinguishable or indistinguishable shortly.

We will begin by assuming that the physical behaviour of a many-particle system is governed by its Schrödigner equation. For our purposes, the Schrödinger equation is a partial differential equation which should be satisfied by the wave-function of a particle or a system of particles.

For example, for a single particle moving in three-dimensional space, the Schrödinger equation is

$$i\hbar\frac{\partial}{\partial t}\psi(\vec{x}, t) = \left[-\frac{\hbar^2}{2m}\vec{\nabla}^2 + V(\vec{x})\right]\psi(\vec{x}, t) \tag{2.1}$$

Here, m is the mass of the particle, $\hbar = h/2\pi$ where h is Planck's constant, $\vec{x} = (x^1, x^2, x^3)$ is a possible position of the particle (we will normally use Cartesian coordinates),

$$\vec{\nabla} \equiv \left(\frac{\partial}{\partial x^1}, \frac{\partial}{\partial x^2}, \frac{\partial}{\partial x^3}\right) \tag{2.2}$$

is the differential operator which is used to take the gradient, divergence or curl of various functions and $V(\vec{x})$ is the potential energy due to the interaction of the particle with its environment when the particle is located at position $\vec{x}$.

The wave-function $\psi(\vec{x}, t)$ is the probability amplitude, from which we would get the probability

$$P(\vec{x}, t) \equiv \psi^\dagger(\vec{x}, t)\psi(\vec{x}, t)\, d^3x \tag{2.3}$$

that, at time t, the particle will occupy an infinitesimally sized volume, d^3x, centred at the position $\vec{x}$. The function $\psi(\vec{x}, t)$ takes values in the complex numbers and $\psi^\dagger(\vec{x}, t)$ is its complex conjugate. The wave-function is usually normalized so that

$$\int d^3x\, \psi^\dagger(\vec{x}, t)\psi(\vec{x}, t) = 1 \tag{2.4}$$

which is equivalent to the statement that the probability of finding the particle somewhere is equal to one. Here, the integral is over the possible positions of the particle. If the set of possible positions is a space with a boundary, the Schrödinger equation must be solved with the appropriate boundary conditions. For simplicity, we will usually take the positions to be anywhere in open, infinite Euclidean space. In that case, the boundary condition is simply that the wave-function be square integrable, that is, that the normalization integral in equation (2.4) exists.

2.2.1 Identical and Indistinguishable Particles

The quantum mechanics of a single particle which we have reviewed above is easily
generalized to the situation where there are two or more particles. The Schrödinger
equation contains the Hamiltonian which is the sum of the energies of the particles
which we can write as

$$H = \sum_{i=1}^{N} \frac{\vec{p}^2}{2m} + V(\vec{x}_1, \vec{x}_2, \ldots, \vec{x}_N)$$

Here, the kinetic energy of each particle is $\frac{\vec{p}^2}{2m}$ where $\vec{p}$ is its linear momentum.
We have also added a potential energy function which we are assuming depends only
on the positions, $\vec{x}_1, \ldots, \vec{x}_N$ of the N particles.

We are going to assume that the particles are identical. We assume that they
have the same masses and other physical attributes, for example, they have identical
interactions with the environment and with each other. Their potential energy function
must therefore be a completely symmetric function of their positions,

$$V(\vec{x}_1, \vec{x}_2, \ldots, \vec{x}_N) = V(\vec{x}_{P(1)}, \vec{x}_{P(2)}, \ldots, \vec{x}_{P(N)})$$

where

$$(1, 2, \ldots, N) \rightarrow (P(1), P(2), \ldots, P(N))$$

is a permutation of the integers $(1, 2, \ldots, N)$. There are $N!$ distinct permutations of
N objects, including the identity permutation $(1, 2, \ldots, N) \rightarrow (1, 2, \ldots, N)$. Given
the Hamiltonian, we can form the Schrödinger equation by replacing the momentum
$\vec{p}_i \rightarrow -i\hbar\vec{\nabla}_i$. The resulting N-particle Schrödinger equation is

$$i\hbar\frac{\partial}{\partial t}\psi(\vec{x}_1, \vec{x}_2, \ldots, \vec{x}_N, t) =$$
$$\left(-\sum_{i=1}^{N} \frac{\hbar^2 \vec{\nabla}_i^2}{2m} + V(\vec{x}_1, \vec{x}_2, \ldots, \vec{x}_N)\right) \psi(\vec{x}_1, \vec{x}_2, \ldots, \vec{x}_N, t) \qquad (2.5)$$

with the boundary condition that the wave-function should be normallizable, so that

$$\int d^3x_1 d^3x_2 \ldots d^3x_N \psi^\dagger(\vec{x}_1, \vec{x}_2, \ldots, \vec{x}_N, t)\psi(\vec{x}_1, \vec{x}_2, \ldots, \vec{x}_N, t) = 1 \qquad (2.6)$$

Now, we can observe that this N-particle Schrödinger equation and its normalization condition have a high degree of symmetry. The fact that the particles are iden-
tical implies that, if we find one solution of the equation, $\psi(\vec{x}_1, \vec{x}_2, \ldots, \vec{x}_N, t)$, then
we can immediately find another solution by permuting the labels on coordinates,
that is, if $\psi(\vec{x}_1, \vec{x}_2, \ldots, \vec{x}_N, t)$ solves (2.5), then so does $\psi(\vec{x}_{P(1)}, \vec{x}_{P(2)}, \ldots, \vec{x}_{P(N)}, t)$
for any permutation P. Then we could ask the question as to whether we have found

multiple solutions of the same equation, that is, whether $\psi(\vec{x}_1, \vec{x}_2, ..., \vec{x}_N, t)$ and $\psi(\vec{x}_{P(1)}, \vec{x}_{P(2)}, ..., \vec{x}_{P(N)}, t)$ are the same quantum state or whether they are distinct quantum states. The wave-functions for distinct quantum states are linearly independent and one could imagine a situation where, by taking permutations of the coordinates, we generate $N!$ distinct linearly independent wave-functions by permuting the labels on the coordinates. And we might wonder whether nature really works this way, that is, whether there is really such a big multiplicity of the quantum states of a many-particle system. The answer seems to be "no", it doesn't.

It is our good fortune that nature always chooses to reduce the multiplicity of multi-particle states in one of two ways. Elementary particles in nature are indistinguishable, as well as being identical. They come in two kinds. One kind is bosons, whose wave-functions are completely symmetric so that

$$\text{bosons}: \quad \psi(\vec{x}_{P(1)}, \vec{x}_{P(2)}, ..., \vec{x}_{P(N)}, t) = \psi(\vec{x}_1, \vec{x}_2, ..., \vec{x}_N, t) \qquad (2.7)$$

and the particles are said to obey "Bose–Einstein statistics". The other kind are fermions, whose wave-functions are completely anti-symmetric,

$$\text{fermions}: \quad \psi(\vec{x}_{P(1)}, \vec{x}_{P(2)}, ..., \vec{x}_{P(N)}, t) = (-1)^{\deg(P)}\psi(\vec{x}_1, \vec{x}_2, ..., \vec{x}_N, t) \quad (2.8)$$

where $\deg(P)$ is the degree of the permutation, the number of exchanges of neighbouring numbers that is needed to restore $(P(1), P(2), \ldots, P(N))$ to $(1, 2, \ldots, N)$. Fermions are said to obey "Fermi-Dirac statistics".[1]

The probability

$$P(\vec{x}_1, \vec{x}_1, \ldots, \vec{x}_N, t) = |\psi(\vec{x}_1, \vec{x}_2, ..., \vec{x}_N, t)|^2 \, d^3x_1 d^3x_2 \ldots d^3x_N \qquad (2.9)$$

is completely symmetric for either fermions or bosons and it is interpreted as the probability of finding a particle in volume d^3x_1 at $\vec{x}_1$, another particle in d^3x_2 at $\vec{x}_2$ and so on with no reference as to which particle is which.

The potential energy function $V(\vec{x}_1, \vec{x}_2, ..., \vec{x}_N)$ represents the potential energy of a particle located at position $\vec{x}_1$ due to the environment of the particle and due to the presence of other particles located at positions $\vec{x}_2, \ldots, \vec{x}_N$. In most applications, it is useful to represent the potential energy energy function $V(\vec{x}_1, ..., \vec{x}_N)$ as the sum of single-particle potential energies, plus a sum of two-particle energies, three particle energies and so on

$$V(\vec{x}_1, ..., \vec{x}_N) = \sum_{i=1}^{N} V(\vec{x}_i) + \sum_{i<j=1}^{N} V(\vec{x}_i, \vec{x}_j) + \sum_{i<j<k=1}^{N} V(\vec{x}_i, \vec{x}_j, \vec{x}_k) + \ldots$$

---

[1] We should comment that, if the dimension of space were one or two, rather than three, there are some exotic alternatives to bosons and fermions called anyons. Their existence is due to the more complicated topology of particle histories in lower dimensional spaces. This fascinating topic is beyond the scope of this book where we will focus on bosons and fermions.

where the two, three and higher particle interactions are symmetric functions of their arguments, for example, $V(\vec{x}_i, \vec{x}_j) = V(\vec{x}_j, \vec{x}_i)$.

There are many interesting physical systems where the single-particle potentials can be ignored and the interaction is entirely due to a two-body potential which depends only on the distance between the particles. For example, a system of identical charged particles with a mutual Coulomb interaction has

$$V_{\text{Coulomb}}(\vec{x}_1, ..., \vec{x}_N) = \sum_{i<j=1}^{N} \frac{e^2}{4\pi\epsilon_0 |\vec{x}_i - \vec{x}_j|}$$

Another example is a simple contact interaction

$$V_{\text{contact}}(\vec{x}_1, ..., \vec{x}_N) = \sum_{i<j=1}^{N} \lambda\delta(\vec{x}_i - \vec{x}_j)$$

where $\delta(\vec{x}_i - \vec{x}_j)$ is the three-dimensional Dirac delta function with the property $\int d^3y\delta(\vec{x} - \vec{y})f(\vec{y}) = f(\vec{x})$. We will often restrict our considerations to a special case such as these.

We have discussed bosons and fermions which have symmetric or anti-symmetric wave-functions. Those wave-functions are also functions of time. Since the Hamiltonian for a system of identical particles is symmetric under any permutation of the particle labels, $\vec{x}_1 \dots , \vec{x}_N \rightarrow \vec{x}_{P(1)}. \dots \vec{x}_{P(N)}$ and since the Hamiltonian determines the time evolution, once we make the wave-function symmetric or anti-symmetric at a given time, it must be so for all times. We will see an explicit example of this in the next section in the case of free particles where the Schrödinger equation is solvable and we can find the wave-function at any time if we know it at an initial time. It is easy to see that if the initial wave-function is symmetric or anti-symmetric, then it remains so thereafter.

2.2.2   The Example of Weakly Interacting Particles

As illustration of many-particle quantum mechanics, let us consider the example where the potential energy vanishes so that the Schrödinger equation is

$$i\hbar\frac{\partial}{\partial t}\psi(\vec{x}_1, \dots , \vec{x}_N, t) = -\sum_{i=1}^{N} \frac{\hbar^2}{2m}\vec{\nabla}_i^2\, \psi(\vec{x}_1, \dots , \vec{x}_N, t) \qquad (2.10)$$

This equation is solved by plane waves. This follows from the property of a plane wave

$$\vec{\nabla}\, e^{i\vec{p}\cdot\vec{x}} = i\vec{p}\, e^{i\vec{p}\cdot\vec{x}}$$

Here p_i is a real number with dimensions of inverse distance. It is called the wave-number. The plane wave

$$e^{i \sum_j \vec{p}_j \cdot \vec{x}_j - i\omega(p)t}$$

solves equation (2.10) if the frequency $\omega(p)$ depends on the wave-vectors $\vec{p}_1, \ldots \vec{p}_N$ as

$$\omega(p) = \frac{1}{\hbar} \sum_{j=1}^{N} \frac{\hbar^2 \vec{p}_j^2}{2m}$$

A general solution of the Schrödinger equation is then gotten by taking a super-position of these plane waves which, since the wave-vectors vary continuously, is in the form of an integral

$$\psi(\vec{x}_1, \ldots, \vec{x}_N, t) = \int d^3 p_1 \ldots d^3 p_N e^{i \sum_{j=1}^{N} \left(\vec{p}_j \cdot \vec{x}_j - \frac{\hbar \vec{p}_j^2}{2m} t \right)} \psi_0(\vec{p}_1, \ldots, \vec{p}_N) \quad (2.11)$$

The function $\psi_0(\vec{p}_1, \ldots, \vec{p}_N)$, which serves as the coefficients in the superposition of plane waves above, can be obtained from initial data. If we know the wave-function at an initial time, say we know that at $t = 0$ it has the form $\psi_0(\vec{x}_1, \ldots, \vec{x}_N)$, then by setting $t = 0$ in equation (2.11) we find

$$\psi_0(\vec{x}_1, \ldots, \vec{x}_N) = \int d^3 p_1 \ldots d^3 p_N e^{i \sum_{j=1}^{N} \vec{p}_j \cdot \vec{x}_j} \psi_0(\vec{p}_1, \ldots, \vec{p}_N)$$

and we see that $\psi_0(\vec{p}_1, \ldots, \vec{p}_N)$ is obtained by the Fourier transform

$$\psi_0(\vec{p}_1, \ldots, \vec{p}_N) = \int \frac{d^3 x_1}{(2\pi)^3} \cdots \frac{d^3 x_N}{(2\pi)^3} e^{-i\vec{p}_1 \cdot \vec{x}_1 - \cdots - i\vec{p}_N \cdot \vec{x}_N} \psi_0(\vec{x}_1, \ldots, \vec{x}_N)$$

$$(2.12)$$

This gives us a general solution of the Schrödinger equation for N non-interacting identical particles. We can then apply it to bosons or fermions by symmetrizing or anti-symmetrizing it in its position variables. It is easy to see from our development above, and we have already observed that it is a general fact, that if the initial wave-function $\psi_0(\vec{x}_1, \ldots, \vec{x}_N)$ is symmetric or anti-symmetric, then the general solution $\psi(\vec{x}_1, \ldots, \vec{x}_N, t)$ is symmetric or anti-symmetric.

By plugging (2.12) into equation (2.11) and integrating over the wave-numbers, we find the general solution of the Schrödinger equation in a single formula

$$\psi(\vec{x}_1, \ldots, \vec{x}_N, t)$$

$$= \int d^3 y_1 \ldots d^3 y_N \left[\frac{im}{2\pi\hbar t} \right]^{\frac{3N}{2}} e^{\frac{im}{2\hbar t} \sum_{j=1}^{N} (\vec{x}_j - \vec{y}_j)^2} \psi(\vec{y}_1, \ldots, \vec{y}_N, 0) \quad (2.13)$$

Specifying and putting an initial wave-function into the equation above and then doing the $3N$ integrations obtains the wave-function at any other time. This is an explicit general solution for the wave-function of N non-interacting particles.

2.2.3 Hamiltonian and Stationary States

We can present the Schrödinger equation (2.10) as a time-independent equation by making the ansatz

$$\psi(\vec{x}_1, \ldots, \vec{x}_N, t) = e^{-iEt/\hbar}\psi_E(\vec{x}_1, \ldots, \vec{x}_N)$$

Then equation (2.10) implies that

$$E\psi_E(\vec{x}_1, \ldots, \vec{x}_N) = \left[-\sum_{i=1}^{N} \frac{\hbar^2\vec{\nabla}_i^2}{2m} + V(\vec{x}_1, ..., \vec{x}_N) \right] \psi_E(\vec{x}_1, \ldots, \vec{x}_N) \quad (2.14)$$

The solution of this equation with the appropriate boundary conditions should give us the wave-functions and the energies E of stationary states. The differential operator on the right-hand-side of equation (2.14) is the N-particle Hamiltonian

$$H = \left[-\sum_{i=1}^{N} \frac{\hbar^2\vec{\nabla}_i^2}{2m} + V(\vec{x}_1, ..., \vec{x}_N) \right] \quad (2.15)$$

and the equation itself is an equation that eigenstates and eigenvalues of the Hamiltonian should satisfy. The solution $\psi_E(\vec{x}_1, \ldots, \vec{x}_N)$ is called an "eigenstate" or "eigenvector" of the Hamiltonian and E is the "eigenvalue" which is associated with it. wave-functions with different energy eigenvalues are orthogonal,

$$\int d^3x_1 \ldots d^3x_N \psi_E^\dagger(\vec{x}_1, \ldots, \vec{x}_N)\psi_{E'}(\vec{x}_1, \ldots, \vec{x}_N) = \delta_{EE'} \quad (2.16)$$

We have used a Kronecker delta function in the right-hand-side of this equation, but note that for a continuum spectrum the states could require a continuum normalization with Dirac delta functions.

Generally, equation (2.14) is difficult to solve when the interaction potential is non-trivial. In fact, there are very few examples of interaction potentials where one can solve for the wave-functions or the energies exactly. One of them is the case of free particles, when the potential is zero. In that case, as we have already noted in the previous section, the explicit wave-function can be found. It is simply constructed from plane-waves,

$$\psi_E(\vec{x}_1, \ldots, \vec{x}_N) = \frac{1}{(2\pi)^{\frac{3N}{2}}} \exp\left(i\sum_{i=1}^{N} \vec{p}_i \cdot \vec{x}_i \right) \quad (2.17)$$

and the energy eigenvalue is

$$E = \sum_{i=1}^{N} \frac{\hbar^2 \vec{p}_i^2}{2m} \tag{2.18}$$

If the particles which obey the Schrödinger equation are bosons or fermions it would be necessary to symmetrize or anti-symmetrize the wave-function.

In the case of bosons the energy eigenstate would be gotten by a total symmetrization of the solution that we have found in equation (2.17),[2]

$$\psi_{\vec{p}_1\ldots\vec{p}_N}(\vec{x}_1, \ldots, \vec{x}_N) = \frac{1}{(2\pi)^{\frac{3N}{2}} N!} \sum_{P} \exp\left(i \sum_{j=1}^{N} \vec{p}_j \cdot \vec{x}_{P(j)}\right) \tag{2.19}$$

where the sum on the right-hand-side is over the $N!$ distinct permutations of $(1, \ldots, N)$. This wave-function is an eigenstate of the Hamiltonian with the energy eigenvalue given in equation (2.18). The normalization integral is

$$\int d^3x_1 \ldots d^3x_N \, \psi_{\vec{p}_1\ldots\vec{p}_N}(\vec{x}_1, \ldots, \vec{x}_N) \psi_{\vec{q}_1\ldots\vec{q}_N}(\vec{x}_1, \ldots, \vec{x}_N)$$

$$= \frac{1}{N!} \sum_{P} \delta(\vec{p}_1 - \vec{q}_{P(1)}) \delta(\vec{p}_2 - \vec{q}_{P(2)}) \ldots \delta(\vec{p}_N - \vec{q}_{P(N)}) \tag{2.20}$$

where $\delta(\vec{p})$ is the three-dimensional Dirac delta function.

For fermions, the wave-function would be gotten by a total anti-symmetrization of the solution that we have found in equation (2.17) above,

$$\psi_{\vec{p}_1\ldots\vec{p}_N}(\vec{x}_1, \ldots, \vec{x}_N) = \frac{1}{(2\pi)^{\frac{3N}{2}} N!} \sum_{P} (-1)^{\deg(P)} \exp\left(i \sum_{j=1}^{N} \vec{p}_j \cdot \vec{x}_{P(j)}\right) \tag{2.21}$$

This wave-function is also an eigenstate of the Hamiltonian with the energy eigenvalue given in equation (2.18). The normalization integral for fermion wave-functions is

$$\int d^3x_1 \ldots d^3x_N \, \psi_{\vec{p}_1\ldots\vec{p}_N}(\vec{x}_1, \ldots, \vec{x}_N) \psi_{\vec{q}_1\ldots\vec{q}_N}(\vec{x}_1, \ldots, \vec{x}_N)$$

$$= \frac{1}{N!} \sum_{P} (-1)^{\deg(P)} \delta(\vec{p}_1 - \vec{q}_{P(1)}) \delta(\vec{p}_2 - \vec{q}_{P(2)}) \ldots \delta(\vec{p}_N - \vec{q}_{P(N)}) \tag{2.22}$$

---

[2] The factor of $N!$ in the normalization in this equation is strictly correct only when no two of the $\vec{p}_i$'s are equal. If this were not the case, we would replace $\frac{1}{N!}$ by $\frac{r_1! r_2! \ldots}{N!}$ when there are r_1 occurrences of $\vec{p}_1$, r_2 occurrences of $\vec{p}_2$, etc.

The fermion wave-function in equation (2.21) has the interesting feature that the right-hand-side vanishes if any two of the wave-vectors $\vec{p}_1, \ldots, \vec{p}_N$ are equal or if any two of the positions $\vec{x}_1, \ldots, \vec{x}_N$ are equal. This is equivalent to the Pauli exclusion principle, the statement that two fermions can never occupy the same quantum state. Note that this feature is absent from the boson wave-function in equation (2.19).

It is sometimes convenient to present the totally anti-symmetric fermion wave-function as a determinant,

$$\psi_{\vec{p}_1 \ldots \vec{p}_N}(\vec{x}_1, \ldots, \vec{x}_N) = \frac{1}{N!} \det \left[\frac{\exp\left(i\,\vec{p}_i \cdot \vec{x}_j\right)}{(2\pi)^{\frac{3}{2}}} \right] \tag{2.23}$$

where we treat $\frac{\exp(i\,\vec{p}_i \cdot \vec{x}_j)}{(2\pi)^{\frac{3}{2}}}$ as an $N \times N$ matrix whose rows are labeled by i and whose columns are labeled by j and the wave-function is proportional to the determinant of this matrix, called the "Slater determinant". This determinant will vanish if $\vec{p}_i = \vec{p}_j$ or if $\vec{x}_k = \vec{x}_\ell$ for any pairs (i, j) or (k, ℓ).

2.2.4   Particles with Spin

There is one elaboration which we should discuss before proceeding to develop our current discussion further. That is the issue of spin. If we want to describe realistic many-particle systems of atoms or electrons, the particles in question generally have spin and their wave-functions must carry an index to label their spin state. To describe these, we add an index to the total wave-function for each particle, so that the wave-function is

$$\psi_{\sigma_1 \sigma_2, \ldots, \sigma_N}(\vec{x}_1, \vec{x}_2, \ldots, \vec{x}_N, t)$$

For spin J, the indices σ_i each run over $2J + 1$ values $\sigma_i = -J, -J + 1, \ldots, J - 1, J$ which correspond to the spin states of a single particle. The wave-function of a system of identical particles must then be either symmetric or anti-symmetric under simultaneous permutations of the spin and position variables of the particles. Generally, bosons have integer spins and fermions have half-odd integer spin. In summary, for bosons, J is an integer and

$$\psi_{\sigma_1 \ldots, \sigma_N}(\vec{x}_1, \ldots, \vec{x}_N, t) = \psi_{\sigma_{P(1)} \ldots, \sigma_{P(N)}}(\vec{x}_{P(1)}, \ldots, \vec{x}_{P(N)}, t) \,, \; \forall P$$

For fermions, J is a half-odd-integer and

$$\psi_{\sigma_1 \ldots, \sigma_N}(\vec{x}_1, \ldots, \vec{x}_N, t) = (-1)^{\deg(P)} \psi_{\sigma_{P(1)} \ldots, \sigma_{P(N)}}(\vec{x}_{P(1)}, \ldots, \vec{x}_{P(N)}, t) \,, \; \forall P$$

where, when we implement the permutation, we permute both the spin and the position labels. The Hamiltonian can also have spin-dependent interactions. In that case, the potential energy, as well as being a function of particle positions, is generally matrix-like object which operates on spin indices. In the most general case, the potential energy term in the Schrödinger equation would be

$$\sum_{\rho_1...\rho_N=-J}^{J} V_{\sigma_1...\sigma_N}^{\;\;\;\;\;\;\rho_1...\rho_N}(\vec{x}_1, ..., \vec{x}_N)\psi_{\rho_1...\rho_N}(\vec{x}_1, ..., \vec{x}_N)$$

For two-body interactions, the two-body potential gets spin indices as $V_{\rho_i\rho_j}^{\sigma_i\sigma_j}(\vec{x}_i, \vec{x}_j)$ and its operation on the wave-function is

$$\sum_{i<j}\sum_{\rho_i\rho_j=-J}^{J} V_{\sigma_i\sigma_j}^{\;\;\;\rho_i\rho_j}(\vec{x}_i, \vec{x}_j)\psi_{\sigma_1...\rho_i...\rho_j...\sigma_N}(\vec{x}_1, ..., \vec{x}_N, t)$$

We will see shortly that this sort of interaction is very easy to implement in second quantization.

If an interaction is spin-independent so that it does not depend on the spin states of the particles, we can simply omit the indices and the interaction term in the Schrödinger equation would be

$$V(\vec{x}_1, ..., \vec{x}_N)\psi_{\sigma_1...\sigma_N}(\vec{x}_1, ..., \vec{x}_N, t)$$

A typical spin-dependent interaction two-body interaction is the spin-spin interaction which, for spin $J = \frac{1}{2}$ particles, has the fomr

$$V_{\rho_i\rho_j}^{\sigma_i\sigma_j}(\vec{x}_i, \vec{x}_j) = \frac{\hbar^2}{4}\vec{\sigma}_{\rho_i}^{\sigma_i} \cdot \vec{\sigma}_{\rho_j}^{\sigma_j} \, v(\vec{x}, \vec{y}) \tag{2.24}$$

where we have introduced the three Pauli matrices as

$$\sigma^1 = \begin{bmatrix} 0 & 1 \\ 1 & 0 \end{bmatrix}, \quad \sigma^2 = \begin{bmatrix} 0 & -i \\ i & 0 \end{bmatrix}, \quad \sigma^3 = \begin{bmatrix} 1 & 0 \\ 0 & -1 \end{bmatrix}, \tag{2.25}$$

We will sometimes denote the three Pauli matrices as a three-component vector $\vec{\sigma}$.

The expectation value of a component of the spin is simply the expectation value of the appropriate component of the spin matrices for each particle. For a spin $J = \frac{1}{2}$ particle it is

$$\langle \Sigma^a \rangle = \int dx_1 ... dx_N \, \Sigma^a(\vec{x}_1, ..., \vec{x}_N, t)$$

$$\Sigma^a(\vec{x}_1, ..., \vec{x}_N, t)$$

$$= \sum_{i=1}^{N} \psi^{\sigma_1...\rho_i...\sigma_N\dagger}(\vec{x}_1, ..., \vec{x}_N, t)\frac{\hbar\sigma_{\rho_i}^{a\;\tau_i}}{2}\psi_{\sigma_1...\tau_i...\sigma_N}(\vec{x}_1, ..., \vec{x}_N, t) \tag{2.26}$$

Here and in the following, we use the Einstein summation convention for repeated up and down indices. In each term on the right-hand-side, each of the indices $\sigma_1, \sigma_2, \ldots, \sigma_{i-1}, \sigma_{i+1}, \ldots, \sigma_N$ and τ_i and ρ_i are assumed to be summed from $-J$ to J. We have omitted the summation symbols

$$\sum_{\sigma_1=-J}^{J} \cdots \sum_{\sigma_{i-1}=-J}^{J} \sum_{\sigma_{i+1}=-J}^{J} \cdots \sum_{\sigma_N=-J}^{J} \sum_{\tau_i=-J}^{J} \sum_{\rho_i=-J}^{J}$$

2.3   Second Quantization in the Schrödinger Picture

Second quantization is a technique which summarizes the many-particle quantum mechanical problem contained in (2.5), together with either Bose or Fermi statistics in an elegant way.

To implement second quantization, we begin by constructing an abstract basis for the states of the N-particle system. For this purpose, we define the Schrödinger field operator, $\psi(\vec{x})$, which depends on one position variable, $\vec{x}$. In spite of the use of the symbol ψ, this operator should not be confused with a wave-function, it is an operator whose important property is that it obeys the commutation relations which will be listed in equations (2.27) and (2.28) or (2.29) and (2.30) which we will discuss in detail in the following. There is one such operator for each different kind of identical particle, for example in a gas of electrons where the electron can exist in two spin states, the field operator would have the spin index, $\psi_\sigma(\vec{x})$ with $\sigma = -\frac{1}{2}, \frac{1}{2}$ labelling the spin. If we describe a plasma that is made up of ionized hydrogen, that is, electrons and protons, we would use two field operators, $\psi_\sigma^{\text{el}}(\vec{x})$ to describe the electron and $\psi_\sigma^{\text{pr}}(\vec{x})$ to describe the proton.

We shall also need the Hermitian conjugate of the field operator, $\psi^{\dagger\sigma}(\vec{x})$. This should be regarded as the Hermitian conjugate of the operator $\psi_\sigma(\vec{x})$ in the sense that, eventually, when we understand how to find matrix elements of it in a space of quantum states, those matrix elements obey

$$(< \chi|\psi_\sigma(\vec{x})|\tilde{\chi} >)^* = < \tilde{\chi}|\psi^{\dagger\sigma}(\vec{x})|\chi >$$

For bosons, we postulate that these operators satisfy the commutation relations

$$\left[\psi_\sigma(\vec{x}), \psi^{\dagger\rho}(\vec{y}) \right] = \delta_\sigma^\rho \delta(\vec{x} - \vec{y}), \tag{2.27}$$

$$\left[\psi_\sigma(\vec{x}), \psi_\rho(\vec{y}) \right] = 0, \quad \left[\psi^{\dagger\sigma}(\vec{x}), \psi^{\dagger\rho}(\vec{y}) \right] = 0 \tag{2.28}$$

where, as usual, the square bracket denotes a commutator ($[A, B] = AB - BA$). In the case of fermions, the commutators should be replaced by anti-commutators so that the operators satisfy the anti-commutation relations

$$\left\{ \psi_\sigma(\vec{x}), \psi^{\dagger\rho}(\vec{y}) \right\} = \delta_\sigma^\rho \delta(\vec{x} - \vec{y}), \tag{2.29}$$

$$\left\{ \psi_\sigma(\vec{x}), \psi_\rho(\vec{y}) \right\} = 0, \quad \left\{ \psi^{\dagger\sigma}(\vec{x}), \psi^{\dagger\rho}(\vec{y}) \right\} = 0 \tag{2.30}$$

We use the curly brackets to denote an anti-commutator, ($\{A, B\} = AB + BA$).

The operators $\psi_\sigma(\vec{x})$ and $\psi^{\dagger\sigma}(\vec{x})$ can be thought of as annihilation and creation operators for a particle located at point $\vec{x}$ and in spin state σ. To see this, consider the following construction. We begin with a specific quantum state which we shall call the "empty vacuum" $|0>$. It is the state where there are no particles at all. It is annihilated by the operators $\psi_\sigma(\vec{x})$ for all values of the position $\vec{x}$ and spin label σ,

$$\psi_\sigma(\vec{x})|0>= 0 \quad \forall \vec{x}, \sigma \tag{2.31}$$

The adjoint of the above statement is that the Hermitian conjugate and the dual state to the empty vacuum also have the property

$$<0|\psi^{\dagger\sigma}(\vec{x}) = 0 \quad \forall \vec{x}, \sigma \tag{2.32}$$

and we take the empty vacuum to be normalized,

$$<0|0>= 1 \tag{2.33}$$

Then, we create particles which occupy the distinct points $\vec{x}_1, \dots, \vec{x}_N$ and are in spin states $\sigma_1, \dots, \sigma_N$ by repeatedly operating $\psi^{\dagger\sigma_i}(\vec{x}_i)$ on the vacuum state

$$|\vec{x}_1, \dots, \vec{x}_N, >^{\sigma_1 \dots \sigma_N} = \frac{1}{\sqrt{N!}} \psi^{\dagger\sigma_1}(\vec{x}_1) \dots \psi^{\dagger\sigma_N}(\vec{x}_N)|0> \tag{2.34}$$

Since the operators $\psi^{\sigma_i\dagger}(\vec{x}_i)$ either commute or anti-commute with each other, $|\vec{x}_1, \dots, \vec{x}_N >^{\sigma_1 \dots \sigma_N}$ is automatically either totally symmetric or totally anti-symmetric under permutations of the position coordinates and spins, appropriate for bosons or fermions, respectively.

Similarly,

$$<\vec{x}_1, \dots, \vec{x}_N|_{\sigma_1 \dots \sigma_N} = \frac{1}{\sqrt{N!}} <0|\psi_{\sigma_N}(\vec{x}_N) \dots \psi_{\sigma_1}(\vec{x}_1) \tag{2.35}$$

The inner product for bosons can easily be found using the commutator algebra (2.27) and (2.28), the properties (2.31), (2.32) and (2.33)

$$<\vec{x}_1, \dots, \vec{x}_N|_{\sigma_1 \dots \sigma_N} |\vec{y}_1, \dots, \vec{y}_N, >^{\rho_1 \dots \rho_N}$$
$$= \frac{1}{N!} \sum_P \delta(\vec{x}_1 - \vec{y}_{P(1)})\delta_{\sigma_1}^{\rho_{P(1)}} \dots \delta(\vec{x}_N - \vec{y}_{P(N)})\delta_{\sigma_N}^{\rho_{P(N)}} \tag{2.36}$$

Similarly, the inner product for fermions is

$$<\vec{x}_1, \dots, \vec{x}_N|_{\sigma_1 \dots \sigma_N} |\vec{y}_1, \dots, \vec{y}_N, >^{\rho_1 \dots \rho_N}$$
$$= \frac{1}{N!} \sum_P (-1)^{\deg(P)}\delta(\vec{x}_1 - \vec{y}_{P(1)})\delta_{\sigma_1}^{\rho_{P(1)}} \dots \delta(\vec{x}_N - \vec{y}_{P(N)})\delta_{\sigma_N}^{\rho_{P(N)}} \tag{2.37}$$

We can form a density operator

$$\rho(\vec{x}) \equiv \psi^{\dagger\sigma}(\vec{x})\psi_\sigma(\vec{x}) \qquad (2.38)$$

Using the algebra of the field operators given in equations (2.27) and (2.28) or (2.29) and (2.30) we can show that

$$\rho(\vec{x}) \, |\vec{x}_1, \dots, \vec{x}_N >^{\sigma_1 \dots \sigma_N} = \left[\sum_{j=1}^{N} \delta(\vec{x} - \vec{x}_j) \right] |\vec{x}_1, \dots, \vec{x}_N >^{\sigma_1 \dots \sigma_N} \qquad (2.39)$$

The state $|\vec{x}_1, \dots, \vec{x}_N, >^{\sigma_1 \dots \sigma_N}$ is therefore an eigenstate of the density operator. The eigenvalue is a sum of Dirac delta functions which are localized at the positions $\vec{x}_1, \dots, \vec{x}_N$. This is the sense in which we would say that $|\vec{x}_1, \dots, \vec{x}_N >^{\sigma_1 \dots \sigma_N}$ is the state with particles located at the positions $\vec{x}_1, \dots, \vec{x}_N$. They also have spin states $\sigma_1, \dots, \sigma_N$.

In second quantization, the vectors $|\vec{x}_1, \dots, \vec{x}_N, >^{\sigma_1 \dots \sigma_N}$ are used to construct a state of the quantum system in the following way. We take a time-dependent superposition of the basis vectors $|\vec{x}_1, \dots, \vec{x}_N >^{\sigma_1 \dots \sigma_N}$ as

$$|\psi(t) >= \int d^3x_1 \dots d^3x_N \psi_{\sigma_1 \dots \sigma_N}(\vec{x}_1, \dots, \vec{x}_N, t)|\vec{x}_1, \dots, \vec{x}_N, >^{\sigma_1 \dots \sigma_N} \qquad (2.40)$$

This is a candidate for the quantum state of the N-particle system. The function $\psi_{\sigma_1 \dots \sigma_N}(\vec{x}_1, \dots, \vec{x}_N, t)$ which we use in this superposition will turn out to be the wave-function which satisfies the N-particle Schrödinger equation. Because the basis vectors $|\vec{x}_1, \dots, \vec{x}_N >^{\sigma_1 \dots \sigma_N}$ are totally symmetric or anti-symmetric, the function $\psi_{\sigma_1 \dots \sigma_N}(\vec{x}_1, \dots, \vec{x}_N, t)$ is automatically totally symmetric or antisymmtric in the positions and spins appropriate for a system of bosons or fermions.

First of all, let us observe that there is a one-to-one correspondence between the state vectors $|\psi(t) >$ and the wave-functions $\psi_{\sigma_1 \dots \sigma_N}(\vec{x}_1, \dots, \vec{x}_N, t)$. If we have a wave-function, $\psi_{\sigma_1 \dots \sigma_N}(\vec{x}_1, \dots, \vec{x}_N, t)$, we simply form the corresponding $|\psi(t) >$ by taking the integrals in equation (2.40). If, on the other hand, we are given $|\psi(t) >$, we can find the wave-function which corresponds to it by taking the inner product,

$$\psi_{\sigma_1 \dots \sigma_N}(\vec{x}_1, \dots, \vec{x}_N, t) =< \vec{x}_1, \dots, \vec{x}_N|_{\sigma_1 \dots \sigma_N} |\psi(t) > \qquad (2.41)$$

The state $|\psi(t) >$ has the same normalization as the wave-function $\psi_{\sigma_1 \dots \sigma_N}(\vec{x}_1, \dots, \vec{x}_N, t)$ which can be taken as unity

$$< \psi(t)|\psi(t) >$$
$$= \int d^3x_1 \dots d^3x_N \psi^{\dagger\sigma_1 \dots \sigma_N}(\vec{x}_1, \dots, \vec{x}_N, t)\psi_{\sigma_1 \dots \sigma_N}(\vec{x}_1, \dots, \vec{x}_N, t)$$
$$= 1 \qquad (2.42)$$

The N-particle Schrödinger equation is equivalent to an equation for $|\psi(t)>$. It is easy to see that the N-particle wave-function $\psi_{\sigma_1\ldots\sigma_N}(\vec{x}_1,\ldots,\vec{x}_N,t)$ satisfies the N-particle Schrödinger equation (2.5) if and only if the state $|\psi(t)>$ obeys the equation

$$i\hbar\frac{d}{dt}|\psi(t)> = H|\psi(t)> \qquad (2.43)$$

where the second quantized Hamiltonian operator is

$$
\begin{aligned}
H = &\int d^3x\left[\frac{\hbar^2}{2m}\vec{\nabla}\psi^{\dagger\sigma}(\vec{x})\cdot\vec{\nabla}\psi_\sigma(\vec{x}) + \psi^{\dagger\sigma}(\vec{x})V_\sigma{}^\tau(\vec{x})\psi_\tau(\vec{x})\right]\\
&+\frac{1}{2}\int d^3x d^3y\,\psi^{\dagger\sigma}(\vec{x})\psi^{\dagger\rho}(\vec{y})V_{\sigma\rho}^{\tilde{\sigma}\tilde{\rho}}(\vec{x},\vec{y})\psi_{\tilde{\rho}}(\vec{y})\psi_{\tilde{\sigma}}(\vec{x})\\
&+\frac{1}{6}\int d^3x d^3y d^3z\,\psi^{\dagger\sigma}(\vec{x})\psi^{\dagger\rho}(\vec{y})\psi^{\dagger\tau}(\vec{z})V_{\sigma\rho\tau}^{\tilde{\sigma}\tilde{\rho}\tilde{\tau}}(\vec{x},\vec{y},\vec{z})\psi_{\tilde{\tau}}(\vec{z})\psi_{\tilde{\rho}}(\vec{y})\psi_{\tilde{\sigma}}(\vec{x})\\
&+\ldots
\end{aligned}
\qquad (2.44)
$$

In H we have allowed for arbitrary spin dependence of the one-, two-, and three-body interactions, with the higher interactions following the same pattern. We have allowed for the most general spin-dependent interaction. In most applications we will consider interactions which are significantly simpler than this most general possibility, for example, where the single-particle potential is a spin-independent constant, $V_\sigma{}^\tau(\vec{x}) = \text{constant}\cdot\delta_\sigma{}^\tau$, the two-body potential is a spin-independent contact interaction, $V_{\sigma\rho}^{\tilde{\sigma}\tilde{\rho}}(\vec{x},\vec{y}) = \lambda\delta(\vec{x}-\vec{y})\delta_\sigma^{\tilde{\sigma}}\delta_\rho^{\tilde{\rho}}$ and the higher interactions absent.

One can construct an initial state $|\psi(0)>$ using the initial many-particle wave-function $\psi_{\sigma_1\ldots\sigma_N}(\vec{x}_1,\ldots,\vec{x}_N,t=0)$ as

$$|\psi(0)> = \frac{1}{\sqrt{N!}}\int d^3x_1\ldots d^3x_N\psi_{\sigma_1\ldots\sigma_N}(\vec{x}_1,\ldots,\vec{x}_N,0)|\vec{x}_1,\ldots,\vec{x}_N,>^{\sigma_1\ldots\sigma_N}$$

$$(2.45)$$

The state at later times is then uniquely determined by (2.43) which has the formal solution

$$|\psi(t)> = e^{-iHt/\hbar}|\psi(0)>$$

and the wave-function at any time that can be extracted from it by taking the inner product,

$$\psi_{\sigma_1\ldots\sigma_N}(\vec{x}_1,\ldots,\vec{x}_N,t) = <\vec{x}_1,\ldots,\vec{x}_N|_{\sigma_1\ldots\sigma_N}|\psi(t)>$$

and it must coincide with the solution of the many-body Schrödinger equation (2.5).

The second quantized Schrödinger equation (2.43) does not contain the explicit information that there are N particles. There is an elegant way to add this information. We have shown in the discussion around equation (2.39) that the state

$|\vec{x}_1 \ldots \vec{x}_N >^{\sigma_1 \ldots \sigma_N}$ that we have constructed should be thought of as the quantum mechanical state where the N particles can be found occupying the positions $\vec{x}_1, \ldots, \vec{x}_N$ and the spin states $\sigma_1 \ldots \sigma_N$, respectively. We noted that $|\vec{x}_1 \ldots \vec{x}_N >^{\sigma_1 \ldots \sigma_N}$ is an eigenstate of the density operator, $\rho(\vec{x}) = \psi^{\sigma \dagger}(\vec{x})\psi_\sigma(\vec{x})$ which was defined in equation (2.39).

The total number of particles can be measured by the number operator which is an integral over space of the density operator,

$$\mathcal{N} = \int d^3x \, \rho(\vec{x}) = \int d^3x \, \psi^{\dagger \sigma}(\vec{x})\psi_\sigma(\vec{x}) \qquad (2.46)$$

Given that $|\vec{x}_1 \ldots \vec{x}_N >^{\sigma_1 \ldots \sigma_N}$ are eigenstates of the density operator, they must also be eigenstates of the number operator, $\mathcal{N}$, with eigenvalue N, the number of particles,

$$\mathcal{N} \, |\vec{x}_1 \ldots \vec{x}_N >^{\sigma_1 \ldots \sigma_N} = N \, |\vec{x}_1 \ldots \vec{x}_N >^{\sigma_1 \ldots \sigma_N} \qquad (2.47)$$

(This can easily be checked explicitly using the definition of $|\vec{x}_1 \ldots \vec{x}_N >^{\sigma_1 \ldots \sigma_N}$ as being formed by operating the fields $\psi^{\dagger \sigma}(\vec{x})$ on the empty vacuum and then using the commutation or anti-commutation relations for the operators $\psi_\sigma(\vec{x})$ and $\psi^{\dagger \sigma}(\vec{x})$.) As a result, the state, $|\psi(t) >$, which we have constructed as a superposition of the states $|\vec{x}_1 \ldots \vec{x}_N >^{\sigma_1 \ldots \sigma_N}$ must be an eigenstate of the particle number operator with eigenvalue N,

$$\mathcal{N} \, |\psi(t) > = N \, |\psi(t) >$$

The Hamiltonian commutes with the number operator,

$$[\mathcal{N}, H] = 0 \qquad (2.48)$$

This can easily be checked explicitly using the algebra for the operators $\psi_\sigma(\vec{x})$ and $\psi^{\dagger \sigma}(\vec{x})$. The result is that the number operator $\mathcal{N}$ is time-independent. The number operator $\mathcal{N}$ and the Hamiltonian H can have simultaneous eigenvalues. Also, the total number of particles can be fixed at an initial time and it will be preserved by the time evolution of the system.

In summary, we have two equivalent formulations of the same theory of N identical and indistinguishable particles. The mathematical problem of solving the second-quantized operator equation (2.43) is identical in all respects to the mathematical problem of solving the many-particle Schrödinger equation (2.5), they are solved when we find the state $|\psi(t) >$ or, equivalently, the wave-function $\psi_{\sigma_1 \ldots \sigma_N}(\vec{x}_1, \ldots, \vec{x}_N, t)$. Solving the second quantized theory is at this point the apparently more abstract problem. In the next chapter, we will study some explicit solutions of it and we will see that it can have some advantages.

2.4   Second Quantization in the Heisenberg Picture

The second quantized formulation of the many-particle problem is essentially the definition of a quantum field theory. The quantum fields are the field operators $\psi_\sigma(\vec{x})$ and $\psi^{\dagger\sigma}(\vec{x})$. Note that they do not depend on time and play a rather simple role in defining the possible quantum states. The time-dependent state $|\psi(t)>$ contains all of the nontrivial information about the time evolution of the system. This is a quantum field theory in the Schrödinger picture of quantum mechanics where operators are time independent and the states carry the time dependence.

The Heisenberg picture is an alternative formulation of quantum mechanics which is equivalent to the Schrödinger picture. It is related to the Schrödingier picture that we have developed so far by a time dependent unitary transformation which acts on the operators and state vectors. The unitary transformation begins with the observation that, if we know the state of the system at an initial time, say at $t = 0$, we can find a formal solution of the equation of motion for the state vector,

$$|\psi(t)\rangle = e^{-iHt/\hbar}\,|\psi(0)\rangle$$

which uses a unitary operator that is obtained by exponentiating the Hamiltonian, $\exp(-iHt/\hbar)$. We can thus set the state vector to its initial condition (assuming $t = 0$ is where we must impose an initial condition) by a unitary transformation. Going to the Heisenberg picture simply does this unitary transformation to all of the states and all of the operators to get an equivalent description of the theory where the operators are time-dependent and the states are independent of time. The unitary transformation of the operators is

$$\psi'_\sigma(\vec{x}, t) = e^{iHt/\hbar}\psi_\sigma(\vec{x})e^{-iHt/\hbar}, \quad \psi^{\dagger\sigma'}(\vec{x}, t) = e^{iHt/\hbar}\psi^{\dagger\sigma}(\vec{x})e^{-iHt/\hbar} \qquad (2.49)$$

In the Heisenberg picture, the states are time independent. For a given physical situation the quantum state is simply given by the initial state of the system. The operators, on the other hand, become time dependent and it is their time dependence which carries the information of the time evolution of the quantum system. In quantum field theory, particularly the relativistic quantum field theory which we shall study later on, the equations of motion are more commonly presented in the Heisenberg picture.

Unlike the time-independent operators $\psi_\sigma(\vec{x})$ and $\psi^{\dagger\sigma}(\vec{x})$ which we introduced in order to construct second quantization, the Heisenberg picture operators $\psi'_\sigma(\vec{x}, t)$ and $\psi^{\dagger\sigma'}(\vec{x}, t)$ now depend on time and their time dependence contains dynamical information, which is determined by equations (2.49). This information can also be given as a differential equation, the Heisenberg equation of motion, which can be obtained by taking a time derivative of equations (2.49).

$$i\hbar\frac{\partial}{\partial t}\psi_\sigma(\vec{x}, t) = \left[\psi_\sigma(\vec{x}, t), H\right], \quad i\hbar\frac{\partial}{\partial t}\psi^{\dagger\rho}(\vec{x}, t) = \left[\psi^{\dagger\rho}(\vec{x}, t), H\right] \qquad (2.50)$$

These are the usual algebraic operator equations which are meant to be solved to find the time dependence of the operators in the Heisenberg picture. In equation (2.50), and elsewhere when it is clear from the context, we will drop the prime notation from the Heisenberg picture fields. They are still distinguished from the Schrödinger picture fields in that they are time dependent and the Schrödinger picture fields are not.

The Heisenberg picture field operators have an equal-time commutator algebra, which can be obtained from (2.27) and (2.28) or (2.29) and (2.30) by multiplying from the left and right by $e^{-iHt/\hbar}$ and $e^{iHt/\hbar}$, respectively. This leads to the canonical equal-time commutation relations for bosons,

$$\left[\psi_\sigma(\vec{x}, t), \psi^{\dagger\rho}(\vec{y}, t)\right] = \delta_\sigma^\rho \delta(\vec{x} - \vec{y}) \tag{2.51}$$

$$\left[\psi_\sigma(\vec{x}, t), \psi_\rho(\vec{y}, t)\right] = 0, \quad \left[\psi^{\dagger\sigma}(\vec{x}, t), \psi^{\dagger\rho}(\vec{y}, t)\right] = 0 \tag{2.52}$$

or the canonical equal-time anti-commutation relations for fermions,

$$\left\{\psi_\sigma(\vec{x}, t), \psi^{\dagger\rho}(\vec{y}, t)\right\} = \delta_\sigma^\rho \delta(\vec{x} - \vec{y}), \tag{2.53}$$

$$\left\{\psi_\sigma(\vec{x}, t), \psi_\rho(\vec{y}, t)\right\} = 0, \quad \left\{\psi^{\dagger\sigma}(\vec{x}, t), \psi^{\dagger\rho}(\vec{y}, t)\right\} = 0 \tag{2.54}$$

At this point the reader should take careful note of the fact that this algebra holds only when the times in both of the operators are the same.

The time-derivative of the time-dependent field $\psi_\sigma(\vec{x}, t)$ can be computed from the Heisenberg equation of motion (2.50) using the equal time commutation relations. It is given by an equation which looks like a non-linear generalization of the Schrödinger equation

$$\left(i\hbar\frac{\partial}{\partial t} + \frac{\hbar^2}{2m}\vec{\nabla}^2\right)\psi_\sigma(\vec{x}, t) = V_\sigma^\rho(\vec{x})\psi_\rho(\vec{x}, t)$$

$$+ \int d^3y\, V_{\sigma\sigma'}^{\rho\rho'}(\vec{x}, \vec{y})\psi^{\dagger\sigma'}(\vec{y}, t)\psi_{\rho'}(\vec{y}, t)\psi_\rho(\vec{x}, t) + \dots \tag{2.55}$$

We have found our third presentation of the same N-particle quantum mechanics problem. It can be written down as the field equation (2.55) of a quantum field theory plus the equal-time commutation relations (2.51) and (2.52) or anti-commutation relations (2.53) and (2.54). The role of the commutation or anti-commutation relations is to define the quantized fields as operators. The field equation (2.55) contains the non-relativistic wave operator

$$i\hbar\frac{\partial}{\partial t} + \frac{\hbar^2}{2m}\vec{\nabla}^2$$

and the non-linear terms in the field equation arise from interactions between the particles. We could fix the total number of particles by requiring that states are eigenvectors of the number operator $\mathcal{N}$ with eigenvalue N. This is compatible with the field equation when $\frac{d}{dt}\mathcal{N} = 0$, which is the case for the class of theories that we are considering where $\mathcal{N}$ commutes with the Hamiltonian.

Solving the quantum field theory that is defined by equations (2.55), (2.51) and (2.52) or equations (2.55), (2.53) and (2.54) must be completely equivalent to solving the N-particle Schrödinger equation (2.5) and adapting the solution to the description of bosons or fermions. In the next chapter we will study how we can solve this quantum field theory in the simple case where the interactions are sufficiently weak that they can be ignored.

Degenerate Fermi and Bose Gases

3

In this chapter, we will study the Heisenberg representation quantum field theories of non-relativistic many-particle systems what we developed in the previous chapter in the limit where the volume is very large, the density is finite and the inter-particle interactions are weak. This will introduce the idea of Fermi energy and Fermi surface for fermions, the concept of particles and holes, and it will allow us to study some of the properties of a weakly interacting Fermi gas. We will also study many boson systems where we will introduce the concept of a Bose condensate and we will examine the low energy excitations of a weakly interacting system of bosons.

3.1 The Limit of Weakly Interacting Particles

As an example of our use of a quantum field theory to describe a quantum mechanical system with many identical particles, let us examine the special case where the particles interact with each other so weakly that, to a first approximation, we can ignore the interactions entirely. In the field equation which we discussed in the previous chapter, this happens when we can ignore the terms containing the interaction potential $V(\vec{x}_1, \ldots, \vec{x}_N)$. We will assume that the particles are either bosons or fermions. The beginning of our development applies equally well to both cases. However, as we shall see later, there are dramatic differences between the low energy states of a system of fermions and a system of bosons.

We have already solved the problem of the N-particle Schrödinger equation for non-interacting particles in Sect. 2.2.2. In this section, we will seek the corresponding solution of the second quantized theory in the Heisenberg picture. In that picture, the quantum field is a space and time-dependent operator which obeys the wave equation

$$\left(i\hbar\frac{\partial}{\partial t} + \frac{\hbar^2}{2m}\vec{\nabla}^2\right)\psi_\sigma(\vec{x}, t) = 0 \tag{3.1}$$

It also obeys, for bosons, the equal time commutation relations

$$\left[\psi_\sigma(\vec{x}, t), \psi^{\dagger\rho}(\vec{y}, t)\right] = \delta_\sigma^\rho \delta^3(\vec{x} - \vec{y}) \tag{3.2}$$

$$\left[\psi_\sigma(\vec{x}, t), \psi_\rho(\vec{y}, t)\right] = 0, \quad \left[\psi^{\dagger\sigma}(\vec{x}, t), \psi^{\dagger\rho}(\vec{y}, t)\right] = 0 \tag{3.3}$$

or, for fermions, the equal time anti-commutation relations

$$\left\{\psi_\sigma(\vec{x}, t), \psi^{\dagger\rho}(\vec{y}, t)\right\} = \delta_\sigma^\rho \delta^3(\vec{x} - \vec{y}) \tag{3.4}$$

$$\left\{\psi_\sigma(\vec{x}, t), \psi_\rho(\vec{y}, t)\right\} = 0, \quad \left\{\psi^{\dagger\sigma}(\vec{x}, t), \psi^{\dagger\rho}(\vec{y}, t)\right\} = 0 \tag{3.5}$$

Here, we have retained the spin indices, σ, ρ, ... to denote the spin state. If the spin is $\frac{1}{2}$, as it is for an electron, these indices each run over the two values, $-\frac{1}{2}$, $\frac{1}{2}$, denoting the two spin states. For a particle with spin J, such as a spin-J atom, this index will run over $2J + 1$ values, $-J$, $-J + 1$, ..., J which label the $2J + 1$ different spin states.

We will take equations (3.1), (3.2) and (3.3) or equations (3.1), (3.4) and (3.5) as the definition of the quantum field theory and, in the following, we will proceed to find a solution of it.

Let us begin by solving the field equation (3.1). We will assume that the coordinates $\vec{x}$ take values on infinite three-dimensional Euclidean space with Cartesian coordinates $\vec{x} = (x^1, x^2, x^3)$. The field equation is a linear partial differential equation which we can easily solve by noting that it is solved by a plane wave

$$\left(i\hbar \frac{\partial}{\partial t} + \frac{\hbar^2}{2m} \vec{\nabla}^2\right) e^{i\vec{k}\cdot\vec{x} - i\frac{\hbar k^2}{2m}t} = 0 \tag{3.6}$$

A general solution of the wave equation is therefore a superposition of plane waves

$$\psi_\sigma(x, t) = \int \frac{d^3k}{(2\pi)^{\frac{3}{2}}} e^{i\vec{k}\cdot\vec{x} - i\frac{\hbar k^2}{2m}t} \alpha_\sigma(\vec{k}) \tag{3.7}$$

Since $\psi_\sigma(x, t)$ is an operator, the object in the integrand of this general solution, $\alpha_\sigma(\vec{k})$, must also be an operator. As well as solving the wave equation, which equation (3.7) accomplishes, we must determine the properties $\alpha_\sigma(\vec{k})$, so that the field $\psi_\sigma(\vec{x}, t)$ in equation (3.7) satisfies the commutation relations (3.2) and (3.3) or the anti-commutation relations (3.4) and (3.5).

It will indeed satisfy those relations if $\alpha_\sigma(\vec{k})$ and $\alpha^{\dagger\sigma}(\vec{k})$ themselves obey commutation relations

$$\left[\alpha_\sigma(\vec{k}), \alpha^{\dagger\rho}(\vec{p})\right] = \delta_\sigma^\rho \delta^3(\vec{k} - \vec{p}) \tag{3.8}$$

$$\left[\alpha_\sigma(\vec{k}), \alpha_\rho(\vec{p})\right] = 0, \quad \left[\alpha^{\dagger\sigma}(\vec{k}), \alpha^{\dagger\rho}(\vec{p})\right] = 0 \tag{3.9}$$

for bosons or anti-commutation relations

$$\left\{\alpha_\sigma(\vec{k}), \alpha^{\dagger\rho}(\vec{p})\right\} = \delta_\sigma^\rho \delta^3(\vec{k} - \vec{p}) \tag{3.10}$$

$$\left\{\alpha_\sigma(\vec{k}), \alpha_\rho(\vec{p})\right\} = 0, \quad \left\{\alpha^{\dagger\sigma}(\vec{k}), \alpha^{\dagger\rho}(\vec{p})\right\} = 0 \tag{3.11}$$

for fermions. To see this explicitly for the case of bosons, consider

$$\left[\psi_\sigma(\vec{x}, t), \psi^{\dagger\rho}(\vec{y}, t)\right]$$

$$= \int \frac{d^3k}{(2\pi)^{\frac{3}{2}}} \int \frac{d^3p}{(2\pi)^{\frac{3}{2}}} e^{i\vec{k}\cdot\vec{x} - i\frac{\hbar k^2}{2m}t} e^{-i\vec{p}\cdot\vec{y} + i\frac{\hbar p^2}{2m}t} \left[\alpha_\sigma(\vec{k}), \alpha^{\dagger\rho}(\vec{p})\right]$$

$$= \int \frac{d^3k}{(2\pi)^{\frac{3}{2}}} \int \frac{d^3p}{(2\pi)^{\frac{3}{2}}} e^{i\vec{k}\cdot\vec{x} - i\frac{\hbar k^2}{2m}t} e^{-i\vec{p}\cdot\vec{y} + i\frac{\hbar p^2}{2m}t} \delta_\sigma^\rho \delta^3(\vec{k} - \vec{p})$$

$$= \delta_\sigma^\rho \int \frac{d^3k}{(2\pi)^3} e^{i\vec{k}\cdot(\vec{x}-\vec{y})} = \delta_\sigma^\rho \, \delta^3(\vec{x} - \vec{y})$$

where we have used the fact that the Fourier transform of the Dirac delta function is given by the formula

$$\int \frac{d^3k}{(2\pi)^3} e^{i\vec{k}\cdot(\vec{x}-\vec{y})} = \delta^3(\vec{x} - \vec{y})$$

Then, in order to complete the solution of the problem, we must find the vector space on which the operators $\alpha_\sigma(\vec{k})$ and $\alpha^{\dagger\sigma}(\vec{k})$ act. This is straightforward, and similar to what we have done in the previous section for the Schrödinger picture field operators (in fact these are the Fourier transform of the field operators at time $t = 0$ where they should be identical to those Schrödinger picture operators). We begin with the state where there are no particles at all, the empty vacuum $|0>$, which has the properties

$$\alpha_\sigma(\vec{k})|0> = 0, \quad <0|\alpha^{\dagger\sigma}(\vec{k}) = 0, \quad \forall \vec{k}, \sigma, \quad <0|0> = 1$$

Then, we construct the multi-particle states by repeatedly operating on the vacuum state with the creation operator $\alpha^{\dagger\sigma}(\vec{k})$,

$$|\vec{k}_1, \vec{k}_2, \ldots, \vec{k}_N >^{\sigma_1\sigma_2\ldots\sigma_N} = \frac{1}{\sqrt{N!}} \alpha^{\dagger\sigma_1}(\vec{k}_1) \alpha^{\dagger\sigma_2}(\vec{k}_2) \ldots \alpha^{\dagger\sigma_N}(\vec{k}_N) |0\rangle \tag{3.12}$$

Because the creation operators either commute with each other for bosons or they anti-commute with each other for fermions, the states in equation (3.12) are either totally symmetric or totally anti-symmetric, respectively, under permuting indices of the wave-numbers and the spins. For a permutation P of the numbers $\{1, 2, \ldots, N\}$, and for bosons,

$$|\vec{k}_{P(1)}, \vec{k}_{P(2)}, \ldots, \vec{k}_{P(N)} >^{\sigma_{P(1)}\sigma_{P(2)}\ldots\sigma_{P(N)}} = |\vec{k}_1, \vec{k}_2, \ldots, \vec{k}_N >^{\sigma_1\sigma_2\ldots\sigma_N}$$

or for fermions

$$|\vec{k}_{P(1)}, \vec{k}_{P(2)}, \ldots, \vec{k}_{P(N)}>^{\sigma P(1) \sigma P(2) \cdots \sigma P(N)} = (-1)^{\deg(P)} |\vec{k}_1, \vec{k}_2, \ldots, \vec{k}_N>^{\sigma_1 \sigma_2 \ldots \sigma_N}$$

The dual states are

$$< \vec{k}_1, \vec{k}_2, \ldots, \vec{k}_N|_{\sigma_1 \sigma_2 \ldots \sigma_N} = \frac{1}{\sqrt{N!}} < 0|\alpha_{\sigma_N}(\vec{k}_N) \ldots \alpha_{\sigma_2}(\vec{k}_2)\alpha_{\sigma_1}(\vec{k}_1)$$

and

$$< \vec{k}_1, \vec{k}_2, \ldots, \vec{k}_N|_{\sigma_1 \sigma_2 \ldots \sigma_N} \; |\vec{p}_1, \vec{p}_2, \ldots, \vec{p}_{N'}>^{\rho_1 \rho_2 \cdots \rho_{N'}}$$
$$= \frac{\delta_{NN'}}{N!} \sum_P \delta(\vec{k}_1 - \vec{p}_{P(1)}) \ldots \delta(\vec{k}_N - \vec{p}_{P(N)})\delta_{\sigma_1}^{\rho P(1)} \ldots \delta_{\sigma_N}^{\rho P(N)}$$

for bosons or

$$< \vec{k}_1, \vec{k}_2, \ldots, \vec{k}_N|_{\sigma_1 \sigma_2 \ldots \sigma_N} \; |\vec{p}_1, \vec{p}_2, \ldots, \vec{p}_{N'}>^{\rho_1 \rho_2 \cdots \rho_{N'}}$$
$$= \frac{\delta_{NN'}}{N!} \sum_P (-1)^{\deg(P)}\delta(\vec{k}_1 - \vec{p}_{P(1)}) \ldots \delta(\vec{k}_N - \vec{p}_{P(N)})\delta_{\sigma_1}^{\rho P(1)} \ldots \delta_{\sigma_N}^{\rho P(N)}$$

for fermions.

When we plug the solution (3.7) into the Hamiltonian (2.44), we get

$$H_0 = \int d^3x \frac{\hbar^2}{2m} \vec{\nabla}\psi^{\dagger\sigma}(x, t) \cdot \vec{\nabla}\psi_\sigma(\vec{x}, t) \tag{3.13}$$

$$= \int d^3k \sum_{\sigma=-J}^{J} \frac{\hbar^2\vec{k}^2}{2m} \alpha^{\dagger\sigma}(\vec{k})\alpha_\sigma(\vec{k}) \tag{3.14}$$

Here we have used the subscript on H_0 to denote the Hamiltonian specialized to the case of non-interacting particles. Note that we use the time-dependent operators in the integrand of H_0 in equation (3.13). We can do this since the Hamiltonian must be time-independent. Indeed, we see in equation (3.14) that when we substitute the solution of the field equation into the integrand we obtain a quantity which does not depend on the time. Also, the number operator is

$$\mathcal{N} = \int d^3k \psi^{\dagger\sigma}(\vec{x})\psi_\sigma(\vec{x}) \tag{3.15}$$

$$= \int d^3k \sum_{\sigma=-J}^{J} \alpha^{\dagger\sigma}(\vec{k})\alpha_\sigma(\vec{k}) \tag{3.16}$$

Again, we see that the result is a combination of operators which does not depend on time.

The quantum states that we have constructed are eigenstates of both the Hamiltonian and the number operator,

$$H_0|\vec{k}_1,\vec{k}_2,\dots,\vec{k}_N>^{\sigma_1\sigma_2\dots\sigma_N} = \sum_{i=1}^{N}\frac{\hbar^2\vec{k}_i^2}{2m}|\vec{k}_1,\vec{k}_2,\dots,\vec{k}_N>^{\sigma_1\sigma_2\dots\sigma_N}$$

$$\mathcal{N}\,|\vec{k}_1,\vec{k}_2,\dots,\vec{k}_N>^{\sigma_1\sigma_2\dots\sigma_N} = N\,|\vec{k}_1,\vec{k}_2,\dots,\vec{k}_N>^{\sigma_1\sigma_2\dots\sigma_N}$$

Note that the energy of a basis state is given by the sum of the energies of the particles in the state. This is a result of the fact that the particles are not interacting, so their total energy is just the sum of their individual energies. There individual energies are entirely due to their kinetic energy, which for a non relativistic particle is $\vec{p}^2/2m$ where, where the particle momentum is given in terms of the wavenumber by $\vec{p}=\hbar\vec{k}$.

The states $|\vec{k}_1,\vec{k}_2,\dots,\vec{k}_N>^{\sigma_1\sigma_2\dots\sigma_N}$ are a convenient basis for the time-independent states of the Heisenberg picture. It is easy to see how the time-dependent operators act on these states,

$$\psi_\sigma(\vec{x},t)|\vec{k}_1,\vec{k}_2,\dots,\vec{k}_N>^{\sigma_1\sigma_2\dots\sigma_N}$$

$$= \sum_{i=1}^{N}\frac{e^{-i\vec{k}_i\cdot\vec{x}+i\frac{\hbar\vec{k}_i^2}{2m}t}}{(2\pi)^{3/2}\sqrt{N}}\delta_\sigma^{\sigma_i}|\vec{k}_1,\dots,\vec{k}_{i-1},\vec{k}_{i+1},\dots,\vec{k}_N>^{\sigma_1\dots\sigma_{i-1}\sigma_{i+1}\dots\sigma_N} \qquad (3.17)$$

$$\psi^{\dagger\rho}(\vec{x},t)|\vec{k}_1,\vec{k}_2,\dots,\vec{k}_N>^{\sigma_1\sigma_2\dots\sigma_N}$$

$$= \sqrt{N+1}\int\frac{d^3p}{(2\pi)^{3/2}}e^{-i\vec{p}\cdot\vec{x}+i\frac{\hbar\vec{p}^2}{2m}t}|\vec{p},\vec{k}_1,\vec{k}_2,\dots,\vec{k}_N>^{\rho\sigma_1\sigma_2\dots\sigma_N} \qquad (3.18)$$

for bosons and, for fermions, equation (3.17) should be replaced by

$$\psi_\sigma(\vec{x},t)|\vec{k}_1,\vec{k}_2,\dots,\vec{k}_N>^{\sigma_1\sigma_2\dots\sigma_N}$$

$$= \sum_{i=1}^{N}\frac{e^{-i\vec{k}_i\cdot\vec{x}+i\frac{\hbar\vec{k}_i^2}{2m}t}}{(2\pi)^{3/2}\sqrt{N}}\delta_\sigma^{\sigma_i}(-1)^{i-1}|\vec{k}_1,\dots,\vec{k}_{i-1},\vec{k}_{i+1},\dots,\vec{k}_N>^{\sigma_1\dots\sigma_{i-1}\sigma_{i+1}\dots\sigma_N}$$

$$(3.19)$$

This gives a complete solution of the Heisenberg picture of the quantum field theory. Observables should be constructed from the fields $\psi_\sigma(\vec{x},t)$ and $\psi^{\dagger\rho}(\vec{x},t)$ and we can then use the above equations which tell us how the fields act on states to compute expectation values of the observables. We have already done this when the observable is the energy.

We can also use the states $|\vec{k}_1,\dots,\vec{k}_N>^{\sigma_1\dots\sigma_N}$ to find a general solution for the time-dependent state which is the solution of the quantum field theory in the Schrödinger picture,

$$|\psi(t)\rangle = \int d^3k_1\dots d^3k_N\, e^{-i\sum_{i=1}^{N}\frac{\hbar\vec{k}_i^2}{2m}t}\,\psi_{\sigma_1\dots\sigma_N}(\vec{k}_1,\dots,\vec{k}_N)|\vec{k}_1,\dots,\vec{k}_N>^{\sigma_1\dots\sigma_N}$$

$$(3.20)$$

The function $\psi_{\sigma_1...\sigma_N}(\vec{k}_1, \ldots, \vec{k}_N)$ could be obtained by doing a Fourier transform of the initial N-particle wave-function $\psi_{\sigma_1...\sigma_N}(\vec{x}_1, \ldots, \vec{x}_N, t = 0)$.

3.2   Degenerate Fermi Gas

In the previous section, we have solved the field equation and constructed the states for an N-particle system as it is described by second quantization in the Heisenberg picture. We did so in a space with no boundaries, that is, we considered a system where the volume is infinite and the number of particles N is finite. For many physical applications we need to study a system where the density, rather than the number of particles, is finite. This could be accomplished by considering the system of N particles confined to a space of finite volume V so that the average density N/V is fixed. We might then want to study aspects of such a system which do not depend on the shape or size of the finite volume or the boundary conditions which would have to be imposed in order to correctly define the Schrödinger equation in a space of finite volume. We might imagine doing this by taking a limit where the volume and the number of particles both go to infinity while the particle density N/V remains fixed. This is sometimes called the thermodynamic limit. The thermodynamic limit can give a very good description of the bulk properties of a system, particularly when the number of degrees in the system is very large and when their typical motion does not encounter the boundaries. For example, the electronic properties of the metal copper, which is a good conductor, are to a certain approximation described by assuming that it contains a gas of weakly interacting electrons. A cubic centimetre of copper would contain of order 10^{23} such electrons, so that the average distance between electrons is of order 10^{-8} cm. The quantum wave-length of the quantum state of a typical conduction electron in copper is also of order 10^{-8} cm. Even a millimetre sized piece of copper is macroscopic. The approximation where it has infinite size is very convenient for describing those properties of the material which don't depend on the nature of its surfaces, conductivity, or specific heat being examples. Our field theoretical formulation of the many particle system is particularly useful in that it allows us to take the "thermodynamic limit" in an elegant way.

3.2.1   The Ground State $|\mathcal{O}>$

When particles are fermions, their many-particle wave-function must be anti-symmetric. This is reflected in second quantization by the fact that the operators which create the fermions anti-commute with each other. Then the algebra of creation operators,

$$\alpha^{\dagger\sigma}(\vec{k})\alpha^{\dagger\sigma'}(\vec{k}') = -\alpha^{\dagger\sigma'}(\vec{k}')\alpha^{\dagger\sigma}(\vec{k})$$

implies

$$\left[\alpha^{\dagger\sigma}(\vec{k})\right]^2 |0> = 0$$

We cannot create a state where two fermions have the same spin σ and the same wave-vector $\vec{k}$. This means that, in any quantum state of a many-fermion system, the fermions must have distinct wave-vectors and spins. Then the lowest energy state (or ground state) of a system of free fermions must be gotten by populating the lowest possible energy states. These are the states where the total energy $E = \sum_{i=1}^{N} \frac{\hbar^2 k_i^2}{2m}$ is minimal. They occupy those values of $\vec{k}$ which are closest to the origin of the wave-vector space. They lie in the interior of a solid sphere, called the Fermi sphere, of radius k_F, called the Fermi wave-vector. Formally, we can then construct the lowest energy state of a many-fermion system by populating the states with $|\vec{k}| \le k_F$ and leaving the states with $|\vec{k}| > k_F$ empty,

$$|\mathcal{O}> = \frac{1}{c} \left[\prod_{|\vec{k}| \le k_F} \prod_{\sigma=-J}^{J} \alpha^{\dagger \sigma}(\vec{k}) \right] |0> \qquad (3.21)$$

The constant c should be chosen so that the state is normalized. Here, we have denoted the lowest energy state of the many-fermion system by $|\mathcal{O}>$ which is distinct from the empty vacuum $|0>$. We will generally call the lowest energy state of a quantum field theory the "vacuum" and we will denote it by $|\mathcal{O}>$.

In principle, we need to deal with the fact that in our infinite volume system, the product over $\vec{k}$'s in equation (3.21) is over a continuously infinite set. Of course, these are an artifact of our use of an infinite volume system. We could be more careful with this definition. However, we shall only need certain properties of the vacuum $|\mathcal{O}>$ and we can specify these properties directly in infinite volume. We shall assume that $|\mathcal{O}>$ is normalized,

$$< \mathcal{O}|\mathcal{O}> = 1 \qquad (3.22)$$

In this state, all of the states with wave-vector smaller that k_F are full

$$\alpha^{\dagger \rho}(\vec{k})|\mathcal{O}> = 0, \quad |\vec{k}| \le k_F \qquad (3.23)$$

All of the states with wave-vectors greater than k_F are empty

$$\alpha_\sigma(\vec{k})|\mathcal{O}> = 0, \quad |\vec{k}| > k_F \qquad (3.24)$$

The maximum wave-number k_F is called the Fermi wave-number. The Fermi momentum is $\hbar k_F$ and the Fermi energy is

$$\epsilon_F = \frac{\hbar^2 k_F^2}{2m} \qquad (3.25)$$

The boundary of the set of occupied states in wave-vector space, those states with $|\vec{k}| = k_F$ and which have the Fermi energy is called the Fermi surface.

3.2.2   Particles and Holes

Equations (3.22), (3.23) and (3.24) define the state $|\mathcal{O}>$ of the many-fermion system at finite density. It is illuminating to relabel the annihilation and creation operators as follows:

$$a_\sigma(\vec{k}) = \alpha_\sigma(\vec{k}), \ a^{\dagger\sigma}(\vec{k}) = \alpha^{\dagger\sigma}(\vec{k}), \ |\vec{k}| > k_F \qquad (3.26)$$

$$b^\sigma(\vec{k}) = \alpha^{\dagger\sigma}(-\vec{k}), \ b^\dagger_\sigma(\vec{k}) = \alpha_\sigma(-\vec{k}), \ |\vec{k}| \le k_F \qquad (3.27)$$

We now have two sets of creation and annihilation operators with the anti-commutator algebra

$$\left\{a_\sigma(k), a^{\dagger\rho}(q)\right\} = \delta^\rho_\sigma \delta(\vec{k} - \vec{q}) \qquad (3.28)$$

$$\left\{a_\sigma(k), a_\rho(q)\right\} = 0, \ \left\{a^{\dagger\sigma}(k), a^{\dagger\rho}(q)\right\} = 0 \qquad (3.29)$$

$$\left\{b^\sigma(k), b^\dagger_\rho(q)\right\} = \delta^\sigma_\rho \delta(\vec{k} - \vec{q}) \qquad (3.30)$$

$$\left\{b^\sigma(k), b^\rho(q)\right\} = 0, \ \left\{b^\dagger_\sigma(k), b^\dagger_\rho(q)\right\} = 0 \qquad (3.31)$$

$$\left\{a_\sigma(k), b^\rho(q)\right\} = 0, \ \left\{a^{\rho\dagger}(k), b^\rho(q)\right\} = 0 \qquad (3.32)$$

$$\left\{a_\sigma(k), b^\dagger_\rho(q)\right\} = 0, \ \left\{a^{\rho\dagger}(k), b^\dagger_\rho(q)\right\} = 0 \qquad (3.33)$$

The reader should note well that we have not introduced any new concept here. We have simply re-labeled some of the same creation and annihilation operators that we had previously defined. The reason for this re-labeling was so that, using the definition in equation (3.21), we can see that the vacuum is annihilated by the new annihilation operators,

$$a_\sigma(\vec{k})|\mathcal{O}>= 0, \ <\mathcal{O}|a^{\dagger\sigma}(\vec{k}) = 0, \ \forall \vec{k}, \sigma \qquad (3.34)$$

$$b^\rho(\vec{k})|\mathcal{O}>= 0, \ <\mathcal{O}|b^\dagger_\rho(\vec{k}) = 0, \ \forall \vec{k}, \rho \qquad (3.35)$$

$$<\mathcal{O}|\mathcal{O}>= 1 \qquad (3.36)$$

Then, there are apparently two types of excitations of this ground state. One is created by $a^{\dagger\sigma}(\vec{k})$. It is a particle with a wave-vector $|\vec{k}| > k_F$ that is outside of the Fermi surface. The other is created by $b^\dagger_\sigma(\vec{k})$. It is called a hole and it is gotten by annihilating a particle which is inside of the Fermi surface, $|\vec{k}| \le k_F$ and which is already contained in $|\mathcal{O}>$. In terms of these particle and hole creation and annihilation operators, the field operator has the form

$$\psi_\sigma(x, t) = \int_{k>k_F} d^3k \frac{e^{i\vec{k}\cdot\vec{x} - i\frac{\hbar k^2}{2m}t}}{(2\pi)^{\frac{3}{2}}} a_\sigma(\vec{k}) +$$

$$+ \int_{k<k_F} d^3k \frac{e^{-i\vec{k}\cdot\vec{x} - i\frac{\hbar k^2}{2m}t}}{(2\pi)^{\frac{3}{2}}} b^\dagger_\sigma(\vec{k}) \qquad (3.37)$$

which is identical to the expression in equation (3.7) with the re-labeling of the creation and annihilation operators in (3.26) and (3.27).

We can plug the expression for the field operator in equation (3.37) into the particle number operator and the Hamiltonian to find the expressions

$$\mathcal{N} = N + \int_{k>k_F} d^3k \, a^{\dagger\sigma}(\vec{k})a_\sigma(\vec{k}) - \int_{k<k_F} d^3k \, b_\sigma^{\dagger}(\vec{k})b^{\sigma}(\vec{k}) \qquad (3.38)$$

$$N = \rho V \qquad (3.39)$$

$$H_0 = E_0 + \int_{k>k_F} d^3k \frac{\hbar^2\vec{k}^2}{2m} a^{\dagger\sigma}(\vec{k})a_\sigma(\vec{k}) - \int_{k<k_F} d^3k \frac{\hbar^2\vec{k}^2}{2m} b_\sigma^{\dagger}(\vec{k})b^{\sigma}(\vec{k}) \qquad (3.40)$$

$$E_0 = U = uV \qquad (3.41)$$

where $N = \rho V$ is the particle number with ρ the density and $U = uV$ is the total internal energy with u the energy density of the Fermi gas. We will discuss these quantities in more detail shortly.

We note that creating a particle by operating $a^{\dagger\sigma}(\vec{k})$ on $|\mathcal{O}>$ adds one to the particle number whereas creating a hole by operating $b_\sigma^{\dagger}(\vec{k})$ subtracts one from the particle number,

$$\mathcal{N} |\mathcal{O}> = N |\mathcal{O}> \qquad (3.42)$$

$$\mathcal{N} \, a^{\dagger\sigma}(\vec{k})|\mathcal{O}> = (N+1) \, a^{\dagger\sigma}(\vec{k})|\mathcal{O}> \qquad (3.43)$$

$$\mathcal{N} \, b_\sigma^{\dagger}(\vec{k})|\mathcal{O}> = (N-1) \, b_\sigma^{\dagger}(\vec{k})|\mathcal{O}> \qquad (3.44)$$

It is a little disquieting that particles and holes also seem to have positive and negative energies, respectively. If this were the correct description of them, we could lower the energy by creating holes. We are prevented from doing this by the fact that the total particle number is fixed to be N, so every time we create a hole, we must also create a particle in order to conserve N. We will return to this issue, but for now, we could fix this apparent problem is by adding a constant to the single-particle energy so that it is defined with respect to the Fermi energy, $\frac{\hbar^2k^2}{2m} \rightarrow \left(\frac{\hbar^2k^2}{2m} - \epsilon_F\right)$. This would also add a constant, $-\epsilon_F N$ to the total energy of the N-particle system. The net effect is the replacement of the Hamiltonian in the above formula as

$$H_0 \rightarrow H_0' \equiv H_0 - \epsilon_F\mathcal{N} \qquad (3.45)$$

$$= E_0' + \int_{k>k_F} d^3k \left(\frac{\hbar^2\vec{k}^2}{2m} - \epsilon_F\right) a^{\dagger\sigma}(\vec{k})a_\sigma(\vec{k})$$

$$+ \int_{k<k_F} d^3k \left(\epsilon_F - \frac{\hbar^2\vec{k}^2}{2m}\right) b_\sigma^{\dagger}(\vec{k})b^{\sigma}(\vec{k})$$

$$E_0' = U - \epsilon_F N = (u - \epsilon_F\rho)V$$

Now, both particles and holes have positive energy relative to the vacuum state.

$$[H_0' - E_0']\, a^{\dagger\sigma}(\vec{k})|\mathcal{O}> = \frac{\hbar^2(\vec{k}^2 - k_F^2)}{2m} a^{\dagger\sigma}(\vec{k})|\mathcal{O}>,\ \ |\vec{k}| > k_F \qquad (3.46)$$

$$[H_0' - E_0']\, b_\sigma^\dagger(\vec{k})|\mathcal{O}> = \frac{\hbar^2(k_F^2 - \vec{k}^2)}{2m} b_\sigma^\dagger(\vec{k})|\mathcal{O}>,\ \ k_F > |\vec{k}| \qquad (3.47)$$

Thus, we see that, when the single-particle energy is measured relative to the Fermi energy, both particles and holes are positive energy excitations. This is obvious for a particle where we were adding a positive energy excitation to the many-particle system. When we create a hole, we are subtracting a particle which has negative energy. This is equivalent to adding positive energy. We will discuss the physical interpretation of the modified Hamiltonian H_0' and the vacuum energy that is found using it, $E_0' = U - \epsilon_F N$ in the next section.

We can find the particle number density $\rho = N/V$ as a function of the Fermi wave-number by taking the expectation value of the density operator in the vacuum state,

$$\rho = <\mathcal{O}|\rho(\vec{x}, t)|\mathcal{O}> = <\mathcal{O}|\psi^{\dagger\sigma}(\vec{x}, t)\psi_\sigma(\vec{x}, t)|\mathcal{O}>$$

$$= (2J + 1) \int\limits_{k \le k_F} \frac{d^3k}{(2\pi)^3} \cdot 1 = \frac{(2J + 1)k_F^3}{6\pi^2}$$

We can solve this equation to determine Fermi wave-number k_F and the Fermi energy ϵ_F in terms of the density,

$$k_F = \left(\frac{6\pi^2\rho}{2J + 1}\right)^{\frac{1}{3}},\quad \epsilon_F = \frac{\hbar^2}{2m}\left(\frac{6\pi^2\rho}{2J + 1}\right)^{\frac{2}{3}}$$

We will sometimes denote the ground state energy E_0 as $E_0 = U = uV$ in analogy with the thermodynamic internal energy which is denoted by U. It is obtained by taking the the expectation value of the Hamiltonian in the vacuum state, $U = <\mathcal{O}|H_0|\mathcal{O}>$, and the internal energy density u is equal to

$$u = \frac{\hbar^2}{2m} <\mathcal{O}|\vec{\nabla}\psi^{\dagger\sigma}(\vec{x}, t) \cdot \vec{\nabla}\psi_\sigma(\vec{x}, t)|\mathcal{O}> = \frac{(2J + 1)\hbar^2}{20\pi^2 m}\left(\frac{6\pi^2\rho}{2J + 1}\right)^{\frac{5}{3}} \qquad (3.48)$$

where we have written it in terms of the particle density.

3.2.3 The Grand Canonical Free Energy

In the previous section, we have observed that the particle and hole energies can both be made positive if we take the Hamiltonian of the system to be $H_0' = H_0 - \epsilon_F N$ rather than H_0. This has the simple effect of adding a constant to what we call the single-fermion energy, replacing $\frac{\hbar^2 k^2}{2m}$ by $\frac{\hbar^2 k^2}{2m} - \epsilon_F$. In this section we will discuss a way in which this change has an interesting thermodynamic interpretation. We emphasize that we are not doing thermodynamics, as the temperature of our system is always zero. However the analogy is useful.

In many practical circumstances in the quantum field theory of fermions, rather than fixing the total number of particles, as we have been doing so far, it is useful to consider an open system where particles can enter and leave the system. In this case, the energy which is minimized in order to find the ground state of the system is not the lowest eigenvalue of the Hamiltonian operator H_0 but the minimal eigenvalue of the quantity

$$H_0' = H_0 - \mu \mathcal{N} \tag{3.49}$$

where μ is a parameter called the chemical potential. It is clear why this must be the case. If fermions are free to come and go from the system, and if we wanted to minimize $< \mathcal{O}|H_0|\mathcal{O} >$, its minimum would be the state with no fermions at all. If we form the combination $H_0 - \mu \mathcal{N}$, where μ is positive, the second term favours states with particles and a mutual minimum of the two terms in

$$< \mathcal{O}|H_0'|\mathcal{O} > = < \mathcal{O}|H_0|\mathcal{O} > -\mu < \mathcal{O}|\mathcal{N}|\mathcal{O} >$$

should be a state with a nonzero number of particles. To see this, assume that $|\mathcal{O} >$ contains particles occupying states up to a Fermi wavenumber k_F. Then

$$< \mathcal{O}|H_0'|\mathcal{O} > = < \mathcal{O}|H_0|\mathcal{O} > -\mu < \mathcal{O}|\mathcal{N}|\mathcal{O} >$$
$$= V \left[\frac{(2J+1)\hbar^2 k_F^5}{20\pi^2 m} - \mu \frac{(2J+1)k_F^3}{6\pi^2} \right]$$

We regard μ as a fixed parameter and k_F as a variable. As a function of k_F the above expression has a minimum where

$$\frac{d}{dk_F} \left[\frac{(2J+1)\hbar^2 k_F^5}{20\pi^2 m} - \mu \frac{(2J+1)k_F^3}{6\pi^2} \right] = 0$$

which is solved by $k_F = \sqrt{2m\mu/\hbar^2}$. Thus k_F is determined once μ is specified. Then, we can find the value of μ which is needed to get a specific value of k_F, it is

$$\mu = \frac{\hbar^2 k_F^2}{2m} \equiv \epsilon_F$$

That is, to achieve a vacuum with a specific k_F, we should use a chemical potential which is equal to the Fermi energy. We note that this identification of μ and the Fermi energy is special for our non-interacting system at zero temperature. If either interactions or temperature were present, μ would differ from ϵ_F. We also note that, when $\mu = \epsilon_F$, the Hamiltonian H_0' which we used above in equation (3.49) is identical to the one which we used in equation (3.45) of the previous section to fix the problem that holes appeared to have negative energy.

The combination

$$\Phi = U - \mu N \tag{3.50}$$

is equal to the zero temperature limit of the grand canonical free energy in thermodynamics. It is this free energy which should be minimized in order to find the state of a system where particles are allowed to enter and leave.

If we use H_0' to generate the time evolution of the fields, the field equation becomes

$$\left(i\hbar \frac{\partial}{\partial t} + \frac{\hbar^2}{2m} \vec{\nabla}^2 + \mu \right) \psi_\sigma(\vec{x}, t) = 0 \tag{3.51}$$

The solution of this field equation (3.51), after setting $\mu = \epsilon_F$, is

$$\psi_\sigma(x, t) = \int_{k>k_F} d^3k \frac{e^{i\vec{k}\cdot\vec{x} - i\frac{\hbar(\vec{k}^2 - k_F^2)}{2m}t}}{(2\pi)^{\frac{3}{2}}} a_\sigma(\vec{k})$$

$$+ \int_{k<k_F} d^3k \frac{e^{-i\vec{k}\cdot\vec{x} + i\frac{\hbar(k_F^2 - \vec{k}^2)}{2m}t}}{(2\pi)^{\frac{3}{2}}} b_\sigma^\dagger(\vec{k}) \tag{3.52}$$

The chemical potential has the a thermodynamic definition,

$$\mu = \left. \frac{\partial u}{\partial \rho} \right|_V$$

where u is the internal energy which we found in equation (3.53) in the previous section. Using equation (3.53), we find that, for our example of non-interacting fermions at zero temperature,

$$\mu = \frac{\hbar^2 k_F^2}{2m} = \epsilon_F$$

that is, the chemical potential is equal to the Fermi energy.

The chemical potential has the statistical mechanics interpretation as the energy that is gained by adding a single particle to the system. In this example, this is clearly equal to the Fermi energy, since if we add one particle, we must add it with an energy greater than or equal to the Fermi energy.

The grand canonical potential $\Phi = \phi V$ is given by the expectation value of the Hamiltonian H' in equation (3.49), which we can find as

$$\phi = u - \mu\rho = -\frac{(2J+1)\hbar^2}{30\pi^2 m}\left(\frac{6\pi^2\rho}{2J+1}\right)^{\frac{5}{3}} \tag{3.53}$$

Notice that, if we use the thermodynamic definition of the pressure of the Fermi gas is given by

$$P = -\frac{\partial}{\partial V}(uV)_N = \frac{(2J+1)\hbar^2}{30\pi^2 m}\left(\frac{6\pi^2\rho}{2J+1}\right)^{\frac{5}{3}} \tag{3.54}$$

we see that, the grand canonical potential for a fee Fermi gas is given by

$$\Phi = -PV$$

and the grand canonical potential density is given by the negative of the pressure, $\phi = -P$.

3.3 Degenerate Bose Gas

Now, if instead of fermions, we examine the states of bosons, we find that the many-particle state at first sight has a much simpler structure. Arbitrarily many particles can occupy the lowest energy state, so the ground state of a system of free bosons is formally given by the state

$$|\mathcal{O}> = \frac{1}{\sqrt{N!}}\left(\alpha^\dagger(\vec{k}=0)\right)^N |0> \tag{3.55}$$

This state is an eigenstate of the Hamiltonian, H_0 defined in equations (3.13) and (3.14), with eigenvalue equal to zero and it is an eigenstate of the particle number operator, $\mathcal{N}$ in equations (3.15) and (3.16), with eigenvalue equal to N. For simplicity, here and in the remainder of this chapter we will assume that the bosons are spinless, that is $J = 0$. It is straightforward to modify what we do here in order to study bosons with spin.

The state where the bosons have a macroscopic occupation of a single eigenstate of the Hamiltonian, usually the ground state, is called a "Bose–Einstein condensate". The expression that we have written in equation (3.55) is not a normalizable state. We could form a normalizable by considering the smeared creation operator

$$\alpha^\dagger(f) \equiv \int d^3k f(\vec{k})\alpha^\dagger(\vec{k}), \quad \int d^3k |f(\vec{k})|^2 = 1$$

where $f(\vec{k})$ is a function which which is concentrated near the origin $\vec{k} \approx 0$. A concrete example would be

$$f(\vec{k}) = \left(\frac{2}{\pi M^2}\right)^{3/4} e^{-\vec{k}^2/M^2}, \text{ as } M \to 0$$

and the normalizable state

$$|\mathcal{O}> = \frac{1}{\sqrt{N!}} \left(a^\dagger(f)\right)^N |0> \tag{3.56}$$

The energy of this state is

$$U = < \mathcal{O}|H_0|\mathcal{O} > = \frac{3\hbar^2 M^2}{4m} \rho V$$

and the internal energy density is

$$u = \frac{3\hbar^2 M^2}{4m} \rho$$

which can be made arbitrarily small by scaling $M \to 0$. In that limit, the chemical potential

$$\mu = \frac{\partial U}{\partial N}\Big|_V = \frac{3\hbar^2 M^2}{4m}$$

goes to zero and the pressure vanishes

$$P = -\frac{\partial U}{\partial V}\Big|_N = 0$$

In this discussion, Bose condensation is the macroscopic occupation of a single energy level for the bosons. The vanishing chemical potential implies, moreover, that in an open system, particles would be free to come and go without any cost in energy. This is consistent with the vanishing of the pressure. This means that the lowest energy state of a gas of bosons is an extremely degenerate state – the states in equation (3.55) have the same energy for all values of N.

In an open system, there is also the possibility that the vacuum is not an eigenstate of the total particle number, but it is a superposition of states with different particle numbers. An example of such a state is

$$|\mathcal{O}> = \sum_{N=0}^{\infty} \frac{c_N}{\sqrt{N!}} \left[\left(a^\dagger(f)\right)^N\right] |0> \tag{3.57}$$

which is a low energy state when $f \to 1$ and it is properly normalized, $< \mathcal{O}|\mathcal{O} > = 1$, when $\sum_{N=0}^{\infty} |c_N|^2 = 1$. A diagnostic for this situation, where the ground state of a system of bosons is a superposition of states with different particle numbers is the expectation value of the field operator. For the state in equation (3.57) the expectation value is

$$< \mathcal{O}|\psi(\vec{x},t)|\mathcal{O}> = \sum_{N=0}^{\infty} c_N^* c_{N+1} \sqrt{(N+1)} f(\vec{x},t) \qquad (3.58)$$

$$f(\vec{x},t) = \int d^3k \frac{e^{i\vec{k}\cdot\vec{x} - i\frac{\hbar k^2}{2m}t}}{(2\pi)^{\frac{3}{2}}} f(\vec{k}) \qquad (3.59)$$

What is required is that at least one pair (c_N, c_{N+1}) are nonzero.[1]

When there are so many possible states of the form (3.57) that minimize the energy of a non-interacting Bose gas, any interaction which resolves the degeneracy, no matter how weak, will have a profound effect on the properties of the system. Of course any realistic system has interactions and true Bose condensates are relatively rare. On the other hand, some of the symptoms of the high degree of degeneracy like the occurrence of states of the form (3.57), is more common. It leads to the phenomenon called superfluidity.

In the following will introduce a weak repulsive interaction and a positive chemical potential for the bosons. We will also consider an open system. We will argue that such a system is a superfluid. The interaction will be a two-body contact interaction whose potential energy has the form

$$V(\vec{x}_1, ..., \vec{x}_N) = \sum_{i<j=1}^{N} \lambda \delta(\vec{x}_i - \vec{x}_j)$$

With this potential, particles interact only when they occupy the same position. The interaction is repulsive when λ is positive and it is weak when λ is small. Of course, λ is a constant with dimensions, so to say that λ is small means that it is smaller than other quantities with the same dimensions that we could make out of the other parameters of the theory. When written in terms of the s-wave scattering length, a, the interaction constant is $\lambda = \frac{4\pi a \hbar^2}{m}$ and the criterion for weak coupling is $a\rho^{\frac{1}{3}} << 1$, that is, the particle density is less than the inverse of the scattering volume, or that the average spacing between particles is much larger than the scattering length.

With the chemical potential and the two-body interaction, the second quantized Hamiltonian has the form

$$H' = \int d^3x \left[\frac{\hbar^2}{2m} \vec{\nabla}\psi^\dagger(x) \cdot \vec{\nabla}\psi(x) - \mu\psi^\dagger(x)\psi(x) + \frac{\lambda}{2}\psi^\dagger(x)\psi^\dagger(x)\psi(x)\psi(x) \right]$$

$$(3.60)$$

---

[1]   More exotic possibilities are possible, for example if $< \mathcal{O}|\psi(\vec{x},t)|\mathcal{O}> = 0$ but $< \mathcal{O}|[\psi(\vec{x},t)]^k|\mathcal{O}> \neq 0$ which would require that at least one pair (c_N, c_{N+k}) is nonzero. These are interesting from a formal point of view, but we will restrict our discussion here to the simplest possibility, the question whether or not $< \mathcal{O}|\psi(\vec{x},t)|\mathcal{O}>$ vanishes.

With this Hamiltonian, none of our formulations of the many-particle problem are exactly solvable. For example, the field equation and commutation relations are now

$$\left(i\hbar \frac{\partial}{\partial t} \psi(\vec{x}, t) + \frac{\hbar^2}{2m} \vec{\nabla}^2 + \mu \right) \psi(\vec{x}, t) = \lambda \psi^\dagger(\vec{x}, t) \psi^2(\vec{x}, t) \tag{3.61}$$

$$\left[\psi(\vec{x}, t), \psi^\dagger(\vec{y}, t) \right] = \delta(\vec{x} - \vec{y}) \tag{3.62}$$

$$\left[\psi(\vec{x}, t), \psi(\vec{y}, t) \right] = 0, \quad \left[\psi^\dagger(\vec{x}, t), \psi^\dagger(\vec{y}, t) \right] = 0 \tag{3.63}$$

The equal-time commutation relations are as they were before, but the field equation is nonlinear and, unlike the free field equation that we analyzed when the interaction was absent, a complete set of exact solutions are not known. The best that we will be able to do is perturbation theory. We will return to this after the next section where we review the Landau argument as to what to expect for the excitations of a superfluid.

3.3.1   Landau's Criterion for Superfluidity

A superfluid is a collection of particles which behave as a fluid with vanishing viscosity. It can flow past barriers or through capillaries without friction or dissipation. Superfluidity occurs in two isotopes helium-3 and helium-4 when they are liquefied by cooling them to cryogenic temperatures.

There is a beautiful argument due to Landau which relates the superfluid property that the fluid flow is non-dissipative to the spectrum of the small oscillations of the fluid, the so-called quasiparticles. Let us briefly review Landau's argument, which is very elegant.

Imagine that a macroscopic piece of superfluid of mass M is flowing with velocity $\vec{v}$ through a pipe. We will assume that the mechanism for dissipation of the fluid flow is by the interaction of the fluid with the pipe which results in the creation of a disturbance in the fluid which we shall call a quasiparticle. Let us assume that the momentum of such a quasiparticle is $\vec{p}$ and its energy is $\omega(p)$. Let us assume that we can study the fluid flow during a time interval which is so short that only one such quasiparticle is created. The question is whether quasiparticle creation will happen at all. We assume that it will readily happen if it lowers the total energy of the system.

Landau's argument relies on a change of reference frames, from the rest frame of the fluid to the rest frame of the pipe, and the transformation of some kinematic quantities when we go between these reference frames. He assumes that this transformation is governed by non-relativistic Galilean relativity where it can be implemented by a Galilean transformation. The assumption is that the dynamics of the fluid are Galilean invariant.

To begin, let us review some facts about Galilean relativity. (There will be a more fullsome presentation of Galilean relativity in the next chapter.) In classical Newtonian mechanics, the momentum and the energy of an object of mass M that is moving with velocity $\vec{v}$ are given by

$$\vec{P} = M\vec{v}, \quad E = \frac{1}{2}Mv^2, \tag{3.64}$$

respectively. According to Galilean relativity, if we view the same particle from a different reference frame, one which is moving with velocity $\vec{V}$ with respect to the first frame, the momentum and energy will appear to be

$$\vec{P}' = M(\vec{v} - \vec{V}) = \vec{P} - M\vec{V} \tag{3.65}$$

$$E' = \frac{1}{2}M(\vec{v} - \vec{V})^2 = E - \vec{P} \cdot \vec{V} + \frac{1}{2}MV^2 \tag{3.66}$$

Equations (3.65) and (3.66) tell us how to transform the momentum and the energy when we view the system in a new reference frame which is moving with velocity $\vec{V}$.

Now, let us consider the situation where a large piece of the fluid of total mass M is flowing through the pipe with uniform velocity $\vec{V}$ with respect to the pipe. We begin by viewing the fluid in its own rest frame. In its rest frame, it has vanishing velocity and momentum and it has energy E_0, the ground state energy of the static fluid.

When the fluid flow produces a quasiparticle with momentum $\vec{p}$ and energy $\omega(p)$, the total momentum in the rest frame of the fluid is that of the quasiparticle, $\vec{p}$, and the total energy is that of the fluid at rest plus the energy of the quasiparticle, $E_0 + \omega(p)$.

Now, let us examine this system in the rest frame of the pipe. We do this by using the transformation of Galilean relativity in equations (3.65) and (3.66) to go to from the rest frame of the fluid to the rest frame of the pipe, a Galilean boost by velocity $-\vec{V}$. The total momentum and energy in the rest frame of the pipe are then, according to the rules of Galilean relativity,

$$\vec{P}' = \vec{p} - M\vec{V} \tag{3.67}$$

$$E' = E_0 + \omega(p) - \vec{p} \cdot \vec{V} + \frac{1}{2}MV^2 \tag{3.68}$$

We should compare this with the same motion but where no quasiparticle is produced. In that case the momentum and energy in the rest frame of the pipe would be

$$\tilde{\vec{P}}' = -M\vec{V} \tag{3.69}$$

$$\tilde{E}' = E_0 + \frac{1}{2}M\vec{V}^2 \tag{3.70}$$

This process of producing a quasiparticle will proceed if it is energetically favourable. This is so if the energy of the state where the quasiparticle was produced is less than the energy of the state where it was not produced, that is if, in the rest frame of the pipe, $E' \leq \tilde{E}'$ or

$$E_0 + \omega(p) - \vec{p} \cdot \vec{v} + \frac{1}{2}Mv^2 \leq E_0 + \frac{1}{2}Mv^2 \qquad (3.71)$$

or if

$$\omega(p) \leq \vec{p} \cdot \vec{v} \text{ at least for some values of } \vec{p} \qquad (3.72)$$

This can happen when

$$v \geq v_c \equiv \text{minimum of } \frac{\omega(p)}{p} \qquad (3.73)$$

The last inequality tells us two important things. One is that dissipation is allowed only when the fluid velocity exceeds a minimum critical velocity, v_c. The other is, that there is a superfluid at all only when $v_c > 0$ and it gives us a way of determining v_c.

This critical velocity could vanish, for example, if $\omega(p) \sim p^2$, as it does for a normal fluid. On the other hand, if $\omega(p) \sim v_s|\vec{p}|$ for small $p = |\vec{p}|$, as it would be for a sound wave, with v_s the sound velocity, the critical velocity would be $v_c = v_s$ and it would be non-zero and dissipation would not occur for fluid flows with velocities smaller v_c. Landau's criterion for a superfluid is that $v_c > 0$. It also relates v_c to the spectrum of quasiparticles.

3.3.2   Vacuum Expectation Value

As we have stated above, for an open system of bosons, the ground state need no longer be an eigenstate of particle number. It can be a superposition of states with different particle numbers as in equation (3.57). We have also emphasized there that a diagnostic of this property of the ground state is the expectation value of the field operator itself in the ground state,

$$\eta(\vec{x}, t) = <\mathcal{O}|\psi(\vec{x}, t)|\mathcal{O}> \qquad (3.74)$$

We will show that this is indeed a typical occurrence in a weakly interacting Bose gas.

When the field operator has an expectation value, as in equation (3.74) , it will turn out to be useful to separate the field operator into a classical part and quantum part,

$$\psi(\vec{x}, t) = \eta(\vec{x}, t) + \tilde{\psi}(\vec{x}, t) \qquad (3.75)$$

where we define this separation so that the quantum part has vanishing vacuum expectation value

$$< \mathcal{O}|\tilde{\psi}(\vec{x}, t)|\mathcal{O} >= 0 \tag{3.76}$$

The quantum part must satisfy the equal-time commutation relations which are obtained by plugging equation (3.74) into equations (3.62) and (3.63)

$$\left[\tilde{\psi}(\vec{x}, t), \tilde{\psi}^{\dagger}(\vec{y}, t) \right] = \delta(\vec{x} - \vec{y}) \tag{3.77}$$

$$\left[\tilde{\psi}(\vec{x}, t), \tilde{\psi}(\vec{y}, t) \right] = 0, \quad \left[\tilde{\psi}^{\dagger}(\vec{x}, t), \tilde{\psi}^{\dagger}(\vec{y}, t) \right] = 0 \tag{3.78}$$

We are going to do a computation which uses perturbation theory. It turns out to be an asymptotic of expansion the fields, both their classical and quantum parts, in the interaction strength λ as follows:

$$\eta(\vec{x}, t) = \frac{1}{\sqrt{\lambda}} \eta^{(-\frac{1}{2})}(\vec{x}, t) + \delta\eta(\vec{x}, t) \tag{3.79}$$

$$\tilde{\psi}(\vec{x}, t) = \tilde{\psi}^{(0)}(\vec{x}, t) + \delta\tilde{\psi}(\vec{x}, t) \tag{3.80}$$

$$\delta\eta(\vec{x}, t) \text{ is of order } \lambda^{\frac{1}{2}} \text{ and higher} \tag{3.81}$$

$$\delta\tilde{\psi}(\vec{x}, t) \text{ is of order } \lambda \text{ and higher} \tag{3.82}$$

Let us begin by studying the field equation. If we plug (3.74) in the form of the expansions in equations (3.79) and (3.80) into the field equation (3.61) and if we set the coefficients of each power of λ to zero, we get a hierarchy of coupled equations. The first two equations in the hierarchy are

$$\left(i\hbar\frac{\partial}{\partial t} + \frac{\hbar^2}{2m}\vec{\nabla}^2 + \mu \right) \eta^{(-\frac{1}{2})}(\vec{x}, t) = |\eta^{(-\frac{1}{2})}(\vec{x}, t)|^2 \eta^{(-\frac{1}{2})}(\vec{x}, t) \tag{3.83}$$

$$\left(i\hbar\frac{\partial}{\partial t} + \frac{\hbar^2}{2m}\vec{\nabla}^2 + \mu - 2|\eta^{(-\frac{1}{2})}(\vec{x}, t)|^2 \right) \tilde{\psi}^{(0)}(\vec{x}, t)$$

$$- [\eta^{(-\frac{1}{2})}(\vec{x}, t)]^2 \tilde{\psi}^{(0)\dagger}(\vec{x}, t) = 0 \tag{3.84}$$

In the first equation (3.83) we see that $\eta^{(-\frac{1}{2})}(\vec{x}, t)$ obeys the classical field equation. Then, the second equation (3.84) is a linear partial differential equation for $\tilde{\psi}^{(0)}(\vec{x}, t)$. It depends on the solution, $\eta^{(-\frac{1}{2})}(\vec{x}, t)$, of the first equation. The remainder of the equations in the hierarchy then determine the higher order corrections, $\delta\eta(\vec{x}, t)$ and $\delta\tilde{\psi}(\vec{x}, t)$.

As well, plugging the expansions into the equal-time commutation relations tells us that the solutions of equation (3.84) must be operators whose vacuum expectation value vanishes, $< \mathcal{O}|\tilde{\psi}^{(0)}(\vec{x}, t)|\mathcal{O} >= 0$ and which obey the equal-time commutators

$$\left[\tilde{\psi}^{(0)}(\vec{x}, t), \tilde{\psi}^{(0)\dagger}(\vec{y}, t)\right] = \delta(\vec{x} - \vec{y}) \tag{3.85}$$

$$\left[\tilde{\psi}^{(0)}(\vec{x}, t), \tilde{\psi}^{(0)}(\vec{y}, t)\right] = 0, \quad \left[\tilde{\psi}^{(0)\dagger}(\vec{x}, t), \tilde{\psi}^{(0)\dagger}(\vec{y}, t)\right] = 0 \tag{3.86}$$

We begin by looking for solutions of the classical field equation (3.83). To examine possible homogeneous stationary states of the system we begin by looking for solutions where $\eta^{(-\frac{1}{2})}$ is a constant. We find that there are two such solutions. First of all

$$\eta^{(-\frac{1}{2})} = 0 \tag{3.87}$$

is always a solution. Then, we see that, when μ is positive (and we have already assumed that λ is positive), there is a family of solutions

$$\eta^{(-\frac{1}{2})} = \sqrt{\mu}e^{i\chi} \tag{3.88}$$

parameterized by the angle $0 \leq \chi < 2\pi$.

We must choose the solution which minimizes the energy. We will see shortly that, for very weak interactions, the vacuum expectation value of the Hamiltonian, given in equation (3.60) is dominated by what is obtained by simply replacing the field operators there by their leading order classical part, $\eta^{(-\frac{1}{2})}$. Upon doing that, we see that when $\mu > 0$ and $\lambda > 0$ the non-trivial solutions (3.88) indeed have the lower energy. Moreover, they have the same energy for all values of the angle χ. This is a result of a symmetry of the theory which we will discuss in more detail shortly. A result of the symmetry will be that any value of χ which we choose when we pick a solution will lead to physics identical to that of any other value of χ. The fact is, to proceed, we must choose one. Let us therefore choose $\chi = 0$. The we take $\eta^{(-\frac{1}{2})} = \sqrt{\mu}$ and the field $\psi(\vec{x}, t)$ has a nonzero vacuum expectation value.

Then, at very weak coupling, the particle density and particle number are gotten by plugging the classical field (3.88) into the density operator and taking the vacuum expectation value,

$$\rho = < \mathcal{O}|\rho(\vec{x}, t)|\mathcal{O} >= \eta^{\dagger}\eta + < \mathcal{O}|\tilde{\psi}^{\dagger}(\vec{x}, t)\tilde{\psi}(\vec{x}, t)|\mathcal{O} > \tag{3.89}$$

$$= \frac{\mu}{\lambda} + \dots \tag{3.90}$$

where the ellipses denote corrections or order λ^0. If we plug our expansion into the Hamiltonian, we can see that when μ and λ are positive, the non-zero solution in equation (3.88) is indeed the solution with lower energy and we find the expression

$$H' = -\frac{\mu^2}{2\lambda}V + \text{constant}$$

$$+ \int d^3x \left\{ \frac{\hbar^2}{2m} \vec{\nabla}\tilde{\psi}^{(0)\dagger}(\vec{x}, t) \cdot \vec{\nabla}\tilde{\psi}^{(0)}(\vec{x}, t) + \mu\psi^{(0)\dagger}(\vec{x}, t)\psi^{(0)}(\vec{x}, t) \right.$$

$$\left. + \frac{\mu}{2}[\tilde{\psi}^{(0)\dagger}(\vec{x}, t)]^2 + [\tilde{\psi}^{(0)}(\vec{x}, t)]^2 \right\} + \mathcal{O}(\lambda) \tag{3.91}$$

The first term on the right-hand-side of equation (3.91) is the leading contribution to the vacuum free energy[2]

$$\Phi = -\frac{\mu^2}{2\lambda}V + \ldots = -\frac{\lambda\rho^2}{2}V + \ldots$$

The second, constant term which follows it in equation (3.91) comes from a correction to η which is of order $\sqrt{\lambda}$. Though finding it is a straightforward exercise, highly recommended for the enterprising student, we will not do it here, so we do not have control over this second term. We do know that it is a number, not an operator. It will contribute to the vacuum free energy at order λ^0 but it will not influence the spectrum of small excitations of the Bose fluid which will be described by the terms in the second line of equation (3.91). That term is, in these leading orders of perturbation theory, quadratic in the field operators. It contains, as well as the free particle energy other terms which are proportional to the chemical potential and which summarize the interaction of the particles with the condensate. The remaining terms, denoted by the ellipses in equation (3.91), are all of higher orders in λ and we will ignore them here, with the assumption that they are not important in the weak coupling, small λ regime. The leading order in those terms turns out to be of order λ.

We can easily diagonalize the leading terms in the approximation to the Hamiltonian that is given by the quadratic terms in the second line of equation (3.91). The eigenstates of that operator will give us the energies of elementary excitations of this system accurately at this order in perturbation theory. The field operator at time $t = 0$ is given by

$$\tilde{\psi}(\vec{x}, 0) = \int \frac{d^3k}{(2\pi)^3} e^{i\vec{k}\cdot\vec{x}} \alpha(\vec{k})$$

with the commutation relations (3.85) and (3.86) satisfied when

$$\left[\alpha(\vec{k}), \alpha(\vec{\ell})\right] = 0 = \left[\alpha^\dagger(\vec{k}), \alpha^\dagger(\vec{\ell})\right], \quad \left[\alpha(\vec{k}), \alpha^\dagger(\vec{\ell})\right] = \delta^3(\vec{k} - \vec{\ell})$$

---

[2] Remeber that we are discussing an open system where the Hamiltonian that we are using has the form $H' = H - \mu\mathcal{N}$ and has expectation value interpreted as the grand canonical free energy of the state, the quantity which we called Φ in the discussion of the degenerate Fermi gas.

The Hamiltonian in terms of creation and annihilation operators is

$$H' = -\frac{\mu^2}{2\lambda}V + \text{constant}$$
$$+ \int d^3k \left\{ \left(\frac{\hbar^2 k^2}{2m} + \mu \right) \alpha^\dagger(\vec{k})\alpha(\vec{k}) + \frac{\mu}{2}[\alpha^\dagger(\vec{k})\alpha^\dagger(-\vec{k}) + \alpha(-\vec{k})\alpha(\vec{k})] \right\}$$
$$+ \mathcal{O}(\lambda) \tag{3.92}$$

This Hamiltonian no longer has the form of an energy like it did for noninteracting particles where the integrand had the single-particle energy $\left[\frac{\hbar^2 k^2}{2m} - \mu \right]$ times the operator $\alpha^\dagger(\vec{k})\alpha(\vec{k})$ which counts the number of excitations with wave-number $\vec{k}$. To diagonalize the Hamiltonian in equation (3.92), we shall find a change of variables in order to get it back into such a form. To this purpose, consider the transformation

$$\begin{bmatrix} a(\vec{k}) \\ a^\dagger(-k) \end{bmatrix} = \begin{bmatrix} \cosh\phi & \sinh\phi \\ \sinh\phi & \cosh\phi \end{bmatrix} \begin{bmatrix} \alpha(k) \\ \alpha^\dagger(-k) \end{bmatrix} \tag{3.93}$$

and its inverse

$$\begin{bmatrix} \alpha(\vec{k}) \\ \alpha^\dagger(-k) \end{bmatrix} = \begin{bmatrix} \cosh\phi & -\sinh\phi \\ -\sinh\phi & \cosh\phi \end{bmatrix} \begin{bmatrix} a(k) \\ a^\dagger(-k) \end{bmatrix} \tag{3.94}$$

where ϕ is a function of $|\vec{k}|$. This is called a Bogoliubov transformation. Its specific form with hyperbolic functions obeying $\cosh^2\phi - \sinh^2\phi = 1$ is designed to preserve the commutation relations so that the new variables also obey

$$\left[a(\vec{k}), a(\vec{\ell}) \right] = 0 = \left[a^\dagger(\vec{k}), a^\dagger(\vec{\ell}) \right], \quad \left[a(\vec{k}), a^\dagger(\vec{\ell}) \right] = \delta^3(\vec{k} - \vec{\ell})$$

We assume that ϕ is a real function of $|\vec{k}|$. When we substitute the transformation in equation (3.94) into the Hamiltonian and require that the Hamiltonian is diagonal, we obtain

$$\tanh 2\phi = \frac{\mu}{\frac{\hbar^2 k^2}{2m} + \mu}$$

and we find that the Hamiltonian is

$$H' = -\frac{\mu^2}{2\lambda}V + \text{constant}$$
$$+ \int d^3k E(k) a^\dagger(\vec{k})a(\vec{k}) + \mathcal{O}(\lambda) \tag{3.95}$$

where the new energies, which we shall call quasiparticle energies, are

$$E(k) = \sqrt{ \left(\frac{\hbar^2 k^2}{2m} + \mu \right)^2 - \mu^2 } \tag{3.96}$$

The vacuum state of the Bose gas now obeys

$$a(\vec{k})|\mathcal{O}>= 0, \quad <\mathcal{O}|a^{\dagger}(\vec{k}), \forall \vec{k}, \quad <\mathcal{O}|\mathcal{O}>= 1 \tag{3.97}$$

and a basis for the space of quantum states is gotten by operating quasiparticle creation operators on the vacuum,

$$|\mathcal{O}>, \; a^{\dagger}(\vec{k})|\mathcal{O}>, \; \frac{1}{\sqrt{n!}}a^{\dagger}(\vec{k}_1)\ldots a^{\dagger}(\vec{k}_n)|\mathcal{O}>, \ldots$$

These are states of the Bose fluid with additional quasiparticle excitations.

High energy quasi-particles, those with large wave-numbers, behave like free particles, they have $E(k) \sim \frac{\hbar^2 \vec{k}^2}{2m}$. On the other hand, at low energies, that is for small wave-number $|\vec{k}|$, the quasiparticle behaves like a sound wave,

$$E(k) \sim \hbar v_S |\vec{k}| \tag{3.98}$$

By Landau's criterion, the critical velocity of the superfluid is given by

$$v_c = \text{ minimum of } \frac{1}{\hbar}\frac{E(k)}{k} = \sqrt{\frac{\lambda \rho}{m}} = v_S \tag{3.99}$$

This is a beautiful confirmation of Landau's general argument for the quasi-particle spectrum in a superfluid.

This leading order in the vacuum energy is equal to the grand canonical free energy which has density

$$\phi = \frac{\Phi}{V} = -\frac{\hat{\mu}^2}{2\lambda} + \ldots \tag{3.100}$$

where corrections, represented by the three dots, begin at order λ^0. If we recall the thermodynamic formulae

$$\rho V = -\frac{\partial \Phi}{\partial \mu}\bigg|_V, \quad P = -\frac{\partial \Phi}{\partial V}\bigg|_\mu$$

we can find an expression for the chemical potential and the pressure in terms of the density,

$$\mu = \lambda \rho + \ldots \tag{3.101}$$

$$P = \frac{\mu^2}{2\lambda} + \ldots = \frac{\lambda \rho^2}{2} + \ldots \tag{3.102}$$

which agrees with our direct computation of the density in equation (3.90) and it shows us that the pressure is a monotonically increasing function of the density, as we expect it to be

The speed of sound can be found from finding an expression for the velocity of a compressional wave as the derivative of the pressure by the mass density[3]

$$v_s^2 = \left. \frac{\partial P}{\partial (m\rho)} \right|_S = \frac{\mu}{m} = \frac{\lambda \rho}{m}$$

What we obtain agrees with the critical velocity and the sound velocity which we found in equation (3.99).

3.4   Spontaneous Symmetry Breaking

The field equation (3.61) and the commutation relations (3.62) and (3.63) which define the many-particle system as a quantum field theory have an important symmetry under changing the phase of the field operator,

$$\psi(\vec{x}, t) \to \tilde{\psi}(\vec{x}, t) = e^{i\chi} \psi(\vec{x}, t) \tag{3.103}$$

$$\psi^\dagger(\vec{x}, t) \to \tilde{\psi}^\dagger(\vec{x}, t) = \psi^\dagger(\vec{x}, t) e^{-i\chi} \tag{3.104}$$

where χ is a constant. This is a symmetry in the sense that, if $\psi(\vec{x}, t)$ is a solution of the field equations and commutation relations, then so is its transform, $\tilde{\psi}(\vec{x}, t) = e^{i\chi} \psi(\vec{x}, t)$.

In the weak coupling limit, where we give the field a classical and quantum part, $\psi(\vec{x}, t) = \eta + \tilde{\psi}(\vec{x}, t)$, we see that the energy is dominated by the classical part

$$\frac{1}{V} <O|H'|O> \approx = \frac{\lambda}{2} \left(\eta^* \eta - \frac{\mu}{\lambda} \right)^2 - \frac{\mu^2}{2\lambda} + \dots \tag{3.105}$$

This energy density for a constant expectation value η which is depicted in Fig. 3.1.

The symmetry is visible in the diagram as the rotation symmetry about the vertical axis. However, a solution must lie in a minimum of this energy function. The set of solutions itself forms an orbit of the symmetry transformation, the ring of minima in the figure. However, when we choose one of the minima, $\eta = \sqrt{\mu\lambda} + \dots$ and we look at the Hamiltonian for the quantum field $\tilde{\psi}(\vec{x}, t)$,

$$H' = \dots + \int d^3x \left\{ \frac{\hbar^2}{2m} \vec{\nabla} \tilde{\psi}^\dagger(\vec{x}, t) \cdot \vec{\nabla} \tilde{\psi}(\vec{x}, t) + \mu \psi^\dagger(\vec{x}, t) \psi(\vec{x}, t) \right.$$
$$\left. + \frac{\mu}{2} \left[\tilde{\psi}^\dagger(\vec{x}, t) \tilde{\psi}^\dagger(\vec{x}, t) + \tilde{\psi}(\vec{x}, t) \tilde{\psi}(\vec{x}, t) \right] \right\} + \dots \tag{3.106}$$

---

[3] The partial derivative must be taken with the entropy held constant. Here the temperature held at zero. This is equivalent to holding the entropy constant.

Fig. 3.1 The energy density is plotted on the vertical axis versus the real and imaginary parts of η on the horizontal axes. The symmetry is under rotations about the vertical axis. The extremum at $\eta = 0$ is a local maxima. There is a ring of minima located at $\eta = e^{i\chi}\sqrt{\frac{\mu}{\lambda}}$ with $0 \le \chi < 2\pi$

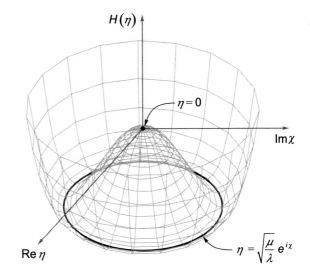

The phase symmetry is nowhere to be seen. Its manifestation is the sound wave excitation which corresponds to a fluctuation of the phase of ψ which ends up being the quasi-particle. It is easiest to see after presenting the Hamiltonian as

$$H' = \ldots +$$
$$+ \int d^3x \frac{1}{2} \left[\tilde{\psi}^\dagger(x,t), \tilde{\psi}(\vec{x},t) \right] \begin{bmatrix} -\frac{\hbar^2 \vec{\nabla}^2}{2m} + \mu & \mu \\ \mu & -\frac{\hbar^2 \vec{\nabla}^2}{2m} + \mu \end{bmatrix} \begin{bmatrix} \tilde{\psi}(\vec{x},t) \\ \tilde{\psi}^\dagger(\vec{x},t) \end{bmatrix} + \ldots$$
$$(3.107)$$

we see that a fluctuation of the phase $\begin{bmatrix} \tilde{\psi}(\vec{x},t) \\ \tilde{\psi}^\dagger(\vec{x},t) \end{bmatrix} \sim \begin{bmatrix} i \\ -i \end{bmatrix}$ has vanishing energy.
This is directly related to the fact that, for any solution occupying one point on the ring of minima of the energy function depicted in Fig. 3.1, there is a flat direction in the energy landscape. In this case, the result is the quasi-particle whose frequency goes to zero as its wave-number goes to zero,

$$\lim_{\vec{k}\to 0} \omega(\vec{k}) = 0$$

In a quantum mechanical system with a finite number of particles and in a space of finite volume, the ground state would not be degenerate, as it is here. In this system the degeneracy survives because, in a sense, the system is a quantum rotor with an infinite moment of inertia, so the rotor gets stuck at one position. This phenomenon is called spontaneous symmetry breaking. The quasiparticle which results from the existence of the flat direction is called a Goldstone boson.

The Action Principle and Noether's Theorem

In the previous chapters, we have formulated the quantum mechanical many-particle system as a quantum field theory. The quantum field theory consisted of a field equation, which was a non-linear partial differential equation, in our example,[1]

$$\left(i\hbar\frac{\partial}{\partial t} + \frac{\hbar^2}{2m}\vec{\nabla}^2\right)\psi_\sigma(\vec{x},t) = \lambda\psi^{\dagger\rho}(\vec{x},t)\psi_\rho(\vec{x},t)\psi_\sigma(\vec{x},t) \tag{4.1}$$

and equal-time commutation relations

$$\left[\psi_\sigma(\vec{x},t), \psi^{\dagger\rho}(\vec{y},t)\right] = \delta_\sigma^\rho\delta(\vec{x}-\vec{y}), \tag{4.2}$$

$$\left[\psi_\sigma(\vec{x},t), \psi_\rho(\vec{y},t)\right] = 0, \quad \left[\psi^{\dagger\sigma}(\vec{x},t), \psi^{\dagger\rho}(\vec{y},t)\right] = 0 \tag{4.3}$$

or equal-time anti-commutation relations

$$\left\{\psi_\sigma(\vec{x},t), \psi^{\dagger\rho}(\vec{y},t)\right\} = \delta_\sigma^\rho\delta(\vec{x}-\vec{y}), \tag{4.4}$$

$$\left\{\psi_\sigma(\vec{x},t), \psi_\rho(\vec{y},t)\right\} = 0, \quad \left\{\psi^{\dagger\sigma}(\vec{x},t), \psi^{\dagger\rho}(\vec{y},t)\right\} = 0 \tag{4.5}$$

In this chapter, we shall examine an alternative way of encoding the information that is contained in the field equation and commutation relations. The essential information that is contained in equations (4.1) and (4.2)–(4.3) or in equations (4.1) and

---

[1] Here, and in the following, we will consider the special case of a contact interaction. Everything that we say can easily be generalized to more complicated interactions.

© The Author(s), under exclusive license to Springer Nature Singapore Pte Ltd. 2023
G. W. Semenoff, *Quantum Field Theory*, Graduate Texts in Physics,
https://doi.org/10.1007/978-981-99-5410-0_4

(4.4)–(4.5) can be extracted from the Lagrangian density

$$\mathcal{L}(\vec{x}, t) = \frac{i\hbar}{2}\left[\psi^{\dagger\sigma}(\vec{x}, t)\frac{\partial}{\partial t}\psi_\sigma(\vec{x}, t) - \frac{\partial}{\partial t}\psi^{\dagger\sigma}(\vec{x}, t)\psi_\sigma(\vec{x}, t)\right]$$
$$- \frac{\hbar^2}{2m}\vec{\nabla}\psi^{\dagger\sigma}(\vec{x}, t) \cdot \vec{\nabla}\psi_\sigma(\vec{x}, t) - \frac{\lambda}{2}\left(\psi^{\dagger\sigma}(\vec{x}, t)\psi_\sigma(\vec{x}, t)\right)^2 \qquad (4.6)$$

and the action, which is the integral over time and space of the Lagrangian density,

$$\mathbf{S}[\psi_\sigma, \psi^{\dagger\sigma}] = \int dt \int d^3x \ \mathcal{L}(\vec{x}, t) \qquad (4.7)$$

To be logically precise here, the fields $\psi_\sigma(\vec{x}, t)$ and $\psi^{\dagger\sigma}(\vec{x}, t)$ occurring in the Lagrangian density (4.6) and the action (4.7) are classical fields. A classical field is simply a function of the space and time coordinates which obeys certain boundary conditions. As is often done, we will use the same notation to denote these classical fields as we do for the quantum field operators of the corresponding quantum field theory.

The idea here is that the Lagrangian density and action are those of a classical field theory which is a dynamical theory which determines the behaviour of the classical fields. The Lagrangian is constructed in such a way that the equation of motion for the classical field theory is identical in form to the field equation (4.1) of the quantum field theory so that, modulo operator ordering, we can deduce the field equation for the quantum field theory by simply replacing the classical fields by the field operators. The only place where operator ordering is an issue in equation (4.1) is in the interaction term on the right-hand-side. Of course, we already know what the ordering should be. But in a case where we did not know we would have to find a way to deal with the ordering ambiguity. There is really no systematic way to do this. It is usually done from case to case. It will turn out to be easy to find the appropriate ordering or to at least parameterize the ordering ambiguity in all of the relatively simple quantum field theories that we study. This allows us to deduce the field equations of quantum field theory from the field equations of classical field theory. Then the classical field theory also has a Poisson bracket structure for its dynamical variables which we can lift to find the commutation relations for the field operators of the quantum field theory. This will also be relatively straightforward and free of ambiguities for the simple field theories that we will study.[2]

Aside from an elegant re-encoding of the information of the quantum field theory, from equations (4.1) and (4.2)–(4.3) or (4.1) and (4.4)–(4.5) into the somewhat

---

[2] Strictly speaking, what we have said in the paragraph above applies only to bosons. If our theory, once it is eventually used as a quantum field theory, will describe fermions rather than bosons, the fields $\psi_\sigma(\vec{x}, t)$ and $\psi^{\dagger\sigma}(\vec{x}, t)$ are slightly more complicated objects, in that they must anti-commute with each other.[3] They are still not operators in the sense that the fermion quantum fields are operators. Rather, they simply should simply be viewed as functions which anti-commute as, for example,

more compact expression $\mathcal{L}(\vec{x}, t)$ in equation (4.6) and (4.7), the classical fields and their action will play an important role in the functional integral formulation of the quantum field theory. In the functional formulation, the computation of the expectation value of an operator in the vacuum state takes the form

$$< \mathcal{O}|\psi^{\dagger\sigma}(\vec{x}, t)\psi_\sigma(\vec{x}, t)|\mathcal{O}> = \frac{\int [d\psi_\sigma(\vec{x}, t)d\psi^{\dagger\rho}(\vec{x}, t)]e^{\frac{i}{\hbar}S[\psi, \psi^\dagger]}\psi^{\dagger\sigma}(\vec{x}, t)\psi_\sigma(\vec{x}, t)}{\int [d\psi_\sigma(\vec{x}, t)d\psi^{\dagger\rho}(\vec{x}, t)]e^{\frac{i}{\hbar}S[\psi, \psi^\dagger]}}$$

The left-hand-side of the above equation is entirely quantum mechanical. It is the expectation value in a quantum state of an operator that is made from the field operators of a quantum field theory. On the other hand, in some sense, the right-hand-side is entirely classical. It contains an integral over all of the possible classical fields that satisfy appropriate boundary conditions. The weight of each classical field configuration in the integral is the exponential of the classical action, $\exp\left(\frac{i}{\hbar}S[\psi, \psi^\dagger]\right)$. We will return to this in some detail in subsequent chapters.

4.1 The Action

The Lagrangian density and the action are important objects in classical field theory.

The action in equation (4.7) is given by an integral over the space and time coordinates of the Lagrangian density given in equation (4.6). The Lagrangian density is a function of the classical fields and their derivatives. An example of a Lagrangian density is our example in equation (4.6) which depends on the classical fields and their first derivatives, the quantities

$$\psi_\sigma(\vec{x}, t), \ \frac{\partial}{\partial t}\psi_\sigma(\vec{x}, t), \ \nabla\psi_\sigma(\vec{x}, t), \ \psi^{\sigma\dagger}(\vec{x}, t), \ \frac{\partial}{\partial t}\psi^{\sigma\dagger}(\vec{x}, t), \ \nabla\psi^{\sigma\dagger}(\vec{x}, t)$$

$$\psi_\sigma(\vec{x}, t)\psi^{\dagger\rho}(\vec{y}, t') = -\psi^{\dagger\rho}(\vec{y}, t')\psi_\sigma(\vec{x}, t)$$
$$\psi_\sigma(\vec{x}, t)\psi_\rho(\vec{y}, t') = -\psi_\rho(\vec{y}, t')\psi_\sigma(\vec{x}, t)$$
$$\psi^{\dagger\sigma}(\vec{x}, t)\psi^{\dagger\rho}(\vec{y}, t') = -\psi^{\dagger\rho}(\vec{y}, t')\psi^{\dagger\sigma}(\vec{x}, t)$$

Clearly, objects which anti-commute with each other in this way are not ordinary functions. In fact a deeper study of the mathematical nature and properties of these objects is an interesting one. However, we will not pursue it here as we shall not require much of the details, other than a few simple rules which will allow us to use them for a few specific things. We will develop those rules as we need them. The fact that the objects anti-commute is the first such rule. The second rule concerns complex conjugation, where, when we conjugate a product of anti-commuting fields, we also reverse the order so that, for example,

$$\left(\psi_\sigma(\vec{y}, t)\, \psi_\rho(\vec{y}, t')\right)^\dagger = \psi^{\dagger\rho}(\vec{y}, t')\, \psi^{\dagger\sigma}(\vec{x}, t)$$

This, for example, makes the density $\psi^{\dagger\sigma}(\vec{x}, t)\psi_\sigma(\vec{x}, t)$ real whereas if we did not reverse the order upon conjugation it would be imaginary.

The action $\mathbf{S}[\psi_\sigma, \psi^{\dagger\rho}]$ is a functional. A functional is a mathematical object which maps functions onto numbers. In this case, we would plug some classical fields $\psi_\sigma(\vec{x}, t)$ and $\psi^{\dagger\rho}(\vec{x}, t)$ into $\mathcal{L}(\vec{x}, t)$ given in equation (4.6). Then we perform the integrals in equation (4.7) to obtain $\mathbf{S}[\psi_\sigma, \psi^{\dagger\rho}]$. The result is a number, in this case a real number.[4]

4.1.1   The Euler–Lagrange Equations

The action principle states that the action functional, considered as a mapping of classical fields obeying the appropriate boundary conditions to the real numbers, is stationary when it is evaluated on the field configurations which obey the classical equations of motion, that is, the classical field equation.

Let us use this principle to find the equations of motion which correspond to a given action and Lagrangian density. Then we consider the classical fields $\psi_\sigma(\vec{x}, t)$ and $\psi^{\dagger\sigma}(\vec{x}, t)$ and some other classical fields which differ from them by an infinitesimal amount,

$$\psi_\sigma(\vec{x}, t) + \delta\psi_\sigma(\vec{x}, t), \psi^{\dagger\sigma}(\vec{x}, t) + \delta\psi^{\dagger\sigma}(\vec{x}, t)$$

where $\delta\psi_\sigma(\vec{x}, t)$ and $\delta\psi^{\dagger\sigma}(\vec{x}, t)$ are functions of infinitesimal magnitude but arbitrary profile. The fields $\psi^{\dagger\sigma}(\vec{x}, t)$ and $\delta\psi^{\dagger\sigma}(\vec{x}, t)$ are the complex conjugates of $\psi_\sigma(\vec{x}, t)$ and $\delta\psi_\sigma(\vec{x}, t)$, respectively. We define the variation of the action by the difference

$$\delta\mathbf{S}[\psi_\sigma, \psi^{\dagger\sigma}] \equiv \mathbf{S}[\psi_\sigma + \delta\psi_\sigma, \psi^{\dagger\sigma} + \delta\psi^{\dagger\sigma}] - S[\psi_\sigma, \psi^{\dagger\sigma}] \qquad (4.8)$$

where, on the right-hand-side, we keep only terms of first order in $\delta\psi_\sigma(\vec{x}, t)$ and $\delta\psi^{\dagger\sigma}(\vec{x}, t)$.

We say that the action is stationary at the location $(\psi_\sigma(\vec{x}, t), \psi^{\dagger\sigma}(\vec{x}, t))$ in the space of all possible fields if the variation of the action defined in equation (4.8) vanishes there, that is, if

$$\delta\mathbf{S}[\psi_\sigma, \psi^{\dagger\sigma}] = 0 \qquad (4.9)$$

---

[4] For fermions, the situation is a little more complicated as the functional of anti-commuting functions is not just a number but it is itself an algebraic entity. Again, we will not take the time to define it here, as only some operational aspects of dealing with anti-commuting functions will be required in the following and we will introduce them when they are needed.

To linear order,

$$
\delta S = \int dt \int d^3 x \left[\delta\psi_\sigma(\vec{x}, t) \frac{\partial \mathcal{L}}{\partial \psi_\sigma(\vec{x}, t)} + \delta\left(\frac{\partial}{\partial t}\psi_\sigma(\vec{x}, t)\right) \frac{\partial \mathcal{L}}{\partial(\frac{\partial}{\partial t}\psi_\sigma(\vec{x}, t))} \right.
$$
$$
+ \delta(\nabla_a\psi_\sigma(\vec{x}, t)) \frac{\partial \mathcal{L}}{\partial(\nabla_a\psi_\sigma(\vec{x}, t))} + \delta\psi^{\dagger\sigma}(\vec{x}, t) \frac{\partial \mathcal{L}}{\partial\psi^{\dagger\sigma}(\vec{x}, t)}
$$
$$
\left. + \delta\left(\frac{\partial}{\partial t}\psi^{\dagger\sigma}(\vec{x}, t)\right) \frac{\partial \mathcal{L}}{\partial(\frac{\partial}{\partial t}\psi^{\dagger\sigma}(\vec{x}, t))} + \delta(\nabla_a\psi^{\dagger\sigma}(\vec{x}, t)) \frac{\partial \mathcal{L}}{\partial(\nabla_a\psi^{\dagger\sigma}(\vec{x}, t))} \right]
$$
$$
(4.10)
$$

In the equation above, we have assumed that the Lagrangian density $\mathcal{L}(\vec{x}, t)$ depends on the fields and their first derivatives, that is, on the variables

$$
\psi_\sigma(\vec{x}, t), \ \frac{\partial}{\partial t}\psi_\sigma(\vec{x}, t), \ \nabla_a\psi_\sigma(\vec{x}, t), \ \psi^{\dagger\rho}(\vec{x}, t), \ \frac{\partial}{\partial t}\psi^{\dagger\rho}(\vec{x}, t), \ \nabla_a\psi^{\dagger\rho}(\vec{x}, t)
$$

but otherwise it is quite general, although we shall often refer back to our specific example given in equation (4.6). The idea here is that, if we fix the space and time coordinates to a specific value, we must treat each of $\psi_\sigma(\vec{x}, t)$, $\frac{\partial}{\partial t}\psi_\sigma(\vec{x}, t)$ and $\nabla_a\psi_\sigma(\vec{x}, t)$ and their complex conjugates $\psi^{\dagger\sigma}(\vec{x}, t)$, $\frac{\partial}{\partial t}\psi^{\dagger\sigma}(\vec{x}, t)$ and $\nabla_a\psi^{\dagger\sigma}(\vec{x}, t)$ as independent variables and this allows us to write the variation of the Lagrangian density in terms of ordinary partial derivatives, as we have done in equation (4.10). The partial derivative with respect to each variable is taken while holding all of the other variables fixed.[5] The variation of the derivatives of the functions are defined as the derivatives of the variations, so that

$$
\delta\left(\frac{\partial}{\partial t}\psi_\sigma(\vec{x}, t)\right) \equiv \frac{\partial}{\partial t}\delta\psi_\sigma(\vec{x}, t), \ \delta\left(\frac{\partial}{\partial t}\psi^{\dagger\rho}(\vec{x}, t)\right) \equiv \frac{\partial}{\partial t}\delta\psi^{\dagger\rho}(\vec{x}, t) \qquad (4.12)
$$

---

[5] In the case of fermions, for fixed $\vec{x}$ and t, we must treat $\psi_\sigma(\vec{x}, t)$, $\frac{\partial}{\partial t}\psi_\sigma(\vec{x}, t)$, $\nabla_a\psi_\sigma(\vec{x}, t)$, $\psi^{\dagger\sigma}(\vec{x}, t)$, $\frac{\partial}{\partial t}\psi^{\dagger\sigma}(\vec{x}, t)$ and $\nabla_a\psi^{\dagger\sigma}(\vec{x}, t)$ as independent anti-commuting numbers. In addition, derivatives by anti-commuting numbers must also be anti-commuting entities. For example,

$$
\frac{\partial}{\partial\psi_\sigma(\vec{x}, t)} \frac{\partial}{\partial\psi_\sigma(\vec{x}', t')} = 0, \ \frac{\partial}{\partial\psi_\sigma(\vec{x}, t)} \frac{\partial}{\partial(\frac{\partial}{\partial t}\psi_\rho(\vec{x}, t))} = -\frac{\partial}{\partial(\frac{\partial}{\partial t}\psi_\rho(\vec{x}, t))} \frac{\partial}{\partial\psi_\sigma(\vec{x}, t)}, \ \text{etc.} \quad (4.11)
$$

Moreover, variables and derivatives by the variables also anti-commute with each other. For example,

$$
\frac{\partial}{\partial\psi_\sigma(\vec{x}, t)} \left(\psi_\rho(\vec{x}, t)f(\psi, \psi^\dagger)\right) = \delta_\rho^\sigma f(\psi, \psi^\dagger) - \psi_\rho(\vec{x}, t)\frac{\partial}{\partial\psi_\sigma(\vec{x}, t)}f(\psi, \psi^\dagger)
$$
$$
\frac{\partial}{\partial\psi_\sigma(\vec{x}, t)} \left(\psi^{\dagger\rho}(\vec{x}, t)f(\psi, \psi^\dagger)\right) = -\psi^{\dagger\rho}(\vec{x}, t)\frac{\partial}{\partial\psi_\sigma(\vec{x}, t)}f(\psi, \psi^\dagger)
$$

These rules should be sufficient for defining the variation of the action in the case of fermions.

$$\delta(\nabla_a \psi_\sigma(\vec{x}, t)) \equiv \nabla_a(\delta \psi_\sigma(\vec{x}, t)), \ \delta(\nabla_a \psi^{\dagger \rho}(\vec{x}, t)) \equiv \nabla_a(\delta \psi^{\dagger \rho}(\vec{x}, t)) \qquad (4.13)$$

Then, using the product rule for taking a derivative of a product of functions, for example,

$$\frac{\partial}{\partial t} \left(\delta \psi_\sigma(\vec{x}, t) \frac{\partial \mathcal{L}}{\partial (\frac{\partial}{\partial t} \psi_\sigma(\vec{x}, t))} \right)$$

$$= \left(\frac{\partial}{\partial t} \delta \psi_\sigma(\vec{x}, t) \right) \frac{\partial \mathcal{L}}{\partial (\frac{\partial}{\partial t} \psi_\sigma(\vec{x}, t))} + \delta \psi_\sigma(\vec{x}, t) \frac{\partial}{\partial t} \left(\frac{\partial \mathcal{L}}{\partial (\frac{\partial}{\partial t} \psi(\vec{x}, t))} \right)$$

we rewrite the expression for the variation of the action in equation (4.10) as

$$\delta \mathbf{S} = \int dt \int d^3 x \cdot$$

$$\left\{ \delta \psi_\sigma(\vec{x}, t) \left[\frac{\partial \mathcal{L}}{\partial \psi_\sigma(\vec{x}, t)} - \frac{\partial}{\partial t} \frac{\partial \mathcal{L}}{\partial (\frac{\partial}{\partial t} \psi_\sigma(\vec{x}, t))} - \nabla_a \frac{\partial \mathcal{L}}{\partial (\nabla_a \psi_\sigma(\vec{x}, t))} \right] \right.$$

$$+ \delta \psi^{\dagger \sigma}(\vec{x}, t) \left[\frac{\partial \mathcal{L}}{\partial \psi^{\dagger \sigma}(\vec{x}, t)} - \frac{\partial}{\partial t} \frac{\partial \mathcal{L}}{\partial (\frac{\partial}{\partial t} \psi^{\dagger \sigma}(\vec{x}, t))} - \nabla_a \frac{\partial \mathcal{L}}{\partial (\nabla_a \psi^{\dagger \sigma}(\vec{x}, t))} \right]$$

$$+ \frac{\partial}{\partial t} \left(\delta \psi_\sigma(\vec{x}, t) \frac{\partial \mathcal{L}}{\partial (\frac{\partial}{\partial t} \psi_\sigma(\vec{x}, t))} \right) + \nabla_a \left(\delta \psi_\sigma(\vec{x}, t) \frac{\partial \mathcal{L}}{\partial (\nabla_a \psi_\sigma(\vec{x}, t))} \right)$$

$$+ \frac{\partial}{\partial t} \left(\delta \psi^{\dagger \sigma}(\vec{x}, t) \frac{\partial \mathcal{L}}{\partial (\frac{\partial}{\partial t} \psi^{\dagger \sigma}(\vec{x}, t))} \right) + \nabla_a \left(\delta \psi^{\dagger \sigma}(\vec{x}, t) \frac{\partial \mathcal{L}}{\partial (\nabla_a \psi^{\dagger \sigma}(\vec{x}, t))} \right) \right\}$$

$$(4.14)$$

We shall call the right-hand-side of equation (4.14) the variation of the action. Gauss's theorem, applied in four-dimensional space-time, can be used to rewrite the last two lines in equation (4.14) as surface integrals. Each of these terms are the four-dimensional volume integral of a total divergence, for example, from the third line of (4.14),

$$\frac{\partial}{\partial t} \left(\delta \psi_\sigma(\vec{x}, t) \frac{\partial \mathcal{L}}{\partial (\frac{\partial}{\partial t} \psi_\sigma(\vec{x}, t))} \right) + \nabla_a \left(\delta \psi_\sigma(\vec{x}, t) \frac{\partial \mathcal{L}}{\partial (\nabla_a \psi_\sigma(\vec{x}, t))} \right) \qquad (4.15)$$

is such a four-divergence. Gauss's theorem allows us to rewrite its space-time volume integral as a surface integral at the boundaries of space and time. We shall assume that the boundary conditions for the functions $\psi_\sigma(\vec{x}, t)$, $\psi^{\dagger \sigma}(\vec{x}, t)$, $\delta \psi_\sigma(\vec{x}, t)$ and $\delta \psi^{\dagger \sigma}(\vec{x}, t)$ are such that the surface terms that are generated in this way all vanish.

Then we can drop the total divergence terms from the variation of the action to get

$$\delta S = \tag{4.16}$$

$$\int dt \int d^3x \left\{ \delta\psi_\sigma(\vec{x}, t) \left[\frac{\partial \mathcal{L}}{\partial \psi_\sigma(\vec{x}, t)} - \frac{\partial}{\partial t} \frac{\partial \mathcal{L}}{\partial(\frac{\partial}{\partial t}\psi_\sigma(\vec{x}, t))} - \nabla_a \frac{\partial \mathcal{L}}{\partial(\nabla_a \psi_\sigma(\vec{x}, t))} \right] \right.$$

$$\left. + \delta\psi^{\dagger\sigma}(\vec{x}, t) \left[\frac{\partial \mathcal{L}}{\partial \psi^{\dagger\sigma}(\vec{x}, t)} - \frac{\partial}{\partial t} \frac{\partial \mathcal{L}}{\partial(\frac{\partial}{\partial t}\psi^{\dagger\sigma}(\vec{x}, t))} - \nabla_a \frac{\partial \mathcal{L}}{\partial(\nabla_a \psi^{\dagger\sigma}(\vec{x}, t))} \right] \right\}.$$

$$\tag{4.17}$$

We are interested in specific elements of the set of all possible functions $\psi_\sigma(\vec{x}, t)$ and $\psi^{\dagger\sigma}(\vec{x}, t)$ where the action $S[\psi_\sigma, \psi^{\dagger\sigma}]$ is stationary. The action is stationary when the terms linear in the variations, which we have found in equation (4.17), vanish. This must be so for any profile of the infinitesimal functions $\delta\psi_\sigma(\vec{x}, t)$ and $\delta\psi^{\dagger\sigma}(\vec{x}, t)$. This requires that the coefficients of these functions under the integrations in equation (4.17) must vanish. This gives us a set of differential equations which the classical field must obey, that is, the classical field equations (4.18) and (4.19) below. They are called the Euler–Lagrange equations

$$\frac{\partial \mathcal{L}}{\partial \psi_\sigma(\vec{x}, t)} - \frac{\partial}{\partial t} \frac{\partial \mathcal{L}}{\partial(\frac{\partial}{\partial t}\psi_\sigma(\vec{x}, t))} - \nabla_a \frac{\partial \mathcal{L}}{\partial(\nabla_a \psi_\sigma(\vec{x}, t))} = 0 \tag{4.18}$$

$$\frac{\partial \mathcal{L}}{\partial \psi^{\dagger\sigma}(\vec{x}, t)} - \frac{\partial}{\partial t} \frac{\partial \mathcal{L}}{\partial(\frac{\partial}{\partial t}\psi^{\dagger\sigma}(\vec{x}, t))} - \nabla_a \frac{\partial \mathcal{L}}{\partial(\nabla_a \psi^{\dagger\sigma}(\vec{x}, t))} = 0 \tag{4.19}$$

Application of the Euler–Lagrange equations (4.18) and (4.19) to the action (4.7) yields equations for the classical fields which is identical to the one for the quantum fields in equation (4.1) plus an equation which is its complex conjugate. This gives us the classical field equations. As well as the field equation, there are boundary conditions, which must be compatible with the boundary conditions which were used to eliminate boundary terms that were encountered when finding the linear variation of the action.

4.2   Canonical Momenta, Poisson Brackets and Commutation Relations

Beyond the field equation, which is the Euler–Lagrange equations, the other data that we need in order to define the quantum field theory are the equal-time commutation relations. The form that these must take are also encoded in the action. In this non-relativistic field theory, the Lagrangian density is linear in the time derivative of the field. For this reason, it is easiest to think of the Lagrangian density as being a function on the phase space of the mechanical system, that is, it is a function of the generalized coordinates and momenta, rather than generalized coordinates and velocities. The analog in classical mechanics, where q_i are the generalized coordinates and p_i are the canonical momenta,[6] is the classical action on phase space

$$S = \int dt\, L(q(t), p(t)), \quad L = p_i(t)\frac{d}{dt}q_i(t) - H(q(t), p(t)) \tag{4.20}$$

L is the Lagrangian and the phase space function $H(q, p)$ is the Hamiltonian. The momenta and coordinates have the Poisson bracket

$$\{q_i, p_j\}_{PB} = \delta_{ij}, \quad \{q_i, q_j\}_{PB} = 0, \quad \{p_i, p_j\}_{PB} = 0 \tag{4.21}$$

which can be read from the first, linear in time derivatives term in the Lagrangian, $L = p_i(t)\,\delta_{ij}\,\frac{\partial}{\partial t}q_j(t) + \dots$. The classical field theory Lagrangian density (4.6) has a form analogous to this plus a total derivative,

$$\mathcal{L} = i\hbar\psi^{\dagger\sigma}(\vec{x}, t)\frac{\partial}{\partial t}\psi_\sigma(\vec{x}, t) - \mathcal{H}(\psi, \psi^\dagger) - \frac{\partial}{\partial t}\left(\frac{i\hbar}{2}\psi^{\dagger\sigma}(\vec{x}, t)\psi_\sigma(\vec{x}, t)\right) \tag{4.22}$$

and the total time derivative can be removed by a canonical transformation.[7] Then, we would identify the generalized coordinate as the field $\psi_\sigma(\vec{x}, t)$ and the canonical which is conjugate to this generalized coordinate as being equal to the coefficient of its time derivative in the Lagrangian density, that is, $i\hbar\psi^{\dagger\sigma}(\vec{x}, t)$. The Poisson brackets for the classical field theory are then the obvious generalization of those of classical mechanics, that is,

$$\{\psi_\sigma(\vec{x}, t), i\hbar\psi^{\dagger\rho}(\vec{y}, t)\}_{PB} = \delta_\sigma^\rho \delta(\vec{x} - \vec{y}), \tag{4.23}$$

$$\{\psi_\sigma(\vec{x}, t), i\hbar\psi_\rho(\vec{y}, t)\}_{PB} = 0, \quad \{\psi^{\dagger\sigma}(\vec{x}, t), i\hbar\psi^{\dagger\rho}(\vec{y}, t)\}_{PB} = 0 \tag{4.24}$$

and, when we quantize, we identify the commutator bracket with $i\hbar$ times the Poisson bracket. This tells us that the commutator in the case of bosons, or anti-commutator in the case of fermions, in the field theory should be

$$\left[\psi_\sigma(\vec{x}, t), \psi^{\dagger\rho}(\vec{y}, t)\right] = \delta_\sigma^\rho \delta^3(\vec{x} - \vec{y}) \tag{4.25}$$

$$\left[\psi_\sigma(\vec{x}, t), \psi_\rho(\vec{y}, t)\right] = 0, \quad \left[\psi^{\dagger\sigma}(\vec{x}, t), \psi^{\dagger\rho}(\vec{y}, t)\right] = 0 \tag{4.26}$$

---

[6] The set of all values of the generalized coordinates and momenta together comprise the phase space.

[7] Even with the total derivative, an equivalent result can be found by analyzing the Lagrangian system as a constrained system where, when the constraints are properly resolved, we would obtain the same bracket for $\psi_\sigma(\vec{x}, t)$ and $\psi^{\rho\dagger}(\vec{x}, t)$. One easy way to see this is to remember that a total time derivative term in a classical action can be removed by a canonical transformation and that the Poisson bracket is left unchanged by canonical transformations.

or

$$\left\{\psi_\sigma(\vec{x}, t), \psi^{\dagger\rho}(\vec{y}, t)\right\} = \delta^\rho_\sigma \delta^3(\vec{x} - \vec{y}) \tag{4.27}$$

$$\left\{\psi_\sigma(\vec{x}, t), \psi_\rho(\vec{y}, t)\right\} = 0, \quad \left\{\psi^{\dagger\sigma}(\vec{x}, t), \psi^{\dagger\rho}(\vec{y}, t)\right\} = 0 \tag{4.28}$$

respectively. These match the equal-time commutation or anti-commutation relations which we have used in our quantum field theories so far.

In addition to the field equation and commutation relations, from equation (4.22) we learn that the Hamiltonian and the Hamiltonian density for our example field theory (4.6) are given by

$$H = \int d^3x \, \mathcal{H}(\psi, \psi^\dagger) \tag{4.29}$$

$$\mathcal{H}(\psi, \psi^\dagger) = \left\{\frac{\hbar^2}{2m}\vec{\nabla}\psi^{\dagger\sigma}(\vec{x}, t) \cdot \vec{\nabla}\psi_\sigma(\vec{x}, t) + \frac{\lambda}{2}\left(\psi^{\dagger\sigma}(\vec{x}, t)\psi_\sigma(\vec{x}, t)\right)^2\right\} \tag{4.30}$$

This agrees with the expression for Hamiltonian which we derived earlier in our study of degenerate Fermi and Bose gases.

4.3 Noether's Theorem

In the last section, we showed how the essential information which appears in the field equations and the commutation relations, if we take those as defining the quantum field theory, is also encoded in the classical action and the action principle. In this section we will demonstrate one of the great benefits of the latter approach. If the field equations can be derived from an action via the action principle, symmetries of the theory lead to conservation laws.

4.3.1 Conservation Laws and Continuity Equations

By a conservation law, we mean an equation of continuity for a density of some physical property of a dynamical system, a generic one of which we will denote by $\mathcal{R}(\vec{x}, t)$ and a current density, which measures the motion of the physical entity. We will denote a generic current density by the symbol $\vec{\mathcal{J}}(\vec{x}, t)$. The physical entity is said to be conserved when the density and current density obey a continuity equation,

$$\frac{\partial}{\partial t}\mathcal{R}(\vec{x}, t) + \vec{\nabla} \cdot \vec{\mathcal{J}}(\vec{x}, t) = 0 \tag{4.31}$$

The implication of the continuity equation can be seen by considering the integral of the density over a region, R, in space, with a fixed value of the time coordinate. This integral produces the time-dependent quantity

$$\mathcal{Q}_R(t) = \int_R d^3x \mathcal{R}(\vec{x}, t) \tag{4.32}$$

We shall sometimes call the quantity $\mathcal{Q}_R(t)$ the "charge" inside the region R. Here, charge is not necessarily electric charge (this discussion will apply to electric charge) but it simply means the amount of the physical entity in question. We will also sometimes call $\mathcal{R}(\vec{x}, t)$ a "charge density".

The derivative of the total charge $\mathcal{Q}_R(t)$ by the time is given by

$$\frac{d}{dt}\mathcal{Q}_R(t) = \int_R d^3x \frac{\partial}{\partial t}\mathcal{R}(\vec{x}, t) \tag{4.33}$$

$$= -\int_R d^3x \vec{\nabla} \cdot \vec{\mathcal{J}}(\vec{x}, t) \tag{4.34}$$

$$= -\oint_{\partial R} d^2\sigma \, \hat{n}(\vec{x}) \cdot \vec{\mathcal{J}}(\vec{x}, t) \tag{4.35}$$

where, in the last line of the above equation, we have used Gauss's theorem to rewrite the volume integral of the divergence of a vector field as a surface integral over the boundary of the volume R, which we denote ∂R, of the normal component of the vector field, in this case the current density.

The formula in equation (4.35) is a statement of charge conservation. It says that the time rate of change of the total charge inside the volume R is equal to the total flux of the current density associated with that charge through the boundaries, ∂R of R. It implies that charge is neither created nor destroyed within R. Any change in the total charge in R is due to charge crossing the boundary of R.

We could then consider the case where R is the entire, infinite, three-dimensional Euclidean space where we have been defining our quantum field theories. In that space, we will normally use boundary conditions such that the current densities that we will consider go to zero sufficiently rapidly at spatial infinity that the surface integral analogous to equation (4.35) vanishes. In that case, the total total amount of charge

$$\mathcal{Q}(t) = \int d^3x \mathcal{R}(\vec{x}, t) \tag{4.36}$$

is independent of the time,

$$\frac{d}{dt}\mathcal{Q} = 0$$

.

4.3.2   Definition of Symmetry

The symmetries which we shall study are those for which there exists the notion of an infinitesimal transformation. There are many examples of such transformations, space and time displacements or rotations being examples that we will study in some detail. An already familiar example is a phase transformation. In this case, $\psi_\sigma(\vec{x}, t) \to \tilde{\psi}_\sigma(\vec{x}, t) = e^{i\chi}\psi_\sigma(\vec{x}, t)$ with infinitesimal χ so that $\delta\psi_\sigma(\vec{x}, t) = i\chi\psi_\sigma(\vec{x}, t)$.

If the transformation is infinitesimal, the fields $\tilde{\psi}_\sigma(\vec{x}, t)$ and $\tilde{\psi}^{\dagger\rho}(\vec{x}, t)$ should undergo an infinitesimal change. This infinitesimal change is implemented by the replacement

$$\psi_\sigma(\vec{x}, t) \to \tilde{\psi}_\sigma(\vec{x}, t) = \psi_\sigma(\vec{x}, t) + \delta\psi_\sigma(\vec{x}, t)$$
$$\psi^{\dagger\rho}(\vec{x}, t) \to \tilde{\psi}^{\dagger\rho}(\vec{x}, t) = \psi^{\dagger\rho}(\vec{x}, t) + \delta\psi^{\dagger\rho}(\vec{x}, t), \qquad (4.37)$$

where $\delta\psi_\sigma(\vec{x}, t)$ and $\delta\psi^{\dagger\rho}(\vec{x}, t)$ are specific infinitesimal variations of the fields which are our candidates for symmetries.

Then we consider the the linear variation of the Lagrangian density $\delta\mathcal{L}(\vec{x}, t)$, written out in detail in equation (4.10) and we specialize it to our specific variations $\delta\psi_\sigma(\vec{x}, t)$ and $\delta\psi^{\dagger\rho}(\vec{x}, t)$ in (4.37). We say that this transformation is a symmetry if, without the use of the equations of motion, with algebra alone, we are able to show that $\delta\mathcal{L}(\vec{x}, t)$ can be assembled into partial derivatives by the space and time coordinates in the form

$$\delta\mathcal{L}(\vec{x}, t) = \frac{\partial}{\partial t}R(\vec{x}, t) + \vec{\nabla} \cdot \vec{J}(\vec{x}, t) \qquad (4.38)$$

In this expression, the quantities $R(\vec{x}, t)$ and $\vec{J}(\vec{x}, t)$ can generally depend on the coordinates $(\vec{x}, t)$, the fields $\psi_\sigma(\vec{x}, t)$ and $\psi^{\dagger\rho}(\vec{x}, t)$ and their derivatives.

The specific form of the quantities $R(\vec{x}, t)$ and $\vec{J}(\vec{x}, t)$ depend on the symmetry transformation in question. We will need their specific forms in order to use Noether's theorem to construct the Noether charge and current densities which we will discuss shortly. First let us pause to examine a few examples of symmetry transformations.

4.3.3   Examples of Symmetries

Our notion of symmetry is best illustrated by examining some specific examples. Consider the Lagrangian density in equation (4.6), which we copy here for the reader's convenience,

$$\mathcal{L}(\psi, \frac{\partial}{\partial t}\psi, \nabla\psi) = \frac{i\hbar}{2}\psi^{\dagger\sigma}(\vec{x}, t)\frac{\partial}{\partial t}\psi_\sigma(\vec{x}, t) - \frac{i\hbar}{2}\frac{\partial}{\partial t}\psi^{\dagger\sigma}(\vec{x}, t)\psi_\sigma(\vec{x}, t)$$
$$-\frac{\hbar^2}{2m}\vec{\nabla}\psi^{\dagger\sigma}(\vec{x}, t) \cdot \vec{\nabla}\psi_\sigma(\vec{x}, t) - \frac{\lambda}{2}\left(\psi^{\dagger\sigma}(\vec{x}, t)\psi_\sigma(\vec{x}, t)\right)^2$$

4.3.3.1 Phase Symmetry

We can see by inspection that the Lagrangian density written above is unchanged if we make the substitution

$$\psi_\sigma(\vec{x}, t) \to \tilde{\psi}_\sigma(\vec{x}, t) = e^{-i\theta}\psi_\sigma(\vec{x}, t)$$
$$\psi^{\dagger\sigma}(\vec{x}, t) \to \tilde{\psi}^{\dagger\sigma}(\vec{x}, t) = e^{i\theta}\psi^{\dagger\sigma}(\vec{x}, t)$$

where θ is a real number. The infinitesimal transformation is

$$\delta\psi_\sigma(\vec{x}, t) = \frac{1}{i\hbar}\psi_\sigma(\vec{x}, t), \ \delta\psi^{\dagger\rho}(\vec{x}, t) = -\frac{1}{i\hbar}\psi^{\dagger\rho}(\vec{x}, t) \tag{4.39}$$

where we have dropped the factor of θ on the right-hand-sides in favour of a factor which we shall find more convenient. Under this transformation,

$$\delta\mathcal{L}(\vec{x}, t) = 0 \tag{4.40}$$

so the above transformation satisfies our definition of a symmetry. In this simple case, $R(\vec{x}, t) = 0$ and $\vec{J}(\vec{x}, t) = 0$.

4.3.3.2 Space and Time-Translation Invariance

A second example is the case of a time translation and a space translation where

$$\psi_\sigma(\vec{x}, t) \rightarrow \tilde{\psi}_\sigma(\vec{x}, t) = \psi_\sigma(\vec{x} + \vec{\epsilon}, t + \epsilon)$$
$$\psi^{\dagger\sigma}(\vec{x}, t) \rightarrow \tilde{\psi}^{\dagger\sigma}(\vec{x}, t) = \psi^{\dagger\sigma}(\vec{x} + \vec{\epsilon}, t + \epsilon)$$

with ϵ an infinitesimal constant and $\vec{\epsilon}$ a constant vector of infinitesimal magnitude. The infinitesimal transformations are obtained by Taylor expansion

$$\tilde{\psi}_\sigma(\vec{x}, t) = \psi_\sigma(\vec{x} + \vec{\epsilon}, t + \epsilon) = \tilde{\psi}_\sigma(\vec{x}, t) + \vec{\epsilon} \cdot \vec{\nabla}\psi_\sigma(\vec{x}, t) + \epsilon\frac{\partial}{\partial t}\psi_\sigma(\vec{x}, t) + \ldots$$

$$\tilde{\psi}^{\dagger\sigma}(\vec{x}, t) = \psi^{\dagger\sigma}(\vec{x}, t + \epsilon) = \tilde{\psi}^{\dagger\sigma}(\vec{x}, t) + \vec{\epsilon} \cdot \vec{\nabla}\psi^{\dagger\sigma}(\vec{x}, t) + \epsilon\frac{\partial}{\partial t}\psi^{\dagger\sigma}(\vec{x}, t) + \ldots$$

so that

$$\delta\psi_\sigma(\vec{x}, t) = \left(\vec{\epsilon} \cdot \vec{\nabla} + \epsilon\frac{\partial}{\partial t}\right)\psi_\sigma(\vec{x}, t), \ \delta\psi^{\dagger\rho}(\vec{x}, t) = \left(\vec{\epsilon} \cdot \vec{\nabla} + \epsilon\frac{\partial}{\partial t}\right)\psi^{\dagger\rho}(\vec{x}, t)$$
$$\tag{4.41}$$

Then, we plug this infinitesimal transformation into the Lagrangian density. In our example, we find, without use of the equations of motion, that

$$\delta\mathcal{L}(\vec{x}, t) = \left(\epsilon\frac{\partial}{\partial t} + \vec{\epsilon} \cdot \vec{\nabla}\right)\mathcal{L}(\vec{x}, t) = \frac{\partial}{\partial t}\left(\epsilon\mathcal{L}(\vec{x}, t)\right) + \vec{\nabla} \cdot \left(\vec{\epsilon}\mathcal{L}(\vec{x}, t)\right) \tag{4.42}$$

which tells us that the transformation is a symmetry and it allows us to identify

$$R(\vec{x}, t) = \epsilon\mathcal{L}(\vec{x}, t), \ \ \vec{J}(\vec{x}, t) = \vec{\epsilon}\mathcal{L}(\vec{x}, t) \tag{4.43}$$

corresponding to the symmetry.

What we have done here can easily be extended to a large class of field theories. Any Lagrangian density which does not have an explicit dependence on the space and time coordinates $\vec{x}$ and t, but which depends on these variables only implicitly through the dependence of the fields and their derivatives on $\vec{x}$ and t, will transform as did the $\mathcal{L}(\vec{x}, t)$ above in equation (4.42) and the quantities $R(\vec{x}, t)$ and $\vec{J}(\vec{x}, t)$ will be those in equation (4.43).

We will return to consider more symmetries after we discuss Noether's theorem.

4.3.4 Proof of Noether's Theorem

We have defined a symmetry as a transformation of the fields under which the transformation of the Lagrangian density can be written in the form given in equation (4.38). This was assumed to be possible with the use of algebra, but without the benefit of the Euler–Lagrange equations of motion. Now we shall assume that, in addition to this fact, the Euler–Lagrange equations are satisfied by the classical fields. We begin with the variation of the Lagrangian density[8]

$$\delta\mathcal{L} \equiv \delta\psi_\sigma \frac{\partial\mathcal{L}}{\partial\psi_\sigma} + \delta\left(\frac{\partial}{\partial t}\psi_\sigma\right)\frac{\partial\mathcal{L}}{\partial(\frac{\partial}{\partial t}\psi_\sigma)} + \delta(\nabla_a\psi_\sigma)\frac{\partial\mathcal{L}}{\partial(\nabla_a\psi_\sigma)}$$

$$+\delta\psi^{\dagger\sigma}\frac{\partial\mathcal{L}}{\partial\psi^{\dagger\sigma}} + \delta\left(\frac{\partial}{\partial t}\psi^{\dagger\sigma}\right)\frac{\partial\mathcal{L}}{\partial(\frac{\partial}{\partial t}\psi^{\dagger\sigma})} + \delta(\nabla_a\psi^{\dagger\sigma})\frac{\partial\mathcal{L}}{\partial(\nabla_a\psi^{\dagger\sigma})}$$

We can reorganize this expression as

$$\delta\mathcal{L} = \delta\psi_\sigma\left\{\frac{\partial\mathcal{L}}{\partial\psi_\sigma} - \frac{\partial}{\partial t}\frac{\partial\mathcal{L}}{\partial(\frac{\partial}{\partial t}\psi_\sigma)} - \nabla_a\frac{\partial\mathcal{L}}{\partial(\nabla_a\psi_\sigma)}\right\}$$

$$+ \delta\psi^{\dagger\sigma}\left\{\frac{\partial\mathcal{L}}{\partial\psi^{\dagger\sigma}} - \frac{\partial}{\partial t}\frac{\partial\mathcal{L}}{\partial(\frac{\partial}{\partial t}\psi^{\dagger\sigma})} - \nabla_a\frac{\partial\mathcal{L}}{\partial(\nabla_a\psi^{\dagger\sigma})}\right\}$$

$$+ \frac{\partial}{\partial t}\left[\left(\delta\psi_\sigma\frac{\partial\mathcal{L}}{\partial(\frac{\partial}{\partial t}\psi_\sigma)} + \delta\psi^{\dagger\sigma}\frac{\partial\mathcal{L}}{\partial(\frac{\partial}{\partial t}\psi^{\dagger\sigma})}\right)\right]$$

$$+ \nabla_a\left[\left(\delta\psi_\sigma\frac{\partial\mathcal{L}}{\partial(\nabla_a\psi_\sigma)} + \delta\psi^{\dagger\sigma}\frac{\partial\mathcal{L}}{\partial(\nabla_a\psi^{\dagger\sigma})}\right)\right]$$

---

[8] We remind the reader that, for the purpose of writing this variational formula, we are treating the fields and their time and space derivatives as independent variables. Moreover, in all cases, the variation of the derivative of a function is equal to the derivative of the variation, $\delta\left(\frac{\partial}{\partial t}\psi\right) = \frac{\partial}{\partial t}(\delta\psi)$, $\delta(\nabla_a\psi) = \nabla_a(\delta\psi)$.

Now, we use the Euler–Lagrange equations to set the first two lines to zero. We obtain

$$
\delta \mathcal{L} = \frac{\partial}{\partial t}\left[\left(\delta\psi_\sigma(\vec{x},t)\frac{\partial\mathcal{L}}{\partial(\frac{\partial}{\partial t}\psi_\sigma(\vec{x},t))} + \delta\psi^{\dagger\sigma}(\vec{x},t)\frac{\partial\mathcal{L}}{\partial(\frac{\partial}{\partial t}\psi^{\dagger\sigma}(\vec{x},t))}\right)\right]
$$
$$
+ \nabla_a\left[\left(\delta\psi_\sigma(\vec{x},t)\frac{\partial\mathcal{L}}{\partial(\nabla_a\psi_\sigma(\vec{x},t))} + \delta\psi^{\dagger\sigma}(\vec{x},t)\frac{\partial\mathcal{L}}{\partial(\nabla_a\psi^{\dagger\sigma}(\vec{x},t))}\right)\right] \quad (4.44)
$$

We have shown that, when the equations of motions are used after the variation of the Lagrangian density, any variation of the Lagrangian density is given by total derivatives.

Then, we can ask what happens when the variation, $\delta\psi_\sigma(\vec{x},t)$ and $\delta\psi^{\dagger\sigma}(\vec{x},t)$, is also a symmetry, that is, when without use of the equations of motion, the variation of the Lagrangian density has the form given in equation (4.38) where it is a total derivatives of the quantities, $R(\vec{x},t)$ and $\vec{J}(\vec{x},t)$. We can equate the two different expressions that we have found for the variation of the Lagrangian density, the one in equation (4.44) and the one in equation (4.38). The result is

$$
\frac{\partial}{\partial t}R(\vec{x},t) + \vec{\nabla}\cdot\vec{J}(\vec{x},t)
$$
$$
= \frac{\partial}{\partial t}\left[\delta\psi_\sigma(\vec{x},t)\frac{\partial\mathcal{L}}{\partial(\frac{\partial}{\partial t}\psi_\sigma(\vec{x},t))} + \delta\psi^{\dagger\sigma}(\vec{x},t)\frac{\partial\mathcal{L}}{\partial(\frac{\partial}{\partial t}\psi^{\dagger\sigma}(\vec{x},t))}\right]
$$
$$
+ \nabla_a\left[\delta\psi_\sigma(\vec{x},t)\frac{\partial\mathcal{L}}{\partial(\nabla_a\psi_\sigma(\vec{x},t))} + \delta\psi^{\dagger\sigma}(\vec{x},t)\frac{\partial\mathcal{L}}{\partial(\nabla_a\psi^{\dagger\sigma}(\vec{x},t))}\right] \quad (4.45)
$$

This equation is a consequence of the two facts, first that the variation is a symmetry and, second that the fields satisfy the equations of motion. By combining the terms in equation (4.45) we obtain the equation of continuity

$$
\frac{\partial}{\partial t}R(\vec{x},t) + \vec{\nabla}\cdot\vec{J}(\vec{x},t) = 0 \quad (4.46)
$$

where the charge density and the current density are given by the expressions

$$
\mathcal{R}(\vec{x},t) = \delta\psi_\sigma(\vec{x},t)\frac{\partial\mathcal{L}}{\partial(\frac{\partial}{\partial t}\psi_\sigma(\vec{x},t))} + \delta\psi^{\dagger\rho}(\vec{x},t)\frac{\partial\mathcal{L}}{\partial(\frac{\partial}{\partial t}\psi^{\dagger\rho}(\vec{x},t))} - R(\vec{x},t) \quad (4.47)
$$
$$
\mathcal{J}^a(\vec{x},t) = \delta\psi_\sigma(\vec{x},t)\frac{\partial\mathcal{L}}{\partial(\nabla_a\psi_\sigma(\vec{x},t))} + \delta\psi^{\dagger\rho}(\vec{x},t)\frac{\partial\mathcal{L}}{\partial(\nabla_a\psi^{\dagger\rho}(\vec{x},t))} - J^a(\vec{x},t)
$$
$$
\quad (4.48)
$$

respectively.

This association of the symmetry with the continuity equation (4.46) is Noether's theorem. The charge density (4.47) and the current density (4.48) are the Noether charge and current densities corresponding to the symmetry. Sometimes they are referred to together as the "Noether current". Our derivation of the conservation law from the existence of a symmetry is the proof of Noether's theorem.

So far, we have been dealing entirely with the classical field theory. When we go to quantum field theory, modulo operator ordering, we will adopt the classical field equations with the classical fields replaced by the operators of the quantum field theory as the field equations which form part of the definition of the quantum field theory. In our non-relativistic example, modulo specifying the ordering of some operators, this would be part of our prescription for recovering the quantum field theory from the classical Lagrangian density. Since the conservation laws for Noether charges and currents are a result of the fact that fields satisfy the field equations, we anticipate that they also survive in the quantum field theory. In most cases that we are interested in, the ordering of operators in the field equations and in the Noether charge and current densities can be specified so that the conservation laws implied by Noether's theorem are indeed inherited by the quantum field theory. Then, these conservation laws, which are operator equations of the form (4.46), (4.47) and (4.48) give us important information about the quantum field theory.[9]

We have discussed Noether's theorem in the context of our non-relativistic quantum field theory. However, it, or straightforward generalizations of it to the case of any field theory where the field equations can be obtained from a Lagrangian density and action by the variational principle. We will make extensive use of this fact later on when we study relativistic field theories.

4.4 Phase Symmetry and the Conservation of Particle Number

Now, let us consider the Lagrangian density (4.6) and the infinitesimal phase transformation

$$\delta\psi = \frac{1}{i\hbar}\psi_\sigma(\vec{x}, t), \quad \delta\psi^{\dagger\rho}(\vec{x}, t) = -\frac{1}{i\hbar}\psi^{\dagger\rho}(\vec{x}, t) \tag{4.49}$$

which we have already identified as a symmetry. When we examined the transformation of the Lagrangian density, we found that $\delta\mathcal{L} = 0$. The quantities $R(\vec{x}, t)$ and $\vec{J}(\vec{x}, t)$ which we would use to construct the Noether charge and current densities are both zero, $R(\vec{x}, t) = 0$, $\vec{J}(\vec{x}, t) = 0$. Then Noether's theorem tells us that the charge density is

---

[9] There are, however, examples of conservation laws that do not survive the transition to quantum field theories. These are associated with phenomena called anomalies and they are generally due to the impossibility of resolving some of the short distance singularities of a quantum field theories in a way which is compatible with the symmetries and their conservation laws.

$$\rho(\vec{x}, t) = \frac{1}{i\hbar} \psi_\sigma(\vec{x}, t) \frac{\partial \mathcal{L}}{\partial(\frac{\partial}{\partial t}\psi(\vec{x}, t))} - \frac{1}{i\hbar} \psi^{\dagger\sigma}(\vec{x}, t) \frac{\partial \mathcal{L}}{\partial(\frac{\partial}{\partial t}\psi^{\dagger\sigma}(\vec{x}, t))} \tag{4.50}$$

and the current density is

$$\mathbf{J}_a(\vec{x}, t) = \frac{1}{i\hbar} \psi_\sigma(\vec{x}, t) \frac{\partial \mathcal{L}}{\partial(\nabla_a \psi_\sigma(\vec{x}, t))} - \frac{1}{i\hbar} \psi^{\dagger\sigma}(\vec{x}, t) \frac{\partial \mathcal{L}}{\partial(\nabla_a \psi^{\dagger\sigma}(\vec{x}, t))} \tag{4.51}$$

These are quite general, in that the above equation applies to any Lagrangian density which has a phase symmetry, and where it is strictly invariant under the phase transformation (and where the Lagrangian density depends on the fields and their first derivatives only). For the Lagrangian density (4.6) which we have been discussing the charge and current densities are

$$\rho(\vec{x}, t) = \psi^{\dagger\sigma}(\vec{x}, t)\psi_\sigma(\vec{x}, t)$$

$$\mathbf{J}_a(\vec{x}, t) = -\frac{i\hbar}{2m} \left[\psi^{\dagger\sigma}(\vec{x}, t)\nabla_a \psi_\sigma(\vec{x}, t) - \nabla_a \psi^{\dagger\sigma}(\vec{x}, t)\,\psi_\sigma(\vec{x}, t) \right] \tag{4.52}$$

Noether's theorem tells us that the charge and current densities must obey the continuity equation

$$\frac{\partial}{\partial t}\rho(\vec{x}, t) + \nabla_a \mathbf{J}^a(\vec{x}, t) = 0 \tag{4.53}$$

The Noether charge is given by

$$\mathcal{N} = \int d^3x \rho(\vec{x}, t) = \int d^3x \psi^{\dagger\sigma}(\vec{x}, t)\psi_\sigma(\vec{x}, t) \tag{4.54}$$

$$\frac{\partial}{\partial t}\mathcal{N} = 0. \tag{4.55}$$

When we take the expressions for $\rho(\vec{x}, t)$ and $\vec{\mathbf{J}}(\vec{x}, t)$ and promote the classical fields to operators so that these quantities are operators in a quantum field theory, we obtain the operator $\mathcal{N}$ whose eigenvalues are the total number of particles and the current density $\vec{\mathbf{J}}(\vec{x}, t)$ which contains information about the transport of particles.

In previous chapters, we observed that the particle number commuted with the Hamiltonian and we expected it to be time-independent. Now, Noether's theorem also tells us that this is so, that the particle number is a conserved Noether charge related to phase symmetry of the Lagrangian density and it will be so for any Lagrangian density which has phase symmetry. Beyond that, it gives us an expression for the particle current density (4.52). If we look back at the details of the derivation of this current density, we see that it only depends on the derivative terms in the Lagrangian density. It therefore has the same form as the expression in equation (4.52) for any interaction potential as long as the potential depended on the fields and not their derivatives. For example, any interaction which depends on the fields only in the combination $\psi^{\dagger\sigma}\psi_\sigma$ has phase symmetry and the interaction current density has the same form as in (4.52).

Non-relativistic Space–Time Symmetries

In the previous chapter, we discussed the action principle, where the data of the quantum field theory is encoded in a Lagrangian density. We learned how to extract the equations of motion and the equal-time commutation relations from knowledge of the Lagrangian density. Then we studied one of the great advantages of the action formalism and describing the field theory using a Lagrangian density. This advantage is the existence of Noether's theorem, which is a systematic way of relating symmetries to conservation laws. We studied the specific example of phase symmetry and we saw that it was associated with the conservation of the total particle number. In this chapter, we will develop further applications of Noether's theorem to the study of symmetries of a non-relativistic system of particles under specific transformations of their space and time coordinates. These will lead to the important conservation laws for energy, momentum and angular momentum.

5.1 Translation Invariance and the Stress Tensor

If the Lagrangian density of a field theory does not contain explicit occurrences of the time or the spatial coordinates, but it depends on the time and the space coordinates only through the implicit dependence of the fields and their derivatives on the these quantities, the field theory that is described by that Lagrangian is symmetric under space and time translations. The infinitesimal space and time translation of the non-relativistic fields that we have been discussing so far are

$$\delta\psi_\sigma(\vec{x}, \vec{t}) = \left(\vec{\epsilon} \cdot \nabla + \epsilon \frac{\partial}{\partial t} \right) \psi_\sigma(\vec{x}, t) \tag{5.1}$$

$$\delta\psi^{\dagger\rho}(\vec{x}, \vec{t}) = \left(\vec{\epsilon} \cdot \nabla + \epsilon \frac{\partial}{\partial t} \right) \psi^{\dagger\rho}(\vec{x}, t) \tag{5.2}$$

© The Author(s), under exclusive license to Springer Nature Singapore Pte Ltd. 2023
G. W. Semenoff, *Quantum Field Theory*, Graduate Texts in Physics,
https://doi.org/10.1007/978-981-99-5410-0_5

Indeed, it is easy to see that, without use of the equations of motion, the Lagrangian density must transform as

$$\delta\mathcal{L}(\vec{x}, t) = \frac{\partial}{\partial t}\left(\epsilon\mathcal{L}(\vec{x}, t)\right) + \vec{\nabla}\cdot\left(\vec{\epsilon}\mathcal{L}(\vec{x}, t)\right) \tag{5.3}$$

Our explicit example in equation (4.6) indeed transforms this way. The fact that the Lagrangian varies as in equation (5.3) tells us that the transformation is a symmetry. It also allows us to identify the two quantities

$$R(\vec{x}, t) = \epsilon\mathcal{L}(\vec{x}, t)$$

$$\vec{J}(\vec{x}, t) = \vec{\epsilon}\mathcal{L}(\vec{x}, t)$$

which are needed to construct the Noether current. The Noether charge and current densities then follow directly from an application of Noether's theorem. They are most usefully presented as a four-by-four array of densities and currents

$$\begin{bmatrix} \mathbf{T}^{tt}(\vec{x}, t) & \mathbf{T}^{tb}(\vec{x}, t) \\ \mathbf{T}^{at}(\vec{x}, t) & \mathbf{T}^{ab}(\vec{x}, t) \end{bmatrix}$$

This object is called the "stress tensor" or the "energy-momentum tensor".[1] It has the components

– The Noether charge density for time translations: $\mathbf{T}^{tt}(\vec{x}, t)$,

– The Noether current density for time translations: $\mathbf{T}^{at}(\vec{x}, t)$,

– The Noether charge density for space translation in "b" direction: $\mathbf{T}^{tb}(\vec{x}, t)$,

– The Noether current density for space translation in "b" direction: $\mathbf{T}^{ab}(\vec{x}, t)$.

---

[1] We have written the tensor with upper indices and the conservation laws below with the derivatives acting with lower indices, reminiscent of relativistic equations. We emphasize that in this non-relativistic theory there is no distinct meaning to upper or lower indices, they are completely equivalent and writing them up or down is simply habit. This will change when we study relativisting field theories where up and down spacetime indices have an important meaning. Here, repeated indices are assumed to be summed over.

Given a Lagrangian density $\mathcal{L}$, the components of the stress tensor given by Noether's theorem as

$$\mathbf{T}^{tt}(\vec{x}, t) = \frac{\partial}{\partial t}\psi_\sigma(\vec{x}, t)\frac{\partial \mathcal{L}}{\partial(\frac{\partial}{\partial t}\psi_\sigma(\vec{x}, t))} + \frac{\partial}{\partial t}\psi^{\dagger\sigma}(\vec{x}, t)\frac{\partial \mathcal{L}}{\partial(\frac{\partial}{\partial t}\psi^{\dagger\sigma}(\vec{x}, t))}$$
$$- \mathcal{L}(\vec{x}, t) \tag{5.4}$$

$$\mathbf{T}^{tb}(\vec{x}, t) = \nabla_b\psi_\sigma(\vec{x}, t)\frac{\partial \mathcal{L}}{\partial(\frac{\partial}{\partial t}\psi_\sigma(\vec{x}, t))} + \nabla_b\psi^{\dagger\sigma}(\vec{x}, t)\frac{\partial \mathcal{L}}{\partial(\frac{\partial}{\partial t}\psi^{\dagger\sigma}(\vec{x}, t))} \tag{5.5}$$

$$\mathbf{T}^{at}(\vec{x}, t) = \frac{\partial}{\partial t}\psi_\sigma(\vec{x}, t)\frac{\partial \mathcal{L}}{\partial(\nabla^a\psi_\sigma(\vec{x}, t))} + \frac{\partial}{\partial t}\psi^{\dagger\sigma}(\vec{x}, t)\frac{\partial \mathcal{L}}{\partial(\nabla^a\psi^{\dagger\sigma}(\vec{x}, t))} \tag{5.6}$$

$$\mathbf{T}^{ab}(\vec{x}, t) = \nabla_b\psi_\sigma(\vec{x}, t)\frac{\partial \mathcal{L}}{\partial(\nabla^a\psi_\sigma(\vec{x}, t))} + \nabla_b\psi^{\dagger\sigma}(\vec{x}, t)\frac{\partial \mathcal{L}}{\partial(\nabla^a\psi^{\dagger\sigma}(\vec{x}, t))}$$
$$- \delta_{ab}\mathcal{L}(\vec{x}, t) \tag{5.7}$$

and, Noether's theorem tells us that in a system which possessed time and space translation invariance, the elements of the stress tensor must obey the conservation laws

$$\frac{\partial}{\partial t}\mathbf{T}^{tt}(\vec{x}, t) + \nabla_a\mathbf{T}^{at}(\vec{x}, t) = 0 \tag{5.8}$$

$$\frac{\partial}{\partial t}\mathbf{T}^{tb}(\vec{x}, t) + \nabla_a\mathbf{T}^{ab}(\vec{x}, t) = 0 \tag{5.9}$$

Explicitly, for the quantum field theory with the Lagrangian density (4.6), the stress tensor is given by

$$\mathbf{T}^{tt}(\vec{x}, t) = \frac{\hbar^2}{2m}\vec{\nabla}\psi^{\dagger\sigma}(\vec{x}, t) \cdot \vec{\nabla}\psi_\sigma(\vec{x}, t) + \frac{\lambda}{2}\left(\psi^{\dagger\sigma}(\vec{x}, t)\psi_\sigma(\vec{x}, t)\right)^2 \tag{5.10}$$

$$\mathbf{T}^{tb}(\vec{x}, t) = -\frac{i\hbar}{2}\left[\nabla_b\psi^{\dagger\sigma}(\vec{x}, t)\psi_\sigma(\vec{x}, t) - \psi^{\dagger\sigma}(\vec{x}, t)\nabla_b\psi_\sigma(\vec{x}, t)\right] \tag{5.11}$$

$$\mathbf{T}^{at}(\vec{x}, t) = -\frac{\hbar^2}{2m}\left(\nabla_a\psi^{\dagger\sigma}(\vec{x}, t)\frac{\partial}{\partial t}\psi_\sigma(\vec{x}, t) + \frac{\partial}{\partial t}\psi^{\dagger\sigma}(\vec{x}, t)\nabla_a\psi(\vec{x}, t)\right) \tag{5.12}$$

$$\mathbf{T}^{ab}(\vec{x}, t) = -\frac{\hbar^2}{2m}\left(\nabla_a\psi^{\dagger\sigma}(\vec{x}, t)\nabla_b\psi_\sigma(\vec{x}, t) + \nabla_b\psi^{\dagger\sigma}(\vec{x}, t)\nabla_a\psi_\sigma(\vec{x}, t)\right)$$
$$+ \delta_{ab}\left(\frac{\hbar^2}{4m}\vec{\nabla}^2\left(\psi^{\sigma\dagger}(\vec{x}, t)\psi_\sigma(\vec{x}, t)\right) - \frac{\lambda}{2}\left(\psi^{\sigma\dagger}(\vec{x}, t)\psi_\sigma(\vec{x}, t)\right)^2\right) \tag{5.13}$$

where we have used the field equation to eliminate the time derivative terms in the Lagrangian density in order to obtain equation (5.13) for $\mathbf{T}^{ab}(\vec{x}, t)$.[2]

The Noether charge and current densities have the physical interpretations:

$\mathbf{T}^{tt}(\vec{x}, t) =$ energy density,

$\mathbf{T}^{at}(\vec{x}, t) =$ flux of energy in the "a"-direction,

$\mathbf{T}^{tb}(\vec{x}, t) =$ density of "b"-component of momentum,

$\mathbf{T}^{ab}(\vec{x}, t) =$ flux of the "b-component of momentum in the "a"-direction.

By flux of a physical entity, we mean the rate of flow of that entity across a surface per unit time and per unit area of the surface. The trace $\mathbf{T}^a{}_a(\vec{x}, t) = -3P$ is related to the thermodynamic pressure, P.

Since $\mathbf{T}^{tt}(\vec{x}, t)$ is interpreted as the energy density, the Noether charge for time-translation symmetry, which is the space integral of $\mathbf{T}^{tt}(\vec{x}, t)$ is interpreted as the total energy. We could call $\mathbf{T}^{tt}(\vec{x}, t)$ the "Hamiltonian density", $\mathcal{H}(\vec{x}, t)$ and its integral over space the "Hamiltonian", H,

$$H = \int d^3x \, \mathbf{T}^{tt}(\vec{x}, t) = \int d^3x \, \mathcal{H}(\vec{x}, t) \tag{5.15}$$

Similarly, the Noether charges for space translation symmetries are the components of the total momentum,

$$P^b = -\int d^3x \, \mathbf{T}^{tb}(\vec{x}, t) \tag{5.16}$$

These Noether charges are time-independent in a field theory which has time and space translation invariance. This time-independence is called conservation of energy and conservation of momentum, respectively.

5.2   Galilean Symmetry

There is an important transformation of the coordinates of a non-relativistic mechanical system which corresponds to transforming to a moving reference frame where the motion is with constant relative velocity. This is the transformation of Galilean relativity which we have already encountered in our discussion of degenerate Bose gases and superfluids in Chap. 2. As an example of Galilean invariance in classical mechanics, Newton's second law of motion for a free particle,

$$m \frac{d^2}{dt^2} \vec{x}(t) = 0 \tag{5.17}$$

---

[2] When we eliminate these time derivatives using the equation of motion, the Lagrangian density becomes

$$\mathcal{L}(\vec{x}, t) = -\frac{\hbar^2}{4m} \vec{\nabla}^2 \left(\psi^{\sigma\dagger}(\vec{x}, t) \psi_\sigma(\vec{x}, t) \right) + \frac{\lambda}{2} \left(\psi^{\sigma\dagger}(\vec{x}, t) \psi_\sigma(\vec{x}, t) \right)^2 \tag{5.14}$$

is invariant under replacement of the position $\vec{x}(t)$ by $\vec{x}(t) + \vec{v}t$ where $\vec{v}$ is a constant vector. The symmetry of the equation of motion under this replacement is a form of non-relativistic relativity called Galilean symmetry. It tells us that the laws of physics, in this case. Newton's second law, hold equally well in either of two different reference frames, the original one, with space coordinates $\vec{x}$ and time coordinate t and a new one with space coordinates $\tilde{\vec{x}} = \vec{x} + \vec{v}t$ and time coordinate $\tilde{t} = t$, that is, where the reference frames are related by the coordinate transformation

$$(\vec{x}, t) \rightarrow (\tilde{\vec{x}}, \tilde{t}) = (\vec{x} + \vec{v}t, t),$$

Now, consider a classical mechanical system of N particles which interact with each other in such a way that their equations of motion are

$$m \frac{d^2}{dt^2} \vec{x}_i(t) = -\sum_{\substack{j=1 \\ j \neq i}}^{N} \vec{\nabla}_i V(\vec{x}_i(t) - \vec{x}_j(t)), \quad i = 1, ..., N \qquad (5.18)$$

where $\vec{x}_i(t)$ is the position of the i'th particle. Their equations of motion contain two-body forces which are obtained by taking the gradients of a two-body potential energy function

$$V(\vec{x}_1, \ldots, \vec{x}_N) = \sum_{i<j=1}^{N} V(\vec{x}_i - \vec{x}_j)$$

where $V(\vec{x} - \vec{y}) = V(\vec{y} - \vec{x})$.

Here, we see that the N equations of motion are also invariant under a Galilean transformation of the particle positions, the replacement

$$(\vec{x}_1(t), \vec{x}_2(t), \ldots, \vec{x}_N(t)) \rightarrow (\vec{x}_1(t) + \vec{v}t, \vec{x}_2(t) + \vec{v}t, \ldots, \vec{x}_N(t) + \vec{v}t) \quad (5.19)$$

The classical mechanics of a free particle or an assembly of free particles interacting by a two-body potential are thus invariant under Galilean transformations. (What is important here is that the potential energy depends only on the interparticle separations, $\vec{x}_i - \vec{x}_j$.) We expect that the quantum mechanics of systems such as these are also invariant.

It is interesting to first ask how this symmetry can be seen in Schrödinger's equation for a single free particle, the quantum mechanical version of the system described by the classical equation (5.17). We expect that the quantum mechanical system is Galilean invariant if the classical one is. The single free particle Schrödinger equation is

$$\left(i\hbar \frac{\partial}{\partial t} + \frac{\hbar^2}{2m} \vec{\nabla}^2 \right) \psi(\vec{x}, t) = 0 \qquad (5.20)$$

We could try to boost this system by substituting $\vec{x} \rightarrow \vec{x} + \vec{v}t$ into the wavefunction. This does not quite work. The wave equation which the "boosted" wave-function $\psi(\vec{x} + \vec{v}t, t)$ satisfies is a little different from the original Schrödinger equation in equation (5.20), it is

$$\left(i\hbar\frac{\partial}{\partial t} - i\hbar\vec{v}\cdot\vec{\nabla} + \frac{\hbar^2}{2m}\vec{\nabla}^2\right)\psi(\vec{x} + \vec{v}t, t) = 0 \qquad (5.21)$$

It has an extra term "$-i\hbar\vec{v}\cdot\vec{\nabla}$". We next observe that this extra term can be removed by an $\vec{x}$-dependent change of phase of the wave-function so that equation (5.21) implies

$$\left(i\hbar\frac{\partial}{\partial t} + \frac{\hbar^2}{2m}\vec{\nabla}^2 + \frac{m}{2}\vec{v}^2\right)\left[e^{-im\vec{v}\cdot\vec{x}/\hbar}\psi(\vec{x} + \vec{v}t, t)\right] = 0 \qquad (5.22)$$

This is still not the same equation as (5.20). It has the extra term, "$\frac{m}{2}\vec{v}^2$". We can remove this term by using another simple change of phase of the wave-function. We get

$$\left(i\hbar\frac{\partial}{\partial t} + \frac{\hbar^2}{2m}\vec{\nabla}^2\right)\left[e^{-im\vec{v}\cdot\vec{x}/\hbar + i\frac{m}{2}\vec{v}^2 t/\hbar}\psi(\vec{x} + \vec{v}t, t)\right] = 0 \qquad (5.23)$$

This equation is now the same as (5.20). Our conclusion is that, if $\psi(\vec{x}, t)$ satisfies the Schrödinger equation (5.20), then the transformed wave-function $e^{-im\vec{v}\cdot\vec{x}/\hbar + i\frac{m}{2}\vec{v}^2 t/\hbar}\psi(\vec{x} + \vec{v}t, t)$ also satisfies the Schrödinger equation (5.20).

This statement, as we have formulated it above, applies to single particle quantum mechanics. We can easily generalize it to the transformation of a many-particle wave-function

$$\psi(\vec{x}_1, ..., \vec{x}_N, t) \rightarrow \tilde{\psi}(\vec{x}_1, ..., \vec{x}_N, t) \qquad (5.24)$$

$$\tilde{\psi}(\vec{x}_1, ..., \vec{x}_N, t) = e^{-im\vec{v}\cdot(\vec{x}_1 + ... + \vec{x}_N)/\hbar + iN\frac{m}{2}\vec{v}^2 t/\hbar}\psi(\vec{x}_1 + \vec{v}t, ..., \vec{x}_N + \vec{v}t, t) \qquad (5.25)$$

It is straightforward to check that the transformed wave-function obeys the many-particle Schrödinger equation of the form given in equation (2.5) whenever the potential energy function $V(\vec{x}_1, ..., \vec{x}_N)$ is a function only of the differences of coordinates $\vec{x}_i - \vec{x}_j$.

We could easily use the transformation in (5.24) to understand how Galilean symmetry works in the quantum field theory formulation of the many-particle problem. However, there is a shortcut to doing this, beginning with free particles. Here, we are actually interested in the classical field theory which yields that quantum field theory when one applies the rules of quantization, as we have discussed them earlier in this chapter. We know that the wave equation which the classical field satisfies, in the case where there are no interactions, is identical in form to the Schrödinger equation,

$$\left(i\hbar\frac{\partial}{\partial t} + \frac{\hbar^2}{2m}\vec{\nabla}^2\right)\psi_\sigma(\vec{x}, t) = 0 \qquad (5.26)$$

This tells us how to do the Galilean transformation of the classical field:

$$\psi_\sigma(\vec{x}, t) \;\rightarrow\; \tilde{\psi}_\sigma(\vec{x}, t) = e^{-im\vec{v}\cdot\vec{x}/\hbar + i\frac{m}{2}\vec{v}^2 t/\hbar}\psi_\sigma(\vec{x} + \vec{v}t, t) \tag{5.27}$$

We know by our discussion above that this must be a symmetry of the non-interacting theory. What remains to check is that it is also a symmetry of a theory with interactions.

To check that this is indeed a symmetry of the classical field theory with interactions, for example the theory with Lagrangian density given in equation (4.6), we consider the infinitesimal transformation

$$\delta\psi_\sigma(\vec{x}, t) = \left(-im\vec{v}\cdot\vec{x}/\hbar + t\vec{v}\cdot\vec{\nabla}\right)\psi_\sigma(\vec{x}, t) \tag{5.28}$$

$$\delta\psi^{\dagger\sigma}(\vec{x}, t) = \left(im\vec{v}\cdot\vec{x}/\hbar + t\vec{v}\cdot\vec{\nabla}\right)\psi^{\dagger\sigma}(\vec{x}, t) \tag{5.29}$$

By plugging this transformation into the specific Lagrangian density (4.6), we see that the variation of the Lagrangian density (without use of the equations of motion) is equal to

$$\delta\mathcal{L} = \vec{\nabla}\cdot(\vec{v}t\mathcal{L}(\vec{x}, t)) \tag{5.30}$$

The transformation in equations (5.28) and (5.29) is therefore a symmetry of the field theory that is defined by the Lagrangian density (4.6). The Noether charge and current densities corresponding to this symmetry are easy to find using equations (4.47) and (4.48) where we take $R(\vec{x}, t) = 0$ and $\vec{J}(\vec{x}, t) = \vec{v}t\mathcal{L}$. Recall that the formal expression for the Noether charge density is

$$\mathcal{B}_b(\vec{x}, t) = \delta_b\psi_\sigma(\vec{x}, t)\frac{\partial\mathcal{L}}{\partial(\partial\psi_\sigma(\vec{x}, t))} + \delta_b\psi^{\dagger\sigma}(\vec{x}, t)\frac{\partial\mathcal{L}}{\partial(\frac{\partial}{\partial t}\psi^{\dagger\sigma}(\vec{x}, t))} - R_b(\vec{x}, t)$$

Using the Galilean boost, there $R_b(\vec{x}, t) = 0$ and the transformations of the fields are the coefficients of v^b in equations (5.28) and (5.29), we find

$$\mathcal{B}_b(\vec{x}, t) = \frac{m}{i\hbar}x_b\psi_\sigma(\vec{x}, t)\frac{\partial\mathcal{L}}{\partial(\frac{\partial}{\partial t}\psi_\sigma(\vec{x}, t))} - \frac{m}{i\hbar}x_b\psi^{\dagger\sigma}(\vec{x}, t)\frac{\partial\mathcal{L}}{\partial(\frac{\partial}{\partial t}\psi^{\dagger\sigma}(\vec{x}, t))}$$
$$+ t\nabla_b\psi_\sigma(\vec{x}, t)\frac{\partial\mathcal{L}}{\partial(\frac{\partial}{\partial t}\psi_\sigma(\vec{x}, t))} + t\vec{\nabla}_b\psi^{\dagger\sigma}(\vec{x}, t)\frac{\partial\mathcal{L}}{\partial(\frac{\partial}{\partial t}\psi^{\dagger\sigma}(\vec{x}, t))}$$

Then, recalling the formal expressions for the phase symmetry Noether current (4.52) and the stress tensor (5.5), we find it most efficient to write the Galilean Noether charge density, and by similar procedure the current density, in terms of components of the stress tensor and the number density and current. The result is

$$\mathcal{B}^b(\vec{x}, t) = t\mathbf{T}^{tb}(\vec{x}, t) + mx^b\,\rho(\vec{x}, t), \tag{5.31}$$

$$\mathcal{B}^{ba}(\vec{x}, t) = t\mathbf{T}^{ab}(\vec{x}, t) + mx^b\,\mathbf{J}^a(\vec{x}, t) \tag{5.32}$$

Here, $(\rho, \vec{\mathbf{J}})$ are the Noether current associated with phase symmetry that we found in equation (4.52). It is straightforward to confirm that the Galilean charge density in (5.31) and the current density in (5.32) obey the continuity equation when the stress tensor is the one corresponding to our example field theory with Lagrangian density (4.6),

$$\frac{\partial}{\partial t} \mathcal{B}^b(\vec{x}, t) + \nabla_a \mathcal{B}^{ba}(\vec{x}, t) = 0$$

Of course, this must be so, as it is implied by Noether's theorem for that Galilean invariant system.

Conversely, suppose that we know that a particular field theory has space- and time-translation invariance and that it therefore has a conserved stress tensor. In addition, it has phase symmetry, so that it has conserved particle number charge and current density. We could use the Noether charge and current densities corresponding to both of those symmetries to form the Galilean Noether charge and current densities that are given in equations (5.31) and (5.32). Then, we can test the conservation law for the Galilean charge and current density,

$$
\begin{aligned}
\frac{\partial}{\partial t} & \left[t \mathbf{T}^{tb}(\vec{x}, t) \; + \; x^b \, m \rho(\vec{x}, t) \right] + \nabla_a \left[t \mathbf{T}^{ab}(\vec{x}, t) \; + \; m x^b \, \mathbf{J}^a(\vec{x}, t) \right] \\
= & \; t \left[\frac{\partial}{\partial t} \mathbf{T}^{tb}(\vec{x}, t) + \nabla_a \mathbf{T}^{ab}(\vec{x}, t) \right] + x^b m \left[\frac{\partial}{\partial t} \rho(\vec{x}, t) + \nabla_a \mathbf{J}^a(\vec{x}, t) \right] \\
& + \mathbf{T}^{tb}(\vec{x}, t) + m \mathbf{J}^b(\vec{x}, t) \\
= & \; \mathbf{T}^{tb}(\vec{x}, t) + m \mathbf{J}^b(\vec{x}, t) \tag{5.33}
\end{aligned}
$$

Here, we have used the conservation law for the stress tensor (5.8) and (5.9) and for the particle current (4.53).

What we find is a condition that a translation and phase symmetric field theory must obey if it is to also have Galilean invariance. This condition states that the stress tensor and particle current must obey the equation

$$-\mathbf{T}^{tb}(\vec{x}, t) \; = \; m \mathbf{J}^b(\vec{x}, t) \tag{5.34}$$

This condition for Galilean invariance equates the momentum density $-\mathbf{T}^{tb}(\vec{x}, t)$ to the particle mass times the particle current density $m \mathbf{J}^b(\vec{x}, t)$. They must be equal if a phase invariant and a space- and time-translation invariant field theory is to also be Galilean invariant. It is easy to check that this is indeed the case for our example, the field theory with Lagrangian density (4.6).

Time translation invariance leads to conservation of energy. Space translation invariance leads to conservation of momentum. The Galilean symmetry also has a dynamical implication. Once the Galilean current in equations (5.31) and (5.32) is

conserved we have the conserved charge given by the volume integral of the charge
density (5.31),

$$Q^b = \int d^3x \, Q^b(\vec{x}, t), \quad \frac{d}{dt} Q^b = 0 \tag{5.35}$$

$$\frac{d}{dt} \int d^3x \left[t \mathbf{T}_{tb}(\vec{x}, t) + x_b \, m\rho(\vec{x}, t) \right] = 0 \tag{5.36}$$

This equation implies

$$\frac{d}{dt} \int d^3x \, m \, \vec{x} \, \rho(\vec{x}, t) = \vec{P} \tag{5.37}$$

where the conserved total momentum is the Noether charge

$$P^b = -\int d^3x \, T^{tb}(\vec{x}, t), \quad \frac{d}{dt} \vec{P} = 0$$

This equation is the simple statement that the time rate of change of the geometrical
centre of mass of the system is equal to the total momentum.

In this and the previous two sections, we have examined some symmetries which
are reasonably common in non-relativistic many-particle systems, phase invariance,
time- and space-translation invariance and Galilean symmetry. Now, we will study
two symmetries which occur much less frequently, but at the same time can have
interesting consequences when they do occur, scale and special Schrödinger symmetry.

5.3   Scale Invariance

In some circumstances, the non-relativistic quantum field theory that we have been
discussing can have a symmetry under scaling of the space and time variables. This
is already so for the non-interacting field theory with the field equation

$$\left(i\hbar \frac{\partial}{\partial t} + \frac{\hbar^2}{2m} \vec{\nabla}^2 \right) \psi_\sigma(\vec{x}, t) = 0 \tag{5.38}$$

which can be derived from the Lagrangian density

$$\mathcal{L}_0(\vec{x}, t) = \frac{i\hbar}{2} \psi^{\dagger\sigma}(\vec{x}, t) \frac{\partial}{\partial t} \psi_\sigma(\vec{x}, t) - \frac{i\hbar}{2} \frac{\partial}{\partial t} \psi^{\dagger\sigma}(\vec{x}, t) \psi_\sigma(\vec{x}, t)$$

$$- \frac{\hbar^2}{2m} \vec{\nabla} \psi^{\dagger\sigma}(\vec{x}, t) \cdot \vec{\nabla} \psi_\sigma(\vec{x}, t) \tag{5.39}$$

Consider the scale transformation[3]

$$\psi_\sigma(\vec{x}, t) \rightarrow \tilde{\psi}_\sigma(\vec{x}, t) = \Lambda^{\frac{d}{2}} \psi_\sigma(\Lambda \vec{x}, \Lambda^2 t) \tag{5.40}$$

$$\psi^{\dagger\sigma}(\vec{x}, t) \rightarrow \tilde{\psi}^{\dagger\sigma}(\vec{x}, t) = \Lambda^{\frac{d}{2}} \psi^{\dagger\sigma}(\Lambda \vec{x}, \Lambda^2 t) \tag{5.41}$$

with Λ is a positive real number and $d = 3$ in three dimensions. We will consider dimensions other than three by retaining d as a parameter. The infinitesimal transformation is

$$\delta\psi_\sigma(\vec{x}, t) = \left(2t \frac{\partial}{\partial t} + \vec{x} \cdot \vec{\nabla} + \frac{d}{2} \right) \psi_\sigma(\vec{x}, t) \tag{5.42}$$

$$\delta\psi^{\dagger\sigma}(\vec{x}, t) = \left(2t \frac{\partial}{\partial t} + \vec{x} \cdot \vec{\nabla} + \frac{d}{2} \right) \psi^{\dagger\sigma}(\vec{x}, t) \tag{5.43}$$

Consider the Lagrangian density that describes non-interacting particles (5.39). With some algebra, and without use of the field equation, we can easily show that, under the scaling transformation in equations (5.42) and (5.43),

$$\delta\mathcal{L}_0 = \frac{\partial}{\partial t} (2t\mathcal{L}_0) + \vec{\nabla} \cdot (\vec{x}\mathcal{L}_0) \tag{5.44}$$

This implies that the non-interacting field theory is scale invariant in any dimensions. We identify

$$R(\vec{x}, t) = 2t\mathcal{L}_0(\vec{x}, t) \tag{5.45}$$

$$\vec{J}(\vec{x}, t) = \vec{x}\mathcal{L}_0(\vec{x}, t) \tag{5.46}$$

and we can use Noether's theorem to form the Noether charge and current densities. The charge density is

$$\mathcal{D}(\vec{x}, t) = \delta\psi_\sigma(\vec{x}, t) \frac{\partial \mathcal{L}_0}{\partial(\frac{\partial}{\partial t}\psi_\sigma(\vec{x}, t))} + \delta\psi^{\dagger\sigma}(\vec{x}, t) \frac{\partial \mathcal{L}_0}{\partial(\frac{\partial}{\partial t}\psi^{\dagger\sigma}(\vec{x}, t))} - R(\vec{x}, t)$$

---

[3] Note that the space and time coordinates to not scale in the same way, in fact $\vec{x} \rightarrow \Lambda\vec{x}$ and $t \rightarrow \Lambda^2 t$. In general $\vec{x} \rightarrow \Lambda\vec{x}$ and $t \rightarrow \Lambda^z t$ where z is called the dynamical critical exponent. For our free non-relativistic field theory, $z = 2$ whereas, for relativistic field theory that we will study in subsequent chapters, $z = 1$.

which, using $R(\vec{x}, t)$ and $\vec{J}(\vec{x}, t)$ as identified above and he transformation in equations (5.42) and (5.43) leads to

$$
\begin{aligned}
\mathcal{D}(\vec{x}, t) = {} & 2t \frac{\partial}{\partial t} \psi_\sigma(\vec{x}, t) \frac{\partial \mathcal{L}_0}{\partial(\frac{\partial}{\partial t} \psi_\sigma(\vec{x}, t))} + 2t \frac{\partial}{\partial t} \psi^{\dagger \sigma}(\vec{x}, t) \frac{\partial \mathcal{L}_0}{\partial(\frac{\partial}{\partial t} \psi^{\dagger \sigma}(\vec{x}, t))} - 2t \mathcal{L}_0 \\
& + \vec{x} \cdot \vec{\nabla} \psi_\sigma(\vec{x}, t) \frac{\partial \mathcal{L}_0}{\partial(\frac{\partial}{\partial t} \psi_\sigma(\vec{x}, t))} + \vec{x} \cdot \vec{\nabla} \psi^{\dagger \sigma}(\vec{x}, t) \frac{\partial \mathcal{L}_0}{\partial(\frac{\partial}{\partial t} \psi^{\dagger \sigma}(\vec{x}, t))} \\
& + \frac{d}{2} \psi_\sigma(\vec{x}, t) \frac{\partial \mathcal{L}_0}{\partial(\frac{\partial}{\partial t} \psi_\sigma(\vec{x}, t))} + \frac{d}{2} \psi^{\dagger \sigma}(\vec{x}, t) \frac{\partial \mathcal{L}_0}{\partial(\frac{\partial}{\partial t} \psi^{\dagger \sigma}(\vec{x}, t))}
\end{aligned}
$$

It is useful to write the first two lines of the above equation in terms of the formal expressions for components of the stress tensor so that it has the form

$$
\begin{aligned}
\mathcal{D}(\vec{x}, t) = {} & 2t \mathbf{T}^{tt}(\vec{x}, t) + x_b \mathbf{T}^{tb}(\vec{x}, t) \\
& + \frac{d}{2} \left[\psi_\sigma(\vec{x}, t) \frac{\partial \mathcal{L}_0}{\partial(\frac{\partial}{\partial t} \psi_\sigma(\vec{x}, t))} + \psi^{\dagger \sigma}(\vec{x}, t) \frac{\partial \mathcal{L}_0}{\partial(\frac{\partial}{\partial t} \psi^{\dagger \sigma}(\vec{x}, t))} \right]
\end{aligned}
$$

By similar reasoning, for the current density,

$$
\begin{aligned}
\mathcal{D}^a(\vec{x}, t) = {} & 2t \mathbf{T}^{at}(\vec{x}, t) + x_b \mathbf{T}^{ab}(\vec{x}, t) \\
& + \frac{d}{2} \left[\psi_\sigma(\vec{x}, t) \frac{\partial \mathcal{L}_0}{\partial(\nabla_a \psi_\sigma(\vec{x}, t))} + \psi^{\dagger \sigma}(\vec{x}, t) \frac{\partial \mathcal{L}_0}{\partial(\nabla_a \psi^{\dagger \sigma}(\vec{x}, t))} \right]
\end{aligned}
$$

Now, the remaining terms depend only on the derivative terms in the Lagrangian density, not the interaction terms. If, for any non-relativistic field theory where the derivative terms in the Lagrangian density are as they appear in the free-field theory governed by $\mathcal{L}_0(\vec{x}, t)$ in (5.39), the second lines in the charge and current densities above are easily evaluated and the result is

$$
\mathcal{D}(\vec{x}, t) = 2t \mathbf{T}^{tt}(\vec{x}, t) + x_b \mathbf{T}^{tb}(\vec{x}, t) \tag{5.47}
$$

$$
\mathcal{D}^a(\vec{x}, t) = 2t \mathbf{T}^{at}(\vec{x}, t) + x_b \mathbf{T}^{ab}(\vec{x}, t) - \frac{d}{2} \frac{\hbar^2}{2m} \nabla^a \rho(\vec{x}, t) \tag{5.48}
$$

The scale invariance of the theory and Noether's theorem imply that this charge and current density obeys the equation of continuity,

$$
\frac{\partial}{\partial t} \mathcal{D}(\vec{x}, t) + \nabla_a \mathcal{D}^a(\vec{x}, t) = 0 \tag{5.49}
$$

Now, if we assume that, for a given field theory, the stress tensor is conserved, that is, the system has time- and space-translation invariance, and we simply form the charge and current densities which are given in equations (5.47) and (5.48) from them, those densities would satisfy the continuity equation (5.49) if the following identity also held

$$2\mathbf{T}^{tt}(\vec{x}, t) + \mathbf{T}_{aa}(\vec{x}, t) - \frac{d}{2}\frac{\hbar^2}{2m}\vec{\nabla}^2 \rho(\vec{x}, t) = 0 \qquad (5.50)$$

Any time and space-translation invariant theory with a conserved stress tensor where the stress tensor also obeys equation (5.50) also has scale symmetry. Note that, for this conservation law, we did not need, rotation symmetry, phase symmetry or Galilean invariance. All that we needed was space and time translation invariance and the stress tensor which follows from it. The last term in (5.50) is special to field theories where the derivative terms in the Lagrangian density are as in $\mathcal{L}_0(\vec{x}, t)$. The stress tensor, on the other hand, could be the one for any field theory which we want to test for scale symmetry.

5.3.1 Improving the Stress Tensor

It is sometimes convenient to consider an "improved" the stress tensor. Improvement takes advantage of an ambiguity in the identification of Noether charge and current densities in order to modify them so that they have favourable properties. We can easily point out the source of the ambiguity in the identification of the Noether current and charge density. It occurs precisely in the manipulations which lead to equation (4.38). There, $R(\vec{x}, t)$ and $\vec{J}(\vec{x}, t)$ are not unique. It is generally possible to find combinations of the fields and their derivatives which we can assemble into charge and current densities and which automatically obey continuity equations. If we could find such combinations, we could add them to $R(\vec{x}, t)$ and $\vec{J}(\vec{x}, t)$ without changing the fact that the transformation is a symmetry. If, in addition, the charge and current densities that are being added are total derivatives which vanish when they are integrated over the volume, they also do not modify the conserved charges. We will see an example shortly.

We will do an improvement for the specific case of the spatial components of the stress tensor $\mathbf{T}^{ab}(\vec{x}, t)$. The purpose is to make equation (5.50) more elegant.

Consider the improved stress tensor

$$\tilde{\mathbf{T}}^{ab}(\vec{x}, t) = \mathbf{T}^{ab}(\vec{x}, t) - \frac{d}{d-1}\frac{\hbar^2}{4m}\left(\delta^{ab}\vec{\nabla}^2 - \nabla^a\nabla^b\right)\rho(\vec{x}, t) \qquad (5.51)$$

where d is the dimension of space and, as usual, $\rho(\vec{x}, t) = \psi^{\sigma\dagger}(\vec{x}, t)\psi_\sigma(\vec{x}, t)$. Since

$$\nabla_a\left[\left(\delta^{ab}\vec{\nabla}^2 - \nabla^a\nabla^b\right)\text{anything}\right] = 0 \qquad (5.52)$$

the divergence of this spatial part of the stress tensor is unchanged

$$\nabla_a\tilde{\mathbf{T}}^{ab}(\vec{x}, t) = \nabla_a\mathbf{T}^{ab}(\vec{x}, t) \qquad (5.53)$$

The additional term in $\tilde{\mathbf{T}}^{ab}(\vec{x}, t)$ therefore does not change the continuity equations, they are still obeyed with the improved tensor,

$$\frac{\partial}{\partial t}\mathbf{T}^{ta}(\vec{x}, t) + \nabla_b\tilde{\mathbf{T}}^{ba}(\vec{x}, t) = 0 \tag{5.54}$$

Also, the quantity that has been added to $\mathbf{T}^{ab}(\vec{x}, t)$ to make $\tilde{\mathbf{T}}^{ab}(\vec{x}, t)$ is a total divergence of derivatives of the density

$$\frac{d}{d-1}\frac{\hbar^2}{4m}\left(\delta^{ab}\vec{\nabla}^2 - \nabla^a\nabla^b\right)\rho(\vec{x}, t) = \nabla_c\left[\frac{d}{d-1}\frac{\hbar^2}{4m}\left(\delta^{ab}\delta^{cd} - \delta^{ac}\delta^{bd}\right)\nabla_d\rho(\vec{x}, t)\right] \tag{5.55}$$

The derivatives $\nabla_d\rho(\vec{x}, t)$ will fall off rapidly at spatial infinity, particularly when the density approaches a constant there. Then, using Gauss's theorem, we see that

$$\int d^3x\,\mathbf{T}^{ab}(\vec{x}, t) = \int d^3x\,\tilde{\mathbf{T}}^{ab}(\vec{x}, t) \tag{5.56}$$

If the spatial part of the stress tensor is symmetric, $\mathbf{T}^{ab}(\vec{x}, t) = \mathbf{T}^{ba}(\vec{x}, t)$ then so is $\tilde{\mathbf{T}}^{ab}(\vec{x}, t) = \tilde{\mathbf{T}}^{ba}(\vec{x}, t)$.

Finally,

$$\tilde{\mathbf{T}}_a^{\ a} = \mathbf{T}_a^{\ a} - d\frac{\hbar^2}{2m}\vec{\nabla}^2\rho(\vec{x}, t) \tag{5.57}$$

where we have written the expression for d space dimensions. The condition for scale invariance becomes

$$2\mathbf{T}^{tt} + \tilde{\mathbf{T}}_a^{\ a} = 0 \tag{5.58}$$

$\tilde{\mathbf{T}}^{ab}$ is called the "improved stress tensor". We can confirm that its use in the other places where we have used $\mathbf{T}^{ab}$ above is also legitimate. For example, in the Galilean Noether current density in equation (5.32) we use $\mathbf{T}^{ab}$ which could equally well be substituted by $\tilde{\mathbf{T}}^{ab}$.

5.3.2   The Consequences of Scale Invariance

The operator equation $2\mathbf{T}^{tt}(\vec{x}, t) + \tilde{\mathbf{T}}_a^{\ a}(\vec{x}, t) = 0$ has interesting consequences. This identity must hold in any scale invariant field theory. Its expectation value must also hold in any state of a scale invariant theory, even when the state itself is not scale invariant. In particular, the ground state $|\mathcal{O}>$ which we have discussed for a weakly interacting Fermi or Bose gas cannot be scale invariant since it contains a finite

density of particles. However, if the theory happened to be scale invariant, we would
have

$$2 < \mathcal{O}|\mathbf{T}^{tt}(\vec{x}, t)|\mathcal{O} > + < \mathcal{O}|\tilde{\mathbf{T}}_a^{\ a}(\vec{x}, t)|\mathcal{O} >= 0 \tag{5.59}$$

Generally, the expectation value of $\mathbf{T}^{tt}(\vec{x}, t)$ is the energy density. Moreover, the
average of the expectation values of the diagonal components of $\tilde{\mathbf{T}}_{ab}(\vec{x}, t)$ is equal
to the pressure. This tells us that, in any state of a scale invariant theory, there is a
relationship between the internal energy density, u and the thermodynamic pressure
P,

$$2u = dP \tag{5.60}$$

where d is the space dimension.

The degenerate Fermi gas is well described by non-interacting fermions which
have a scale invariant dynamics. One can confirm from equations (3.53) and (3.54)
that equation (5.60) holds in that case.

Generally, once we introduce interactions, the system that we are discussing is not
scale invariant. In fact, since for the Lagrangian density (4.6), the improved stress
tensor is

$$\begin{aligned}
\tilde{\mathbf{T}}^{ab} = &-\frac{\hbar^2}{2m}\left(\nabla^a\psi^{\dagger\sigma}(\vec{x}, t)\nabla^b\psi_\sigma(\vec{x}, t) + \nabla^b\psi^{\dagger\sigma}(\vec{x}, t)\nabla^a a\psi_\sigma(\vec{x}, t)\right) \\
&+ \delta^{ab}\left[\vec{\nabla}^2(\psi^{\dagger\sigma}\psi_\sigma) - \frac{\lambda}{2}\left(\psi^{\sigma\dagger}(\vec{x}, t)\psi_\sigma(\vec{x}, t)\right)^2\right] \\
&- \frac{d}{d-1}\frac{\hbar^2}{2m}\left(\delta^{ab}\vec{\nabla}^2 - \nabla^a\nabla^b\right)\left(\psi^{\sigma\dagger}(\vec{x}, t)\psi_\sigma(\vec{x}, t)\right)
\end{aligned} \tag{5.61}$$

where we have used the equation of motion to eliminate the time derivative terms.
The trace condition for scale invariance is

$$2\mathbf{T}^{tt} + \tilde{\mathbf{T}}^a_{\ a} = (2 - d)\frac{\lambda}{2}\left(\psi^{\sigma\dagger}(\vec{x}, t)\psi_\sigma(\vec{x}, t)\right)^2 \tag{5.62}$$

which is non-zero. The system can only have scale invariance if the right-hand-side
vanishes, as an operator. Outside of two dimensions, the interaction is not scale
invariant. It turns out that the apparent scale invariance when $d = 2$ is violated by
a scale anomaly, so even in two dimensions, there is no scale invariance once the
particles interact with each other with generic values of the coupling constant.

There can be some special values of the coupling, called "fixed points" at which
the theory has scale invariance. Of course, free field theory where all of the couplings
vanish is an example. The Feschbach resonance, or unitary point of a cold atom gas
is another example.

5.4 Special Schrödinger Symmetry

Scale invariance is often accompanied by an additional symmetry which is a close analog of the conformal transformations that we will learn about in later chapters. If a field theory has translation invariance, so that it has a conserved stress tensor, and it has scale invariance, so that the stress tensor satisfies $2\mathbf{T}^{tt} + \tilde{\mathbf{T}}_{aa} = 0$, if, in addition it has Galilean symmetry, so that $\mathbf{T}_{ta} = -m\mathbf{J}_a$, then it has another symmetry, called the special Schrödinger symmetry. The special Schrödinger transformation of the fields is

$$\delta\psi_\sigma(\vec{x}, t) = \left(t^2 \frac{\partial}{\partial t} + t\vec{x} \cdot \vec{\nabla} - \frac{im}{\hbar} \frac{\vec{x}^2}{2} + \frac{d}{2} t \right) \psi_\sigma(\vec{x}, t) \tag{5.63}$$

$$\delta\psi^{\dagger\sigma}(\vec{x}, t) = \left(t^2 \frac{\partial}{\partial t} + t\vec{x} \cdot \vec{\nabla} + \frac{im}{\hbar} \frac{\vec{x}^2}{2} + \frac{d}{2} t \right) \psi^{\dagger\sigma}(\vec{x}, t) \tag{5.64}$$

The free field Lagrangian density (5.39) has special Schödinger symmetry. Under the transforms in equations (5.63) and (5.64), $\mathcal{L}_0(\vec{x}, t)$ transforms by the total derivative terms,

$$\delta\mathcal{L}_0(\vec{x}, t) = \frac{\partial}{\partial t} \left(t^2 \mathcal{L}_0(\vec{x}, t) \right) + \vec{\nabla} \cdot (t\vec{x} \mathcal{L}_0(\vec{x}, t))$$

The Noether charge and current densities are constructed in the standard manner. They are related to the stress tensor components and the particle and number current densities as

$$\mathcal{K}(\vec{x}, t) = t^2 \mathbf{T}^{tt}(\vec{x}, t) + tx_b \mathbf{T}_{tb}(\vec{x}, t) + m \frac{x^2}{2} \rho(\vec{x}, t)$$

$$\mathcal{K}_a(\vec{x}, t) = t^2 \mathbf{T}_{at}(\vec{x}, t) + tx_b \mathbf{T}_{ab}(\vec{x}, t) + m \frac{x^2}{2} \mathbf{J}_a(\vec{x}, t) - dt \frac{\hbar^2}{4m} \nabla_a \rho(\vec{x}, t)$$

If the stress tensor is conserved and the number density and current are conserved, the above charge and current densities also obey a continuity equation if the following expression vanishes

$$2t\mathbf{T}^{tt}(\vec{x}, t) + x^b \mathbf{T}_{tb}(\vec{x}, t) + t\mathbf{T}_{aa}(\vec{x}, t) + mx^b \mathbf{J}_b(\vec{x}, t) - td\frac{\hbar^2}{4m}\vec{\nabla}^2 \rho(\vec{x}, t)) = 0$$

Scale and Galilean symmetry are enough to guarantee the above. We conclude that a translation invariant quantum field theory which has a conserved stress tensor associated with the time and space-translation invariance and a conserved particle density and current, and which also also has a Galilean symmetry, so that

$$\mathbf{T}_{tb}(\vec{x}, t) + m\mathbf{J}_b(\vec{x}, t) = 0 \tag{5.65}$$

and a scale symmetry so that

$$2\mathbf{T}^{tt}(\vec{x}, t) + \mathbf{T}_a^{\ a}(\vec{x}, t) - d\frac{\hbar^2}{4m}\vec{\nabla}^2\rho(\vec{x}, t) = 0 \tag{5.66}$$

must also have a conserved current corresponding to the special Schrödinger symmetry. We can also use the improved stress tensor that we found in the previous section. For that case, the Noether current density would be

$$\begin{aligned}
\mathcal{K}_b(\vec{x}, t) &= t^2\mathbf{T}_{bt}(\vec{x}, t) + tx^a\tilde{\mathbf{T}}_{ab}(\vec{x}, t) + m\frac{x^2}{2}\mathbf{J}_b(\vec{x}, t) \\
&\quad - \left(\delta_{ba}\vec{\nabla}^2 - \nabla_b\nabla_a\right)\left(\frac{d}{d-1}\frac{\hbar^2}{4m}x^a\rho(\vec{x}, t)\right)
\end{aligned} \tag{5.67}$$

We see that, re-writing $\mathcal{K}_b(\vec{x}, t)$ with $\tilde{\mathbf{T}}_{ab}$ results in an additional total derivative, automatically conserved term which could just be dropped in order to write improved charge and current densities

$$\mathcal{K}(\vec{x}, t) = t^2\mathbf{T}^{tt}(\vec{x}, t) + tx^a\mathbf{T}_{ta}(\vec{x}, t) + m\frac{x^2}{2}\rho(\vec{x}, t) \tag{5.68}$$

$$\tilde{\mathcal{K}}_b(\vec{x}, t) = t^2\mathbf{T}_{bt}(\vec{x}, t) + tx^a\tilde{\mathbf{T}}_{ab}(\vec{x}, t) + m\frac{x^2}{2}\mathbf{J}_b(\vec{x}, t) \tag{5.69}$$

5.5   Summary

The following is a summary of the symmetry transformations, their Noether charge and current densities—the number current and stress tensor—for the non-relativistic quantum field theory that we have discussed:

Symmetry Transformations

Phase transformation:

$$\delta\psi = \frac{1}{i\hbar}\psi_\sigma(\vec{x}, t), \quad \delta\psi^{\dagger\rho}(\vec{x}, t) = -\frac{1}{i\hbar}\psi^{\dagger\rho}(\vec{x}, t) \tag{5.70}$$

Space translation:

$$\delta^a\psi_\sigma(\vec{x}, \vec{t}) = \nabla^a\psi_\sigma(\vec{x}, t), \quad \delta^a\psi^{\dagger\rho}(\vec{x}, \vec{t}) = \nabla^a\psi^{\dagger\rho}(\vec{x}, t) \tag{5.71}$$

Time translation:

$$\delta\psi_\sigma(\vec{x}, \vec{t}) = \frac{\partial}{\partial t}\psi_\sigma(\vec{x}, t), \quad \delta\psi^{\dagger\rho}(\vec{x}, t) = \frac{\partial}{\partial t}\psi^{\dagger\rho}(\vec{x}, t) \tag{5.72}$$

Space rotation:

$$\delta^{ab}\psi_\sigma(\vec{x},\vec{t}) = (x^a\nabla^b - x^b\nabla^a)\psi_\sigma(\vec{x},t) \tag{5.73}$$

$$\delta^{ab}\psi^{\dagger\rho}(\vec{x},\vec{t}) = (x^a\nabla^b - x^b\nabla^a)\psi^{\dagger\rho}(\vec{x},t) \tag{5.74}$$

Galilean boost:

$$\delta^a\psi_\sigma(\vec{x},t) = \left(-im\vec{x}^a/\hbar + t\nabla^a\right)\psi_\sigma(\vec{x},t) \tag{5.75}$$

$$\delta^a\psi^{\dagger\sigma}(\vec{x},t) = \left(imx^a/\hbar + t\nabla^a\right)\psi^{\dagger\sigma}(\vec{x},t) \tag{5.76}$$

Scale transformation:

$$\delta\psi_\sigma(\vec{x},t) = \left(2t\frac{\partial}{\partial t} + \vec{x}\cdot\vec{\nabla} + \frac{d}{2}\right)\psi_\sigma(\vec{x},t) \tag{5.77}$$

$$\delta\psi^{\dagger\sigma}(\vec{x},t) = \left(2\frac{\partial}{\partial t} + \vec{x}\cdot\vec{\nabla} + \frac{d}{2}\right)\psi^{\dagger\sigma}(\vec{x},t) \tag{5.78}$$

Special Schrödinger's transformation:

$$\delta\psi_\sigma(\vec{x},t) = \left(t^2\frac{\partial}{\partial t} + t\vec{x}\cdot\vec{\nabla} - \frac{im\,\vec{x}^2}{\hbar\,2} + \frac{d}{2}t\right)\psi_\sigma(\vec{x},t) \tag{5.79}$$

$$\delta\psi^{\dagger\sigma}(\vec{x},t) = \left(t^2\frac{\partial}{\partial t} + t\vec{x}\cdot\vec{\nabla} + \frac{im\,\vec{x}^2}{\hbar\,2} + \frac{d}{2}t\right)\psi^{\dagger\sigma}(\vec{x},t) \tag{5.80}$$

Noether Charge and Current Densities

For the Lagrangian density

$$\mathcal{L} = \frac{i\hbar}{2}\psi^{\dagger\sigma}\frac{\partial}{\partial t}\psi_\sigma - \frac{i\hbar}{2}\frac{\partial}{\partial t}\psi^{\dagger\sigma}\psi_\sigma - \frac{\hbar^2}{2m}\vec{\nabla}\psi^{\dagger\sigma}\cdot\vec{\nabla}\psi_\sigma - \frac{\lambda}{2}\left(\psi^{\dagger\sigma}\psi_\sigma\right)^2 \tag{5.81}$$

the number current is

$$\rho(\vec{x},t) = \psi^{\dagger\sigma}(\vec{x},t)\psi_\sigma(\vec{x},t) \tag{5.82}$$

$$\mathbf{J}_a(\vec{x},t) = -\frac{i\hbar}{2m}\left[\psi^{\dagger\sigma}(\vec{x},t)\nabla_a\psi_\sigma(\vec{x},t) - \nabla_a\psi^{\dagger\sigma}(\vec{x},t)\,\psi_\sigma(\vec{x},t)\right] \tag{5.83}$$

and the improved stress tensor is

$$\mathbf{T}^{tt} = \frac{\hbar^2}{2m}\vec{\nabla}\psi^{\dagger\sigma}(\vec{x},t)\cdot\vec{\nabla}\psi_\sigma(\vec{x},t) + \frac{\lambda}{2}\left(\psi^{\dagger\sigma}(\vec{x},t)\psi_\sigma(\vec{x},t)\right)^2 \tag{5.84}$$

$$\mathbf{T}^{tb} = -\frac{i\hbar}{2}\left[\nabla_b\psi^{\dagger\sigma}(\vec{x},t)\psi_\sigma(\vec{x},t) - \psi^{\dagger\sigma}(\vec{x},t)\nabla_b\psi_\sigma(\vec{x},t)\right] \tag{5.85}$$

$$\mathbf{T}^{at} = -\frac{\hbar^2}{2m}\left(\nabla_a\psi^{\dagger\sigma}(\vec{x},t)\frac{\partial}{\partial t}\psi_\sigma(\vec{x},t) + \frac{\partial}{\partial t}\psi^{\dagger\sigma}(\vec{x},t)\nabla_a\psi(\vec{x},t)\right) \tag{5.86}$$

$$\tilde{\mathbf{T}}^{ab} = -\frac{\hbar^2}{2m}\left(\nabla^a\psi^{\dagger\sigma}(\vec{x},t)\nabla^b\psi_\sigma(\vec{x},t) + \nabla^b\psi^{\dagger\sigma}(\vec{x},t)\nabla^a\psi_\sigma(\vec{x},t)\right)$$
$$+ \delta^{ab}\left[\vec{\nabla}^2(\psi^{\dagger\sigma}\psi_\sigma) - \frac{\lambda}{2}\left(\psi^{\sigma\dagger}(\vec{x},t)\psi_\sigma(\vec{x},t)\right)^2\right]$$
$$- \frac{d}{d-1}\frac{\hbar^2}{2m}\left(\delta^{ab}\vec{\nabla}^2 - \nabla^a\nabla^b\right)\left(\psi^{\sigma\dagger}(\vec{x},t)\psi_\sigma(\vec{x},t)\right) \tag{5.87}$$

Conserved Noether Charges
Phase Symmetry: $\exists(\rho,\vec{\mathbf{J}})$ with

$$\frac{\partial}{\partial t}\rho + \vec{\nabla}\cdot\mathbf{J} = 0 \tag{5.88}$$

particle number

$$\mathcal{N} = \int d^d x\,\rho(\vec{x},t) \tag{5.89}$$

Space–time translation symmetry: $\exists(\mathbf{T}^{tt},\mathbf{T}^{tb},\mathbf{T}^{at},\tilde{\mathbf{T}}^{ab})$ such that

$$\frac{\partial}{\partial t}\mathbf{T}^{tt} + \nabla_a\mathbf{T}^{at} = 0, \quad \frac{\partial}{\partial t}\mathbf{T}^{tb} + \nabla_a\tilde{\mathbf{T}}^{ab} = 0 \tag{5.90}$$

energy and momentum

$$H = \int d^d x\,\mathbf{T}^{tt}(\vec{x},t), \quad P^a = -\int d^d x\,\mathbf{T}^{ta}(\vec{x},t) \tag{5.91}$$

Rotation symmetry: $\exists(\mathbf{T}^{tt},\mathbf{T}^{tb},\mathbf{T}^{at},\tilde{\mathbf{T}}^{ab})$ with (5.90) and $\tilde{\mathbf{T}}^{ab} = \tilde{\mathbf{T}}^{ba}$, conserved angular momenta

$$J^{ab} = \int d^d x\left[x^a\mathbf{T}^{tb}(\vec{x},t) - x^b\mathbf{T}^{tb}(\vec{x},t)\right] \tag{5.92}$$

Galilean invariance: $\exists(\rho, \vec{\mathbf{J}})$ and $\exists(\mathbf{T}^{tt}, \mathbf{T}^{tb}, \mathbf{T}^{at}, \tilde{\mathbf{T}}^{ab})$ with (5.88), (5.90) and $\mathbf{T}^{ta} = -m\mathbf{J}^a$

Galilean boost generators

$$\mathcal{B}^b = \int d^d x \left[t\mathbf{T}^{tb}(\vec{x}, t) + mx^b \rho(\vec{x}, t) \right] \tag{5.93}$$

Scale invariance: $\exists(\mathbf{T}^{tt}, \mathbf{T}^{tb}, \mathbf{T}^{at}, \tilde{\mathbf{T}}^{ab})$ with (5.90), $2\mathbf{T}^{tt} + \tilde{\mathbf{T}}_a{}^a = 0$

Dilatation generator

$$\mathcal{D} = \int d^d x \left[2t\mathbf{T}^{tt}(\vec{x}, t) + x_b\mathbf{T}^{tb}(\vec{x}, t) \right] \tag{5.94}$$

Special Schrödinger symmetry: $\exists(\rho, \vec{\mathbf{J}}(\vec{x}, t))$ with (5.88) and
$\exists(\mathbf{T}^{tt}, \mathbf{T}^{tb}, \mathbf{T}^{at}, \tilde{\mathbf{T}}^{ab})$ with (5.90) and $\mathbf{T}^{ta} = -m\mathbf{J}^a$, $2\mathbf{T}^{tt} + \tilde{\mathbf{T}}_a{}^a = 0$,

Conformal generator

$$\mathcal{K} = \int d^d x \left[t^2\mathbf{T}^{tt}(\vec{x}, t) + tx_b\mathbf{T}_{tb}(\vec{x}, t) + m\frac{x^2}{2}\rho(\vec{x}, t) \right] \tag{5.95}$$

Galilei Algebra and Schrödinger Algebra: We could either use the Poisson brackets at the classical level or, once the classical fields are replaced by quantum fields, the commutators of the quantum fields to compute the commutation relation of the Noether charges with the fields and with each other. With the fields, the Noether charges generate the infinitesimal transformations

$$\delta\psi_\sigma(\vec{x}, t) = \frac{1}{i\hbar}\left[\psi_\sigma(\vec{x}, t), \mathcal{R}\right], \quad \delta\psi^{\dagger\sigma}(\vec{x}, t) = \frac{1}{i\hbar}\left[\psi^{\dagger\sigma}(\vec{x}, t), \mathcal{R}\right] \tag{5.96}$$

For the charges, the result is the Galilei algebra

$$\frac{1}{i\hbar}[\mathcal{N}, H] = 0 \quad \frac{1}{i\hbar}[\mathcal{N}, P_a] = 0 \quad \frac{1}{i\hbar}[\mathcal{N}, J_{ab}] = 0 \quad \frac{1}{i\hbar}[\mathcal{N}, \mathcal{B}_a] = 0 \tag{5.97}$$

$$\frac{1}{i\hbar}[P_a, P_b] = 0 \quad \frac{1}{i\hbar}[J_{ab}, H] = 0 \quad \frac{1}{i\hbar}[P_a, H] = 0. \tag{5.98}$$

$$\frac{1}{i\hbar}[J_{ab}, J_{cd}] = \delta_{ad}J_{bc} - \delta_{ac}J_{db} + \delta_{bc}J_{da} - \delta_{bd}J_{ac} \tag{5.99}$$

$$\frac{1}{i\hbar}[J_{ab}, P_c] = \delta_{ac}P_b - \delta_{bc}P_a \quad \frac{1}{i\hbar}[J_{ab}, \mathcal{B}_c] = \delta_{ac}\mathcal{B}_b - \delta_{bc}\mathcal{B}_a \tag{5.100}$$

$$\frac{1}{i\hbar}[\mathcal{B}_a, \mathcal{B}_b] = 0 \quad \frac{1}{i\hbar}[\mathcal{B}_a, H] = P_a \quad \frac{1}{i\hbar}[\mathcal{B}_a, P_b] = -\delta_{ab}m\frac{\mathcal{N}}{2} \tag{5.101}$$

If we are interested in a theory that also has scale invariance, we can add two more charges, the dilatation, and the special Schrödinger operator and the commutators

$$\frac{1}{i\hbar}[\mathcal{D}, H] = 2H, \quad \frac{1}{i\hbar}[\mathcal{D}, P_a] = P_a, \quad \frac{1}{i\hbar}[\mathcal{D}, \mathcal{B}_a] = -\mathcal{B}_a, \quad \frac{1}{i\hbar}[\mathcal{D}, J_{ab}] = 0$$

$$(5.102)$$

$$\frac{1}{i\hbar}[\mathcal{K}, H] = \mathcal{D}, \quad \frac{1}{i\hbar}[\mathcal{K}, P_a] = -\mathcal{B}_a, \quad \frac{1}{i\hbar}[\mathcal{K}, \mathcal{B}_a] = 0, \quad \frac{1}{i\hbar}[\mathcal{K}, J_{ab}] = 0 \quad (5.103)$$

$$\frac{1}{i\hbar}[\mathcal{K}, \mathcal{D}] = 2\mathcal{K} \tag{5.104}$$

which together with the Galilei algebra forms the Schrödinger algebra. The three generators H, $\mathcal{D}$ and $\mathcal{K}$ form an SL(2,R) subalgebra which has had some interesting consequences for the physics of cold atoms.

Space–Time Symmetry and Relativistic Field Theory

6

6.1 Quantum Mechanics and Special Relativity

It is often said that quantum field theory is the natural marriage of Einstein's special theory of relativity and the quantum theory. The point of this section will be to motivate this statement, before we proceed to a closer study of relativity. We will begin with a single free quantum mechanical particle and ask what is wrong with simply assuming that it is quantized in the same way as a single non-relativistic particle, with eigenstates of momentum $|p>$ or position $|x>$ but where its energy spectrum is given by the relativistic expression $E(\vec{p}) = \sqrt{m^2c^4 + c^2\vec{p}^2}$, rather than the non-relativistic $E(\vec{p}) = \frac{\vec{p}^2}{2m}$. After all, for small momenta, the first expression is $E(\vec{p}) \approx mc^2 + \frac{\vec{p}^2}{2m} + \dots$ which is the non-relativistic energy plus a constant, the rest mass.

Let us assume that the particle travels on open, infinite three dimensional space. It is described by its position $\vec{x}$ and momentum $\vec{p}$ which, for the quantum mechanical particle, are operators with the commutation relation

$$\left[\hat{x}^a, \hat{p}_b\right] = i\hbar\delta^a_b$$

Momentum and energy are conserved and the energy and momentum are related by

$$E(\vec{p}) = \sqrt{m^2c^4 + \vec{p}^2c^2} \qquad (6.1)$$

where m is the rest mass of the particle and c is the speed of light. In the quantum mechanics of a single particle, we could consider a quantum state of the particle which is an eigenstate of its linear momentum,

$$\hat{p}^a \, |p\rangle = p^a \, |p\rangle \,, \quad i = 1, 2, 3 \tag{6.2}$$

These eigenstates of momentum have a continuum normalization, so that

$$\langle p|p'\rangle = \delta^3(\vec{p} - \vec{p}\,'). \tag{6.3}$$

This equation is perhaps the first sign of trouble as it is not Lorentz covariant, for that it would have to be $\langle p|p'\rangle = 2E(\vec{p})\delta^3(\vec{p} - \vec{p}\,')$.

We will ignore this problem for the time being and stick with our initial question which asks what is wrong with our minimal modification of non-relativistic quantum mechanics where we simply replace the free particle Hamiltonian by one which gives us the relativistic particle dispersion relation (6.1),

$$H = \frac{\vec{p}^2}{2m} \; \rightarrow \; H = \sqrt{m^2c^4 + \vec{p}^2 c^2} \tag{6.4}$$

Of course, the action of this Hamiltonian is easy to understand when it operates on eigenstates of the momentum,

$$H \, |\vec{p}\rangle = \sqrt{m^2 c^4 + \vec{p}^2 c^2} \, |\vec{p}\rangle \tag{6.5}$$

The Schrödinger equation must be satisfied by the time-dependent state vector, $|\Psi(t)\rangle$,

$$i\hbar \frac{\partial}{\partial t} \, |\Psi(t)\rangle = H \, |\Psi(t)\rangle \tag{6.6}$$

which has the solution

$$|\Psi(t)\rangle = e^{-iHt/\hbar} \, |\Psi(0)\rangle \tag{6.7}$$

The probability amplitude that the particle which begins in momentum eigenstate $|\vec{p}>$ at the initial time evolves to the momentum eigenstate $|\vec{q}>$ after time t has elapsed is given by the matrix element

$$<\vec{q}| \, e^{-iHt/\hbar} \, |\vec{p}> = \; e^{-i\sqrt{m^2c^4 + \vec{p}^2 c^2}\,t/\hbar} \, \delta(\vec{p} - \vec{q}) \tag{6.8}$$

We can also consider eigenstates of the position operator

$$\hat{\vec{x}}|\vec{x}> = \vec{x}|\vec{x}>, \quad <\vec{x}_1|\vec{x}_2> = \delta(\vec{x}_1 - \vec{x}_2)$$

We can then ask about the probability amplitude that the particle begins in a state which is localized at a position $\vec{x}$ and propagates to a position $\vec{y}$. The easiest way is to use completeness of the set of all momentum and position eigenstates, so that the unit operator in Hilbert space is

$$\mathcal{I} = \int d^3p \, |\vec{p}><\vec{p}|, \quad \mathcal{I} = \int d^3x \, |\vec{x}><\vec{x}|$$

and the overlap

$$<\vec{x}|\vec{p}> = \frac{e^{i\vec{p}\cdot\vec{x}/\hbar}}{(2\pi\hbar)^{\frac{3}{2}}}$$

Using these, we get the expression

$$
\begin{aligned}
<\vec{y}| \, e^{-iHt/\hbar} \, |\vec{x}> &= \int d^3q \int d^3p <\vec{y}|\vec{q}><\vec{q}| \, e^{-iHt/\hbar} \, |\vec{p}><\vec{p}|\vec{x}> \\
&= \int d^3q \frac{e^{i\vec{q}\cdot\vec{y}/\hbar}}{(2\pi\hbar)^{\frac{3}{2}}} \int d^3p \frac{e^{-i\vec{p}\cdot\vec{x}/\hbar}}{(2\pi\hbar)^{\frac{3}{2}}} <\vec{q}| \, e^{-iHt/\hbar} \, |\vec{p}> \\
&= \int d^3p \frac{1}{(2\pi\hbar)^3} e^{i\vec{p}\cdot(\vec{y}-\vec{x})/\hbar - i\sqrt{m^2c^4+\vec{p}^2c^2}\,t/\hbar}
\end{aligned}
\tag{6.9}
$$

This result should be interpreted as a probability amplitude.

There are a few problems with it. One is the problem that was pointed out in the discussion following equation (6.3), that we have been using states whose normalization and inner products are not Lorentz covariant. Lorentz covariance would require that this probability amplitude in (6.9), up to a phase, is a function of $c^2t^2 - |\vec{y} - \vec{x}|^2$, and it is not. (Upon close examination, we could see that it transforms like the time component of a four-vector.)

In fact, there is a small modification of (6.9) which is Lorentz invariant and which is closer to the correct answer. The amplitude becomes Lorentz invariant if we include a factor of $1/2E$ in the integrand, to get

$$
\begin{aligned}
<\vec{y}| \, e^{-iHt/\hbar} \, |\vec{x}>_{\text{improved}} \\
= \int d^3p \frac{c}{(2\pi\hbar)^3 2E(\vec{p})} e^{i\vec{p}\cdot(\vec{y}-\vec{x})/\hbar - i\sqrt{m^2c^4+\vec{p}^2c^2}\,t/\hbar}
\end{aligned}
\tag{6.10}
$$

This expression is Lorentz covariant. A way to see this is to use the identity for Dirac delta functions

$$\delta(p_\mu p^\mu + m^2c^2)\theta(p^0) = \frac{\delta(p^0 - \sqrt{\vec{p}^2 + m^2c^2})}{2\sqrt{\vec{p}^2 + m^2c^2}}\tag{6.11}$$

where the left-hand-side is invariant under proper Lorentz transformations (which cannot change the sign of p^0).

$$
\begin{aligned}
<\vec{y}| \, e^{-iHt/\hbar} \, |\vec{x}>_{\text{improved}} \\
= \int dp \frac{1}{(2\pi\hbar)^3} \delta(p_\mu p^\mu + m^2c^2)\theta(p^0) e^{i[\vec{p}\cdot(\vec{y}-\vec{x}) - p^0 ct]/\hbar}
\end{aligned}
\tag{6.12}
$$

where the right-hand-side is now expressed in terms of the Lorentz invariant quantities dp, $p_\mu p^\mu$ and $p_\mu x^\mu$ (these symbols will be defined in subsequent sections of this chapter) and (6.10) is thus manifestly Lorentz invariant.

The "improved" amplitude (6.10) has the interesting feature that, as $t \to 0$,

$$\lim_{t \to 0} < \vec{y} |\, e^{-iHt/\hbar} \,| \vec{x} >_{\text{improved}} \qquad = \int d^3 p \, \frac{c \, e^{i\vec{p}\cdot(\vec{y}-\vec{x})/\hbar}}{(2\pi\hbar)^3 2E(\vec{p})} \qquad (6.13)$$

the right-hand-side is not a delta function in positions as it would be in non-relativistic physics. This is a symptom of the fact that a relativistic particle cannot be localized in a volume that is smaller than its Compton wavelength. Indeed, the integral on the right-hand-side of (6.13) resembles that for a non-relativistic particle in the regime $|\vec{p}| \ll mc$.

Now, we find other fatal difficulty that is shared by both of the expressions for the amplitudes (6.9) and (6.10). One of the postulates of the special theory of relativity states that the speed of light is a maximum speed. However, from either equation (6.9) or its Lorentz invariant fix-up (6.10) we can see that the probability amplitude is nonzero in the causally forbidden region, where $|\vec{y} - \vec{x}| > ct$. There is non-zero probability of the particle moving at a speed that is greater than the speed of light. This behaviour does not fit nature as we currently understand it where, apparently, particles cannot propagate at speeds greater than that of light.

A formal way to see that (6.9) or (6.10) are indeed nonzero in the forbidden region is to consider t where it occurs in either of those equations as a complex variable. Then, both (6.9) and (6.10) are analytic in the lower half of the complex t-plane. When t is real, the expressions are distributions which should be defined by their limits as complex t approaches the real axis from the lower half plane. Given that they are analytic in this domain, they cannot be zero in any extended region of the lower half plane plus the real axis except for discrete points, otherwise they would have to be zero everywhere. This is a fundamental property of analytic functions. The right-hand-sides of (6.9) and (6.10) are definitely not zero for all times. Thus, they cannot be zero in the entire extended region where t is real and $ct < |\vec{y} - \vec{x}|$. They both describe an amplitude which allows propagation faster than the speed of light.

The argument above establishes the point. To see this more explicitly, we can do the integrals for the special case where $m = 0$. We could do this for either of the integrals in (6.9) or (6.10) and the results would be similar. Here, we will concentrate of equation (6.10) which is a little bit simpler. We put $\vec{p} \to \hbar p$ and we go to spherical polar coordinates where $\int d^3 p \ldots = \int_0^\infty p^2 dp \int_0^{2\pi} \int_0^\pi d\phi \sin\theta d\theta \ldots$ and where we assume that θ is the angle between the vectors $\vec{p}$ and $\vec{x}_f - \vec{x}_i$. The integral becomes

$$< \vec{y} |\, e^{-iHt/\hbar} \,| \vec{x} > = \frac{1}{4\pi^2} \int_0^\infty p\, dp \int_{-1}^1 d\cos\theta \, e^{ip(|\vec{y}-\vec{x}|\cos\theta - ct)} \qquad (6.14)$$

$$= \frac{1}{4\pi^2 i |\vec{y}-\vec{x}|} \int_0^\infty dp \left[e^{ip(|\vec{y}-\vec{x}|-ct)} - e^{ip(-|\vec{y}-\vec{x}|-ct)} \right]$$

$$= \lim_{\epsilon \to 0^+} \frac{1}{4\pi^2 |\vec{y} - \vec{x}|} \left[\frac{1}{|\vec{y} - \vec{x}| - ct + i\epsilon} + \frac{1}{|\vec{y} - \vec{x}| + ct - i\epsilon} \right]$$

$$= \lim_{\epsilon \to 0^+} \frac{1}{4\pi^2 |\vec{y} - \vec{x}|} \left[\frac{\mathcal{P}}{|\vec{y} - \vec{x}| - ct} + \frac{\mathcal{P}}{|\vec{y} - \vec{x}| + ct} \right.$$

$$\left. - i\pi\delta(|\vec{y} - \vec{x}| - ct) + i\pi\delta(|\vec{y} - \vec{x}| + ct) \right]$$

$$= \lim_{\epsilon \to 0^+} \frac{1}{2\pi^2} \left[-i\pi\delta(|\vec{y} - \vec{x}|^2 - c^2 t^2)\epsilon(t) + \frac{\mathcal{P}}{|\vec{y} - \vec{x}|^2 - c^2 t^2} \right] \qquad (6.15)$$

where

$$\epsilon(t) = \theta(t) - \theta(-t) = \text{sign}(t)$$

We have used the identity

$$\lim_{\epsilon \to 0^+} \frac{1}{x - i\epsilon} = \frac{\mathcal{P}}{x} + i\pi\delta(x)$$

where $\mathcal{P}/x$ is the principal value distribution. Also, for the Dirac delta function

$$\delta(t^2 - a^2) = \frac{1}{2|a|} \left(\delta(t - a) + \delta(t + a) \right)$$

$$\delta(t^2 - a^2)\epsilon(t) = \frac{1}{2|a|} \left(\delta(t - |a|) - \delta(t + |a|) \right)$$

In equation (6.15), we see that the wave-function of a massless particle spreads in two ways. The first is a wave which travels at the speed of light and is therefore confined to the light cone - where $|\vec{y} - \vec{x}| = ct$. The second is a principle value distribution which is non-zero everywhere except $|\vec{y} - \vec{x}| = ct$, including in the forbidden region where $|\vec{y} - \vec{x}| > ct$. This latter spreading of the wave packet violates causality. It tells us that, in our quantum mechanical system, the result of a measurement of the position of the particle at position $\vec{y}$ after time $t < \frac{1}{c}|\vec{y} - \vec{x}|$ would indeed be possible. The particle could be observed as travelling faster than light. This would certainly seem to be incompatible with the principles of the special theory of relativity where objects are restricted to having sub-luminal speeds (see Fig. 6.1).

Now that we have found a difficulty with causality, we need to find a way to resolve it. We will resolve it by going beyond single-particle quantum mechanics to an extended theory where there is another process which competes with the one that we have described. The total amplitude will then be the sum of the amplitudes for the two processes and we will rely on destructive interference of the amplitudes to solve our problem, that is, to make the probability of detecting the particle identically zero in the entire forbidden region $|\vec{y} - \vec{x}| > ct$.

To include the second process, we will begin by framing the first process, the one we have discussed so far, as the following thought experiment. An observer, whom

Fig. 6.1 The wave packet is initially localized at $\vec{x}$ and as time evolves it spreads in such a way that there is a nonzero amplitude for detecting it in the vicinity of point $\vec{y}$. If it is detected at $\vec{y}$, since $|ct| < |\vec{y} - \vec{x}|$, its classical velocity would be greater than that of light

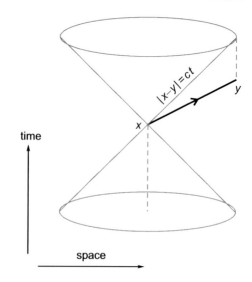

we shall call Alice, is located at position $\vec{x}$ and prepares and releases the particle in the quantum state which is localized at $\vec{x}$.[1]

The particle which Alice releases is then allowed to evolve by its natural time evolution, the one which we have described above, so that after time t, its quantum amplitude for being localized near the position $\vec{y}$ is given by equation (6.9) or its relativistic fix-up (6.10). Then, after the time t has elapsed, another observer, Bob, who is located at point $\vec{y}$ does an experiment to detect the particle. Of course, in a given experiment, Bob might or might not find the particle. But, given that the particle has non-zero amplitude to propagate there, if Alice and Bob repeat this experiment sufficiently many times, Bob will eventually detect the particle at $\vec{y}$. The result of the experiment is to collapse the particle's wave function to one which is localized at $\vec{y}$. The amplitude for the particle to propagate to $\vec{y}$ is given by (6.9) or (6.10). If this were all there is to it, the result of the experiments would occasionally violate causality.

The second process that we will super-impose with the one that we have described above will require other states to be introduced. It then clearly involves an extension of single particle quantum mechanics. In the second process, the attempt by Bob, the observer who is located at $\vec{y}$, to observe a particle's position creates a pair consisting of a particle and an anti-particle. The position measure collapses the wave function of the particle into the position eigenstate localized at $\vec{y}$, the position which was the final state of the particle in the first experiment (see Fig. 6.2).

The anti-particle is interpreted as a particle which moves backward in time, from time t to time 0. After time $-t$ it has an amplitude to arrive at position $\vec{x}$ where it

---

[1] If, as we have stated, it is not possible for Alice to completely localize the particle at $\vec{x}$, it is sufficient to have her release it in a region that is localized near $\vec{x}$ and of size scale much larger than the Compton wave-length but still much smaller than $|\vec{y} - \vec{x}|$.

Fig. 6.2 We should add to
the amplitude for the particle
to travel from $\vec{y}$ to $\vec{x}$ the
amplitude for another
process as in Fig. 6.1, the
amplitude that a
particle-anti-particle pair is
created at $\vec{y}$, the particle
continues forward in time as
it did in the first process, the
anti-particle propagates
backward in time and
annihilates the particle which
was prepared in the state
localized at $\vec{x}$

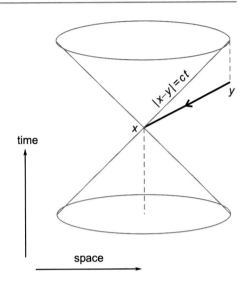

annihilates the particle that Alice, the observer at $\vec{x}$, has prepared in the localized
state. To an impartial observer, the result of this second process is the precisely same
as that of the first process, it appears that a particle begins in a state localized at x
and after a time t it is detected in a state localized at $\vec{y}$. The amplitude for the second
process is similar but not identical to that of the first process, due to the fact that the
antiparticle propagates backward in time. It is

$$< \vec{y}|e^{iHt/\hbar}|\vec{x}>_{\text{antiparticle}}=$$
$$= -\int d^3p\, \frac{c}{(2\pi\hbar)^3 2E(\vec{p})} e^{-i\vec{p}\cdot(\vec{y}-\vec{x})/\hbar + i\sqrt{m^2c^4+\vec{p}^2c^2}\,t/\hbar} \qquad (6.16)$$

where we have adjusted the sign in front of the expression. The total amplitude is
the sum of amplitudes of the two processes,

$$< \vec{y}|e^{-iHt/\hbar}|\vec{x}>_{\text{particle}} + < \vec{y}|e^{iHt/\hbar}|\vec{x}>_{\text{antiparticle}}$$
$$= \int \frac{d^3p\, c}{(2\pi\hbar)^3 2E(\vec{p})} \left\{ e^{\frac{i}{\hbar}\left[\vec{p}\cdot(\vec{y}-\vec{x})-\sqrt{m^2c^2+\vec{p}^2}\,ct\right]} - e^{-\frac{i}{\hbar}\left[\vec{p}\cdot(\vec{y}-\vec{x})-\sqrt{m^2c^2+\vec{p}^2}\,ct\right]} \right\} \qquad (6.17)$$

Now, we can see that the right-hand-side of the above formula is imaginary, the
real part of the two integrals cancels. This is enough to produce the destructive
interference that we want. For example, the result for a massless particle that we
found in equation (6.15) then becomes

$$< \vec{y}|e^{-iHt/\hbar}|\vec{x}>_{\text{particle}} + < \vec{y}|e^{iHt/\hbar}|\vec{x}>_{\text{antiparticle}}$$
$$= \lim_{\epsilon\to 0^+} \frac{-i}{\pi}\delta(|\vec{y}-\vec{x}|^2 - c^2t^2)\epsilon(t) \qquad (6.18)$$

We see in equation (6.18) that the principal value part of the expression in (6.15), which was nonzero outside of the light cone, has canceled. What remains describes the wave function of the initial particle spreading along its light cone, as we might expect for a massless particle, which travels at the speed of light. We have demonstrated that the anti-particle amplitude can be tuned in order to restore causality of our initial result (6.10). Similar can be done for the other option (6.9) with the same result.

The upshot of the above development is that a correct treatment of a quantum mechanical particle which also obeys the laws of special relativity requires more than just single particle quantum mechanics. The resolution of the difficulty that we have suggested needs an anti-particle. Relativistic quantum field theory will supply us with an anti-particle. Another lesson is that the properties of the anti-particle must be finely tuned to be very similar to that of the particle. Otherwise the exact cancellation of the amplitude outside of the light cone would not happen. We will eventually see that this fine-tuning is generally a result of the relativistic wave equations which replace the Schrödingier equation. They have both positive and negative energy solutions which we shall interpret as belonging to the particle and the anti-particle that the wave equation simultaneously describes. We will put off further discussion of this fact until we study relativistic fields and their wave equations.

6.2   Coordinates

The non-relativistic classical and quantum fields $\psi_\sigma(\vec{x}, t)$ which we have dealt with so far are functions of both the time, t, and the space coordinates, labeled by the vector $\vec{x}$. So far, $\vec{x}$ label points in three dimensional Euclidean space and time is parameterized by the variable t. In the following, when we proceed to study relativistic field theories, we will find it convenient to think of space and time coordinates from a unified point of view and include time to form a four-vector $(ct, \vec{x})$. The time is defined with a factor of the speed of light, so that if t is measured in time units, the $x^0 = ct$ is measured in distance units. Points in the four-dimensional space-time are called events.

In the remainder of this chapter, we will introduce some of the notation which we will use to describe the relevant properties of spacetime when we are discussing relativistic field theories. We will also introduce scalar, vector and tensor fields. Then, we will discuss the symmetries of space-time and of our four-dimensional Minkowski space in particular.

For the most part we will be interested in infinite flat four-dimensional Minkowski space. However, at the outset, it is useful to briefly consider more general space-time. In any spacetime geometry, our basic need is a coordinate system which labels the events of the space-time. A coordinate system assigns a unique sequence of real numbers $\left(x^0, x^1, x^2, \ldots, x^{D-1}\right)$ to each event in space-time. The number of entries in the sequence, D, is the dimension of the space-time. We will usually deal with the physical case of four dimensions. The four real numbers $\left(x^0, x^1, x^2, x^3\right)$ contain the four bits of data that are necessary for locating an event. The component $x^0 = ct$

is associated with time, the other three components x^1, x^2, x^3 are said to be the spatial coordinates. For short, we denote the array (x^0, x^1, x^2, x^3) by an indexed object, x^μ, where the index μ runs over the values $\mu = 0, 1, 2, 3$. Each distinct event in space-time should be associated with a distinct set of four numbers. Conversely, each distinct set of four numbers should label a unique event. Note that the position of the index of x^μ is up. In the following, this will be important. An up index will be different from, and must be distinguished from, a down index. Also, we will use the convention that an index which is a Greek letter, such as $\mu, \nu, \lambda, \sigma, \rho, \alpha, \beta, \ldots$, typically runs over the range 0, 1, 2, 3 and is used to denote a four-component object, whereas an index which is a letter $a.b, c, \ldots$ runs over the range 1, 2, 3 and is used to denote the three spatial dimensions. We will sometimes denote the spatial part of x^μ by x^a or $\vec{x}$ and the time component as x^0 or ct. We will sometimes consider dimensions D other than four. In that case, there is always one time dimension and $D - 1$ space dimensions.

A useful idea is that of changing between different coordinate systems. To some extent the labelling of events in space-time is arbitrary. As well as the coordinate system with labels x^μ, we could use an alternative, say one with some different labels, $\tilde{x}^\mu$. To be precise, if the four numbers (x^0, x^1, x^2, x^3) label a specific event in space-time in the old coordinate system, the same event has the label $(\tilde{x}^0, \tilde{x}^1, \tilde{x}^2, \tilde{x}^3)$ in the new coordinate system. We could build up a dictionary for translating between the old and new coordinate systems. This dictionary is encoded in transformation functions $\tilde{x}^\mu(x)$. These are four functions, $(\tilde{x}^0(x), \tilde{x}^1(x), \tilde{x}^2(x), \tilde{x}^3(x))$, each one a function of four variables, $x^\mu = (x^0, x^1, x^2, x^3)$. If we plug the old coordinates of a space-time event, x^μ, into these functions, they give us the new coordinates, $\tilde{x}^\mu$, of the same event. We assume that such a coordinate transformation is invertible. This means that, if we know the functions $\tilde{x}^\mu(x)$, we could at least in principle find the inverse transformation $x^\mu(\tilde{x})$. We also assume that we can take derivatives of the transformation functions so that

$$\frac{\partial \tilde{x}^\mu}{\partial x^\nu} \equiv \partial_\nu \tilde{x}^\mu, \quad \frac{\partial x^\mu}{\partial \tilde{x}^\nu} \equiv \tilde{\partial}_\nu x^\nu \tag{6.19}$$

are both non-singular 4×4 matrices, at least in the ranges of coordinates of interest.

Note that we have defined derivatives, $\partial_\mu \equiv \frac{\partial}{\partial x^\mu}$, with a down index. The difference between down indices and up indices occurs in the way in which the objects carrying the indices transform under a general coordinate transformation, for example, the four-gradient, ∂_μ, which has a down index, transforms as

$$\tilde{\partial}_\mu = \frac{\partial}{\partial \tilde{x}^\mu} = \frac{\partial x^\nu}{\partial \tilde{x}^\mu} \frac{\partial}{\partial x^\nu} = \frac{\partial x^\nu}{\partial \tilde{x}^\mu} \partial_\nu$$

where we have used the chain rule for differentiation. We also remind the reader that we are using the Einstein summation convention by which, unless it is explicitly stated otherwise, repeated up and down indices are assumed to be summed over their range. Thus,

$$\frac{\partial x^\nu}{\partial \tilde{x}^\mu} \frac{\partial}{\partial x^\nu} \equiv \sum_{\nu=0}^{D-1} \frac{\partial x^\nu}{\partial \tilde{x}^\mu} \frac{\partial}{\partial x^\nu}$$

An infinitesimal increment of the coordinates, dx^μ, has an up index and transforms as

$$d\tilde{x}^\mu = \frac{\partial \tilde{x}^\mu}{\partial x^\nu} dx^\nu$$

Take careful note that this transformation is different from that of ∂_μ, in summary,

$$\tilde{\partial}_\mu = \frac{\partial x^\nu}{\partial \tilde{x}^\mu} \partial_\nu, \quad d\tilde{x}^\mu = \frac{\partial \tilde{x}^\mu}{\partial x^\nu} dx^\nu \tag{6.20}$$

This equation establishes the rules for how objects with up indices and objects with down indices transform when we do a general coordinate transformation.

We will often see expressions where an up index on one vector is set equal to a down index on another vector and then the value of the index is summed over all values that the index can take. This usually creates an object with a simpler transformation law. For example, consider

$$d\tilde{x}^\mu \tilde{\partial}_\mu = \frac{\partial \tilde{x}^\mu}{\partial x^\rho} dx^\rho \frac{\partial x^\sigma}{\partial \tilde{x}^\mu} \partial_\sigma = dx^\sigma \partial_\sigma$$

In the above equation, we have used (6.20) and also the chain rule of differential calculus,

$$\frac{\partial \tilde{x}^\mu}{\partial x^\rho} \frac{\partial x^\sigma}{\partial \tilde{x}^\mu} = \frac{\partial x^\sigma}{\partial x^\rho} = \delta^\sigma{}_\rho$$

As a result we see that $dx^\sigma \partial_\sigma$ is left unchanged by the coordinate transformation. Here, $\delta^\sigma{}_\rho$ is the Kronecker delta symbol, which is equal to one when the up and down indices are equal, $\rho = \sigma$, and zero otherwise. If we use the rules in equation (6.20) to transform it, we see that it is an invariant object,

$$\tilde{\delta}^\sigma{}_\rho = \frac{\partial x^\mu}{\partial \tilde{x}^\rho} \frac{\partial \tilde{x}^\sigma}{\partial x^\nu} \delta^\nu{}_\mu = \frac{\partial x^\mu}{\partial \tilde{x}^\rho} \frac{\partial \tilde{x}^\sigma}{\partial x^\mu} = \delta^\sigma{}_\rho$$

It will often be useful to consider infinitesimal coordinate transformations. An infinitesimal transformation is one where the new coordinates differ from the old coordinates by an infinitesimal amount, which can be encoded in four functions $f^\mu(x)$ of infinitesimal magnitude and arbitrary profile, so that

$$\tilde{x}^\mu = x^\mu + f^\mu(x) \tag{6.21}$$

To linear order in infinitesimals, it is easy to find the inverse of this transformation

$$x^\mu = \tilde{x}^\mu - f^\mu(\tilde{x}) \tag{6.22}$$

For these infinitesimal transformations,

$$\frac{\partial \tilde{x}^\mu}{\partial x^\nu} = \delta^\mu{}_\nu + \partial_\nu f^\mu(x), \quad \frac{\partial x^\mu}{\partial \tilde{x}^\nu} = \delta^\mu{}_\nu - \partial_\nu f^\mu(x) \tag{6.23}$$

where we have written the right-hand-sides only to the linear order in the infinitesimal transformation.

6.3   Scalars, Vectors, Tensors

So far, the only structure which we have given space-time is the existence of a coordinate system and the possibility of transforming between different coordinate systems. This is already sufficient structure to define some fields. A relativistic field is a function of the space-time coordinates that transforms in a certain way.

The simplest example of a field is a scalar field. A scalar field is a function of the space-time coordinates whose values at particular space-time events specify the values of some physical quantity. The field should have the same value at the same event of space-time when that event is described in any coordinate system. If x^μ and $\tilde{x}^\mu$ are coordinates which label the same event in two different coordinate systems, and the scalar field has functional form $\phi(x)$ in the x^μ coordinates and $\tilde{\phi}(\tilde{x})$ in the $\tilde{x}^\mu$ coordinates, then the statement that the scalar field has the same value at the same event of space-time in the two coordinate systems gives the scalar field transformation law,

$$\tilde{\phi}(\tilde{x}) = \phi(x) \tag{6.24}$$

In terms of infinitesimal transformations (6.21), equation (6.24) is

$$\tilde{\phi}(\tilde{x}) = \phi(x) + \delta\phi(x) + f^\mu(x)\partial_\mu\phi(x) + \ldots = \phi(x)$$

In the above equation, we see that $\tilde{\phi}(\tilde{x})$ differs from $\phi(x)$ in two ways. First of all, its functional form changes. This is $\delta\phi(x)$. Secondly, the coordinates which it depends on change, this is the term $f^\mu(x)\partial_\mu\phi(x)$. Canceling the untransformed $\phi(x)$ from each side of the above equation, we obtain, to linear order, the transformation law for the scalar field,

$$\delta\phi(x) = -f^\lambda(x)\partial_\lambda\phi(x) \tag{6.25}$$

Like an increment of the coordinates, dx^μ, a vector field $V^\mu(x)$ has a direction at a given space-time point. The components of the vector field, $V^\mu(x)$ transform in a similar way,

$$\tilde{V}^\mu(\tilde{x}) = \frac{\partial \tilde{x}^\mu}{\partial x^\nu} V^\nu(x) \tag{6.26}$$

and, for an infinitesimal transformation,

$$\delta V^\mu(x) = -f^\lambda(x)\partial_\lambda V^\mu(x) + \partial_\lambda f^\mu(x)V^\lambda(x) \tag{6.27}$$

We could also consider a vector field with a lower index, $A_\mu(x)$ which transforms like the gradient operator $\tilde{\partial}_\mu = \frac{\partial x^\nu}{\partial \tilde{x}^\mu} \partial_\nu$,

$$\tilde{A}_\mu(\tilde{x}) = \frac{\partial x^\nu}{\partial \tilde{x}^\mu} A_\nu(x) \tag{6.28}$$

and the infinitesimal transformation

$$\delta A_\mu(x) = -f^\lambda(x)\partial_\lambda A_\mu(x) - \partial_\mu f^\lambda(x) A_\lambda(x) \tag{6.29}$$

By similar reasoning, a tensor field with any number of up and down indices, $T^{\mu_1\cdots\mu_k}{}_{\nu_1\ldots\nu_\ell}(x)$, has the transformation law,

$$\tilde{T}^{\mu_1\cdots\mu_k}{}_{\nu_1\ldots\nu_\ell}(\tilde{x}) = \frac{\partial \tilde{x}^{\mu_1}}{\partial x^{\rho_1}} \cdots \frac{\partial \tilde{x}^{\mu_k}}{\partial x^{\rho_k}} \frac{\partial x^{\sigma_1}}{\partial \tilde{x}^{\nu_1}} \cdots \frac{\partial x^{\sigma_\ell}}{\partial \tilde{x}^{\nu_\ell}} T^{\rho_1\cdots\rho_k}{}_{\sigma_1\ldots\sigma_\ell}(x) \tag{6.30}$$

and

$$\begin{aligned}
\delta T^{\mu_1\cdots\mu_k}{}_{\nu_1\ldots\nu_\ell}(x) &= -f^\lambda(x)\partial_\lambda T^{\mu_1\cdots\mu_k}{}_{\nu_1\ldots\nu_\ell}(\tilde{x}) \\
&+ \partial_\lambda f^{\mu_1}(x) T^{\lambda\cdots\mu_k}{}_{\nu_1\ldots\nu_\ell}(\tilde{x}) + \ldots + \partial_\lambda f^{\mu_k}(x) T^{\mu_1\cdots\lambda}{}_{\nu_1\ldots\nu_\ell}(\tilde{x}) \\
&- \partial_{\nu_1} f^\lambda(x) T^{\mu_1\cdots\mu_k}{}_{\lambda\ldots\nu_\ell}(\tilde{x}) - \ldots - \partial_{\nu_\ell} f^\lambda(x) T^{\mu_1\cdots\mu_k}{}_{\nu_1\ldots\lambda}(\tilde{x})
\end{aligned} \tag{6.31}$$

Any physical scalar, vector or tensor field should transform in the way which we have outlined if they are to have physical meaning.

When we set the indices in a pair equal, where one is an upper index and one is a lower index, and then we sum over all values of the index, the transformation law acting on those indices cancel. For example, the composite of two vector fields

$$V^\mu(x) A_\mu(x)$$

transforms like a scalar field

$$\delta\left[V^\mu(x) A_\mu(x)\right] = -f^\lambda(x)\partial_\lambda\left[V^\mu(x) A_\mu(x)\right]$$

6.4   The Metric

Now that we have introduced coordinates of space-time, we must discuss how some fundamental quantities, like time and distance, for example, are to be computed. All of the information that is needed about the geometry of space-time is encoded in a symmetric two-index tensor field called the metric, $g_{\mu\nu}(x)$. The metric obeys $g_{\mu\nu}(x) = g_{\nu\mu}(x)$ and it transforms like a tensor field with lower indices

$$\tilde{g}_{\mu\nu}(\tilde{x}) = \frac{\partial x^\rho}{\partial \tilde{x}^\mu} \frac{\partial x^\sigma}{\partial \tilde{x}^\nu} g_{\rho\sigma}(x) \tag{6.32}$$

The metric is usually assumed to be non-singular, so that it can be inverted and its inverse is denoted by the same symbol, but with up-indices, $g^{\mu\nu}(x)$, so that

$$g^{\mu\nu}(x)g_{\nu\lambda}(x) = \delta^{\mu}{}_{\lambda}, \quad g_{\mu\nu}(x)g^{\nu\lambda}(x) = \delta^{\lambda}{}_{\mu} \tag{6.33}$$

The inverse of the metric transforms like a tensor with two up-indices,

$$\tilde{g}^{\mu\nu}(\tilde{x}) = \frac{\partial \tilde{x}^{\mu}}{\partial x^{\rho}} \frac{\partial \tilde{x}^{\nu}}{\partial x^{\sigma}} g^{\rho\sigma}(x) \tag{6.34}$$

(This is consistent with the fact that the matrix $\frac{\partial \tilde{x}^{\mu}}{\partial x^{\rho}}$ is the inverse of the matrix $\frac{\partial x^{\rho}}{\partial \tilde{x}^{\mu}}$. It is also consistent with that fact that, as we shall learn shortly, $g^{\mu\nu}(x)$ and $g_{\mu\nu}(x)$ can be used to raise and lower indices on a tensor.)

The infinitesimal transformations of the metric and its inverse are

$$\delta g_{\mu\nu}(x) = -f^{\lambda}(x)\partial_{\lambda}g_{\mu\nu}(x) - \partial_{\mu}f^{\lambda}(x)g_{\lambda\nu}(x) - \partial_{\nu}f^{\lambda}(x)g_{\mu\lambda}(x) \tag{6.35}$$

$$\delta g^{\mu\nu}(x) = -f^{\lambda}(x)\partial_{\lambda}g^{\mu\nu}(x) + \partial_{\lambda}f^{\mu}(x)g^{\lambda\nu}(x) + \partial_{\lambda}f^{\nu}(x)g^{\mu\lambda}(x) \tag{6.36}$$

The metric is what gives a space-time its structure. It can be used to compute the proper time that elapses as a trajectory is traversed. Given an infinitesimal increment of the coordinates, dx^{μ}, the increment of proper time is defined by

$$-c^2 d\tau^2 = g_{\mu\nu}(x)dx^{\mu}dx^{\nu}$$

(The minus sign on the left-hand-side of this equation is a matter of convention.). Generally, dx^{μ} are measured in units of distance and $g_{\mu\nu}(x)$ is dimensionless. so we need to have a constant with the units of velocity if $d\tau$ is to have units of time. The constant that we have written there is c, the speed of light. This proper time is the time which elapses on a clock which moves with an object along a trajectory. Here, the trajectory goes from x^{μ} to $x^{\mu} + dx^{\mu}$ and it is given by the parametric equation

$$x^{\mu}(s) = x^{\mu} + s dx^{\mu}, \ 0 \le s \le 1$$

The proper time is a well-defined physical quantity and it should not depend on the coordinate system which is used. This is guaranteed by the coordinate transformation of the increment dx^{μ} and the metric (6.32) which combine to

$$-c^2 d\tilde{\tau}^2 = \tilde{g}_{\mu\nu}(\tilde{x})d\tilde{x}^{\mu}d\tilde{x}^{\nu} = g_{\mu\nu}(x)dx^{\mu}dx^{\nu} = -c^2 d\tau^2$$

We observe that the metric can be used to raise and lower indices. If we take a vector field, $V^{\mu}(x)$, we can create a vector with a lower index by contracting it with the metric tensor

$$V_{\mu}(x) = g_{\mu\nu}(x)V^{\nu}(x)$$

and raise the index with the inverse of the metric,

$$V^{\mu}(x) = g^{\mu\nu}(x)V_{\nu}(x)$$

We can use the coordinate transformation laws for the metric and for vector fields to see that, indeed, $V^{\mu}(x)$ and $V_{\mu}(x)$ transform as vector fields with an upper and a lower index, respectively, when the metric and its inverse transform like tensors of the appropriate type.

For the most part, we will not be interested in general space-times, but will focus on Minkowski space. Minkowski space is defined as that space-time where one can find a coordinate system in which the metric has the special form

$$\eta_{\mu\nu} = \begin{bmatrix} -1 & 0 & 0 & 0 \\ 0 & 1 & 0 & 0 \\ 0 & 0 & 1 & 0 \\ 0 & 0 & 0 & 1 \end{bmatrix}, \quad \eta^{\mu\nu} = \begin{bmatrix} -1 & 0 & 0 & 0 \\ 0 & 1 & 0 & 0 \\ 0 & 0 & 1 & 0 \\ 0 & 0 & 0 & 1 \end{bmatrix}, \tag{6.37}$$

$$\eta^{\mu\nu}\eta_{\nu\lambda} = \delta^{\mu}{}_{\lambda}, \quad \eta_{\mu\nu}\eta^{\nu\lambda} = \delta_{\mu}{}^{\lambda} \tag{6.38}$$

where we denote this special metric by the symbol $\eta_{\mu\nu}$ and its inverse by $\eta^{\mu\nu}$. Minkowski space is also called "flat space" where flatness refers to its geometry. Also, we generally take the range of the coordinates x^{μ} of Minkowski space to be the entire real numbers, $(-\infty, \infty)$.

6.5   Symmetry of Space–Time

Now that we have introduced the concept of metric, we can discuss the idea of a symmetry of a space-time. We define a symmetry transformation of space-time as a general coordinate transformation under which the metric remains unchanged, $\delta g_{\mu\nu}(x) = 0$ where $\delta g_{\mu\nu}(x)$ is given in equation (6.35). This can be so only for certain special infinitesimal transformations, $\tilde{x}^{\mu} = x^{\mu} + \hat{f}^{\mu}(x)$. We will use a hat on a vector field which corresponds to a symmetry, $\hat{f}^{\mu}(x)$, in order to distinguish it from a general coordinate transformation which we shall still denote by $f^{\mu}(x)$. The condition that the coordinate transformation does not change the metric gives us a partial differential equation which the four functions $\hat{f}^{\mu}(x)$ must obey,

$$\partial_{\mu}\hat{f}^{\lambda}(x)g_{\lambda\nu}(x) + \partial_{\nu}\hat{f}^{\lambda}(x)g_{\mu\lambda}(x) + \hat{f}^{\lambda}(x)\partial_{\lambda}g_{\mu\nu}(x) = 0 \tag{6.39}$$

This equation is called the Killing equation. The solutions, $\hat{f}^{\mu}(x)$, of the Killing equation are called Killing vectors. Each linearly independent Killing vector generates a symmetry of space-time. Different space-times can have different symmetries, varying both in the number and the nature of the symmetry transformations and the Killing vectors associated with them. As one can imagine, a generic space-time might have no symmetry at all. There turns out to be a maximum number of symmetries that a space-time can have, in D-dimensions it is $D(D + 1)/2$. The space-time that we will be the most interested in, four-dimensional Minkowski space, whose metric we wrote down in equation (6.37) and on which we elaborate in the next section, is one of the maximally symmetric four dimensional spaces. We therefore expect, and will confirm that it has ten Killing vectors.

6.6   The Symmetries of Minkowski Space

The Killing equation for Minkowski space is obtained by plugging the Minkowski metric from equation (6.37) into the generic Killing equation (6.39) to get

$$\partial_\mu \hat{f}_\nu(x) + \partial_\nu \hat{f}_\mu(x) = 0 \qquad (6.40)$$

where we have defined

$$\hat{f}_\mu(x) \equiv \eta_{\mu\nu} \hat{f}^\nu(x) \qquad (6.41)$$

The ten solutions of this equation are:

1. four constants $\hat{f}^\mu = c^\mu$ corresponding to constant translations of the space-time coordinates
2. $\hat{f}^\mu = \omega^\mu_{\ \nu} x^\nu$ with constants $\omega_{\mu\nu} = -\omega_{\nu\mu}$, where $\omega_{\mu\nu} = \eta_{\mu\rho}\omega^\rho_{\ \nu}$. There are six independent components of this 4×4 anti-symmetric matrix which correspond to three infinitesimal spatial rotations and three infinitesimal Lorentz boosts.

It is easy to verify that these are solutions by plugging them into the equation (6.40) and seeing that it is indeed solved. The rotations and boosts which are generated by the six linearly independent Killing vectors of the kind $\hat{f}^\mu(x) = \omega^\mu_{\ \nu} x^\nu$ are generally called Lorentz transformations. When we study a field theory which also has these symmetries we will say that it is Lorentz invariant. If we add on the translations as well, the transformations generated by all ten of the linearly independent Killing vectors in the set $f^\mu(x) = \left(c^\mu, \omega^\mu_{\ \nu} x^\nu\right)$ are called the Poincare transformations. A field theory that has these symmetries will be said to be Poincare symmetric.

Given that we have found the infinitesimal transformations, it is easy to find the general form of finite Poincare transformations. As an infinitesimal transformation, $x^\mu \rightarrow x^\mu + \hat{f}^\mu(x)$ with $f^\mu(x) = \left(c^\mu, \omega^\mu_{\ \nu} x^\nu\right)$ is a combination of a translation of x^μ and a linear transformation on the components of x^μ. If we think of building up a finite transformation from doing a large number if infinitesimal transformations, we can see that the the finite transformation will still be a translation of the components of x^μ and a linear transformation on those components. We can therefore make an ansatz for a finite transformation which is a combination of the two,

$$x^\mu \rightarrow \tilde{x}^\mu = \Lambda^\mu_{\ \nu} x^\nu + c^\mu \qquad (6.42)$$

where c^μ are constants (which are no longer infinitesimal) and $\Lambda^\mu_{\ \nu}$ is a constant 4×4 matrix. We can then require that the finite coordinate transformation is a symmetry of the Minkowski metric. If we do this, we see that the above linear transformation is indeed a symmetry of the Minkowski metric when the matrix Λ^μ_ν obeys the equation

$$\Lambda^\rho_{\ \mu} \Lambda^\sigma_{\ \nu} \eta_{\rho\sigma} = \eta_{\mu\nu} \tag{6.43}$$

We can check that this is consistent with what we know about the infinitesimal transformations by examining the limit where Λ corresponds to an infinitesimal transformation. In this limit, it should be very close to the identity. If we look for a Λ^μ_ν which is close to the identity,

$$\Lambda^\mu_{\ \nu} \approx \delta^\mu_{\ \nu} + \omega^\mu_{\ \nu} \tag{6.44}$$

where $\omega_{\mu\nu}$ is infinitesimal and we require that Λ^μ_ν satisfy equation (6.43) to the first order in $\omega_{\mu\nu}$ we see that $\omega_{\mu\nu}$ must be a real anti-symmetric 4×4 matrix. Then if we use our infinitesimal transformation (6.44) in equation (6.42) we see that we simply recover an the infinitesimal Lorentz transformation of the coordinate.

We should note that there are solutions of the equation (6.43) for Λ^μ_ν which can never be close to the identity. To see this, we take a determinant of both sides of equation (6.43) and we use the property of determinants that the determinant of a product of two matrices is a product of the determinants, we see that

$$\det \left[\Lambda^\rho_{\ \mu} \eta_{\rho\sigma} \Lambda^\sigma_{\ \nu} \right] = \det \left[\eta_{\mu\nu} \right]$$
$$\det \left[\Lambda^t \eta \Lambda \right] = \det \left[\eta \right] \tag{6.45}$$

where Λ^t is the transpose of the matrix Λ (and we recall that $\det \Lambda^t = \det \Lambda$) to conclude that

$$[\det \Lambda]^2 = 1 \tag{6.46}$$

and that there are thus two kinds of Λ^μ_ν, those with

$$\det \Lambda = +1 \tag{6.47}$$

and those with

$$\det \Lambda = -1 \tag{6.48}$$

One kind has determinant equal to 1 and the other kind has determinant equal to -1.

The Lorentz transformations with determinant -1 reverse the orientation of the coordinate system – they contain a reflection as well as a Lorentz transformation and they therefore do not have an infinitesimal version. The set of all Lorentz transformations with determinant -1 and the set with determinant $+1$ have the same size, in fact it is easy to see there is a one-to-one mapping between them. If Λ has determinant one, $\Pi\Lambda$ where Π is the matrix

$$\Pi = \begin{bmatrix} 1 & 0 & 0 & 0 \\ 0 & -1 & 0 & 0 \\ 0 & 0 & -1 & 0 \\ 0 & 0 & 0 & -1 \end{bmatrix}$$

has determinant 1. If the matrix Λ^{μ}_{ν} is a Lorentz transform, then it is easy to confirm that the product $\Pi^{\mu}_{\rho}\Lambda^{\rho}_{\nu}$ is also a Lorentz transformation in that it obeys equation (6.43). Also, taking its determinant, we see that

$$\det[\Pi\Lambda] = \det[\Pi]\det[\Lambda] = -\det[\Lambda]$$

and multiplying by Π changes the sign of the determinant. This gives a way of mapping back and forth between Lorentz transformations where the determinant is $+1$ and -1 and establishes a one-to-one mapping between them.

We can also see by taking the 00 component of equation (6.43), that

$$(\Lambda^0_0)^2 = 1 + \sum_a (\Lambda^a_0)^2 \geq 1 \tag{6.49}$$

and the set of all Λ's decomposes into two distinct subsets in another way, the subset with $\Lambda^0_0 > 0$ and the subset $\Lambda^0_0 < 0$. In the first case, under the Lorentz transformation

$$\tilde{x}^0 = \Lambda^0_0 x^0 + \Lambda^0_a x^a \tag{6.50}$$

and since, by equation (6.49), $|\Lambda^0_0| > \sqrt{\sum_a(\Lambda^a_0)^2}$, and if $|x^0| > |\vec{x}|$, the transformations with $\Lambda^0_0 > 0$ do not change the sign of the time coordinate, were by a similar argument, those with $\Lambda^0_0 < 0$ do change the sign of the time coordinate. Again, it is easy to see that the subset of Lorentz transformations with $\Lambda^0_0 < 0$ can never be close to the identity.

For some applications, we will want to restrict our attention to those Lorentz transformations which can be close to the identity. Those have $\det \Lambda = 1$ and $\Lambda^0_0 > 0$. This subset are sometimes called the proper, orthochronous Lorentz transformations.

If A^{μ} and B^{ν} are two four-vectors, that is, two entities which transform like

$$A^{\mu} \rightarrow \tilde{A}^{\mu} = \Lambda^{\mu}_{\rho}A^{\rho}, \ B^{\nu} \rightarrow \tilde{B}^{\nu} = \Lambda^{\nu}_{\sigma}A^{\sigma} \tag{6.51}$$

then the combination which we will denote

$$A \cdot B \equiv AB \equiv A_{\mu}B^{\mu} \equiv A^{\mu}\eta_{\mu\nu}B^{\nu} \tag{6.52}$$

transforms like a scalar, that is, it is invariant under a Lorentz transformation

$$\tilde{A} \cdot \tilde{B} = A \cdot B \tag{6.53}$$

We distinguish this four-vector "dot product" from the three dimensional, spatial one that we have been using so far by the absence of arrows over A and B. A spatial dot product would read $\vec{A} \cdot \vec{B}$. Since coordinates transform like four-vectors,

$$x^2 \equiv x_\mu x^\mu \equiv \eta_{\mu\nu} x^\mu x^\nu \tag{6.54}$$

is invariant under Lorentz transformations. This means that displacements in Minkowski space (possible values of x^μ) come in three distinct kinds

- timelike: $x^2 < 0$
- spacelike: $x^2 > 0$
- lightlike or null: $x^2 = 0$

These properties are preserved by any Lorentz transformation.

The transformations for vector fields can be a little more general than the one that we have used so far. We call a vector A^μ a pseudo-vector if it transforms as

$$A^\mu \; \rightarrow \; \tilde{A}^\mu = [\det \Lambda] \Lambda^\mu{}_\nu A^\nu \tag{6.55}$$

Then, if A^μ is a pseudo-vector and if B^ν is a vector, $A \cdot A$ is Lorentz invariant, we might call it a scalar and $A \cdot B$ is a pseudoscalar in that

$$\tilde{A} \cdot \tilde{B} = [\det \Lambda] A \cdot B \tag{6.56}$$

6.7   Natural Units

We have defined the time coordinate of spacetime to be $x^0 = ct$. This definition effectively means that x^0 is measured in distance units. If x^0 were one meter, it corresponds to the interval of time that it takes light to move one meter in vacuum where the dielectric constant is equal to one. This convention is convenient as it renders our discussions less cluttered with factors of c. We could take this one step further and simply declare that x^0 is the time, measured in distance units, to be converted to conventional time units by introducing a factor of c.

To simplify our notation further, we can also take a system of units where Planck's constant $\hbar$ is set equal to one. In such a system of units, energy and frequency as well as momentum and wave-number are measured in the same units. Then, an inverse centimetre, for example, is both the energy and the frequency of a light wave which has wave-length one centimetre.

When $\hbar = 1 = c$, our equations are written more elegantly. Moreover, these constants can always be restored if we need them to quote physical quantities in a specific system of measure by simple dimensional analysis.

6.8   Relativistic Fields

We have taken the point of view that Poincare symmetries are, first of all, symmetries of Minkowski space, entirely independent of other considerations. Once we have a symmetric spacetime, we will attempt to embed a dynamical system, a quantum field theory, into that spacetime in such a way that the quantum field theory also respects the symmetries of Minkowski space. This will mean that the quantum field theory will also be symmetric under Poincare transformations. Such a field theory is what we mean by the term relativistic quantum field theory. In order to formulate a relativistic field theory, we will have to study how fields transform. We have already discussed how scalar, vector and tensor fields transform under coordinate transformations. It is natural that these same fields will then inherit their transformations law under Poincare transformations, which we will now restrict our attention to, from the way in which they transform under general coordinate transformations.

Under a general Poincare transformation, a scalar field transforms as it did under a general coordinate transformation, but with the coordinate transformation specialized to be a combination of a Lorentz transformation and a translation. The coordinate transformation is given by

$$x^\mu \;\to\; \tilde{x}^\mu \;=\; \Lambda^\mu_{\ \nu} x^\nu + c^\mu \tag{6.57}$$

where $\Lambda^\mu_{\ \nu}$ satisfies equation (6.43). The coordinate transformation law for a scalar field, applied to the Poincare transformation, tells us that

$$\tilde{\phi}(\Lambda^\mu_{\ \nu} x^\nu + c^\mu) = \phi(x) \tag{6.58}$$

This equation is often written in an alternative form by inverse transforming x^μ. The transformation law is then

$$\phi(x) \;\to\; \tilde{\phi}(x^\mu) \;=\; \phi([\Lambda^{-1}]^\mu_{\ \nu}(x^\nu - c^\nu)) \tag{6.59}$$

This defines how a scalar field transforms under the Poincare transformation. We could also define a pseudoscalar field as one which has the transformation law

$$\chi(x) \;\to\; \tilde{\chi}(x^\mu) \;=\; [\det \Lambda]\, \chi([\Lambda^{-1}]^\mu_{\ \nu}(x^\nu - c^\nu)) \tag{6.60}$$

If the Lorentz transformation is an infinitesimal one, that is, if

$$\Lambda^\mu_{\ \nu} = \delta^\mu_{\ \nu} + \omega^\mu_{\ \nu} \tag{6.61}$$

with infinitesimal $\omega^\mu_{\ \nu}$, and if the translation, c^μ, is infinitesimal, and if we define $\delta\phi(x)$ by

$$\tilde{\phi}(x) = \phi(x) + \delta\phi(x) \tag{6.62}$$

the transformation of the scalar field in equation (6.60) implies

$$\delta\phi(x) = -[c^\mu + \omega^\mu{}_\nu x^\nu]\partial_\mu\phi(x) \tag{6.63}$$

The same formula applies to the infinitesimal transformation of a pseudo-scalar field.

Similarly, for the case of a vector field, the rule for how it transforms under a general coordinate transformation implies that, under a Poincare transformation,

$$V^\lambda(x) \;\rightarrow\; \tilde{V}^\lambda(x^\mu) = \Lambda^\lambda{}_\rho V^\rho([\Lambda^{-1}]^\mu{}_\nu(x^\nu - c^\nu)) \tag{6.64}$$

We could also define a pseudo-vector field as a field which transforms as

$$A^\lambda(x) \;\rightarrow\; \tilde{A}^\lambda(x^\mu) \;=\; [\det \Lambda]\, \Lambda^\lambda{}_\rho A^\rho([\Lambda^{-1}]^\mu{}_\nu(x^\nu - c^\nu)) \tag{6.65}$$

Under an infinitesimal Poincare transformation, a vector field or a pseudo-vector field transform as

$$\delta V^\lambda(x^\mu) = -[c^\mu + \omega^\mu{}_\nu x^\nu]\partial_\mu V^\lambda(x) + \omega^\lambda{}_\rho V^\rho(x) \tag{6.66}$$

A vector field with a lower index transforms as

$$\delta V_\lambda(x^\mu) = -[c^\mu + \omega^\mu{}_\nu x^\nu]\partial_\mu V_\lambda(x) - \omega^\rho{}_\lambda V_\rho(x) \tag{6.67}$$

Note that this latter transformation law can be gotten from equation (6.66) by lowering the index on the vector field using the Minkowski metric, so that $V_\lambda(x) \equiv \eta_{\lambda\sigma} V^\sigma(x)$.

The Real Scalar Quantum Field Theory

7

In this chapter we will examine an example of a relativistic quantum field theory. It contains a single, real scalar field and it will have spacetime symmetry under Poincare transformations. We will do this, for the moment, without physical context. In the following chapter we will discuss a scenario where a theory of real scalar fields is emergent.

7.1 Constructing a Relativistic Lagrangian Density

A relativistic field theory is one where the field equations and the equal-time commutation relations have relativistic symmetry, that is, where they are invariant under Poincare transformations. This symmetry is guaranteed if the Lagrangian density from which the field equations and the commutation relations are derived has Poincare symmetry. It is easy to see that this will be the case if the Lagrangian density itself transforms like a scalar field.

Consider an infinitesimal Poincare transformation

$$x^\mu \to \tilde{x}^\mu = x^\mu + \hat{f}^\mu(x)$$

where $\hat{f}^\mu(x)$ is a Killing vector of Minkowski space. If, upon properly transforming the fields that are contained in the Lagrangian density using their transformation laws, the variation of the Lagrangian density is the same as that for a scalar field,

$$\delta \mathcal{L}(x) = -\hat{f}^\mu(x)\partial_\mu \mathcal{L}(x) \tag{7.1}$$

then, due to the fact that, for a Killing vector of Minkowski space

$$\partial_\mu \hat{f}^\mu(x) = 0$$

the transformation of the Lagrangian density can be written as

$$\delta\mathcal{L}(x) = \partial_\mu\left[-\hat{f}^\mu(x)\mathcal{L}(x)\right] \tag{7.2}$$

The spacetime transformation satisfies our criterion for a symmetry.

The guideline for constructing a Lagrangian density for a relativistic quantum field theory is then straightforward. We need to ensure that the Lagrangian density will transform like a scalar field. This will be so if each term in the Lagrangian density transforms like a scalar field. This is particularly simple in the Lagrangian density for a single scalar field, $\phi(x)$, where the field transforms as

$$\delta\phi(x) = -\hat{f}^\mu(x)\partial_\mu\phi(x) \tag{7.3}$$

There, it is easy to see that $\phi^n(x)$ transforms like a scalar field when $\phi(x)$ does. Terms containing derivatives of the scalar field will transform like a scalar field when all of the four-vector indices of the derivative operators are paired and summed over in Lorentz invariant so that there are no free indices left. The Minkowski space metric can also be used in this pairing. An example of a term with derivatives which transforms like a scalar is

$$\partial_\mu\phi(x)\eta^{\mu\nu}\partial_\nu\phi(x) = \partial_\mu\phi(x)\partial^\mu\phi(x)$$

Important constraints in constructing an acceptable Lagrangian density for a quantum field theory will come from the physical and mathematical consistency of the quantum field theory itself. Locality and causality of the field theory and its interactions are important and, amongst other things, they require that the individual terms in the Lagrangian density are constructed from products of the fields and their derivatives all of which are evaluated at the same spacetime point. In addition, the number of derivatives should be finite. This criterion ensures that we do not build acausal behaviour or action at a distance into our field theory from the get-go. We have been following this criterion in our discussion so far and all of our relativistic field theories will conform to it.

Beyond locality, the most important criteria are renormalizability of the quantum field theory and unitarity of its time evolution. We have not developed enough formalism to discuss either of these with any degree of rigour. Suffice it to say that they are both important for the mathematical definition of the quantum field theory and for the quantum states of the field theory to have a sensible interpretation as probability amplitudes and also a sensible time evolution. Moreover, they are both highly constraining. For example, if we demand renormalizability and Poincare symmetry and if we also impose a discrete symmetry under the transformation $\phi(x) \to -\phi(x)$, the most general acceptable Lagrangian density for a scalar field in four spacetime dimensions is given by

$$\mathcal{L}(x) = -\frac{1}{2}\partial_\mu\phi(x)\partial^\mu\phi(x) - \frac{m^2}{2}\phi^2(x) - \frac{\lambda}{4!}\phi^4(x) \tag{7.4}$$

Renormalizability excludes the addition of higher powers $\phi^n(x)$ with $n > 4$ and any terms like $\partial_\mu \phi(x)\partial^\mu \phi(x)\ \phi^n(x)$ with $n > 0$ or any terms with more than two derivatives. This has the interesting implication that any relativistic quantum field theory of a single scalar field in four dimensional spacetime with a $\phi \to -\phi$ symmetry can have only three independent parameters. Its most general Lagrangian density is the one in equation (7.4) which contains the two parameters m^2 and λ. We could obtain a third parameter by a rescaling $\phi(x) \to Z^{\frac{1}{2}}\phi(x)$ of the scalar field itself. This gives the quantum field theory remarkable predictive power. Any modelling of a physical system with this real scalar field theory would be a three parameter fit.

If we did not impose the $\phi \to -\phi$ symmetry, the only possible terms we could add are $\phi(x)$ and $\phi^3(x)$ which would introduce two additional parameters. Also, the reader should be aware that the criterion of renormalizability differs in different dimensions and our discussion here has been specialized to four dimensional Minkowski space. We will learn more about the constraint of renormalizability later.

In the discussion above, we have assumed that the degree of freedom of interest is a single real scalar field $\phi(x)$ and we have constructed a Lagrangian density (7.4) which is constrained by the criterion of renormalizability and an additional assumed $\phi(x) \to -\phi(x)$ symmetry. By construction, it has Poincare symmetry in that, under the transformation (7.3) it transforms as the total derivative in equation (7.2). We can now proceed to study this field theory in more detail.

7.2   Field Equation and Commutation Relations

The Euler–Lagrange equations applied to the scalar field theory and using relativistic notation have the rather elegant form

$$\frac{\partial \mathcal{L}(x)}{\partial \phi(x)} - \partial_\mu \frac{\partial \mathcal{L}(x)}{\partial (\partial_\mu \phi(x))} = 0 \tag{7.5}$$

Applying them to the Lagrangian density in equation (7.4) results in the field equation

$$\left(\partial^2 - m^2\right)\phi(x) = \frac{\lambda}{3!}\phi^3(x) \tag{7.6}$$

The left-hand-side of this equation is linear in the field and it has a wave operator $\left(\partial^2 - m^2\right)$, sometimes called the Klein-Gordon operator, acting on the field. The right-hand-side is nonlinear. The nonlinear term is what allows the fields, or the particles which are their quanta, to interact.

The operator nature of $\phi(x)$ is determined by its equal-time commutation relations. To find out what the equal time commutation relations should be, we need to examine the terms in the Lagrangian density which contain time derivatives of the fields. In equation (7.4) these terms are

$$\mathcal{L}(x) = \frac{1}{2}\frac{\partial \phi(x)}{\partial x^0}\frac{\partial \phi(x)}{\partial x^0} + \dots \tag{7.7}$$

Unlike our non-relativistic field theories, where the Lagrangian density was linear in the time derivatives, this Lagrangian density is quadratic in $\frac{\partial\phi(x)}{\partial x^0}$.

Our field theory Lagrangian density is the field theory analog of a particle mechanics Lagrangian of the form

$$L(\dot{q}(t), q(t)) = \frac{1}{2}\dot{q}^2(t) - V(q(t))$$

where $q(t)$ is a generalized coordinate and $\dot{q}(t) \equiv dq(t)/dt$ is the velocity corresponding to that coordinate. In the particle mechanics, the canonical momentum is given by

$$p = \frac{\partial L}{\partial \dot{q}} = \dot{q}$$

To quantize the mechanical theory, we would promote the coordinate and the momentum to operators which have the commutation relation

$$[q, p] = i\hbar$$

In the field theory case, the generalize coordinate is the field $\phi(x)$ whose time derivatives appear in the Lagrangian density as displayed in equation (7.7). By direct analogy with particle mechanics, the canonical momentum corresponding to the field is then identified as the partial derivative of the field by the time coordinate, $\frac{\partial\phi(x)}{\partial x^0}$. Then, also analogously to the particle mechanics case, we see that the non-trivial equal-time commutation relations between the field and its time derivative are,

$$\left[\phi(x), \frac{\partial}{\partial y^0}\phi(y)\right]\delta(x^0 - y^0) = i\hbar\delta^{(4)}(x - y) \tag{7.8}$$

$$[\phi(x), \phi(y)]\delta(x^0 - y^0) = 0, \quad \left[\frac{\partial}{\partial x^0}\phi(x), \frac{\partial}{\partial y^0}\phi(y)\right]\delta(x^0 - y^0) = 0 \tag{7.9}$$

The field equation (7.6) and the commutation relations (7.8) and (7.9) define a quantum field theory. The right-hand side of equation (7.8) has the factor $\hbar$. In natural units, which we will use most of the time, $\hbar = 1$. Notice that, in this particular case, there are no operator ordering ambiguities involved in promoting the classical fields to the field operators of the quantum field theory.

7.3   Noether's Theorem and Poincare Symmetry

It is interesting to study the implication of Noether's theorem for this scalar field theory. The fact that, according to equation (7.2), the infinitesimal Poincare transformation is a symmetry of the theory allows us to deduce the conserved Noether current density corresponding to it. To do this, we implement Noether's theorem. A transformation of the scalar field

$$\phi(x) \rightarrow \tilde{\phi}(x) = \phi(x) + \delta\phi(x)$$

is a symmetry of the field theory if, without use of the equations of motion, we can show that

$$\delta\mathcal{L}(x) = \partial_\mu R^\mu(x)$$

Then Noether's theorem tells us that

$$\mathcal{J}^\mu(x) = \delta\phi(x)\frac{\partial\mathcal{L}(x)}{\partial(\partial_\mu\phi(x))} - R^\mu(x)$$

is a conserved current in that, once the equations of motion are taken into account,

$$\partial_\mu\mathcal{J}^\mu(x) = 0$$

Since we have constructed the relativistic field theory so that, under the spacetime transformation of the scalar field in equation (7.3), the Lagrangian density varies as in equation (7.1), the transformation is a symmetry of the theory and the conserved Noether current corresponding to it is

$$\mathcal{J}^\mu_{\hat{f}}(x) = \delta\phi(x)\frac{\partial\mathcal{L}(x)}{\partial(\partial_\mu\phi(x))} + \hat{f}^\mu(x)\mathcal{L}(x) \tag{7.10}$$

Evaluating the right-hand-side of this equation results in

$$\mathcal{J}^\mu_{\hat{f}}(x) = -\mathbf{T}^{\mu\nu}_0(x)\hat{f}_\nu(x) \tag{7.11}$$

where the stress tensor is

$$\mathbf{T}^{\mu\nu}_0(x) = \partial^\mu\phi(x)\partial^\nu\phi(x) - \eta^{\mu\nu}\left[\frac{1}{2}\partial_\mu\phi(x)\partial^\mu\phi(x) + \frac{m^2}{2}\phi^2(x) + \frac{\lambda}{4!}\phi^4(x)\right] \tag{7.12}$$

Noether's theorem tells us that

$$\partial_\mu\mathcal{J}^\mu_{\hat{f}}(x) = \partial_\mu\left(-\mathbf{T}^{\mu\nu}_0(x)\hat{f}_\nu(x)\right) = 0 \tag{7.13}$$

for any of the ten Killing vectors $\hat{f}^\mu(x)$. Equivalently, we can see that

$$\begin{aligned}
0 &= \partial_\mu\left(-\mathbf{T}^{\mu\nu}_0(x)\hat{f}_\nu(x)\right) \\
&= -\left[\partial_\mu\mathbf{T}^{\mu\nu}_0(x)\right]\hat{f}_\nu(x) - \frac{1}{4}\left[\mathbf{T}^{\mu\nu}_0(x) + \mathbf{T}^{\nu\mu}_0(x)\right]\left[\partial_\mu\hat{f}_\nu(x) + \partial_\nu\hat{f}_\mu(x)\right] \\
&\quad - \frac{1}{4}\left[\mathbf{T}^{\mu\nu}_0(x) - \mathbf{T}^{\nu\mu}_0(x)\right]\left[\partial_\mu\hat{f}_\nu(x) - \partial_\nu\hat{f}_\mu(x)\right]
\end{aligned} \tag{7.14}$$

is satisfied if $\hat{f}^\mu(x)$ obeys the Killing equation for Minkowski space–time and if the stress tensor is conserved and symmetric,

$$\partial_\mu \mathbf{T}_0^{\mu\nu}(x) = 0, \quad \mathbf{T}_0^{\nu\mu}(x) = \mathbf{T}_0^{\mu\nu}(x) \tag{7.15}$$

Of course, these properties can also be confirmed by taking the divergence of the explicit form of $\mathbf{T}_0^{\mu\nu}(x)$ in equation (7.12) and then using the equation of motion (7.6). Note that we have denoted the stress tensor $\mathbf{T}_0^{\mu\nu}(x)$ with a subscript 0 since later on, when we discuss scale and conformal symmetry, we will find a need to improve it. Until then, this stress tensor will suffice for our discussion of Poincare symmetry.

The conserved Noether charges for space and time translations can be assembled into the energy-momentum four-vector,

$$P^\mu = \int d^3x \, \mathbf{T}_0^{0\mu}(x) \tag{7.16}$$

Here, P^0 is the Hamiltonian, given by the integral

$$P^0 = \int d^3x \, \mathbf{T}_0^{00}(x) \tag{7.17}$$

$$= \int d^3x \left\{ \frac{1}{2}\left(\frac{\partial\phi(x)}{\partial x^0}\right)^2 + \frac{1}{2}\vec{\nabla}\phi(x)\cdot\vec{\nabla}\phi(x) + \frac{m^2}{2}\phi^2(x) + \frac{\lambda}{4!}\phi^4(x) \right\} \tag{7.18}$$

and the momentum operator is

$$P^a = \int d^3x \, \mathbf{T}_0^{0a}(x) \tag{7.19}$$

$$= -\int d^3x \frac{\partial\phi(x)}{\partial x^0} \nabla^a \phi(x) \tag{7.20}$$

The Noether charge corresponding to Lorentz transformations is

$$M^{\mu\nu} = \int d^3x \left[\mathbf{T}_0^{0\mu}(x)x^\nu - \mathbf{T}_0^{0\nu}(x)x^\mu \right] \tag{7.21}$$

We can easily use the equal-time commutation relations, and the fact that these Noether charges are time-independent, so that their time arguments can be adjusted to any value, to show that the commutators of the fields with the Noether charges actually generate the infinitesimal transformations of the fields,

$$i\left[P_\mu, \phi(x)\right] = \partial_\mu \phi(x) \tag{7.22}$$

$$i\left[M^{\nu\mu}, \phi(x)\right] = \left(x^\mu \partial^\nu - x^\nu \partial^\mu\right)\phi(x) \tag{7.23}$$

What is more, the commutators of the charges can also be found

$$i\left[P^\mu, P^\nu\right] = 0 \tag{7.24}$$

$$i\left[M^{\mu\nu}, P^\lambda\right] = \eta^{\mu\lambda} P^\nu - \eta^{\nu\lambda} P^\mu \tag{7.25}$$

$$i\left[M^{\mu\nu}, M^{\rho\sigma}\right] = \eta^{\mu\rho} M^{\nu\sigma} - \eta^{\mu\sigma} M^{\nu\rho} - \eta^{\nu\rho} M^{\mu\sigma} + \eta^{\nu\sigma} M^{\mu\rho} \tag{7.26}$$

This set of commutation relations is called the Poincare algebra.

7.4   Correlation Functions of the Real Scalar Field

In this section we will discuss correlation functions of the scalar field theory. These functions contain valuable information about the structure of the theory and they will eventually be our primary tool for studying that structure. Here, we will present some definitions and some simple results.

A correlation function of quantum fields is an expectation value of a product of field operators where, generally, each of the operators has a different space–time argument. We will consider such correlations in the ground state of the system, the vacuum $|\mathcal{O}>$. As usual, the vacuum is defined as the lowest eigenstate of the Hamiltonian operator P^0. This contains the assumption that the Hamiltonian has a lowest energy eigenstate and that it is non-degenerate.

The vacuum expectation value of a product of n scalar fields, each situated at one of n different spacetime points,

$$< \mathcal{O}|\phi(x_1)\phi(x_2)\ldots\phi(x_n)|\mathcal{O} > \tag{7.27}$$

is called the Wightman n point function.

Of most use to us, particularly later on when we discuss perturbation theory, will be time-ordered correlations such as

$$\Gamma(x_1, x_2, ..., x_n) \equiv < \mathcal{O}|T\phi(x_1)\phi(x_2)\ldots\phi(x_n)|\mathcal{O} > \tag{7.28}$$

when we generally call the n point function.

Time ordering is signified by the presence of the symbol T inside the bracket in equation (7.28). This symbol indicates that the operators which occur immediately to the right of it are to be put in a particular order. They must be ordered in such a way that their time arguments have decreasing value as we follow the operators from left to right. That is, the T operation specifies the ordering of operators which occur to the right of it in the following way.

$$T\phi(x_1)\phi(x_2)\ldots\phi(x_n) = \phi(x_{P(1)})\phi(x_{P(2)})\ldots\phi(x_{P(n)})$$

if $x^0_{P(1)} > x^0_{P(2)} > ... > x^0_{P(n)}$ and where $\{P(1), P(2), ..., P(n)\}$ is a permutation of $\{1, 2, ..., n\}$.

Sometimes it is useful to use the Heaviside step function to indicate the time ordering, for example, for $n = 2$,

$$\mathcal{T}\phi(x_1)\phi(x_2) = \theta(x_1^0 - x_2^0)\phi(x_1)\phi(x_2) + \theta(x_2^0 - x_1^0)\phi(x_2)\phi(x_1) \qquad (7.29)$$

The time ordered function has the distinct advantage that it is symmetric under the interchange of the arguments of the fields, for example

$$\mathcal{T}\phi(x_1)\phi(x_2) = \mathcal{T}\phi(x_2)\phi(x_1)$$

It is symmetric because the time ordering symbol $\mathcal{T}$ will order the operators in the same way, irrespective of the order in which they appear in the above expression. This time-ordering operation and its properties should be familiar from time-dependent perturbation theory in quantum mechanics where it is widely used.

The 2 point function of the scalar field will be important to us in subsequent developments and we shall denote it by a special symbol

$$D(x_1, x_2) \equiv \Gamma(x_1, x_2) \equiv \ <\mathcal{O}|\mathcal{T}\phi(x_1)\phi(_2)|\mathcal{O}> \qquad (7.30)$$

$$= \int \frac{dk}{(2\pi)^4} e^{-k(x-y)} D(k^2) \qquad (7.31)$$

(Here we have anticipated the fact that the Fourier transform depends only on k^2.)

7.5   The Free Scalar Field

In the free field limit of our scalar field theory, that is, the limit where we set the coupling constant λ to zero, it is exactly solvable and it is useful to study the solution. The field equation (7.6) with λ put to zero reduces to the relativistic wave equation

$$\left(\partial^2 - m^2\right)\phi(x) = 0 \qquad (7.32)$$

which must be solved in such a way that the field obeys the equal-time commutation relations in equations (7.8)–(7.9).

The general solution of equation (7.32) is obtained by noting that the wave equation is solved by plane waves,

$$\left(\partial^2 - m^2\right) e^{ik_\mu x^\mu} = 0 \text{ when } k^0 = \pm E(\vec{k}), \ \ E(\vec{k}) = \sqrt{\vec{k}^2 + m^2}$$

There are plane wave solutions with both positive and negative energies or positive and negative frequencies. When we quantized non-relativistic fermions, we interpreted the negative frequency solutions as holes in the Fermi sea. For the scalar field, which describes bosons, there is no such interpretation. We will use the negative frequency solutions in a different way. We will superpose the negative and positive frequency solutions in such a way that the field operator is Hermitian, $\phi(x) = \phi^\dagger(x)$.

This is the quantum mechanical analog of the the fact that the classical scalar field that we are discussing is real.

With this in mind, we write the general solution of the wave equation as a superposition of the plane waves as

$$\phi(x) = \int \frac{d^3k}{\sqrt{(2\pi)^3 2E(\vec{k})}} \left[e^{ik_\mu x^\mu} a(\vec{k}) + e^{-ik_\mu x^\mu} a^\dagger(\vec{k}) \right] \tag{7.33}$$

$$k^0 = E(\vec{k})$$

The normalization factors inside the integration in equation (7.33) are determined so that the scalar field obeys the commutation relations (7.8)–(7.9) when the commutators of creation and annihilation operators are

$$\left[a(\vec{k}), a^\dagger(\vec{k}') \right] = \delta(\vec{k} - \vec{k}'), \quad \left[a(\vec{k}), a(\vec{k}') \right] = 0, \quad \left[a^\dagger(\vec{k}), a^\dagger(\vec{k}') \right] = 0 \tag{7.34}$$

The vacuum state (which will turn out to be the lowest energy state) is defined as the state which is annihilated by all of the annihilation operators

$$a(\vec{k})|\mathcal{O} >= 0, \quad <\mathcal{O}|a^\dagger(\vec{k}) = 0, \quad \forall \vec{k}, \quad <\mathcal{O}|\mathcal{O} >= 1$$

and a basis for multi-particle states is created by operating with creation operators on the vacuum,

$$\left\{ |\mathcal{O}>, \; a^\dagger(\vec{k})|\mathcal{O}>, \; ..., \; \frac{1}{\sqrt{n!}} a^\dagger(\vec{k}_1)...a^\dagger(\vec{k}_n)|\mathcal{O}>, \; ... \right\} \tag{7.35}$$

This set of states is a basis for the Fock space of quantum states of the real scalar field.

With our solution for the field operator in equation (7.33) we can form the four-momentum operator of the free field theory P^μ from the Noether charges in equations (7.17) and (7.19). The result is

$$P^0 = \int d^3k \, E(\vec{k}) \, a^\dagger(\vec{k}) a(\vec{k}) \; + \; \text{constant} \tag{7.36}$$

$$\vec{P} = \int d^3k \, \vec{k} \, a^\dagger(\vec{k}) a(\vec{k}) \tag{7.37}$$

We can use equation (7.36) to confirm that the vacuum is indeed the state of this theory with the smallest value of the energy. The other states in the basis (7.35) are eigenstates of the four-momentum

$$P^\mu \frac{1}{\sqrt{n!}} a^\dagger(\vec{k}_1)...a^\dagger(\vec{k}_n)|\mathcal{O}>= \left[\sum_{i=1}^{n} (E(\vec{k}_i), \vec{k}_i) \right] \frac{1}{\sqrt{n!}} a^\dagger(\vec{k}_1)...a^\dagger(\vec{k}_n)|\mathcal{O}> \tag{7.38}$$

These equations tell us that we have found all of the eigenstates and eigenvalues of the Hamiltonian and momentum operators. The eigenvalues are just the sums of the energies and the momenta of assemblies of particles. Since the energies are additive, it is clear that these particles do not interact with each other. Since the states $\frac{1}{\sqrt{n!}}a^{\dagger}(\vec{k}_1)...a^{\dagger}(\vec{k}_n)|\mathcal{O}>$ are completely symmetric under permutations of the individual wave-vectors $\vec{k}_1, ..., \vec{k}_n$, this theory describes non-interacting bosons. In fact, these bosons are spinless relativistic particles.

For free field theory we will call the 2 point function the propagator and we will give it a special symbol

$$\Delta(x, y) = \lim_{\lambda \to 0} D(x, y) \tag{7.39}$$

or, in momentum space, we will show in the discussion below that

$$\Delta(k^2) = \lim_{\lambda \to 0} D(k^2) \tag{7.40}$$

$$= \frac{-i}{k^2 + m^2 - i\epsilon} \tag{7.41}$$

$\Delta(x, y)$ and $\Delta(k^2)$ will play an important role in perturbation theory. In equation (7.41) we have anticipated the form of the Fourier transform of $\Delta(x, y)$. We will derive this equation shortly.

To study the 2 point function, we take the definition of the time-ordered product of two fields in equation (7.29), plug in the expansion in equation (7.33) and take the vacuum expectation value. The result is the expression

$$\Delta(x, y) \equiv < \mathcal{O}|T\phi(x)\phi(y)|\mathcal{O} > \tag{7.42}$$

$$= \theta(x^0 - y^0) \int d^3k \frac{e^{i\vec{k}\cdot(\vec{x}-\vec{y}) - iE(k)(x^0-y^0)}}{(2\pi)^3 2E(k)}$$

$$+ \theta(y^0 - x^0) \int d^3k \frac{e^{-i\vec{k}\cdot(\vec{x}-\vec{y}) + iE(k)(x^0-y^0)}}{(2\pi)^3 2E(k)} \tag{7.43}$$

We can simplify the form of this formula. For this purpose, we use Cauchy's integral formula[1] to show that

$$\theta(x^0 - y^0)e^{-iE(k)(x^0-y^0)} = -\lim_{\epsilon \to 0} \int \frac{dk^0}{2\pi i} \frac{e^{-ik^0(x^0-y^0)}}{k^0 - E(k) + i\epsilon} \tag{7.44}$$

---

[1] Cauchy's formula considers an integral of a holomorphic function $f(z)$ along a closed contour in the complex plane. It states that, of the function has simple poles at points z_i in the interior of the contour,

$$\oint_C dz f(z) = 2\pi i \sum_i \text{Res } f(z) \text{ at } z_i$$

where, near the pole at z_i, $f(z) \sim [\text{Res } f(z) \text{ at } z_i]/(z - z_i)$.

where ϵ is an infinitesimal positive real number.[2] Similarly

$$\theta(y^0 - x^0)e^{iE(\vec{k})(x^0-y^0)} = \lim_{\epsilon \to 0} \int \frac{dk^0}{2\pi i} \frac{e^{-ik^0(x^0-y^0)}}{k^0 + E(\vec{k}) - i\epsilon} \tag{7.45}$$

Putting the terms together, we have

$$\Delta(x, y) = -\int \frac{dk}{(2\pi)^4} \left[\frac{e^{i\vec{k}\cdot(\vec{x}-\vec{y})-ik^0(x^0-y^0)}}{2i\,E(k)(k^0 - E(\vec{k}) + i\epsilon)} - \frac{e^{i\vec{k}\cdot(\vec{x}-\vec{y})-ik^0(x^0-y^0)}}{2i\,E(k)(k^0 + E(\vec{k}) - i\epsilon)} \right] \tag{7.46}$$

or, upon combining the integrands,

$$\Delta(x, y) = -i \int \frac{dk}{(2\pi)^4} \frac{e^{ik_\mu(x-y)^\mu}}{k_\mu k^\mu + m^2 - i\epsilon} \tag{7.47}$$

If we operate the wave operator on the above expression for the 2 point function

$$(-\partial^2 + m^2)\Delta(x, y) = -i \int \frac{dk}{(2\pi)^4} (-\partial^2 + m^2) \frac{e^{ik_\mu(x-y)^\mu}}{k_\mu k^\mu + m^2 - i\epsilon}$$

$$= -i \int \frac{dk}{(2\pi)^4} (k_\mu k^\mu + m^2) \frac{e^{ik_\mu(x-y)^\mu}}{k_\mu k^\mu + m^2 - i\epsilon}$$

$$= -i \int \frac{dk}{(2\pi)^4} e^{ik_\mu(x-y)^\mu} = -i\delta(x - y)$$

we see that it is proportional to a Green function, that is

$$\left(-\partial^2 + m^2\right)\Delta(x, y) = -i\delta(x - y) \tag{7.48}$$

---

[2] To understand this formula, we observe that, the integral over k^0 in equation (7.44) is a line integral along the real axis in the complex k^0-plane. The integrand, $-\frac{1}{2\pi i}\frac{e^{-ik^0(x^0-y^0)}}{k^0-E(\vec{k})+i\epsilon}$, is regarded as a function of the complex variable k^0. It has a pole at $k^0 = E(\vec{k}) - i\epsilon$ which is in the lower half of the complex plane. The residue at the pole is $-\frac{1}{2\pi i}e^{(-iE(\vec{k})-\epsilon)(x^0-y^0)}$. If $x^0 - y^0 > 0$, the factor in the integrand $e^{-ik^0(x^0-y^0)}$ goes to zero on the half-circle at infinity of the lower half-plane. The line integral along this half-circle is thus zero and it can be added to the integral in equation (7.44) without changing the value of the integral. The resulting integral is a line integral over a contour which encloses the entire lower half-plane including the position of the pole. The orientation of the contour is clockwise, thus the minus sign. It can then be evaluated using Cauchy's integral formula, which evaluates it as $-2\pi i$ times the residue of the pole which is enclosed by the contour. Here, the pole is at $k^0 = E(\vec{k}) - i\epsilon$ and the net result is $e^{(-iE(\vec{k})-\epsilon)(x^0-y^0)}$ which is the value of the left-hand-side of (7.44) when $x^0 - y^0 > 0$. On the other hand, if $x^0 - y^0 < 0$, the integrand goes to zero at the boundaries of the upper half-plane and, completing the contour there and using Cauchy's integral formula we obtain zero. This also agrees with the left-hand-side of equation (7.44).

A Green function for a wave operator must always be defined using a boundary condition. The "$i\epsilon$" which appears in the denominator of the integrand in (7.47) is where the information that $\Delta(x, y)$ must be the time ordered Green function is input.

To get an alternative check on the validity of equation (7.48), we can use the field equation and commutation relations to obtain a formula for the 2 point function. To begin, let us take its first time derivative,

$$
\frac{\partial}{\partial x^0} \Delta(x, y) = < \mathcal{O}|T \frac{\partial}{\partial x^0} \phi(x)\phi(y)|\mathcal{O} > + \frac{d}{dx^0}\left[\theta(x^0 - y^0)\right] < \mathcal{O}|\phi(x)\phi(y)|\mathcal{O} >
$$

$$
+ \frac{d}{dx^0}\left[\theta(y^0 - x^0)\right] < \mathcal{O}|\phi(y)\phi(x)|\mathcal{O} >
$$

$$
= < \mathcal{O}|T \frac{\partial}{\partial x^0} \phi(x)\phi(y)|\mathcal{O} > + \delta(x^0 - y^0) < \mathcal{O}|\left[\phi(x), \phi(y)\right]|\mathcal{O} >
$$

$$
= < \mathcal{O}|T \frac{\partial}{\partial x^0} \phi(x)\phi(y)|\mathcal{O} >
$$

where we have used the formula $\frac{d}{dx}\theta(x) = \delta(x)$ for the derivative of the Heavyside step function and then the equal-time commutation relation (7.9). We see that the first time derivative actually commutes with the time ordering.

Then, consider the second time derivative,

$$
\frac{\partial^2}{\partial x^{0^2}} \Delta(x, y) = < \mathcal{O}|T \frac{\partial^2}{\partial x^{0^2}} \phi(x)\phi(y)|\mathcal{O} >
$$

$$
+ \left[\frac{d}{dx^0}\theta(x^0 - y^0)\right] < \mathcal{O}|\frac{\partial}{\partial x^0}\phi(x)\phi(y)|\mathcal{O} >
$$

$$
+ \left[\frac{d}{dx^0}\theta(y^0 - x^0)\right] < \mathcal{O}|\phi(y)\frac{\partial}{\partial x^0}\phi(x)|\mathcal{O} >
$$

$$
= < \mathcal{O}|T \frac{\partial^2}{\partial x^{0^2}} \phi(x)\phi(y)|\mathcal{O} > + \delta(x^0 - y^0) < \mathcal{O}|\left[\frac{\partial}{\partial x^0}\phi(x), \phi(y)\right]|\mathcal{O} >
$$

$$
= < \mathcal{O}|T \frac{\partial^2}{\partial x^{0^2}} \phi(x)\phi(y)|\mathcal{O} > -i\delta(x - y)
$$

$$
= \left(\vec{\nabla}^2 - m^2\right) < \mathcal{O}|T\phi(x)\phi(y)|\mathcal{O} > -i\delta(x - y)
$$

where we have used the field equation and the equal-time commutation relation (7.8). This yields a derivation of equation (7.48). Thus, the 2 point function of the free scalar field is proportional to a Green function for the wave-operator $-\partial^2 + m^2$.

We have found the 2 point correlation function for the free scalar field. We could proceed to find all of the higher multi-point functions in a similar way by plugging the expression for the scalar field in terms of creation and annihilation operators in equation (7.33) and evaluating the matrix element directly. This is straightforward but we will put off further discussion until we study functional techniques where we will find an even easier way to get all of the n point functions of the free scalar field.

7.6   Consequences of Spacetime Symmetry

In contrast with the free field theory of the previous section where we have an exact solution of the theory, we know very little about the quantum states or other properties of the interacting quantum field theory that is described the the Lagrangian density in equation (7.4). Unlike the free field theory where we have creation and annihilation operators which can be used to create a basis of states from the vacuum and were we have a good understanding of the matrix elements of important operators in those basis states, once we turn on the $-\frac{\lambda}{4!}\phi^4(x)$ interaction term in the Lagrangian density, the field equation becomes nonlinear and we do not have a general solution available to us.

We have emphasized that, via Noether's theorem, symmetries of the quantum field theory imply conservation laws. Assuming that the conservation laws survive the passage from classical to quantum field theory, they become exact statements about the behaviour of the quantum field theory. In this section we are going to explore some of the ways that this information can be used to deduce properties of the solution of the quantum field theory if it exists, even when we do not know the solution explicitly. We will do this in the context of the scalar field theory with Lagrangian density (7.4). In that theory, the only continuous symmetry is the relativistic Poincare invariance which resulted in conservation of the stress tensor $\mathbf{T}^{\mu\nu}(x)$ and the Noether charges P^μ and $M^{\mu\nu}$ as time-independent operators which have the algebraic properties summarized in equations (7.22)–(7.26).

Of course, in order to progress at all, we shall have to make some assumptions about the quantum field theory. We will assume that P^μ and $M^{\mu\nu}$ are Hermitian operators. We will assume that the operator P^μ have eigenvalues and eigenstates which are contained in the Hilbert space of quantum states of the quantum field theory.

We will assume that the quantum field theory has a vacuum state $|\mathcal{O}>$ which, as we have been doing so far, is defined to be that eigenstate of the Hamiltonian, P^0, with minimum eigenvalue. We can, for the sake of our arguments here, set this eigenvalue to zero so that $P^0|\mathcal{O}> = 0$.

We will further assume that the quantum states other than the vacuum can be organized so that they are eigenstates of the full energy and momentum four-vector, P^μ, and that the vacuum state is an eigenstate of these operators with zero eigenvalue,

$$P^\mu|\mathcal{O}> = 0 \tag{7.49}$$

We shall also assume that both spacetime translations and Lorentz transformations are implemented by the action of unitary operators on the space of states. Let us begin with a discussion of how this works for space and time translations. An iteration of the commutator of the energy-momentum operator with the field $\phi(x)$ in equation (7.22) yields, for the k'th iteration

$$\partial^{\mu_1}\partial^{\mu_2}...\partial^{\mu_k}\,\phi(x) = i^k\left[P^{\mu_1},\left[P^{\mu_2},\ldots\left[P^{\mu_k},\phi(x)\right]...\right]\right] \tag{7.50}$$

We can use this equation in the Taylor expansion of the field about some point, say the point $x^\mu = 0$,

$$\phi(x) = \sum_{k=0}^{\infty} \frac{1}{k!} x_{\mu_1} x_{\mu_2} \cdots x_{\mu_k} \partial^{\mu_1} \partial^{\mu_2} \dots \partial^{\mu_k} \phi(x) \Big|_{x=0} \qquad (7.51)$$

$$= \sum_{k=0}^{\infty} \frac{1}{k!} i^k x_{\mu_1} x_{\mu_2} \cdots x_{\mu_k} \left[P^{\mu_1}, \left[P^{\mu_2}, \dots \left[P^{\mu_k}, \phi(0) \right] \dots \right] \right] \qquad (7.52)$$

We could compare the above equation with the Taylor expansion of the expression $e^{i P_\mu x^\mu} \phi(0) e^{-i P_\nu x^\nu}$ in x^μ and conclude that

$$\phi(x) = e^{i P_\mu x^\mu} \phi(0) e^{-i P_\nu x^\nu} \qquad (7.53)$$

Since P_μ are Hermitian operators, $e^{i P_\mu a^\mu}$ is a unitary operator. It is the unitary operator which implements the spacetime translation $x^\mu \to x^\mu + a^\mu$,

$$\phi(x + a) = e^{i P_\mu a^\mu} \phi(x) e^{-i P_\nu a^\nu} \qquad (7.54)$$

If it acts on operators as in equation (7.54), it must act on states as

$$|\psi> \to e^{i P_\mu a^\mu} |\psi> \qquad (7.55)$$

We have assumed equation (7.49) for the vacuum state, that it is an eigenstate of P^μ with zero eigenvalue. For the unitary operator that generates a spacetime translation, this implies that the vacuum state is strictly invariant,

$$e^{i P_\mu a^\mu} |\mathcal{O}> = |\mathcal{O}> \qquad (7.56)$$

Similarly, we shall assume that the vacuum state is invariant under Lorentz transformations. Invariance under infinitesimal Lorentz transformations requires

$$M^{\mu\nu} |\mathcal{O}> = 0 \qquad (7.57)$$

We postulate the existence of a unitary operator, $U(\Lambda)$, which operates on the states of the quantum field theory and implements the Lorentz transformation which corresponds to the coordinate transformation

$$x^\mu \to \tilde{x}^\mu = \Lambda^\mu_{\ \nu} x^\nu \qquad (7.58)$$

Since it is unitary, $U^\dagger(\Lambda) U(\Lambda) = \mathcal{I}$, where $\mathcal{I}$ is the unit operator acting on quantum states. Also, $U(1) = \mathcal{I}$ and $U(\Lambda^{-1}) = U^\dagger(\Lambda)$.

The unitary operator $U(\Lambda)$ must implement a Lorentz transformation of the scalar field,

$$\tilde{\phi}(x) = U(\Lambda) \phi(x) U^\dagger(\Lambda) = \phi(\Lambda^{-1} x) \qquad (7.59)$$

If the transformation is infinitesimal, so that

$$\Lambda^{\mu}_{\ \nu} = \delta^{\mu}_{\ \nu} + \omega^{\mu}_{\ \nu}$$

with $\omega^{\mu}_{\ \nu}$ infinitesimal,

$$U(1 + \omega) = \mathcal{I} + i\omega_{\mu\nu}M^{\mu\nu} + \ldots \tag{7.60}$$

Lorentz invariance of the vacuum is equivalent to the expression

$$U(\Lambda)\,|\mathcal{O}> \ = \ |\mathcal{O}> \tag{7.61}$$

for any Λ.

We will make the further assumption that the rest of the quantum states, beyond the vacuum, can be organized into states which are eigenstates of the energy-momentum operator. So a generic state $|p>$ must obey

$$P^{\mu}\,|p> \ = \ p^{\mu}\,|p> \tag{7.62}$$

Generally, the energy and momentum of these states should be time-like, that is, $p^2 = p_{\mu}p^{\mu} < 0$ and the total energy should be positive, $p^0 > 0$.

Finally, we assume that the unitary operators which implements a Lorentz transformation acts on states which are eigenstates of the four-momentum by Lorentz transforming the momentum,

$$\mathcal{U}(\Lambda)|p> = |\Lambda p> \tag{7.63}$$

Let us now make use of the Poincare invariance of the quantum field theory. Consider the expectation value of the product of two field operators,

$$< \mathcal{O}|\phi(x)\phi(y)|\mathcal{O} >$$

which we have called the Wightman 2 point function.

It is easy to see that, as a consequence of translation invariance, the Wightman function depends only on the difference of the coordinates, $x_1^{\mu} - x_2^{\mu}$. For this, we substitute equation (7.53) into the expectation value to get

$$< \mathcal{O}|\phi(x_1)\phi(x_2)|\mathcal{O} > = < \mathcal{O}|e^{iPx_1}\phi(0)e^{iP(x_2-x_1)}\phi(0)e^{-iPx_2}|\mathcal{O} >$$

We use the fact that $< \mathcal{O}|e^{iPx_1} = < \mathcal{O}|$ and $e^{-iPx_2}|\mathcal{O} > = |\mathcal{O} >$ to rewrite the above equation as

$$< \mathcal{O}|\phi(x_1)\phi(x_2)|\mathcal{O} > = < \mathcal{O}|\phi(0)e^{iP(x_2-x_1)}\phi(0)|\mathcal{O} >$$

We see that, as a consequence of translation symmetry, the Wightman function depends on the coordinates only in the combination $x_1^{\mu} - x_2^{\mu}$.

Now let us study the consequence of Lorentz invariance. We use the invariance of the vacuum, $< \mathcal{O}|U^{-1}(\Lambda) =< \mathcal{O}|$ and $U(\Lambda)|\mathcal{O} >= |\mathcal{O} >$ and also $U(\Lambda)U^{-1}(\Lambda) = \mathcal{I}$ to get

$$< \mathcal{O}|\phi(x_1)\phi(x_2)|\mathcal{O} >=< \mathcal{O}|U^{-1}(\Lambda)\phi(x_1)U(\Lambda)U^{-1}(\Lambda)\phi(x_2)U(\Lambda)|\mathcal{O} >$$

Then we use the equation (7.59) in the form $U^{-1}(\Lambda)\phi(x_2)U(\Lambda) = \phi(\Lambda x)$ to get the equation

$$< \mathcal{O}|\phi(x_1)\phi(x_2)|\mathcal{O} >=< \mathcal{O}|\phi(\Lambda x_1)\phi(\Lambda x_2)|\mathcal{O} >$$

When it is combined with translation invariance, Lorentz invariance shows us that $< \mathcal{O}|\phi(x_1)\phi(x_2)|\mathcal{O} >$ can depend only the Lorentz invariant combination on $(x_1 - x_2)^2$.

Thus the spacetime symmetries tell us that $< \mathcal{O}|\phi(x_1)\phi(x_2)|\mathcal{O} >$ which would nominally be a function of eight variables, the components of x_1^μ and x_2^μ, is a function of only one variable, $(x_1 - x_2)^2$.

7.7   Spectral Theorem

We can extend our discussion of the section above by adding one ingredient. Once the states of the quantum field theory can always be written in combinations where they are eigenstates of the energy and momentum four-vector as in equation (7.62), we assume that there is a complete set of states and we can write a completeness sum as

$$\mathcal{I} = \sum_p |p >< p| \tag{7.64}$$

The summation sign in the above equation is formal in the sense that it represents summations and integrations over all of the states with each value of the energy and momentum p^μ and then a sum or integral over the allowed values of p^μ itself. We will not have to worry about the precise nature of these summations and integrations here, we are only going to use the very reasonable postulate that the decomposition of the unit vector in the form (7.64) exists.

We begin by following the process in the previous section of applying our assumption that the quantum field theory is translation invariant and that the vacuum state has zero energy and momentum to write the Wightman function as

$$< \mathcal{O}|\phi(x)\phi(y)|\mathcal{O} >=< \mathcal{O}|\phi(0)e^{-i P_\mu(x-y)^\mu}\phi(0)|\mathcal{O} > \tag{7.65}$$

Then, we insert a complete set of states into the middle of the second expression to get

$$< \mathcal{O}|\phi(x)\phi(y)|\mathcal{O} >= \sum_p < \mathcal{O}|\phi(0)|p >< p|e^{-i P_\mu(x-y)^\mu}\phi(0)|\mathcal{O} > \tag{7.66}$$

Next, we use fact that $|p>$ is an eigenstate of P_μ so that

$$< p|e^{-iP_\mu(x-y)^\mu} =< p|e^{-ip_\mu(x-y)^\mu}$$

to get

$$< \mathcal{O}|\phi(x)\phi(y)|\mathcal{O} >= \sum_p |< p|\phi(0)|\mathcal{O} >|^2\, e^{-ip_\mu(x-y)^\mu} \qquad (7.67)$$

Then, finally, we do the trick of inserting unity in the form of

$$1 = \int dq\, \delta^{(4)}(p-q)$$

into the summation to get

$$< \mathcal{O}|\phi(x)\phi(y)|\mathcal{O} >= \int dq \left\{ \sum_p |< p|\phi(0)|\mathcal{O} >|^2 \delta^{(4)}(p-q) \right\} e^{iq_\mu(x-y)^\mu}$$
$$(7.68)$$

We have already shown in the previous section that the left-hand-side of the above equation is a function of the coordinates in only the one combination, $(x-y)^2$. This actually implies that its Fourier transform, the quantity in the brackets on the right-hand-side of the above equation, is a function of the four-momentum q^μ only in the one combination q^2. This is indeed so, with one additional caveat, the fact that the energies are positive, that is, $q^0 > 0$.

Then, motivated by this last expression, we define the spectral function by

$$\sum_p |< p|\phi(0)|\mathcal{O} >|^2 \delta^{(4)}(p-q) = \frac{1}{(2\pi)^3}\theta(q^0)\rho(q^2) \qquad (7.69)$$

where we have used the fact mentioned above. Due to Lorentz invariance, $< p|\phi(0)|\mathcal{O} >$ depends only on p^2 and it is nonzero only when $p^2 \le 0$ and $p^0 \ge 0$.

Using the spectral function we can write the Wightman function as

$$< \mathcal{O}|\phi(x)\phi(y)|\mathcal{O} >= \int dq\, \frac{1}{(2\pi)^3}\theta(q^0)\rho(q^2)\, e^{iq_\mu(x-y)^\mu} \qquad (7.70)$$

$$= \int d\mu^2 \rho(\mu^2) \int dq\, \frac{1}{(2\pi)^3}\theta(q^0)\delta(q^2-\mu^2)\, e^{iq_\mu(x-y)^\mu} \qquad (7.71)$$

$$= \int d\mu^2 \rho(\mu^2)\, W_0(x,y;\mu^2) \qquad (7.72)$$

where we have noted the fact that

$$W_0(x,y;\mu^2) = \int dq\, \frac{1}{(2\pi)^3}\theta(q^0)\delta(q^2-\mu^2)\, e^{iq_\mu(x-y)^\mu} \qquad (7.73)$$

is the Wightman function for a free scalar field with mass μ.

The same arguments as we presented above for the Wightman function can equally well be used on the time-ordered 2 point function. The result is

$$D(q^2) = \int d\mu^2 \rho(\mu^2) \frac{-i}{q^2 + \mu^2 - i\epsilon} \tag{7.74}$$

which is called the spectral representation of the 2 point function. This equation has many interesting implications. One is that the asymptotic form of the 2 point function must obey

$$\lim_{q^2 \to \infty} D(q^2) \sim \frac{1}{q^2} \tag{7.75}$$

We can also use the fact that

$$\frac{-i}{q^2 + \mu^2 - i\epsilon} = \frac{-i\mathcal{P}}{q^2 + \mu^2} + \pi\delta(q^2 + \mu^2)$$

to find the spectral function in terms of $D(q^2)$. If μ^2 is real, then

$$\rho(\mu^2) = \frac{1}{2\pi}\left[D(-\mu^2) + D^*(-\mu^2)\right] \tag{7.76}$$

For free field theory, $\rho(\mu^2) = \delta(\mu^2 - m^2)$ and the spectral representation simply reproduces the free field propagator.

It is also useful to think of $D(q^2)$ defined in equation (7.74) as a function of a complex variable $z = -q^2$. The spectral representation then tells us that $D(z)$ is analytic everywhere in the complex plane which is away from the positive real axis. What is more, it tells us about the nature of the singularities of $D(z)$ that occur on the positive real axis. They occur for those values of $z = \mu^2$ where $\rho(\mu^2)$ is nonzero. In free field theory this only happens at one point, as $\rho(\mu^2) = \delta(\mu^2 - m^2)$ where m is the mass of the particle and $D(z)$ has a simple pole at $z = m^2$.

In an interacting theory we know that $\rho(\mu^2)$ is nonzero for those values of μ^2 which are equal to eigenvalues of of the operator $-P^2$. From the form of the spectral function in equation (7.69) we can see that, as well as being an eigenvalue of $-P^2$, the corresponding eigenvector must have a nonzero inner product with the state which is gotten by acting on the vacuum with the field operator, $\phi(x)|O>$.

Generally, since $\rho(\mu^2)$ can be nonzero for the value of $\mu^2 = -p^2$ where p^2 are the eigenvalues of $-P^2$ for some physical states of the theory, the form of the function ρ can come in two kinds. First of all, $-P^2$ could have a discrete spectrum with isolated real positive eigenvalues. In that case, $\rho(\mu^2)$ can contain Dirac delta functions similar to free field theory. Such states have the energy and momentum dispersion relation of a single relativistic particle and the eigenvalue of $-P^2$ is the mass of the particle. Then, there would be states where $-P^2$ has a continuous spectrum. In that case the spectral density $\rho(\mu^2)$ is a continuous function of μ^2 and the spectral representation (7.74) tells us that the 2 point function $D(z)$ has a cut singularity on the real z-axis with discontinuity across the cut related to $\rho(\mu^2)$.

7.8   Normalization of the Spectral Function

The equal-time commutation relation constrains the spectral function of the scalar field. To see this, note that the equal time commutation relation implies

$$
\begin{aligned}
i\delta^{(4)}(x-y) &= <\mathcal{O}|\,[\phi(x),\partial_0\phi(y)]\,|\mathcal{O}> \delta(x^0-y^0) \\
&= [<\mathcal{O}|\phi(x)\partial_0\phi(y)|\mathcal{O}> - <\mathcal{O}|\partial_0\phi(y)\phi(x)|\mathcal{O}>]\delta(x^0-y^0) \\
&= \int d\mu^2\rho(\mu^2)\int dq\,\frac{1}{(2\pi)^3}\theta(q^0)\delta(q^2-\mu^2)\cdot \\
&\qquad \cdot\left\{-iq_0e^{iq_\mu(x-y)^\mu} - iq_0e^{-iq_\mu(x-y)^\mu}\right\}\delta(x^0-y^0) \\
&= i\int d\mu^2\rho(\mu^2)\int dq\,\frac{1}{(2\pi)^3}\theta(q^0)\delta(q^2-\mu^2)2q^0\,e^{-i\vec{q}\cdot(\vec{x}-\vec{y})}\delta(x^0-y^0) \\
&= i\int d\mu^2\rho(\mu^2)\delta(\vec{x}-\vec{y})\delta(x^0-y^0) = i\delta^{(4)}(x-y)\int d\mu^2\rho(\mu^2) \quad (7.77)
\end{aligned}
$$

where we have used the following identity for the Dirac delta function

$$
\delta(q^2-\mu^2)\theta(q^0) = \frac{1}{2q^0}\delta\left(q^0 - \sqrt{\vec{q}^2+\mu^2}\right)
$$

This result is that the spectral function is normalized as

$$
\int_0^\infty d\mu^2\rho(\mu^2) = 1 \tag{7.78}
$$

In renormalized perturbation theory, we will be forced to re-scale the field by a factor of $Z^{\frac{1}{2}}$ which, of course, rescales the commutation relations and it will be the rescaled fields whose 2 point functions we compute. Of course they must have a spectral representation and the normalization of the spectral function is changed to

$$
\int_0^\infty d\mu^2\rho(\mu^2) = Z \tag{7.79}
$$

7.9   Analyticity

The fact that the energy p^0 of a quantum state must be positive, or zero if the state is the vacuum has an interesting consequence for correlation functions. In this section we will see that it implies the analyticity of correlation functions when the time variables are replaced by complex numbers.

To see how this analyticity arises, consider the Wightman n point function function in the scalar field theory,

$$< 0|\phi(\vec{x}_1, x_1^0)\phi(\vec{x}_2, x_2^0)\ldots\phi(\vec{x}_n, x_n^0)|0 > \tag{7.80}$$

We have assumed that the exponential of the four-momentum operator is a unitary operator which generates translations of the space–time coordinates of the fields. For the present discussion, all we need is the specialization of this assumption to time translation,

$$\phi(\vec{x}, x^0) = e^{iP^0 x^0}\phi(\vec{x}, 0)e^{-iP^0 x^0}$$

where P^0 is the Hamiltonian. With this assumption, we can write the Wightman n point function (7.80) as

$$< 0|e^{iP^0 x_1^0}\phi(\vec{x}_1, 0)e^{-iP^0(x_1^0-x_2^0)}\ldots e^{-iP^0(x_{n-1}^0-x_n^0)}\phi(\vec{x}_n, 0)e^{-iP^0 x_n^0}|0 > \tag{7.81}$$

Then, we assume that the vacuum is the eigenstate of the energy operator P^0 the lowest eigenvalue and we can take that eigenvalue as being zero, $P^0|0>=0$. Moreover, the remaining eigenvalues of P^0 are positive, so we can treat the P^0's in the formula as positive numbers. To be concrete, we can insert complete sets of states in between the field operators to get

$$\sum_{p_1\ldots p_{n-1}} e^{-ip_1^0(x_1^0-x_2^0)-\ldots-ip_{n-1}^0(x_{n-1}^0-x_n^0)}.$$
$$\cdot < 0|\phi(\vec{x}_1, 0)|p_1><p_1|\ldots|p_{n-1}><p_{n-1}|\phi(\vec{x}_n, 0)|0 > \tag{7.82}$$

The essential point here is that all of the numbers p_k^0 are positive. As a result, we can replace any of the time differences, say $(x_{j-1}^0 - x_j^0)$ by a complex number with a negative imaginary part and the sum over intermediate energy states would still converge, in fact it would converge better than it did when the time was just a real number. Put in other words, the correlation function in (7.82), viewed as a function of any one of the time differences

$$x_1^0 - x_2^0, x_2^0 - x_3^0, \ldots, x_{n-1}^0 - x_n^0$$

is analytic in the lower half of the complex plane. This applies to any of the time difference variables. Since any difference $x_j^0 - x_\ell^0$ for any $j < \ell$ can be found by adding some number of the differences $(x_{j-1}^0 - x_j^0)$ together, the Wightman n point function, as a function of any of the differences $x_j^0 - x_\ell^0$, with $j < \ell$, is analytic in the lower half-plane. This is the signal, contained in correlation functions, that the system has time translation invariance, that there is a conserved energy associated with this time translation invariance and that the energy of the quantum states has a lower bound and that the vacuum, the external state in the correlation function, is the lowest energy state.

We can apply a similar argument to a time-ordered n point function. There, the time ordering helps us to make a more elegant statement. Using similar reasoning as the above, we can see that the time-ordered n point function

$$< 0|\mathcal{T}\phi(\vec{x}_1, x_1^0)\phi(\vec{x}_2, x_2^0)\ldots\phi(\vec{x}_n, x_n^0)|0 >$$

is analytic in the second and fourth quadrants of the complex plane for any of the differences of the times $x_j^0 - x_\ell^0$ for any j and ℓ, that is, for a given pair of times, in the fourth quadrant of the complex plane

$$\Re x_j^0 - x_\ell^0 > 0, \quad \Im x_j^0 - x_\ell^0 < 0$$

or in the second quadrant

$$\Re x_j^0 - x_\ell^0 < 0, \quad \Im x_j^0 - x_\ell^0 > 0$$

This analyticity translates to an analyticity in the momenta k^0 in the Fourier transform of the correlation function. We will discuss the important concept of the Fourier transform of the n point functions later. Here, let us just consider the time domain Fourier transform,

$$\Gamma(\vec{x}_1, k_1^0, \ldots, \vec{x}_n, k_n^0)(2\pi)\delta(k_1^0 + \ldots + k_n^0)$$
$$= \int dx_1^0 \ldots dx_n^0 e^{ik_1^0 x_1^0 + \ldots + ik_n^0 x_n^0} < 0|\mathcal{T}\phi(\vec{x}_1, x_1^0)\phi(\vec{x}_2, x_2^0)\ldots\phi(\vec{x}_n, x_n^0)|0 > \tag{7.83}$$

In this expression we can use the fact that all of the frequencies k_j^0 must sum to zero to write the exponential in the Fourier transform as

$$\Gamma(\vec{x}_1, k_1^0, \ldots, \vec{x}_n, k_n^0)(2\pi)\delta(k_1^0 + \ldots + k_n^0)$$
$$= \int dx_1^0 \ldots dx_n^0 e^{ik_1^0(x_1^0 - x_n^0) + \ldots + ik_{n-1}^0(x_{n-1}^0 - x_n^0)} < 0|\mathcal{T}\phi(\vec{x}_1, x_1^0)\ldots\phi(\vec{x}_n, x_n^0)|0 > \tag{7.84}$$

The fact that the integrand is analytic in the time differences in the second and fourth quadrants of the complex plane means that we can simultaneously rotate the integration contours $x_j^0 - x_n^0 \to e^{-i\phi}(x_j^0 - x_n^0)$ and the energy variables $k_j^0 \to e^{i\phi}k_j^0$ where $0 \leq \phi \leq \frac{\pi}{2}$ without encountering a singularity. This in turn implies that the function $\Gamma(\vec{x}_1, k_1^0, \ldots, \vec{x}_n, k_n^0)$ remains non-singular when we replace the frequencies by complex numbers as in $\Gamma(\vec{x}_1, e^{i\phi}k_1^0, \ldots, \vec{x}_n, e^{i\phi}k_n^0)$ and we have the freedom of varying the angle ϕ between 0 and $\pi/2$. This analyticity is intimately connected with the procedure called Wick rotation which we shall employ in evaluating Feynman integrals when we discuss perturbation theory.

7.9.1   The Reeh–Schlieder Theorem

The spectral representation which we discussed in a previous section gave us a guide to the information that is contained in the 2 point function. The spectral representation is a measure of the ability of the field operator $\phi(x)$ to create states from the vacuum, when these states are characterized by their energies and momenta.

We might imagine that we could create all possible quantum states by operating with the basic field $\phi(x)$ on the vacuum state some number of times, $\phi(x_1)\phi(x_2)\ldots\phi(x_n)|\mathcal{O}>$ with varying values of x_1, ..., x_n. This reasonable expectation, which can be turned into a theorem which we shall not prove, is contained in the statement that the set of state vectors

$$\phi(x_1)\phi(x_2)\ldots\phi(x_n)|\mathcal{O}>$$

for any number of operations, n, and with the points $x_1, \ldots, x_n$ located anywhere in Minkowski space, contains a complete set of states.[3] The correct terminology is that this is a "dense" set of states in that any quantum state can be approximated to any degree of accuracy by some linear combination of these states and any state at all can be found by a suitably convergent sequences of such states. One thing that this implies is that there is no state which is orthogonal to all of the states that are created by operating $\phi's$, that is, if

$$< \xi|\phi(x_1)\phi(x_2)\ldots\phi(x_n)|\mathcal{O}>= 0, \quad \forall n, \ x_1, ..., x_n$$

then

$$|\xi>= 0$$

the zero vector in the Hilbert space.

The facts of the above paragraph are not be surprising. What is surprising is something that we have already seen in our arguments that relativistic quantum mechanics requires anti-particles. Intuitively, we would expect that the state that we create by operating the field operator at a space–time point, $\phi(x)|\mathcal{O}>$, to be, in some sense, localized there. However, it is a fact that, its overlap with a state that we create by operating the field operator at a different point, $\phi(y)|\mathcal{O}>$ is never zero, even when x and y have space-like separations. The overlap is just the Wightman 2 point function

$$< \mathcal{O}|\phi(y)\phi(x)|\mathcal{O}>$$

---

[3] Here we are assuming that there are no charges (such as topological charges) which are not carried by the field ϕ—otherwise the charged states could not be found in this way. Their existence would require inclusion of other operators which carry the charge.

which we have found a spectral representation for in equation (7.72),

$$< \mathcal{O}|\phi(y)\phi(x)|\mathcal{O} >$$
$$= \int d\mu^2 \rho(\mu^2) \int dq \frac{1}{(2\pi)^3} \theta(q^0)\delta(q^2 - \mu^2)\, e^{iq_\mu(y-x)^\mu} \qquad (7.85)$$

We have already used analyticity to argue that the right-hand-side of this equation, which is a function of $(y - x)^2$, can never vanish for any points where $(y - x)^2$ is a finite number, even when the points are space-like separated. If the points have space-like separations, we could find a reference frame where $x^0 = y^0$ and, in that reference frame,

$$< \mathcal{O}|\phi(y)\phi(x)|\mathcal{O} >= \int d\mu^2 \rho(\mu^2) \int d^3q \frac{e^{i\vec{q}\cdot(\vec{y}-\vec{x})}}{(2\pi)^3 2\sqrt{\vec{q}^2 + \mu^2}} \qquad (7.86)$$

The right-hand-side of this equation is definitely non-zero if $\vec{x} \neq \vec{y}$. This tells us that there is an intrinsic non-locality in quantum field theory, related to the fact that it is not possible to localize a relativistic particle.

The second miracle is that this non-locality does not lead to violations of causality. To see this, we observed that the right-hand-side of equation (7.86) is an even function of $\vec{y} - \vec{x}$ and therefore we must have

$$< \mathcal{O}|\,[\phi(y),\phi(x)]\,|\mathcal{O} >= 0 \ \text{ if } (y - x)^2 > 0 \qquad (7.87)$$

Measurements of the "observables" $\phi(x)$ and $\phi(y)$ do not interfere with each other if x and y are space-like separated points.

This discussion has a generalization called the Reeh–Schlieder theorem. We assume that or scalar field theory is a quantum mechanical system whose states are vectors in a Hilbert space. We will assume that there is a lowest energy state, the vacuum $|\mathcal{O} >$, that the field theory lives in Minkowski space and that it has space–time translation invariance. The theorem states that

$$\phi(x_1)\phi(x_2)...\phi(x_n)|\mathcal{O} >$$

for any n and for $x_1, ..., x_n$ in any open subset of space–time is a "dense" subset of the quantum Hilbert space. Without defining dense, we can simply say that it provides enough states that linear combinations of them can describe any state to any accuracy.

The proof of this theorem is straightforward and it only relies on translation invariance, the implementation of spacetime translations by the unitary operator e^{iPx}, the fact that the vacuum is the minimum energy state and the existence of the complete set of states in the form of the completeness sum

$$\mathcal{I} = \sum_p |p><p|$$

Consider points $x_1, ..., x_n$ which are all located in a region of space–time, $\mathcal{U}$, which is an open set,[4] that is,

$$x_i \in \mathcal{U}, \quad i = 1, 2, ..., n$$

The Reeh–Schlieder Theorem says that by considering all states of the form

$$\phi(x_1)\phi(x_2)\dots\phi(x_n)|\mathcal{O}>, \quad x_1, x_2, ..., x_n \in \mathcal{U}, \quad n = 0, 1, 2, ... \qquad (7.88)$$

and their superpositions we can find all of the quantum states of the quantum field theory.

The proof of this surprising theorem is elementary. Let us assume that it were not the case. Then there would have to be a state $|\zeta>$ which is orthogonal to all of the states of the form (7.88), that is,

$$< \zeta|\phi(x_1)\phi(x_2)\dots\phi(x_n)|\mathcal{O}> = 0, \quad x_i \in \mathcal{U}, \quad n = 0, 1, 2, ... \qquad (7.89)$$

Now, we can work on the quantities $< \zeta|\phi(x_1)\phi(x_2)\dots\phi(x_n)|\mathcal{O}>$ using what we know about spacetime translation invariance. Consider a space–time translation of the last point, x_n, from $x_n \in \mathcal{U}$ to $x_n + zv \in \mathcal{U}$ for z a (perhaps very small) real number and v^μ a time-like four-vector with positive time component, $v^0 > 0$. This should always be possible since $\mathcal{U}$ is an open set. Then

$$< \zeta|\phi(x_1)\phi(x_2)\dots\phi(x_n + vz)|\mathcal{O}>$$
$$= < \zeta|\phi(x_1)\phi(x_2)\dots e^{iP_\mu v^\mu z}\phi(x_n)e^{-iP_\mu v^\mu z}|\mathcal{O}>$$
$$= \sum_p < \zeta|\phi(x_1)\phi(x_2)\dots|p><p|\phi(x_n)|\mathcal{O}> \, e^{iP_\mu v^\mu z}$$

Since both v^μ and p^μ are time-like vectors, $p^0 > 0$ and $v^0 > 0$, we know that $p_\mu v^\mu < 0$. This tells us that, as a function of z, the above expression is analytic in the lower half of the complex z-plane. An elementary application of the edge of the wedge theorem of complex analysis tells us that, if a function is zero for some interval of z on the real axis, and if it is analytic in the lower half plane, it must be zero everywhere in the lower half plane. Moreover, if it is zero in the entire lower half plane, it must also be zero on the entire real axis.

Thus we conclude that, if

$$< \zeta|\phi(x_1)\phi(x_2)\dots\phi(x_{n-1})\phi(x_n)|\mathcal{O}> = 0$$

---

[4] Remember that an example of an open set of real numbers are the real numbers x obeying $a < x < b$ for some $a < b$. An example of an open set $\mathcal{U}$ is the simple generalization of this to a multi-dimensional space.

for $x_1, \ldots, x_n \in \mathcal{U}$, we must also have

$$< \zeta|\phi(x_1)\phi(x_2)\ldots\phi(x_{n-1})\phi(x_n + vz)|\mathcal{O} >= 0$$

for $x_1, \ldots, x_n \in \mathcal{U}$ and for all real values of z.

Now we repeat the argument with another time-like vector v' and another variable z' and iterate to conclude that if

$$< \zeta|\phi(x_1)\phi(x_2)\ldots\phi(x_{n-1})\phi(x_n)|\mathcal{O} >= 0$$

for $x_1, \ldots, x_n \in \mathcal{U}$, we must also have

$$< \zeta|\phi(x_1)\phi(x_2)\ldots\phi(x_{n-1})\phi(x_n + vz + v'z' + v''z'' + v'''z''')|\mathcal{O} >= 0$$

for $x_1, \ldots, x_n \in \mathcal{U}$ and for all real values of z, z', z'', z'''.

We are free to choose the vectors v, v', v'', v''' as long as they are time-like and have positive time component. It is possible to choose a basis for Minkowski space as a set of four such vectors. Let us assume that v, v', v'', v''' is such a basis. Then z, z', z'', z''' can be chosen so that $x_n + vz + v'z' + v''z'' + v'''z'''$ is any point in Minkowski space. Our conclusion is then, if

$$< \zeta|\phi(x_1)\phi(x_2)\ldots\phi(x_{n-1})\phi(x_n)|\mathcal{O} >= 0$$

for $x_1, \ldots, x_n \in \mathcal{U}$, we must also have

$$< \zeta|\phi(x_1)\phi(x_2)\ldots\phi(x_{n-1})\phi(x_n)|\mathcal{O} >= 0$$

for $x_1, \ldots, x_{n-1} \in \mathcal{U}$ and x_n any point in Minkowski space.

Next we consider translation of the last two coordinates,

$$< \zeta|\phi(x_1)\phi(x_2)\ldots\phi(x_{n-1} + vz)\phi(x_n + vz)|\mathcal{O} >$$

with $x_1, \ldots, x_{n-1} \in \mathcal{U}$, x_n any point of Minkowski space v a time-like four-vector with positive time component and z initially small enough that $x_{n-1} + vz \in \mathcal{U}$.

Again we use translation invariance and insert a complete set of states to rewrite the above expression as

$$\sum_{pp'} < \zeta|\phi(x_1)\ldots\phi(x_{n-2})|p > < p|\phi(0)|p' > < p'|\phi(0)|\mathcal{O} > \cdot$$

$$\cdot e^{ip(x_{n-1}+vz) - ip'(x_{n-1}-x_n)}$$

For the same reasoning as above, this quantity is analytic in the lower half of the complex z-plane. It is also zero for some range of real values of z where $x_{n-1} + vz$ is still in the region $\mathcal{U}$. Analyticity then tells us that it must zero everywhere in the lower half-plane and on the entire real axis. We can then repeat this argument with more time-like vectors and show that $x_{n-1} - x_n$ can can be any point in Minkowski space. Since x_n can be anywhere in Minkowski x_{n-1} can be anywhere in Minkowski space. This leads us to the consequence of equation (7.89) and translation invariance, if

$$< \zeta|\phi(x_1)\phi(x_2)\ldots\phi(x_n)|\mathcal{O} >= 0$$

when $x_1, \ldots, x_n \in \mathcal{U}$, it must also be true that

$$< \zeta|\phi(x_1)\phi(x_2)\ldots\phi(x_n)|\mathcal{O} >= 0$$

when $x_1, \ldots, x_{n-2} \in \mathcal{U}$ and x_{n-1} and x_n are any points on Minkowski spacetime.

Iterating this process for the point x_{n-2} and so on leads us to the conclusion that

$$< \zeta|\phi(x_1)\phi(x_2)\ldots\phi(x_n)|\mathcal{O} >= 0$$

when $x_1, \ldots, x_n \in \mathcal{U}$, it must also be true that

$$< \zeta|\phi(x_1)\phi(x_2)\ldots\phi(x_n)|\mathcal{O} >= 0 \qquad (7.90)$$

when $x_1, \ldots, x_n$ are any points on Minkowski spacetime.

Our assumption that the set of states $\phi(x_1)\phi(x_2)\ldots\phi(x_n)|\mathcal{O} >$ for all n and with $x_1, \ldots, x_n$ any points on Minkowski spacetime sweeps out a complete set of states then implies that equation (7.90) then implies that the state $|\zeta >$ is orthogonal to every state in the Hilbert space. This can only be true if

$$|\zeta >= 0$$

the zero vector in Hilbert space.

This then demonstrates that the states in equation (7.89) are "dense" in that any state in the Hilbert space can be approximated by a linear combination of them and gotten exactly by a sequence of such linear combinations with certain convergence properties.

The Reeh–Schlieder theorem has many implications the discussion of which would take us far beyond the scope of this monograph.

7.10 Conformal Symmetry

There is an interesting enhancement of Poincare symmetry which has some important applications in physics. This extended Poincare symmetry is called conformal symmetry. Quantum field theories which possess conformal symmetry are called conformal field theories.

Recall that a symmetry of space–time was defined as a coordinate transformation which leaves the metric unchanged, $\delta g_{\mu\nu}(x) = 0$. Writing this equation for an infinitesimal coordinate transformation led to the Killing equation whose solutions were Killing vectors which in turn are vector fields which generate the symmetries of spacetime. To look for conformal symmetries, we seek coordinate transformations under which the metric is not left unchanged, but under which it can change by a quantity which is proportional to the metric itself. The equation for an infinitesimal coordinate transformation of the metric has the form

$$\delta g_{\mu\nu}(x) = -\hat{f}^\lambda(x)\partial_\lambda g_{\mu\nu}(x) - \partial_\mu \hat{f}^\lambda(x)g_{\lambda\nu}(x) - \partial_\nu \hat{f}^\lambda(x)g_{\mu\lambda}(x)$$
$$= \Omega(x)g_{\mu\nu}(x) \tag{7.91}$$

which, specialized to Minkowski space, and noting that the equation itself determines $\Omega(x)$, gives the conformal Killing equation of Minkowski space,

$$\partial_\mu \hat{f}_\nu(x) + \partial_\nu \hat{f}_\mu(x) - \frac{2}{D}\eta_{\mu\nu}\partial_\lambda \hat{f}^\lambda(x) = 0 \tag{7.92}$$

where D is the dimension of spacetime and, as in the Killing equation for Minkowski space, $\hat{f}_\mu(x) \equiv \eta_{\mu\nu}\hat{f}^\nu(x)$. Solutions of the conformal Killing equation (7.92) are called conformal Killing vectors. Each linearly independent conformal Killing vector corresponds to a conformal symmetry of spacetime.

Every solution of the Killing equation must also be a solution of the conformal Killing equation. Therefore the Killing vectors which correspond to the Poincare symmetries are also conformal Killing vectors. Then, there can be more conformal Killing vectors which are not Killing vectors. In D dimensions, the maximum number of conformal Killing vectors is $(D + 1)(D + 2)/2$ which, for four dimensional Minkowski space, where $D = 4$, is equal to fifteen. Ten of them are the Killing vectors and there must be five more. It is straightforward to find them, so that the total list of conformal Killing vectors is

1. Translations: $\hat{f}^\mu = c^\mu$, constants
2. Lorentz transformations $\hat{f}^\mu(x) = \omega^\mu_{\ \nu}x^\nu$ with $\omega_{\nu\mu} = -\omega_{\mu\nu}$.
3. Scale transformations $\hat{f}^\mu(x) = x^\mu$
4. Conformal transformations $\hat{f}^\mu(x) = 2x^\mu b_\nu x^\nu - b^\mu x^2$ with b^μ a constant vector.

As for the finite transformations, it is clear that the scale transformation is simply

$$x^\mu \to \tilde{x}^\mu = \lambda x^\mu \tag{7.93}$$

for a real number λ. The finite conformal transformation is

$$x^\mu \to \tilde{x}^\mu = \frac{x^\mu - b^\mu x^2}{1 - 2bx + b^2 x^2} \approx x^\mu + 2x^\mu bx - b^\mu x^2 + \ldots \tag{7.94}$$

In order to study the conformal symmetry of a field theory we need to generalize our discussion of coordinate transformations a little. Consider the definition of the transformation of the field as

$$\delta\phi(x) = -\hat{f}^\mu(x)\partial_\mu \phi(x) - \frac{\Delta}{D}\partial_\nu \hat{f}^\nu(x)\phi(x) \tag{7.95}$$

where, in the last term, D is the spacetime dimension and Δ is a real number called the scaling dimension of the field. The transformation of the field in equation (7.95) still works for the Poincare symmetries since, for a Killing vector, $\partial_\mu \hat{f}^\nu(x) = 0$ and the transformation then matches the one that we have used up until now, that given in equation (6.63).

A scale transformation of the scalar field is then

$$\delta\phi(x) = \left(-x^\mu \partial_\mu - \Delta\right)\phi(x) \tag{7.96}$$

and the finite transformation is

$$\phi(x) \to \tilde{\phi}(x) = \lambda^{-\Delta}\phi(\lambda^{-1}x) \tag{7.97}$$

A simple example of a conformal field theory is the free massless scalar field theory with Lagrangian density

$$\mathcal{L}(x) = -\frac{1}{2}\partial_\mu\phi(x)\partial^\mu(x) \tag{7.98}$$

In this field theory, when the spacetime dimension is four, the field $\phi(x)$ has scaling dimension $\Delta = 1$. We can easily show that the modified transformation of the scalar field in equation (7.95) is a symmetry of the Lagrangian density in (7.98). However, there is a shorter way to an acceptable Noether current. We begin the stress tensor that we found for the scalar field in equation (7.12) above, specialized to the field theory with the Lagrangian density in equation (7.98)

$$\mathbf{T}_0^{\mu\nu}(x) = \partial^\mu\phi(x)\partial^\nu\phi(x) - \frac{1}{2}\eta^{\mu\nu}\partial_\lambda\phi(x)\partial^\lambda\phi(x) \tag{7.99}$$

This stress tensor is symmetric

$$\mathbf{T}_0^{\mu\nu}(x) = \mathbf{T}_0^{\nu\mu}(x)$$

It is conserved,

$$\partial_\mu\mathbf{T}_0^{\mu\nu}(x) = 0$$

whenever $\phi(x)$ obeys the equation of motion ,

$$-\partial^2\phi(x) = 0$$

that follows from the Lagrangian density in equation (7.98).

However the trace of $\mathbf{T}_0^{\mu\nu}(x)$ is not zero. If it were traceless, the ansatz for the Noether current for a scale transformation $\mathbf{T}^\mu{}_{\nu 0}(x)x^\nu$, for example, would be conserved. The idea that we will pursue here is to improve the tensor $\mathbf{T}_0^{\mu\nu}(x)$ by adding an automatically conserved, symmetric, total derivative term which renders it traceless. Consider the improvement

$$\mathbf{T}^{\mu\nu}(x) = \partial^{\mu}\phi(x)\partial^{\nu}\phi(x) - \frac{1}{2}\eta^{\mu\nu}\partial_{\lambda}\phi(x)\partial^{\lambda}\phi(x)$$

$$+ \frac{1}{3}\left(\eta^{\mu\nu}\partial^2 - \partial^{\mu}\partial^{\nu}\right)\left(\frac{1}{2}\phi^2(x)\right) \tag{7.100}$$

This tensor differs from $\mathbf{T}_0^{\mu\nu}(x)$ by the last term.

We have improved the stress tensor in such a way that it is now traceless,

$$\mathbf{T}^{\mu}_{\ \mu}(x) = 0 \tag{7.101}$$

Once we have a conserved, symmetric and traceless stress tensor, the current

$$\mathcal{J}^{\mu}_{\hat{f}}(x) = \mathbf{T}^{\mu\nu}(x)\hat{f}_{\nu}(x) \tag{7.102}$$

is conserved for any conformal Killing vector $\hat{f}^{\mu}(x)$.

The Noether charges corresponding to these Noether currents are

$$P^{\mu} = \int d^3x\, \mathbf{T}^{0\mu}(x) \tag{7.103}$$

$$M^{\mu\nu} = \int d^3x \left(\mathbf{T}^{0\mu}(x)x^{\nu} - \mathbf{T}^{0\nu}(x)x^{\mu}\right) \tag{7.104}$$

$$D = \int d^3x\, T^{0\mu}(x)x_{\mu} \tag{7.105}$$

$$K^{\mu} = \int d^3x\, \mathbf{T}^{0\lambda}(x)(x_{\lambda}x^{\mu} - \delta^{\mu}_{\lambda}x^2/2) \tag{7.106}$$

In fact, we can show that the improvement has no effect on P^{μ} or $M^{\mu\nu}$. These quantities are as they were before, when they were constructed from $\mathbf{T}_0^{\mu\nu}(x)$. In addition, we have the additional generators, D of dilitations and K^{μ} of conformal transformations.

These Noether charges $P^{\mu}, M^{\mu\nu}, D, K^{\mu}$ are guaranteed to be independent of time. As a result, the time argument in their integrands can be set equal to a value of convenience. This is useful when we study the commutation relations of these quantities with the fields and with each other. It is possible to use the equal-time commutation relations for the field $\phi(x)$ and its first time derivative to show that the Noether charges generate the transformations in the sense that

$$\frac{1}{i}\left[\phi(x),\, P^{\mu}\right] = \partial^{\mu}\phi(x) \tag{7.107}$$

$$\frac{1}{i}\left[\phi(x),\, M^{\mu\nu}\right] = \left(x^{\mu}\partial^{\nu} - x^{\nu}\partial^{\mu}\right)\phi(x) \tag{7.108}$$

$$\frac{1}{i}[\phi(x),\, D] = (x^{\mu}\partial_{\mu} + 1)\phi(x) \tag{7.109}$$

$$\frac{1}{i}\left[\phi(x),\, K^{\mu}\right] = \left[(x^{\mu}x_{\lambda} - \delta^{\mu}_{\lambda}x^2/2)\partial^{\lambda} + x^{\mu}\right]\phi(x) \tag{7.110}$$

The first two of the equations above are identical to what we had for the Poincare invariant field theory in the previous section in equation (7.23) and they should indeed be present in any Poincare invariant quantum field theory, with the appropriate Poincare transformations of the fields on the right-hand-sides. The second two of the above equations are the extension of the Poincare transformations to the full conformal transformations. The transformations are generated by the Noether charges Δ and K^{μ}. In this quantum field theory, and in many other theories, once the theory is Poincare and scale invariant, it is also conformal invariant and there exists a conserved charge K^{μ} which generates the conformal transformation of the field.

Moreover, the commutators of the generators give an extension of the Poincare algebra, called the conformal algebra

$$\frac{1}{i}\left[P^{\mu},\, P^{\nu}\right] = 0 \tag{7.111}$$

$$\frac{1}{i}\left[M^{\mu\nu},\, P^{\lambda}\right] = \eta^{\mu\lambda}P^{\nu} - \eta^{\nu\lambda}P^{\mu} \tag{7.112}$$

$$\frac{1}{i}\left[M^{\mu\nu},\, M^{\rho\sigma}\right] = \eta^{\mu\rho}M^{\nu\sigma} - \eta^{\mu\sigma}M^{\nu\rho} - \eta^{\nu\rho}M^{\mu\sigma} + \eta^{\nu\sigma}M^{\mu\rho} \tag{7.113}$$

$$\frac{1}{i}\left[M^{\mu\nu},\, K^{\lambda}\right] = \eta^{\mu\lambda}K^{\nu} - \eta^{\nu\lambda}K^{\mu}, \quad \frac{1}{i}\left[M^{\mu\nu},\, D\right] = 0 \tag{7.114}$$

$$\frac{1}{i}\left[D,\, P^{\mu}\right] = -P^{\mu}, \quad \frac{1}{i}\left[D,\, K^{\mu}\right] = K^{\mu}, \quad \frac{1}{i}\left[K^{\mu},\, P^{\nu}\right] = 2D\eta^{\mu\nu} + M^{\mu\nu} \tag{7.115}$$

Again, equations (7.111)–(7.113) are as they were in the Poincare invariant field theory, quoted in equations (7.24)–(7.26) of the previous section and some new commutation relations (7.114)–(7.115) involving the scale and conformal generators have been added. All together, equations (7.111)–(7.115) are called the conformal algebra.

The additional presence of conformal symmetry allows a different approach to the solution of the field theory, using the techniques of conformal field theory. The details of how to proceed are very interesting but they are beyond what we are prepared to discuss here. We will only add a few comments as a motivation for further study of this fascinating subject.

We will generally use the term "operators" to describe products of the basic field $\phi(x)$ with itself and its derivatives all evaluated at the same spacetime point. Examples are[5]

$$\phi(x), :\phi^2(x):, \partial\phi(x), \partial\phi(x)\phi(x), :\phi^3(x):, \ldots$$

We can classify operators as to how they transform under Poincare transformations as scalar, vector or tensor operators. Once we organize the operators in this way, some operators have a simple commutator with the generators of scale and conformal transformations. Such an operator, for example, a scalar operator $O_\Delta(x)$, obeys a generalization of equations (7.109) and (7.110) to

$$\frac{1}{i}[O_\Delta(x), D] = (x^\mu \partial_\mu + \Delta)O_\Delta(x) \tag{7.116}$$

$$\frac{1}{i}\left[O_\Delta(x), K^\mu\right] = \left[(x^\mu x_\lambda - \delta^\mu_\lambda x^2/2)\partial^\lambda + \Delta x^\mu\right]O_\Delta(x) \tag{7.117}$$

where Δ is a constant. Operators that transform in this way are said to be scalar "primary operators" with dimension Δ. We can easily see that in our free scalar field theory, $\phi(x)$ and $:\phi^2(x):$ are examples of such operators with dimensions $\Delta = 1$ and $\Delta = 2$, respectively. There are called scalar operators since, under Poincare transformations, they transform as scalar fields. A slight generalization of the definition in equations (7.116) and (7.117) to consider operators which Poincare transform like vector or tensor fields is straightforward.

As an example of the power of scale symmetry, we recall that an important consequence of Poincare symmetry was the fact that a 2 point function of scalar fields $<O|TO_\Delta(x_1)O_\Delta(x_2)|O>$ can depend on the coordinates only in the one combination $(x_1 - x_2)^2$. Scale invariance has further implications. If we assume that the scale transformation generator annihilates the vacuum, $D|O>= 0$, and take into account the scale transformation of the scalar primary field $O_\Delta(x)_1$ we can show, using equation (7.116) that scale invariance requires

$$<O|TO_\Delta(x_1)O_\Delta(x_2)|O> \; = \; \frac{c(\Delta_1, \Delta_2)}{[(x-y)^2]^{\frac{\Delta_1+\Delta_2}{2}}}$$

The 2 point function is determined up to an overall constant. If, in addition to scale symmetry, $D|O>= 0$, we assume conformal symmetry, $K^\mu|O>= 0$, we can use equation (7.117) to show that

$$<O|TO_\Delta(x)_1 O_\Delta(x_2)|O> \; = \; \frac{c(\Delta_1)\delta_{\Delta_1,\Delta_2}}{[(x-y)^2]^{\frac{\Delta_1+\Delta_2}{2}}} \tag{7.118}$$

---

[5] For an operator like $\phi^2(x)$ we must define it as $\phi^2(x) - <O|\phi^2(x)|O>$. We have used the notation $:\phi^2(x):$ for this operator since in the free field theory that we are considering the subtraction is equivalent to normal ordering the creation and annihilation operators contained in the fields.

that is, the 2 point function is nonzero only if the dimensions of the primary operators are equal.

Moreover, conformal symmetry also implies something for 3 point functions. If we study 3 point functions of scalar primary operators, scale and conformal symmetry implies

$$
< \mathcal{O}|O_{\Delta_1}(x_1)O_{\Delta_2}(x_2)O_{\Delta_3}(x_3)|\mathcal{O} >=
$$
$$
\frac{\lambda(\Delta_1, \Delta_2, \Delta_3)}{(x_1 - x_2)^{\frac{\Delta_1+\Delta_2-\Delta_3}{2}} (x_1 - x_3)^{\frac{\Delta_1+\Delta_3-\Delta_2}{2}} (x_2 - x_3)^{\frac{\Delta_2+\Delta_3-\Delta_1}{2}}} \tag{7.119}
$$

3 point functions of primary operators are fixed up to one overall constant which is a function of the dimensions of the three operators involved. Again, this expression has straightforward generalizations to vector and tensor operators.

In this brief foray into conformal field theory, we have seen in equations (7.118) and (7.119), some of the implications of conformal symmetry. We will leave this as a teaser intended to motivate the reader to a more intensive study of this subject that is far beyond the scope of this monograph.

Emergent Relativistic Symmetry

8

The relativistic symmetry of Minkowski spacetime which we discussed in the previous chapter is usually used to formulate fundamental theories of the elementary particles of nature. The reason is that, as far as elementary particle physics experiments to date are concerned, Poincare invariance seems to be a basic symmetry of nature.

In this chapter we will examine a different source of relativistic symmetry in a quantum field theory, the case where the symmetry is emergent. Emergent symmetry occurs when a subset of the set of all of the dynamical variables in a quantum mechanical system turns out to have different symmetries than the system as a whole. Of course, that subset of the degrees of freedom should also be an important one for the physical properties of the system. Moreover, it must be isolated from the remainder of the system so that its symmetries are not violated by its interactions with that remainder. In as much as it can be isolated, an approximate description of these degrees of freedom can be as a dynamical theory which exhibits the emergent symmetry.

In this chapter, we will outline a few examples of emergent relativistic symmetry. One of our interests in these theories is the fact that it broadens the applications of relativistic quantum field theory from those few very specific models which describe the known elementary particles to other types of quantized fields which are applicable to a wider variety of physical problems. There is also the possibility that the relativistic symmetry that elementary particles seem to possess is itself emergent from a deeper and as yet unknown structure and it is desirable to have some concrete examples of how this might occur.

© The Author(s), under exclusive license to Springer Nature Singapore Pte Ltd. 2023   137
G. W. Semenoff, *Quantum Field Theory*, Graduate Texts in Physics,
https://doi.org/10.1007/978-981-99-5410-0_8

8.1   Phonons

Phonons are the quanta of the vibrations of a solid whose constituent atoms occupy the sites of a lattice. A lattice is a regular array of discrete points. The simple example that we will use is a cubic lattice which has sites at the spatial points

$$\vec{x}_n = n_1 \tilde{c} \hat{e}^1 + n_2 \tilde{c} \hat{e}^2 + n_3 \tilde{c} \hat{e}_3 \tag{8.1}$$

where (n_1, n_2, n_3) are integers, $\hat{e}^1, \hat{e}^2, \hat{e}^3$ are the unit vectors which are parallel to each of the axes of the Cartesian coordinate system and $\tilde{c}$ is the lattice constant, or lattice spacing, the distance between neighbouring sites of the lattice. As each of the n_a sweep over all of the integers, the vectors $\vec{x}_n$ sweep over the Cartesian coordinates of the sites of the infinite cubic lattice.

Now we shall consider a monatomic solid with a single type of atoms occupying the sites of the infinite cubic lattice that we have been discussing. We will take the idealization where each atom interacts with its nearest neighbours only. We will also assume that, when an atom moves away from its equilibrium position, the force on it due to each neighbouring atom is linearly proportional to the distance between it and that atom. This force can be derived from a quadratic potential energy. We shall denote the position of the atom whose equilibrium position would be $\vec{x}_n$ by

$$\vec{x}_n + \vec{\phi}(\vec{x}_n, t)$$

The functions $\vec{\phi}(\vec{x}_n, t)$ which denote the displacement of each atom from it equilibrium position, will be the dynamical variable that we are interested in. We will make the assumption that these displacements from equilibrium are small, that is, $|\vec{\phi}(\vec{x}_n, t)| \ll \tilde{c}$. Then, our model for the total potential energy of the solid assumes that it is the sum of the potential energies that are stored in the interaction of each pair of atoms and that, for each pair, the potential energy is simply proportional to the square of the distance between the atoms,

$$\mathcal{U} = \sum_{n_1,n_2,n_3=-\infty}^{\infty} \sum_{a=1}^{3} \frac{\kappa}{2} \left(\left(x_n + \tilde{c} \hat{e}^a + \vec{\phi}(x_n + \tilde{c} \hat{e}^a, t) \right) - \left(x_n + \vec{\phi}(x_n, t) \right) \right)^2 \tag{8.2}$$

Here, we have taken the potential energy of a bond of length x to be $\frac{\kappa}{2} x^2$. We can simplify the potential energy slightly to get

$$\mathcal{U} = 3V \frac{\kappa}{2} \tilde{c}^2 + \sum_{n_1,n_2,n_3=-\infty}^{\infty} \tilde{c}^3 \sum_{a=1}^{3} \frac{\kappa}{2\tilde{c}} \left(\frac{\vec{\phi}(x_n + \tilde{c} \hat{e}^a, t) - \vec{\phi}(x_n, t)}{\tilde{c}} \right)^2 \tag{8.3}$$

where V is the (infinite) volume of the system. It is easy to see that this potential energy has a minimum which is given by the trivial configuration $\vec{\phi}(x_n, t) = 0$.[1] Now, let us consider the kinetic energy which must be the sum of the kinetic energies of each of the atoms,

$$\mathcal{K} = \sum_{n_1,n_2,n_3=-\infty}^{\infty} \frac{M}{2} \dot{\vec{\phi}}(\vec{x}_n, t)^2 \tag{8.4}$$

Here, M is the mass of each atom and its velocity is the time derivative of its position. The Hamiltonian is the sum of the kinetic and potential energies,

$$H = \mathcal{K} + \mathcal{U} \tag{8.5}$$

$$= \sum_{n_1,n_2,n_3=-\infty}^{\infty} \tilde{c}^3 \left\{ \frac{M}{2\tilde{c}^3} \dot{\vec{\phi}}(\vec{x}_n, t)^2 + \sum_{a=1}^{3} \frac{\kappa}{2\tilde{c}} \left(\frac{\vec{\phi}(x_n + \tilde{c}\hat{e}^a, t) - \vec{\phi}(x_n, t)}{\tilde{c}} \right)^2 \right\} \tag{8.6}$$

The mechanical momentum of an atom is equal to its mass times its velocity, $M\dot{\vec{\phi}}(\vec{x}_n, t)$. To quantize this system, we impose a canonical commutation relation on the positions and the momenta of the atoms, that is, we require

$$\left[\phi^i(\vec{x}_n, t), \dot{\phi}^j(x_{n'}, t) \right] = i\hbar \frac{\tilde{c}^3}{M} \frac{\delta_{n_1 n_1'} \delta_{n_2 n_2'} \delta_{n_3 n_3'}}{\tilde{c}^3} \tag{8.7}$$

$$\left[\phi^i(\vec{x}_n, t), \dot{\phi}^j(x_{n'}, t) \right] = 0, \quad \left[\phi^i(\vec{x}_n, t), \dot{\phi}^j(x_{n'}, t) \right] = 0 \tag{8.8}$$

Now, we are going to assume that the physical effects that we wish to describe take place on length and distance scales that are much larger than the lattice spacing $\tilde{c}$, so that, compared to these distance scales, $\tilde{c}$ is effectively zero. We recognize in our equations above, as $\tilde{c}$ is taken to zero, the definition of the Riemann integral, the definition of derivative and the definition of the Dirac delta function,

---

[1] There is, of course, a lower energy minimum of the potential where $\vec{\phi}(x_n) = -\vec{x}_n$, the entire lattice collapsing to a point, but it violates our assumption that $|\vec{\phi}| << \tilde{c}$. We would assume that when this assumption is violated, other interactions set in which stabilize the lattice configuration. It would also be more realistic to allow the equilibrium bond-length to be nonzero, so that the potential energy would have the form $\frac{\kappa}{2}(x - x_0)^2$. This elaboration, or in fact practically any more realistic estimate of a potential energy function than the one that we have chosen, will not give us a relativistic field theory. We will discuss this once we have found the emergent field theory due to the potential energy function (8.2).

$$\sum_{n_1,n_2,n_3=-\infty}^{\infty} \tilde{c}^3 \implies \int d^3x \tag{8.9}$$

$$\frac{\vec{\phi}(x_n + \tilde{c}\hat{e}^a, t) - \vec{\phi}(x_n, t)}{\tilde{c}} \implies \nabla^a \vec{\phi}(\vec{x}, t) \tag{8.10}$$

$$\frac{\delta_{n_1 n_1'} \delta_{n_2 n_2'} \delta_{n_3 n_3'}}{\tilde{c}^3} \implies \delta^3(\vec{x} - \vec{x}') \tag{8.11}$$

If we define

$$\phi(\vec{x}, t) = \sqrt{\frac{\tilde{c}^3}{M}} \vec{\phi}(\vec{x}, t) \tag{8.12}$$

and the parameter

$$c = \sqrt{\frac{\kappa \tilde{c}^2}{M}} \tag{8.13}$$

we find a Hamiltonian

$$H = \int d^3x \left\{ \frac{1}{2}\dot{\vec{\phi}}(\vec{x}, t)^2 + \frac{c^2}{2} \vec{\nabla}\phi^i(\vec{x}, t) \cdot \vec{\nabla}\phi^i(\vec{x}, t) \right\} \tag{8.14}$$

and the equal time commutation relations

$$\left[\phi^i(\vec{x}, t), \dot{\phi}^j(\vec{y}, t) \right] = i\hbar \delta^{ij} \delta^3(\vec{x} - \vec{x}') \tag{8.15}$$

$$\left[\phi^i(\vec{x}, t)\phi^j(\vec{y}, t) \right] = 0, \quad \left[\dot{\phi}^i(\vec{x}, t), \dot{\phi}^j(\vec{y}, t) \right] = 0 \tag{8.16}$$

The Hamilton equations of motion are

$$\dot{\phi}^i(\vec{x}, t) = \frac{1}{i\hbar} \left[\phi^i(\vec{x}, t), H \right], \quad \ddot{\phi}^i(\vec{x}, t) = \frac{1}{i\hbar} \left[\dot{\phi}^i(\vec{x}, t), H \right] \tag{8.17}$$

The first of these equations confirms the identification of the momentum operator and the second of them yields a wave equation for the field,

$$\left(\frac{1}{c^2} \frac{\partial_2}{\partial t^2} - \vec{\nabla}^2 \right) \phi^i(\vec{x}, t) = 0 \tag{8.18}$$

This is a relativistic wave equation where c is the emergent speed of light. At this point, we might note that this theory is identical to three copies of the scalar field which we studied in some detail in the previous chapter. In fact, in the approximation where we are studying it, the interaction and mass terms in the field equation are absent and this is just three copies of the massless free scalar field which we used in the previous chapter as a simple example of a field theory with conformal symmetry. This is indeed the case. For clarity, we will continue to review some of the details which are important for our discussion of phonons.

The wave equation (8.18) has plane wave solutions

$$\exp\left(i\vec{k}\cdot\vec{x}-ic|\vec{k}|t\right), \quad \exp\left(i\vec{k}\cdot\vec{x}+ic|\vec{k}|t\right)$$

where $\vec{k}$ is the wave-number and $c|\vec{k}|$ or $-c|\vec{k}|$ is the frequency. Note that there is both a positive and negative frequency solution and, to have a complete set of states, we shall need to make use of both of them. We saw something similar to this when we studied non-relativistic fermions. There were positive frequency modes with energies above the Fermi surface (with $\frac{\hbar^2\vec{k}^2}{2m}-\epsilon_F > 0$) which we identified as particles and then there were negative frequency modes with energies below the Fermi surface (with $\frac{\hbar^2\vec{k}^2}{2m}-\epsilon_F < 0$). We relabeled the negative frequency particles as positive frequency holes, and in the quantum field theory they indeed ended up corresponding to positive energy excitations.

Here, in our phonon theory, the phonons are bosons and we have a different use for the negative frequency modes. We use them in equation (8.19) below to make the field operator Hermitian, $\phi^{i\dagger}(\vec{x},t)=\phi^i(\vec{x},t)$. Of course, the phonon field corresponds to a physical displacement of atoms and spatial displacement is a real variable. A real classical variable, one that satisfies $\phi^{i*}(\vec{x},t)=\phi^i(\vec{x},t)$ at the classical level, should correspond to a Hermitian operator in the quantum theory.

It is straightforward to find the general solution of the wave equation by superposing plane waves,

$$\phi^i(\vec{x},t)=\int\frac{d^3k}{\sqrt{(2\pi)^3 2c|\vec{k}|/\hbar}}\left[e^{i\vec{k}\cdot\vec{x}-ic|\vec{k}|t}a^i(\vec{k})+e^{-i\vec{k}\cdot\vec{x}+ic|\vec{k}|t}a^{i\dagger}(\vec{k})\right] \quad (8.19)$$

where we have fixed the relative phases of the first and second term so that the phonon field is Hermitian, $\phi^{i\dagger}(\vec{x},t)=\phi^i(\vec{x},t)$.

The creation and annihilation operators $a^{i\dagger}(\vec{k})$ and $a^i(\vec{k})$ obey the commutation relations

$$\left[a^i(\vec{k}),a^{j\dagger}(\vec{k}')\right]=\delta^{ij}\delta^3(\vec{k}-\vec{k}') \quad (8.20)$$

$$\left[a^i(\vec{k}),a^j(\vec{k}')\right]=0, \quad \left[a^{i\dagger}(\vec{k}),a^{j\dagger}(\vec{k}')\right]=0 \quad (8.21)$$

The operators $a^{i\dagger}(\vec{k})$ and $a^i(\vec{k})$ are creation and annihilation operators for elementary excitations of the lattice which are called phonons. The Hilbert space of phonon states begins with a vacuum which obeys

$$a^i(\vec{k})|\mathcal{O}>=0 \quad <\mathcal{O}|a^{i\dagger}(\vec{k})=0 \; \forall i,\vec{k}, \quad <\mathcal{O}|\mathcal{O}>=1 \quad (8.22)$$

The vacuum is simply that state of the lattice system where none of the atoms are displaced at all. States with coherent small displacements of the atoms are the phonon quantum states that are created by operating creation operators on the vacuum,

$$\left\{ |\mathcal{O}>, a^{i\dagger}(\vec{k})|\mathcal{O}>, \ldots, \frac{1}{\sqrt{n!}} a^{i_1\dagger}(\vec{k}_1) \ldots a^{i_n\dagger}(\vec{k}_n)|\mathcal{O}>, \ldots \right\} \qquad (8.23)$$

The state vectors are the basis vectors for the space of quantum states of our theory of phonons.

The energies of these states are gotten by plugging the solution of the equation of motion into the Hamiltonian operator (8.14) to obtain

$$H = \sum_{i=1}^{3} \int d^3k \; \hbar c |\vec{k}| \; a^{i\dagger}(\vec{k}) a^i(\vec{k}) \; + \; E_0 \qquad (8.24)$$

where the ground state (or zero point) energy comes from arranging the creation and annihilation operators into the order given in (8.24),

$$E_0 = V u_0, \quad u_0 = \frac{3}{2} \int d^3k \; \hbar |c\vec{k}| \qquad (8.25)$$

with V the volume of space and where u_0 is the vacuum energy density. Note that u_0 is given by a divergent integral. We will deal with this shortly.

The energy of a single phonon is given by the relativistic dispersion relation

$$E(k) = \hbar c |\vec{k}| \qquad (8.26)$$

The multi-phonon states are eigenstates of the Hamiltonian

$$H \frac{1}{\sqrt{n!}} a^{i_1\dagger}(\vec{k}_1) \ldots a^{i_n\dagger}(\vec{k}_n)|\mathcal{O}> = E \frac{1}{\sqrt{n!}} a^{i_1\dagger}(\vec{k}_1) \ldots a^{i_n\dagger}(\vec{k}_n)|\mathcal{O}> \qquad (8.27)$$

$$E = \hbar c |\vec{k}_1| + \ldots + \hbar c |\vec{k}_n| + E_0 \qquad (8.28)$$

The eigenvalue of the Hamiltonian is simply given by the sum of the energies of the phonons.

The field equation in (8.18) and the equal-time commutation relations in (8.15) and (8.16) define a quantum field theory. This quantum field theory is very similar to the one which we studied in the previous chapter in the context of spacetime symmetries. They can be derived from an action and Lagrangian density

$$S = \int dt d^3x \mathcal{L}(\vec{x}, t) \qquad (8.29)$$

$$\mathcal{L}(x) = \frac{1}{2} \frac{\partial \phi^i(\vec{x}, t)}{\partial t} \frac{\partial \phi^i(\vec{x}, t)}{\partial t} - \frac{c^2}{2} \nabla_a \vec{\phi}(\vec{x}, t) \cdot \nabla_a \vec{\phi}(\vec{x}, t) \qquad (8.30)$$

where, in this case, the Euler–Lagrange equation is

$$\frac{\partial \mathcal{L}(\vec{x},t)}{\partial \phi^i(\vec{x},t)} = \frac{\partial}{\partial t}\frac{\partial \mathcal{L}(\vec{x},t)}{\partial(\partial \phi^i(\vec{x},t)/\partial t)} + \nabla_a \frac{\partial \mathcal{L}(\vec{x},t)}{\partial(\nabla_a \phi^i(\vec{x},t))} \tag{8.31}$$

Here, we are using the fact that the scalar field has three components to write it as a vector $\vec{\phi}$. Of course, the displacements which we began our discussion of phonons with, and which become the field $\vec{\phi}(x)$ were spatial vectors. Under a rotation of the spatial coordinates, they should transform like a vector. However, when they appear as they do in the Lagrangian density (8.30), there are two symmetries, one is a simple rotation of the components of $\vec{\phi}$, without a rotation of the coordinates, and another is a rotation of the coordinates, without a mixing of the components of $\vec{\phi}$.

The emergence of these two independent symmetries is a symptom of the rather special assumptions that went into the model that we used at the lattice level, for example, the quadratic potentials being of the form $\kappa \vec{x}^2/2$, rather than a more realistic $\kappa(|\vec{x}| - x_0)^2/2$ which has a minimum energy at separation, x_0. If we had used this more realistic term, the Lagrangian density that we have found would have the additional term

$$-\frac{a}{2}\nabla_a \phi^a(\vec{x},t)\nabla_b \phi^b(\vec{x},t) \tag{8.32}$$

The effect of such a term is to make the speed of longitudinal sound waves— displacements of the atoms are in the direction of the propagation of the sound wave— different from the speed of the transverse sound waves—where the displacement of the atoms is perpendicular to the direction of propagation of the wave. In typical monatomic solids at room temperature, the speeds of these two kinds of sound waves are indeed different. The typical difference is between 30 and 50 percent, so it can be significant. As we are interested in emergent relativistic symmetry here, for now, we will ignore the difference and corrections to the Lagrangian density of the form (8.32). Then, the classical and quantum field theory described by the Lagrangian density (8.30) has emergent relativistic symmetry.

8.2   The Debye Theory of Solids

The quantum field theory whose Lagrangian density is described by (8.30) is our model for the long wave-length modes of lattice vibrations. It has one parameter, the emergent speed of light, c, which is given by the basic properties of the system, the mass, M, of the atoms, the spring constant, κ, and the lattice constant, $\tilde{c}$. One might wonder whether this model actually describes the physics of a material. Indeed, the phonons that it describes contribute to the low temperature properties of solids. This was first realized by Debye who succeeded in computing the phonon contribution to the low temperature specific heat of a solid. Indeed, if the solid is an insulator, it gives a reasonable estimate of this quantity, since the lowest energy excitations of an insulating regular solid are typically phonons.

A qualitative derivation of Debye's result is easy. In natural units, the scale invariant field theory describing the phonons has no dimensionful constants at all. Therefore, if we introduce temperature, that temperature would be the only constant with dimensions. Dimensional analysis tells us that the internal energy per unit volume, which has dimensions of energy times the inverse of volume, must scale as the inverse of length to the fourth power. It must therefore depend on temperature as

$$\frac{U}{V} = cT^4$$

where c is a dimensionless constant. The specific heat at constant volume is given by the derivative

$$c_V = \frac{\partial}{\partial T} \frac{U}{V}\bigg|_V = 3cT^3$$

which is cubic in the temperature. The cubic in temperature dependence of this specific heat, which is observed in many solids, was the great success of Debye's idea that quantized phonons are dominantly responsible for the low temperature thermodynamic properties of some solids.[2]

Debye's simple model is slightly more sophisticated than the dimensional analysis which we have discussed above. It introduces a cutoff of the high energy and wave-number components of the phonon field. The purpose of this cutoff is to get the continuum field theory to count the total number of degrees of freedom correctly. Once there is a large wave-number (or large frequency and energy) cutoff, we would expect that the cubic dependence of the specific heat that we have discussed above is still good in the limit where the temperature is much smaller than the cutoff, or in the limit as the cutoff is taken to be infinite. On the other hand, once the temperature is of the same magnitude or even greater than the cutoff, the result should be different. Also, by the general principle of equipartition of energy, we know what the internal energy should be, it should be one unit of $k_B T$ for each degree of freedom. The high temperature limit of the internal energy should then be

$$U = 3k_B T \, N, \quad N = \rho V \quad \rightarrow \quad u = 3\rho k_B T$$

and the specific heat should go to a constant in the high temperature limit,

$$c_V = \frac{\partial u}{\partial T}\bigg|_V = 3\rho k_B$$

Debye's cutoff is determined by a computation of the number of phonon degrees of freedom by recalling the beginnings of the model where the degrees of freedom were

---

[2] These were generally insulating materials. Conductors have free electrons and the low temperature specific heat due to the electrons is linear in the temperature and it is the dominant contribution in conducting solids at low temperatures.

simply one three-dimensional harmonic oscillator per site, so that the total number of oscillators is

$$3\rho V = 3V \int \frac{d^3k}{(2\pi)^3} \theta(k_D^2 - \vec{k}^2) = \frac{3V}{(2\pi)^3} \frac{4}{3}\pi k_D^3$$

where ρ is the number of atoms per unit volume. On the left-hand side of the above equation we have a counting of the total number of oscillators per unit volume times the volume. On the right-hand-side we have the estimate of that same quantity in the continuum field theory where there is one degree of freedom for each cell of phase space of volume $dkdx/2\pi$. To make the number finite, k_D is introduced. It is a cutoff with units of inverse distance. It is called the Debye wavenumber. It can be converted to a frequency by multiplying by the speed of sound to get ck_D and to an energy by further multiplying by $\hbar$ to get the Debye energy $\hbar ck_D$. The above integral tells us that

$$k_D = \left(6\pi^2\rho\right)^{\frac{1}{3}}$$

Then, with this cutoff, we can calculate the thermodynamic internal energy

$$U = 3\int d^3x \int \frac{d^3k}{(2\pi)^3} \theta(k_D^2 - \vec{k}^2)\hbar c|\vec{k}| \frac{1}{e^{\hbar c|\vec{k}|/k_B T} - 1}$$

where we have introduced the thermal distribution function $\frac{1}{e^{\hbar c|\vec{k}|/k_B T} - 1}$ which gives the expectation of the number of excitations with energy $\hbar c|\vec{k}|$ when the excitations are bosons, the temperature is T and k_B is the Boltzmann constant. The equation for the internal energy can be re-organized to get

$$U = V\frac{3\hbar c}{2\pi^2}\left(\frac{k_B T}{\hbar c}\right)^4 \mathcal{D}\left(\frac{\hbar c\left(6\pi^2\rho\right)^{\frac{1}{3}}}{k_B T}\right)$$

where

$$\mathcal{D}(x) = \int\limits_0^x dk \frac{k^3}{e^k - 1}$$

is related to the third Debye function. It has the asymptotic expansions

$$\mathcal{D}(x) = \frac{x^3}{3} - \frac{x^4}{8} + \dots, \quad x << 1$$
$$\mathcal{D}(x) = 6\zeta(4) + \mathcal{O}(x^3 e^{-x}), \quad x >> 1$$

where $\zeta(s) = \sum_1^\infty n^{-s}$ is Riemann's zeta function and $\zeta(4) = \pi^4/90$.

With these asymptotics, the internal energy density is

$$\frac{U}{V} = \frac{9\zeta(4)}{\pi^2(\hbar c)^3}(k_B T)^4 + \mathcal{O}\left(\frac{\hbar c\left(6\pi^2\rho\right)^{\frac{1}{3}}}{k_B T}\right)^3 \exp\left(-\frac{\hbar c\left(6\pi^2\rho\right)^{\frac{1}{3}}}{k_B T}\right) \quad \text{(low } T)$$

$$\frac{U}{V} = 3\rho k_B T + \mathcal{O}(\text{constant}) \quad \text{(high } T)$$

The low temperature limit agrees with our qualitative argument which invoked scale invariance where the internal energy density must vary with temperature like T^4. It also gives us the coefficient of the T^4 behaviour which can be taken as the prediction of the Debye model in this result. The high temperature limit simply agrees with what is expected from the equipartition of energy in classical thermodynamics, every degree of freedom carries internal energy $k_B T$, so the internal energy density is $\frac{U}{V} \to 3\rho k_B T$.

The specific heat per unit volume at constant volume is given by the thermodynamic formula

$$c_V = \frac{\partial}{\partial T}\frac{U}{V}\bigg|_V$$

from which we find the low and high temperature limits as

$$c_V = \frac{\pi^2 k_B^4}{10(\hbar c)^3}T^3 + \dots \qquad\qquad T \ll \frac{\hbar c\left(6\pi^2\rho\right)^{\frac{1}{3}}}{k_B}$$

$$c_V = 3\rho k_B + \dots \qquad\qquad T \gg \hbar c\left(6\pi^2\rho\right)^{\frac{1}{3}}/k_B$$

As we have discussed, the cubic temperature dependence of the low temperature limit is the achievement of the Debye theory. We also get a value for the constant in front of T^3 which is a prediction of the theory. That constant contains only one parameter, the speed of sound, c. The slight elaboration of the model which we will discuss below introduces another parameter, the difference between the speeds of longitudinal and transverse sound waves. With those two parameters, the computation of c_V in the Debye theory gives a beautiful fit to the low temperature specific heat of a number of solids.

Finally, we comment that, if we had considered a more accurate model of the solid, the Lagrangian density which we would have derived would have the form

$$\mathcal{L}(x) = \frac{1}{2}\dot{\phi}^a\dot{\phi}^a - \frac{c^2}{2}\nabla_a\phi^b\nabla_a\phi^b - \frac{a}{2}\nabla_a\phi^a\nabla_b\phi^b \tag{8.33}$$

This field theory has spacetime translation invariance, spatial rotation invariance where

$$\delta\phi^a = (x^c\nabla^d - x^d\nabla^c)\phi^a(x) + \delta^{ca}\phi^d(x) - \delta^{da}\phi^c(\vec{x})$$

This transformation must now simultaneously rotate the components of the field ϕ^a into each other as well as the space coordinates. Finally, it has scale invariance, that is, symmetry under the transformation

$$\delta\phi^a(x) = (x^\mu\partial_\mu + 1)\phi^a(x)$$

In fact, it decomposes into two relativistic field theories which do not interact with each other and which have differing speeds of light. The Lagrangian density can be written

$$\mathcal{L}(x) = \frac{1}{2}\dot{\phi}_T^a\dot{\phi}_T^a - \frac{c^2}{2}\nabla_a\phi_T^b\nabla_a\phi_T^b + \frac{1}{2}\dot{\phi}_L^a\dot{\phi}_L^a - \frac{c^2+a}{2}\nabla_a\phi_L^a\nabla_b\phi_L^b \qquad (8.34)$$

where the transverse and longitudinal parts of the phonon field obey the constraints

$$\vec{\nabla}\times\vec{\phi}_L(x) = 0, \quad \vec{\nabla}\cdot\vec{\phi}_T(x) = 0 \qquad (8.35)$$

The two polarizations of the transverse phonons are relativistic massless scalar fields which travel at the speed of light, c and the one longitudinal polarization is a relativistic massless scalar field with speed of light $\sqrt{c^2+a}$. They are separately Lorentz invariant but any interaction between them would violate symmetry under Lorentz boosts and conformal transformations, but not under spacetime translations, spatial rotations and scale transformations.

8.3   Relativistic Fermions in Graphene

Now, let us turn to another system with emergent special relativity. This occurs in the physics of graphene. Graphene has the remarkable property that, when its valence electrons interact with the graphene lattice, they take on what is for all intents and purposes a relativistic spectrum and they have relativistic dynamics. To a first approximation, they obey a continuum Dirac equation which is appropriate to the two-dimensional material that they inhabit. In the following we will give a basic outline for how this comes about.

Graphene is a planar material consisting of a single sheet of Carbon atoms bound to each other to form a two-dimensional hexagonal lattice. Each Carbon atom has four valence electrons. In each atom, three of the four valence electrons are used to form strong covalent bonds with the three nearest neighbours in the lattice. The fourth electron is less strongly bound and it becomes the degree of freedom which determines the electronic properties of graphene.

The hexagonal lattice of graphene is depicted in Fig. 8.1. It is a combination of two triangular lattices, the A and B sub-lattices, depicted by the black and the white dots in Fig. 8.1, respectively. All of the nearest neighbours of a site on the A sub-lattice are equidistant from it and they are on the B sub-lattice. If we label the positions of the sites of the A sub-lattice as the two-dimensional vectors, $\vec{A}$, the points of the B sub-lattice which are the nearest neighbours of A are $\vec{A}+\vec{s}_1$, $\vec{A}+\vec{s}_2$ and $\vec{A}+\vec{s}_3$.

Fig. 8.1 The hexagonal
lattice of graphene is
depicted with triangular A
and B sub-lattices denoted
by the black and the white
dots, respectively. The
vectors which connect the
sublattices are labeled as
$\vec{s}_1, \vec{s}_2$ and $\vec{s}_3$. Basis vectors
for the A sub-lattice can be
taken to be $\vec{s}_3 - \vec{s}_1$ and
$\vec{s}_2 - \vec{s}_1$

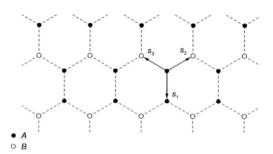

The vectors $\vec{s}_i$ which connect the sub-lattices have magnitude which is the lattice
spacing, α.

We will denote the operators which create and annihilate an electron at position
$\vec{A}$ on sub-lattice A as $a^\dagger_{\vec{A}}$ and $a_{\vec{A}}$, respectively and those which create and annihilate
an electron at position $\vec{A} + \vec{s}_i$ on sub-lattice B as $b^\dagger_{\vec{A}+\vec{s}_i}$ and $b_{\vec{A}+\vec{s}_i}$, respectively.
Being creation and annihilation operators for fermions, they must have the anti-
commutation relations

$$\left\{a_{\vec{A}}, a^\dagger_{\vec{A}'}\right\} = \delta_{\vec{A}\vec{A}'}, \quad \left\{a_{\vec{A}}, a_{\vec{A}'}\right\} = 0, \quad \left\{a^\dagger_{\vec{A}}, a^\dagger_{\vec{A}'}\right\} = 0 \tag{8.36}$$

$$\left\{b_{\vec{A}+\vec{s}_i}, b^\dagger_{\vec{A}'+\vec{s}_j}\right\} = \delta_{\vec{A}+\vec{s}_i.\vec{A}'+\vec{s}_j}, \quad \left\{b_{\vec{A}+\vec{s}_i}, b^\dagger_{\vec{A}'+\vec{s}_j}\right\} = 0 \tag{8.37}$$

and all of the a-operators anti-commute with all of the b-operators.

Much of the physics of graphene is described by a simple tight-binding model.
In this model, as a first approximation, the valence electron of each Carbon atom is
located in a shallow bound state that is localized at the site of the Carbon atom. An
electron can occupy this state and, moreover, it can tunnel from the state at this location to an identical state at the location of one of the other atoms. The tunnelling with
the largest amplitude is between nearest neighbours. The tight-binding Hamiltonian
which describes this tunnelling process is

$$H = \sum_{\vec{A},i} \left(t b^\dagger_{\vec{A}+\vec{s}_i} a_{\vec{A}} + t^* a^\dagger_{\vec{A}} b_{\vec{A}+\vec{s}_i}\right) \tag{8.38}$$

The only parameter in this Hamiltonian is the hopping amplitude, t, which, in
graphene, has magnitude $|t| \sim 2.8\,\mathrm{eV}$. A term in the Hamiltonian, say $b^\dagger_{\vec{A}+\vec{s}_i} a_{\vec{A}}$,
describes an electron tunnelling from the site located at $\vec{A}$ to the nearest neighbour
site located at $\vec{A} + \vec{s}_i$.

To diagonalize the Hamiltonian (8.38), it is most illuminating to go to wave-vector space by using the Fourier transform of the creation and annihilation operators,

$$a_{\vec{A}} = \int_{\Omega_B} \frac{d^2k}{\sqrt{\Omega_B}} e^{i\vec{k}\cdot\vec{A}} a(k), \qquad a_{\vec{A}}^\dagger = \int_{\Omega_B} \frac{d^2k}{\sqrt{\Omega_B}} e^{-i\vec{k}\cdot\vec{A}} a^\dagger(k) \quad (8.39)$$

$$b_{\vec{A}+\vec{s}_i} = \int_{\Omega_B} \frac{d^2k}{\sqrt{\Omega_B}} e^{i\vec{k}\cdot(\vec{A}+\vec{s}_i)} b(k), \qquad b_{\vec{A}+\vec{s}_i}^\dagger = \int_{\Omega_B} \frac{d^2k}{\sqrt{\Omega_B}} e^{-i\vec{k}\cdot(\vec{A}+\vec{s}_i)} b^\dagger(k) \quad (8.40)$$

where the momentum integrals are over the Brillouin zone, which we will define shortly. We have denoted both the Brillouin zone and its volume by Ω_B. The Fourier transforms are normalized so that the non-vanishing anti-commutators are

$$\{a(k), a^\dagger(\ell)\} = \delta^2(\vec{k} - \vec{\ell}), \quad \{b(k), b^\dagger(\ell)\} = \delta^2(\vec{k} - \vec{\ell}) \quad (8.41)$$

The Brillouin zone is by definition the fundamental domain in the space of wave-vectors. It is found by first finding the vectors which generate the lattice. In our case, for the A sub-lattice, these can be taken to be

$$\vec{A}_1 = \vec{s}_3 - \vec{s}_1, \quad \vec{A}_2 = \vec{s}_2 - \vec{s}_1$$

where $\vec{s}_1, \vec{s}_2, \vec{s}_3$ are the three vectors which connect the A and B sub-lattices. They are depicted in Fig. 8.1. The A sublattice is thus given by the set of points

$$n_1\vec{A}_1 + n_2\vec{A}_2, \quad (n_1, n_2) \text{ integers}$$

The $\vec{s}_1, \vec{s}_2, \vec{s}_3$ lie in a plane and have the properties

$$\vec{s}_1 + \vec{s}_2 + \vec{s}_3 = 0 \tag{8.42}$$

$$\vec{s}_i^2 = \alpha^2 \tag{8.43}$$

$$\vec{s}_i \cdot \vec{s}_j = -\frac{1}{2}\alpha^2 \quad i \neq j \tag{8.44}$$

where α is the lattice spacing. (In graphene, $\alpha \sim 1.42$ Å.)

To find the Brillouin zone we need to find the reciprocal (or dual) lattice. The dual lattice basis vectors are defined by

$$\vec{r}_1 \cdot \vec{A}_1 = \vec{r}_1 \cdot (\vec{s}_3 - \vec{s}_1) = 2\pi, \ \vec{r}_1 \cdot \vec{A}_2 = \vec{r}_1 \cdot (\vec{s}_2 - \vec{s}_1) = 0 \tag{8.45}$$

and

$$\vec{r}_2 \cdot \vec{A}_1 = \vec{r}_2 \cdot (\vec{s}_3 - \vec{s}_1) = 0, \ \vec{r}_2 \cdot \vec{A}_2 = \vec{r}_2 \cdot (\vec{s}_2 - \vec{s}_1) = 2\pi \tag{8.46}$$

The explicit solution for these dual lattice vectors are

$$\vec{r}_1 = -\frac{4\pi}{3\alpha^2} (\vec{s}_2 + \vec{s}_1) \tag{8.47}$$

$$\vec{r}_2 = -\frac{4\pi}{3\alpha^2} (\vec{s}_3 + \vec{s}_1) \tag{8.48}$$

Then, the reciprocal (or dual) lattice is the lattice generated by $\vec{r}_1$ and $\vec{r}_2$, that is, the points in the two-dimensional plane that are swept out by the vectors $m_2\vec{r}_1 + m_2\vec{r}_2$ as m_1 and m_2 vary over all of the integers. The Brillouin zone is the unit cell of this dual lattice. We could take it to be the region given by the vectors

$$\Omega_B = \left\{ \vec{k} = k_1\vec{r}_1 + k_2\vec{r}_2 \;\middle|\; -\frac{1}{2} < k_1, k_2 \le \frac{1}{2} \right\}$$

Our Brillouin zone is a parallelogram, whereas the one commonly presented in literature is a hexagon. The dual lattice is indeed a hexagonal lattice and the natural unit cell is indeed a hexagon. However, there is some freedom in choosing the unit cell and the parallelogram that we have chosen is also a possibility. The Brillouin zone for the B sub-lattice is identical to the one for the A sub-lattice. The Brillouin zone contains the wave-number vectors over which the area integrals in the Fourier transforms in equations (8.39) and (8.40) are done. The volume of the Brillouin zone is the volume of the parallellogram spanned by the vectors $\vec{r}_1$ and $\vec{r}_2$, that is $\Omega_B = \frac{8\pi^2}{3\sqrt{3}}\frac{1}{\alpha^2}$.

With the Fourier transforms (8.39) and (8.40) the Hamiltonian has the form

$$H = \int_{\Omega_B} d^3k \left[a^\dagger(k)\; b^\dagger(k)\right] \begin{bmatrix} 0 & \Delta^*(k) \\ \Delta(k) & 0 \end{bmatrix} \begin{bmatrix} a(k) \\ b(k) \end{bmatrix} \tag{8.49}$$

where

$$\Delta(k) = t \sum_i e^{i\vec{k}\cdot\vec{b}_i} \tag{8.50}$$

For each value of $\vec{k}$ in the Brillouin zone, the Hamiltonian acts on the two-component wave-function $\begin{bmatrix} a(k) \\ b(k) \end{bmatrix}$ as the two-by-two matrix

$$h(k) = \begin{bmatrix} 0 & \Delta^*(k) \\ \Delta(k) & 0 \end{bmatrix} \tag{8.51}$$

The possible energy levels of the electrons are then given by the eigenvalues, $E(k)$, of $h(k)$ which are easy to find, they are given by

$$E(k) = \pm|\Delta(k)|$$

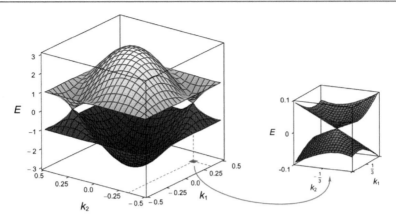

Fig. 8.2 The positive and negative energy bands $\pm E(k)$ of graphene are depicted including the K points where the bands touch at zero energy. There are two K points in the Brillouin zone

We see that the energy spectrum has two bands, a positive and a negative energy band. Moreover, the positive and negative energy states are in one-to-one correspondence with each other: If we find a state with positive energy, so that

$$h(k) \begin{bmatrix} a_E(k) \\ b_E(k) \end{bmatrix} = E(k) \begin{bmatrix} a_E(k) \\ b_E(k) \end{bmatrix}$$

then there is automatically another state with negative energy,

$$h(k) \begin{bmatrix} a_E(k) \\ -b_E(k) \end{bmatrix} = -E(k) \begin{bmatrix} a_E(k) \\ -b_E(k) \end{bmatrix}$$

This one-to-one correspondence between positive and negative energy states is called particle-hole symmetry. This symmetry was absent from our previous examples of non-relativistic many-electron systems that we studied in earlier chapters. Here, it is a special property of the tight-binding lattice model of graphene.

All of the possible energies of each branch of the spectrum, $E(k)$ and $-E(k)$ are found as $\vec{k}$ sweeps over the Brillouin zone. These are depicted in Fig. 8.2. The two branches of the energy spectrum have zeros and come into contact for special values of the wave-vector where $\Delta(k) = 0$ and therefore $E(k) = 0$. (These are called the K-points.) The equation $\Delta(K) = 0$ has a real part and an imaginary part, and it is therefore two real equations. This gives us two equations in two variables, k_1 and k_2, and these equations should generically have solutions at discrete points. Indeed, there are two solutions within the Brillouin zone. These are found by looking for values of $\vec{k}$ which solve the equation

$$0 = \Delta(k)/t = e^{i\vec{k}\cdot\vec{s}_1} + e^{i\vec{k}\cdot\vec{s}_2} + e^{i\vec{k}\cdot\vec{s}_3}$$

or, equivalently, the equation $1 + e^{i\vec{k}\cdot(\vec{s}_2-\vec{s}_1)} + e^{i\vec{k}\cdot(\vec{s}_3-\vec{s}_1)} = 0$ or

$$1 + e^{i\vec{k}\cdot\vec{A}_1} + e^{i\vec{k}\cdot\vec{A}_2} = 0$$

Given that $\vec{k} = k_1\vec{r}_1 + k_2\vec{r}_2$ with $-\frac{1}{2} < k_1, k_2 \leq \frac{1}{2}$, and using equations (8.45) and (8.46), the above equation reads

$$1 + e^{2\pi i k_1} + e^{2\pi i k_2} = 0 \tag{8.52}$$

Remembering that the sum of the cube roots of unity vanish, that is, if $\omega = e^{2\pi i/3}$ then $1 + \omega + \omega^2 = 0$, we see that equation (8.52) has two solutions for (k_1, k_2) in the Brillouin zone,

$$k_1 = \frac{1}{3}, \ k_2 = -\frac{1}{3}, \ \vec{K} = \frac{1}{3}(\vec{r}_1 - \vec{r}_2) = \frac{4\pi}{3\alpha^2}(\vec{s}_3 - \vec{s}_2)$$

and

$$k_1 = -\frac{1}{3}, \ k_2 = \frac{1}{3}, \ \vec{\tilde{K}} = -\vec{K}$$

The K-points, $\vec{K}$ and $\vec{\tilde{K}}$ are points in the Brillouin zone where the positive and negative energy bands touch. These are depicted in Fig. 8.2.

Now, we will use another fact about graphene, that the electronic states of charge-neutral graphene are half-filled. There is one positively charged Carbon ion on each lattice site. It is neutralized by one valence electron, so if the entire material is to be charge neutral, the electrons must have a density of one electron per lattice site. Thus, neutral graphene has a valence electron density of one electron per lattice site. But electrons have spin $\frac{1}{2}$ and they therefore have two spin states. The tight binding model would accommodate two electrons per site, one with each spin polarization. This leaves the single-electron states of neutral graphene half-filled. Given that there is a one-to-one mapping between the positive and negative energy states, the half of the states with the lowest energy are the negative energy states.

Thus, the lowest energy state of charge neutral graphene must have all of the negative energy states filled. The Fermi level is at zero energy, precisely at the K-points, of which there are two in the Brillouin zone.[3]

The ground state of the tight-binding model thus has a Fermi surface consisting of two points, the K-points. We will focus on the very low energy excitations of the system, that is the states of the many-particle system with energies very close to the Fermi energy. This is done by studying excitations whose wave-numbers are very near the K-points, that is, whose wave-numbers are $\vec{K} + \vec{k}$ or else $\vec{\tilde{K}} + \vec{k}$ where in both cases $|\vec{k}| << |\vec{K}| = |\vec{\tilde{K}}|$. Then

$$\Delta(\vec{K} + \vec{k}) = \frac{3}{2}\alpha \left(k_x - i k_y \right) + \dots \tag{8.53}$$

---

[3] There is a spin-orbit interaction for the valence electrons in graphene and this interaction indeed does couple the state of the electron's motion on the lattice to its spin state. However, this interaction, as well as the Zeeman interaction of the electron's magnetic moment with a magnetic field are negligibly small and we shall neglect them entirely in our discussion.

where we have used the explicit form of the lattice vectors $\vec{s}_3 = (0, -\alpha)$ and $\vec{s}_2 - \vec{s}_3 = (-\alpha, 0)$ which is apparent from Fig. 8.1. The ellipses denote orders higher than one in the dimensionless combination $\alpha|\vec{k}|$ and we are assuming that these are negligibly small corrections when the energy is small. Similarly,

$$\Delta(\vec{K} + \vec{k}) = \frac{3}{2}\alpha \left(k_x + ik_y \right) + \dots \tag{8.54}$$

The single-particle Hamiltonians which describe the low energy excitations are then obtained by substituting the expansions in equations (8.53) and (8.54) into the single particle Hamiltonian (8.51) to get

$$h(\vec{K} + \vec{k}) = \frac{3\alpha t}{2} \begin{bmatrix} 0 & k_x - ik_y \\ k_x + ik_y & 0 \end{bmatrix} + \dots$$

$$h(\vec{\tilde{K}} + \vec{k}) = \frac{3\alpha t}{2} \begin{bmatrix} 0 & k_x + ik_y \\ k_x - ik_y & 0 \end{bmatrix} + \dots$$

The two regions where the electrons can have energies near the Fermi level, the regions near the K and $\tilde{K}$ points are called "valleys" to describe the fact that they are local minima in the (positive) energy landscape. The two-component operators for each valley written as

$$\psi_1(k) = \begin{bmatrix} a(K + k) \\ b(K + k) \end{bmatrix}$$

and

$$\psi_2(k) = \begin{bmatrix} b(\tilde{K} + k) \\ a(\tilde{K} + k) \end{bmatrix}$$

where we have arranged the position of the a- and b- operators so that the two ψ's have the same Hamiltonian. As far as our low energy limit is concerned the degrees of freedom in each of the two valleys, described by the fields $\psi_1(k)$ and $\psi_2(k)$ have identical properties. Then, if we also restore the spin degeneracy of the electrons (and if, as we have been assuming, the dynamics is spin-independent) this further doubles the number of fermion operators with identical properties, so we denote the field by $\psi_a(k)$ where $a = 1, 2, 3, 4$. Note that the spinor index, that would indicate that, for each a, $\psi_a(k)$ is a two-component object, has been suppressed here, and we will continue to suppress it in the following. The Hamiltonian which describes our low energy electronic degrees of freedom in graphene is then

$$H = \int_{|\vec{k}| < \Lambda} d^3k \, \psi^{\dagger a}(k) h(k) \psi_a(k) \tag{8.55}$$

$$h(k) = \hbar v_F \begin{bmatrix} 0 & k_x - ik_y \\ k_x + ik_y & 0 \end{bmatrix} + \dots \tag{8.56}$$

where there is an implicit sum over $a = 1, 2, 3, 4$. We have suppressed the spinor indices of $\psi_a(k)$ for each a and will continue to do so in the following. The two-by-two matrix Hamiltonian $h(k)$ acts on these spinor indices. Here, $\Lambda << \frac{1}{\alpha}$ is a cutoff for wave-numbers which is placed in the regime which is much larger that the wave-vectors of the low energy modes that we are interested in but still small enough that the lattice corrections to the linear in wave-vectors approximation to the energy spectrum are vanishingly small. Also, we have renamed the constant

$$\frac{3\alpha|t|}{2} \equiv \hbar v_F$$

where v_F is the speed of the graphene electron. With the values of a and t in graphene, $v_F \sim c/300$ where c is the vacuum speed of light.

The fermion field operators as functions of wave-vectors must obey canonical anti-commutation relations,

$$\left\{ \psi_a(\vec{k}), \psi^{\dagger b}(\vec{\ell}) \right\} = \delta_a^b \delta(\vec{k} - \vec{\ell}) \tag{8.57}$$

in the above equation the spin and valley indices a, b are shown explicitly, but spinor components of the fields are suppressed. The right-hand-side of the commutation relation should therefore be regarded as a unit matrix in those suppressed indices.

Now, we could further model this system by re-introducing high energy states, but not with the graphene energy dispersion relation $E(k) = |\Delta(k)|$, but with relativistic energies $E(k) = \hbar v_F |\vec{k}|$ for the wave-vectors with magnitudes $\Lambda < |\vec{k}| < \infty$. In this free field theory, the low and high energy states are not coupled to each other, so if we agree to only probe the system in a way that excites the low energy states, the inclusion of high energy states will never be noticeable and it becomes a cosmetic improvement. However, interactions will couple high and low energy degrees of freedom and more is needed to understand that our removal of the cutoff is innocuous, indeed the quantum field theory machinery for that is beyond the scope of this monograph. One facet of this improvement is that the appropriately defined fields with space coordinates,

$$\psi_a(\vec{x}) = \int \frac{d^2 k}{2\pi} e^{i\vec{k}\cdot\vec{x}} \psi_a(\vec{k}) \tag{8.58}$$

now satisfy the usual anti-commutation relations. Equation (8.57) can be used to show that

$$\left\{ \psi_a(\vec{x}), \psi^{\dagger b}(\vec{y}) \right\} = \delta_a^b \delta(\vec{x} - \vec{y}) \tag{8.59}$$

If we also remove the cutoff in the Hamiltonian (8.55) and substitute the Fourier transform of the fields, we find

$$H = \hbar v_F \int d^2 x \psi^{\dagger a}(\vec{x}) \begin{bmatrix} 0 & i\partial_x + \partial_y \\ i\partial_x - \partial_y & 0 \end{bmatrix} \psi_a(\vec{x}) \tag{8.60}$$

The equations (8.59) and (8.60) define a quantum field theory in the Schrödinger picture.

The matrix Hamiltonian H in (8.60) is called a Dirac Hamiltonian or Dirac operator. Its generalization to three space dimensions which we will explore in the next chapter turns out to be the correct way to describe a relativistic particle with spin $J = \frac{1}{2}$. Here, we have found a Dirac Hamiltonian which is adapted to one lower dimension. It nonetheless has relativistic symmetry, in fact its quanta all propagate at the same speed, the "graphene speed of light" v_F, similar to what photons do in four dimensions. A photon in vacuum always travels at the same speed regardless of how it was created or what its history is. Within the validity of our modelling, a graphene electron does the same, except instead of c, the speed is $v_F \sim c/300$. The electron travels like a massless relativistic particle. The dispersion relation for a massive relativistic particle would be $E = \sqrt{m^2c^4 + \vec{p}^2c^2}$ which reduces to $E = c|\vec{p}| = \hbar c|\vec{k}|$ when the mass is taken to zero. The graphene electron dispersion relation is identical to this, with c replaced by v_F. Moreover, not only is the dispersion relation that of a relativistic particle, but the matrix structure of the Dirac Hamiltonian in equation (8.60) is what is needed to give the theory relativistic spacetime symmetry in a space–time which has two, rather than three space dimensions.

If we use the Hamiltonian (8.60) to go to the Heisenberg picture and we find that the field equation is the Dirac equation,

$$i\hbar\frac{\partial}{\partial t}\psi_a(x) = \hbar v_F \begin{bmatrix} 0 & i\partial_x + \partial_y \\ i\partial_x - \partial_y & 0 \end{bmatrix} \psi_a(x) \qquad (8.61)$$

(where we use (x) to denote $(\vec{x}, t)$) and the spinor field $\psi_a(x) \equiv \psi_a(\vec{x}, t)$ has the equal-time anti-commutation relations

$$\left\{\psi_a(x), \psi^{\dagger b}(y)\right\} \delta(x^0 - y^0) = \delta_a^b \delta(x - y) \qquad (8.62)$$

where, here, $\delta(x - y)$ is a Dirac delta function in three dimensions. It equates the time as well as the two space coordinates.

Equations (8.61) and (8.62) define a quantum field theory, that of the electron in 2+1 spacetime dimensions. We will study the Dirac field theory in more detail, primarily in four rather than three space–time dimensions, in the next chapter. It will turn out to have the full symmetries of Minkowski spacetime—translation invariance and Lorentz invariance—and when it is massless, as is the case of graphene—these extend to the full conformal symmetry.

The Dirac Field Theory

9

9.1   The Dirac Equation

In our studies of many particle systems, we formulated an approach to their quantum mechanics which led us to the non-relativistic field equation, in the absence of interactions,

$$i\hbar\frac{\partial}{\partial t}\psi_\sigma(\vec{x},t) = \left[-\frac{\hbar^2}{2m}\vec{\nabla}^2 + \epsilon_F\right]\psi_\sigma(\vec{x},t) \tag{9.1}$$

where we have included the Fermi energy ϵ_F as is appropriate for an open system. We will focus on fermions which have spin $J = \frac{1}{2}$ so that the spin index takes on two values, $\sigma = \pm\frac{1}{2}$. In this chapter we shall discuss the equation which replaces this one in a relativistic quantum field theory. From our point of view, the main difference between the two is the space–time symmetry. The field equation above has time and space translation, spatial rotation and Galilean symmetry. We want to replace it with an equation which retains the space and time translation symmetry as well as the rotation invariance and which is symmetric under Lorentz, rather than Galilean transformations.

To seek the appropriate relativistic wave equation, we could recall our discussion of the previous chapters. If we simply postulate a Hamiltonian with a relativistic dispersion relation, a differential operator whose eigenvalues are given by $E(p) = \sqrt{m^2c^4 + c^2\vec{p}^2}$, so that the wave equation (9.1) is replaced by

$$i\hbar\frac{\partial}{\partial t}\psi_\sigma(\vec{x},t) = \sqrt{m^2c^4 - c^2\hbar^2\vec{\nabla}^2}\psi_\sigma(\vec{x},t) \tag{9.2}$$

the resulting theory has difficulties with causality. There is a finite probability of the particle propagating faster than the speed of light. The difficulty lies in the fact

© The Author(s), under exclusive license to Springer Nature Singapore Pte Ltd. 2023
G. W. Semenoff, *Quantum Field Theory*, Graduate Texts in Physics,
https://doi.org/10.1007/978-981-99-5410-0_9

that the "Hamiltonian" operator on the right-hand-side of this equation is not a finite polynomial in derivatives, if one Taylor expands in the momentum squared, all orders up to infinity occur there. Having all orders in derivatives results in the fact that it is not a local differential operator. Dirac found a way to replace this equation by one where the Hamiltonian has the same spectrum, but the operator is a polynomial in derivatives. We also have this goal.

In our discussion of single particle relativistic quantum mechanics, the problems with causality had a potential solution if, besides the particle, the theory contained an anti-particle. Let us address the problems with causality by postulating the existence of an anti-particle which would satisfy and equation similar to (9.2) but with negative energy,

$$i\hbar\frac{\partial}{\partial t}\tilde{\psi}_{\bar{\sigma}}(\vec{x}, t) = -\sqrt{m^2c^4 - c^2\hbar^2\vec{\nabla}^2}\,\tilde{\psi}_{\bar{\sigma}}(\vec{x}, t) \tag{9.3}$$

Assuming that both the particle and the anti-particle occur in the same theory, we could combine the two into the same multi-component field to find

$$i\hbar\frac{\partial}{\partial t}\begin{bmatrix}\psi_\sigma(\vec{x}, t)\\ \tilde{\psi}_{\bar{\sigma}}(\vec{x}, t)\end{bmatrix} = \begin{bmatrix}\sqrt{m^2c^4 - c^2\hbar^2\vec{\nabla}^2} & 0\\ 0 & -\sqrt{m^2c^4 - c^2\hbar^2\vec{\nabla}^2}\end{bmatrix}\begin{bmatrix}\psi_\sigma(\vec{x}, t)\\ \tilde{\psi}_{\bar{\sigma}}(\vec{x}, t)\end{bmatrix} \tag{9.4}$$

If the spin index takes its two values, $\sigma = -\frac{1}{2}, \frac{1}{2}$, the above is an equation for a four-component object which has both positive and negative energy states, both with relativistic dispersion relations. However, it is still not quite right. As we shall see shortly, it would be right if the label σ referred to helicity, rather than spin. (We need the following development to know what helicity is.) This is something that we will learn from the correct equation.

One might wonder whether there is a matrix Hamiltonian, rather than the diagonal one in equation (9.2), which has the same eigenvalues as the one in (9.2), two of the eigenvalues would be $\sqrt{m^2c^2 + c^2\hbar^2\vec{k}^2}$ and two of the eigenvalues would be $-\sqrt{m^2c^2 + c^2\hbar^2\vec{k}^2}$. Moreover, it would be desirable that the matrix Hamiltonian is a polynomial in derivatives. There is no such 2×2 or 3×3 matrix. It is Dirac's great insight that our problem can be solved with a 4×4 matrix. Consider the four Hermitian 4×4 matrices

$$\beta, \; \alpha^1, \; \alpha^2, \; \alpha^3$$

which have the properties

$$\beta = \beta^\dagger, \; \alpha^1 = (\alpha^1)^\dagger, \; \alpha^2 = (\alpha^2)^\dagger, \; \alpha^3 = (\alpha^3)^\dagger$$
$$\beta\beta = 1, \; \beta\alpha^a + \alpha^a\beta = 0, \; \alpha^a\alpha^b + \alpha^b\alpha^a = 2\delta^{ab}1$$

where 1 is the 4×4 unit matrix. (Alternatively, if we consider a set of four matrices with the above properties, there is a way to show that the minimal size of such matrices is 4×4.)

Then, we consider the wave equation

$$i\hbar \frac{\partial}{\partial t} \begin{bmatrix} \psi_1(\vec{x},t) \\ \psi_2(\vec{x},t) \\ \psi_3(\vec{x},t) \\ \psi_4(\vec{x},t) \end{bmatrix} = \left[\beta mc^2 + i\hbar c \vec{\alpha} \cdot \vec{\nabla} \right] \begin{bmatrix} \psi_1(\vec{x},t) \\ \psi_2(\vec{x},t) \\ \psi_3(\vec{x},t) \\ \psi_4(\vec{x},t) \end{bmatrix}$$

The "Dirac Hamiltonian"

$$h_D = \left[\beta mc^2 + i\hbar c \vec{\alpha} \cdot \vec{\nabla} \right] \tag{9.5}$$

is a Hermitian operator. It must have real eigenvalues. What is more, if we square the operator, we can use the properties of the matrices in the Hamiltonian to show that

$$h_D{}^2 = m^2 c^4 - \hbar^2 c^2 \vec{\nabla}^2 \tag{9.6}$$

so, since the eigenvalues of $\vec{\nabla}^2$ are $-\vec{k}^2$, h_D^2 has four degenerate eigenvalues $m^2 c^4 + \hbar^2 c^2 \vec{k}^2$. This tells us that h_D must have eigenvalues which are either $+\sqrt{m^2 c^4 + \hbar^2 c^2 \vec{k}^2}$ or $-\sqrt{m^2 c^4 + \hbar^2 c^2 \vec{k}^2}$. It is also easy to see that there are two of each of these. This is due to the fact that there is a unitary matrix $\beta \alpha^1 \alpha^2 \alpha^3$ which anti-commutes with h_D, so that if $h_D u = \sqrt{m^2 c^4 + \hbar^2 c^2 \vec{k}^2} u$, $h_D[\beta \alpha^1 \alpha^2 \alpha^3]u = -\sqrt{m^2 c^4 + \hbar^2 c^2 \vec{k}^2}[\beta \alpha^1 \alpha^2 \alpha^3]u$, that is, for each positive eigenvalue, there is a negative eigenvalue. Our conclusion is that there are two eigenvalues $+\sqrt{m^2 c^4 + \hbar^2 c^2 \vec{k}^2}$ and two eigenvalues $-\sqrt{m^2 c^4 + \hbar^2 c^2 \vec{k}^2}$.

To make the Dirac equation look more covariant, we can define the matrices

$$\beta = i\gamma^0, \ \alpha = \gamma^0 \vec{\gamma} \tag{9.7}$$
$$\gamma^0 = -\gamma^{0\dagger}, \ \gamma^i = \gamma^{i\dagger} \tag{9.8}$$

which have the algebraic property that

$$\{\gamma^\mu, \gamma^\nu\} = 2\eta^{\mu\nu} \tag{9.9}$$

γ^μ are called the Dirac gamma-matrices, or Dirac matrices. Using them, the Dirac equation is the matrix differential equation

$$\sum_{b=1}^{4} \left[\sum_{\mu=0}^{3} \gamma_{ab}^\mu \partial_\mu + \frac{mc}{\hbar} \delta_{ab} \right] \psi_b(\vec{x},t) = 0 \tag{9.10}$$

or, more elegantly with the indices and summation signs suppressed,

$$\left[\slashed{\partial} + \frac{mc}{\hbar} \right] \psi(x) = 0 \tag{9.11}$$

where we will often use the Feynman slash notation for the relativistic inner product of the Dirac matrices with any other four-vector

$$\slashed{\partial} \equiv \gamma^\mu \partial_\mu, \quad \slashed{A} \equiv \gamma^\mu A_\mu \tag{9.12}$$

We will find it useful to define

$$\bar{\psi}(x) \equiv \psi^\dagger(x)\gamma^0, \quad \psi^\dagger(x) \equiv -\bar{\psi}(x)\gamma^0 \tag{9.13}$$

Using this definition and taking a Hermitian conjugate of the equation in (9.11), we obtain

$$\bar{\psi}(x)\left[-\overleftarrow{\slashed{\partial}} + \frac{mc}{\hbar} \right] = 0 \tag{9.14}$$

where the left oriented arrow above the derivative indicates that it operates on whatever is to the left of it.

The Dirac equation has a structure similar to the Schrödinger wave equation. It can be written as

$$i\hbar \frac{\partial}{\partial t} \psi(x) = i\hbar c\gamma^0 \left[\vec{\gamma} \cdot \vec{\nabla} + \frac{mc}{\hbar} \right] \psi(x) \tag{9.15}$$

with the difference that, what we would call the single-particle Hamiltonian, the quantity h_D in equation (9.5), which we could also write as

$$h_D = i\hbar c\gamma^0 \left[\vec{\gamma} \cdot \vec{\nabla} + \frac{mc}{\hbar} \right] \tag{9.16}$$

We can consider equation (9.15) as the relativistic generalization of the non-relativistic field equation for non-interacting identical particles. It will turn out that this will be correct only when the particles are fermions with spin $J = \frac{1}{2}$. It does not apply to bosons and for fermions with other spins.

In order to define the quantum field theory, we also need the equal-time anti-commutation relations for the field operators. Anticipating that, following the similarity with the nonrelativistic theory that we have been emphasizing in the discussion so far, the Lagrangian will have the time derivative term $\mathcal{L} = i\hbar\psi_a^\dagger(x)\partial_0\psi_a(x) + \ldots$ we can simply take the anti-commutation relations from the non-relativistic theory

$$\left\{ \psi_a(x), \psi_b^\dagger(y) \right\}_{x^0=y^0} = \delta_{ab}\delta(\vec{x} - \vec{y})$$

$$\{ \psi_a(x), \psi_b(y) \}_{x^0=y^0} = 0, \quad \left\{ \psi_a^\dagger(x), \psi_b^\dagger(y) \right\}_{x^0=y^0} = 0 \tag{9.17}$$

Of course, for the quantum field theory to have Lorentz symmetry, the commutation relations must be compatible with Lorentz invariance. We shall return to a discussion of why this is so once we have studied how Lorentz transformations act on the Dirac fields.

We note here that we constructed the Dirac equation so that it has a relativistic spectrum. A short way to see that is to use the anti-commutation algebra of the Dirac matrices to show that

$$(\gamma^\mu \partial_\mu)^2 = \gamma^\mu \gamma^\nu \partial_\mu \partial_\nu = \frac{1}{2}\{\gamma^\mu, \gamma^\nu\}\partial_\mu \partial_\nu = \eta^{\mu\nu}\partial_\mu \partial_\nu = \partial_\mu \partial^\mu \equiv \partial^2$$

Using this identity, we can operate the matrix valued differential operator $(-\gamma^\nu \partial_\nu + mc/\hbar)$ on the Dirac equation from the left to obtain

$$(-\gamma^\nu \partial_\nu + mc/\hbar)(\gamma^\mu \partial_\mu + mc/\hbar)\psi = 0 \rightarrow (-\hbar^2 \partial^2 + m^2 c^2)\psi = 0$$

We see that, if $\psi(x)$ obeys the Dirac equation, it also obeys the relativistic wave equation $(-\hbar^2 \partial^2 + m^2 c^2)\psi(x) = 0$. This implies that the solutions of the Dirac equation also obey this relativistic wave equation, and must therefore propagate like relativistic matter waves.

9.2 Solving the Dirac Equation

The Dirac equation in natural units is

$$\left(\slashed{\partial} + m\right)\psi(x) = 0 \tag{9.18}$$

To see how the Dirac equation is solved, it is useful to choose a specific form for the Dirac matrices. For example, the following matrices obey the correct algebra

$$\gamma^a = \begin{bmatrix} 0 & \sigma^a \\ \sigma^a & 0 \end{bmatrix}, \quad \gamma^0 = \begin{bmatrix} 0 & 1 \\ -1 & 0 \end{bmatrix} \tag{9.19}$$

where we use 1 to denote the 2×2 unit matrix which appears in the upper and lower triangle of γ^0. Also, σ^i are the 2×2 Pauli matrices, which we remind the reader are given by

$$\sigma^1 = \begin{bmatrix} 0 & 1 \\ 1 & 0 \end{bmatrix}, \quad \sigma^2 = \begin{bmatrix} 0 & -i \\ i & 0 \end{bmatrix}, \quad \sigma^3 = \begin{bmatrix} 1 & 0 \\ 0 & -1 \end{bmatrix} \tag{9.20}$$

The Pauli matrices have the properties

$$\sigma^a \sigma^b + \sigma^b \sigma^a = 2\delta^{ab} 1, \quad \sigma^a \sigma^b - \sigma^b \sigma^a = 2i \epsilon^{abc} \sigma^c$$

where ϵ^{abc} is the totally anti-symmetric tensor with $\epsilon^{123} = 1$. It is easy to confirm that the explicit form (9.19) indeed have γ^0 anti-Hermitian, γ^a Hermitian and that they have the correct anti-commutation algebra for Dirac matrices.

With the matrices in (9.19), the Dirac equation is

$$\begin{bmatrix} m & \partial_0 + \vec{\sigma} \cdot \vec{\nabla} \\ -\partial_0 + \vec{\sigma} \cdot \vec{\nabla} & m \end{bmatrix} \begin{bmatrix} u(x) \\ v(x) \end{bmatrix} = 0 \tag{9.21}$$

where we have split the four-component Dirac spinor into $u(x)$ and $v(x)$ which are two 2-component objects.

To solve the differential equation, we use the ansatz

$$\begin{bmatrix} u(x) \\ v(x) \end{bmatrix} = e^{ik^\mu \eta_{\mu\nu} x^\nu} \begin{bmatrix} u(k) \\ v(k) \end{bmatrix} = e^{-ik^0 t + i\vec{k}\cdot\vec{x}} \begin{bmatrix} u(k) \\ v(k) \end{bmatrix} \tag{9.22}$$

so that the Dirac equation becomes two coupled equations for the two-component objects $u(k)$ and $v(k)$,

$$m\, u(k) - i[k^0 - \vec{\sigma} \cdot \vec{k}]\, v(k) = 0 \tag{9.23}$$

$$m\, v(k) + i[k^0 + \vec{\sigma} \cdot \vec{k}]\, u(k) = 0 \tag{9.24}$$

We have now reduced the Dirac equation to two matrix equations. Equation (9.24) determines the 2-component object $v(k)$ in terms of $u(k)$, that is, if $u(k)$ were known, we could determine v as

$$v(k) = -\frac{i}{m}[k^0 + \vec{\sigma} \cdot \vec{k}]\, u(k) \tag{9.25}$$

Plugging this into (9.23) yields the condition

$$(k^0)^2 = \vec{k}^2 + m^2 \tag{9.26}$$

which has two solutions for k^0, $k^0 = \sqrt{\vec{k}^2 + m^2}$ and $k^0 = -\sqrt{\vec{k}^2 + m^2}$, the positive and negative energy solutions, respectively. In the following, we will use the notation where k^0 is the frequency which can be either positive or negative, and $E(\vec{k}) = \sqrt{\vec{k}^2 + m^2}$ is positive, and sometimes abbreviated by E. Then, the two types of solution that we are finding have $k^0 = E(\vec{k})$ and $k^0 = -E(\vec{k})$.

Next, we note that $\vec{\sigma} \cdot \vec{k}$ is a Hermitian matrix which can be diagonalized. Once diagonal, it has real eigenvalues. The eigenstates of this matrix are said to be "eigenstates of helicity". It is left as an exercise to the reader to show that there exist two eigenvalues and two eigenvectors,

$$\vec{\sigma} \cdot \vec{k}\, u_+(k) = |\vec{k}|u_+(k) \tag{9.27}$$

$$\vec{\sigma} \cdot \vec{k}\, u_-(k) = -|\vec{k}|u_-(k) \tag{9.28}$$

$$u_+^\dagger(k)u_+(k) = 1 = u_-^\dagger(k)u_-(k) \tag{9.29}$$

$$u_+^\dagger(k)u_-(k) = 0 = u_-^\dagger(k)u_+(k) \tag{9.30}$$

In addition to the above properties of the eigenvectors, we shall find the following identities very useful

$$u_+(k)u_+^\dagger(k) = \frac{|\vec{k}| + \vec{\sigma} \cdot \vec{k}}{2|\vec{k}|}, \quad u_-(k)u_-^\dagger(k) = \frac{|\vec{k}| - \vec{\sigma} \cdot \vec{k}}{2|\vec{k}|} \qquad (9.31)$$

Then, putting it all together, we have four linearly independent solutions, one for each of the two signs of the helicity and the two signs of the energy k^0. We can take a superposition of these four solutions to obtain the general solution for the Dirac field,

$$\psi(x) = \int d^3k \frac{e^{i\vec{k}\cdot\vec{x} - iEt}}{\sqrt{2}(2\pi)^{\frac{3}{2}}} \left\{ \begin{bmatrix} i\sqrt{1 - \frac{|\vec{k}|}{E}} \, u_+(k) \\ \sqrt{1 + \frac{|\vec{k}|}{E}} \, u_+(k) \end{bmatrix} a_+(\vec{k}) + \begin{bmatrix} i\sqrt{1 + \frac{|\vec{k}|}{E}} \, u_-(k) \\ \sqrt{1 - \frac{|\vec{k}|}{E}} \, u_-(k) \end{bmatrix} a_-(\vec{k}) \right\}$$

$$+ \int d^3k \frac{e^{-i\vec{k}\cdot\vec{x} + iEt}}{\sqrt{2}(2\pi)^{\frac{3}{2}}} \left\{ \begin{bmatrix} i\sqrt{1 - \frac{|\vec{k}|}{E}} \, u_+(k) \\ -\sqrt{1 + \frac{|\vec{k}|}{E}} \, u_+(k) \end{bmatrix} b_+^\dagger(\vec{k}) + \begin{bmatrix} i\sqrt{1 + \frac{|\vec{k}|}{E}} \, u_-(k) \\ -\sqrt{1 - \frac{|\vec{k}|}{E}} \, u_-(k) \end{bmatrix} b_-^\dagger(\vec{k}) \right\} \qquad (9.32)$$

where $E(\vec{k}) = \sqrt{k^2 + m^2}$. Here, as in our previous non-relativistic treatment of fermions, taking negative energy solutions to represent holes, so the annihilation operator for a negative energy particle is replaced by a creation operator for a positive energy hole. The holes will be our anti-particles. This solution will obey the anti-commutation relation for the Dirac field (9.17) if the Fourier coefficients satisfy the non-vanishing anti-commutation relations

$$\left\{ a_+(\vec{k}), a_+^\dagger(\vec{k}') \right\} = \delta(\vec{k} - \vec{k}'), \quad \left\{ a_-(\vec{k}), a_-^\dagger(\vec{k}') \right\} = \delta(\vec{k} - \vec{k}') \qquad (9.33)$$

$$\left\{ b_+(\vec{k}), b_+^\dagger(\vec{k}') \right\} = \delta(\vec{k} - \vec{k}'), \quad \left\{ b_-(\vec{k}), b_-^\dagger(\vec{k}') \right\} = \delta(\vec{k} - \vec{k}') \qquad (9.34)$$

All of other combinations of creation and annihilation operators have vanishing anti-commutators. We can easily check that, with these anti-commutation relations for creation and annihilation operators, the solution in equation (9.32) obeys equations the correct equal-time anti-commutations relations (9.17).

We can form a Hamiltonian for the Dirac theory. We will discuss how it comes about when we analyze the space–time symmetries, in particular, time translation invariance of the Dirac equation. Here, we simply state that the quantum field theory Dirac Hamiltonian can be written as

$$H = i \int d^3x \bar{\psi}(x) \left[\vec{\gamma} \cdot \vec{\nabla} + m \right] \psi(x) \qquad (9.35)$$

Indeed, the commutator of the fields with this Hamiltonian can be used to find the Dirac equation. If we plug our solution (9.32) of the Dirac equation into this quantum field theory Hamiltonian we obtain

$$H = \int d^3k \sqrt{\vec{k}^2 + m^2} \left(a_+^\dagger(\vec{k})a_+(\vec{k}) + a_-^\dagger(\vec{k})a_-(\vec{k}) + b_+^\dagger(\vec{k})b_+(\vec{k}) + b_-^\dagger(\vec{k})b_-(\vec{k}) \right) \qquad (9.36)$$

We can do the same for the number operator, which gives us

$$
\mathcal{N} = \int d^3x \psi^\dagger(x,t)\psi(\vec{x},t)
$$

$$
= \int d^3k \left(a_+^\dagger(\vec{k})a_+(\vec{k}) + a_-^\dagger(\vec{k})a_-(\vec{k}) - b_+^\dagger(\vec{k})b_+(\vec{k}) - b_-^\dagger(\vec{k})b_-(\vec{k}) \right) \quad (9.37)
$$

In both of the above expressions, we have dropped infinite constants. Unlike in the non relativistic theory, the vacuum energy density and the vacuum charge density both contain infinite constants which we have to simply drop in order to have a sensible Hamiltonian and number operator.

We have written down the expressions for the Hamiltonian (9.36) and the number operator (9.37) in terms of the Dirac field by analogy with the same quantities in our non-relativistic field theories of fermions. Later on, we will see that these quantities are indeed the appropriate Noether charges for symmetries of the Dirac field theory.

We see from the expression (9.37), that, in direct analogy to the non relativistic system that we have studied, particles, which are associated with $a_+(\vec{k})$ and $a_+^\dagger(\vec{k})$ contribute positively to the particle number whereas holes, or anti-particles, which are associated with $b_+(\vec{k})$ and $b_+^\dagger(\vec{k})$, have negative particle number. What differs from the non-relativistic theory is the fact that particles and holes have the same energy spectrum. We can think of this as a symptom of the existence of a symmetry of the theory which interchanges the particles and holes, or particle-hole symmetry, or in the relativistic parlance, it interchanges particles and anti-particles and is called charge conjugation symmetry. Since the energy of a single electron, $\sqrt{\vec{k}^2 + m^2}$, can be arbitrarily large, for large values of $|\vec{k}|$, it is also so for holes (or antiparticles). This means that, as we shall see, unlike the Fermi sea of our non-relativistic many-particle theory, in this relativistic analog of it, the "Dirac sea" is infinitely deep. This is what leads to the infinite values of the energy and number densities (which we have dropped). (Recall that in our non-relativistic theories, the energies and the numbers were infinite but the densities were finite and they indeed had a physical interpretation. Here the densities are also infinite.)

Another difference with the non-relativistic theory, where the electron had a state of well-defined spin is that in the relativistic theory, it is the helicity (the states labeled by subscripts $(+)$ and $(-)$), which are important. The helicity can be thought of as the projection of the spin in the direction of motion of the fermion.

We construct the basis of the Fock space beginning with the vacuum $|\mathcal{O}>$ which we assume is normalized,

$$
< \mathcal{O}|\mathcal{O} >= 1 \quad (9.38)
$$

and has the property that it is annihilated by all of the annihilation operators,

$$
a_+(\vec{k})|\mathcal{O} >= 0, \; a_-(\vec{k})|\mathcal{O} >= 0, \; b_+(\vec{k})|\mathcal{O} >= 0, \; b_-(\vec{k})|\mathcal{O} >= 0 \quad (9.39)
$$

for all values of $\vec{k}$. There is the adjoint statement

$$< \mathcal{O}|a_+^\dagger(\vec{k}) = 0, \quad < \mathcal{O}|a_-^\dagger(\vec{k}) = 0, \quad < \mathcal{O}|b_+^\dagger(\vec{k}) >= 0, \quad < \mathcal{O}|b_-^\dagger(\vec{k}) = 0 \quad (9.40)$$

Then, multi particle and anti-particle states are created by operating with creation operators, for example

$$a_+^\dagger(\vec{k}_1)\ldots a_+^\dagger(\vec{k}_m)a_-^\dagger(\vec{k}_1')\ldots a_-^\dagger(\vec{k}_{m'}')b_+^\dagger(\vec{\ell}_1)\ldots b_+^\dagger(\vec{\ell}_n)b_-^\dagger(\vec{\ell}_1')\ldots b_-^\dagger(\vec{\ell}_{n'}')|\mathcal{O}> \tag{9.41}$$

These states are eigenstates of particle number, $\mathcal{N}$, with eigenvalue

$$N = m + m' - n - n' \tag{9.42}$$

and they are eigenstates of the Hamiltonian, H_0, with eigenvalue the total energy,

$$E = \sum_1^m \sqrt{\vec{k}_i^2 + m^2} + \sum_1^{m'} \sqrt{(\vec{k}_i')^2 + m^2} + \sum_1^n \sqrt{\vec{\ell}_i^2 + m^2} + \sum_1^{n'} \sqrt{(\vec{\ell}_i')^2 + m^2} \tag{9.43}$$

We see that the total energy of a system of non-interacting relativistic fermions and anti-fermions is equal to the sum of their individual energies. We also see that, as they did for non-relativistic fermions, the holes or anti-particles contribute negatively to the particle number whereas the particles have positive particle number. Also, both the particles and holes have positive energies.

9.3   Lorentz Invariance of the Dirac Equation

The Dirac equation is clearly invariant under translations of the space–time coordinates. If $\psi(\vec{x}, t)$ is a solution of the Dirac equation, then $\psi(\vec{x} + \vec{a}, t + \tau)$, with constants $\vec{a}$ and τ, is also a solution. What about Lorentz transformations? Of course, we showed that the Dirac equation leads to a relativistic energy spectrum for the particles and anti-particles. One might then anticipate that it is Lorentz invariant. Let confirm this fact explicitly.

Let us begin by reviewing our discussion of how a Lorentz transformation of a scalar field is implemented. Recall that a Lorentz transformation is the linear transformation on the coordinates

$$x^\mu \to \tilde{x}^\mu = \Lambda^\mu{}_\nu x^\nu$$

where the matrices $\Lambda^\mu{}_\nu x^\nu$ satisfy the equation

$$\Lambda^\mu{}_\nu \Lambda^\rho{}_\sigma \eta_{\mu\rho} = \eta_{\nu\sigma}$$

We are often interested in infinitesimal transformations. For a Lorentz transformation

$$\Lambda^{\mu}{}_{\nu} = \delta^{\mu}{}_{\nu} + \omega^{\mu}{}_{\nu} + \dots$$
$$(\Lambda^{-1})^{\mu}{}_{\nu} = \delta^{\mu}{}_{\nu} - \omega^{\mu}{}_{\nu} + \dots$$

In our discussion of coordinate transformations, we have derived the transformation property of the scalar field,

$$\tilde{\phi}(\tilde{x}) = \phi(x)$$

which, for the Lorentz transformation, we can rewrite as

$$\tilde{\phi}(x) = \phi(\Lambda^{-1}x)$$

The infinitesimal transformation is then

$$\delta\phi(x) = -\omega_{\mu\nu}x^{\nu}\partial^{\mu}\phi(x)$$

We expect that, under a Lorentz transformation, the argument of the Dirac field also changes and some vestige of the scalar transformation law should also apply to it. However, the Dirac field has four components and the Lorentz transformation could also mix the components. Thus, we could make the ansatz that the Lorentz transformation of the Dirac field involves multiplication of it by a matrix, as well as a Lorentz transformation of the coordinates,

$$\tilde{\psi}(x) = S(\Lambda)\psi(\Lambda^{-1}x) = \psi(x) - \omega^{\mu}{}_{\nu}x^{\nu}\partial_{\mu}\psi(x) + s\psi(x) + \dots$$

Here,

$$S(\Lambda) = 1 + s + \dots \tag{9.44}$$

is a 4×4 matrix which depends on the Lorentz transformation. Since a Lorentz transformation should be invertible, we expect that S is an invertible matrix, that is, that $\det S \neq 0$. For an infinitesimal transformation, S differs from the identity matrix by an infinitesimal matrix, which we have denoted by s. This transformation is a symmetry of the Dirac equation if the transformed field also satisfies the equation, that is, if we assume that $[\slashed{\partial} + m]\psi(x) = 0$ then, we need

$$0 = [\slashed{\partial} + m]\tilde{\psi}(x)$$
$$= [\slashed{\partial} + m]\psi(x) + [\slashed{\partial} + m]\left[-\omega^{\mu}{}_{\nu}x^{\nu}\partial_{\mu} + s\right]\psi(x)$$
$$= \left[-\omega^{\mu}{}_{\nu}\gamma^{\nu}\partial_{\mu} + [\slashed{\partial}, s]\right]\psi(x) = \left[-\omega^{\mu}{}_{\nu}\gamma^{\nu} + [\gamma^{\mu}, s]\right]\partial_{\mu}\psi(x) = 0$$
$$\text{if } -\omega^{\mu}{}_{\nu}\gamma^{\nu} + [\gamma^{\mu}, s] = 0$$

Thus we see that, in order to show that the Lorentz transformation is a symmetry transformation of the Dirac equation, and to understand how the Dirac field transforms under an infinitesimal Lorentz transformation, we need to find the 4×4 matrix s which is a linear function of the components of the matrix $\omega^\mu{}_\nu$ and which has the property that

$$\left[\gamma^\mu, s\right] = \omega^\mu{}_\nu \gamma^\nu$$

By using the commutator algebra of the Dirac gamma-matrices, we can easily see that this equation is solved by

$$s = \frac{1}{8}\left[\gamma^\rho, \gamma^\sigma\right]\omega_{\rho\sigma}$$

To see this, consider

$$\left[\gamma^\mu, s\right] = \frac{1}{8}\left[\gamma^\mu, \left[\gamma^\rho, \gamma^\sigma\right]\right]\omega_{\rho\sigma} = \frac{1}{4}\left[\gamma^\mu, \gamma^\rho\gamma^\sigma\right]\omega_{\rho\sigma}$$

$$= \frac{1}{4}\left(\{\gamma^\mu, \gamma^\rho\}\gamma^\sigma - \gamma^\rho\{\gamma^\mu, \gamma^\sigma\}\right)\omega_{\rho\sigma}$$

$$= \frac{1}{4}\left(2\eta^{\mu\rho}\gamma^\sigma - \gamma^\rho 2\eta^{\mu\sigma}\right)\omega_{\rho\sigma} = \omega^\mu{}_\sigma\gamma^\sigma$$

Thus, we have found the infinitesimal Lorentz transformation of the Dirac field,

$$\delta\psi(x) = \omega_{\mu\nu}\left[x^\mu\partial^\nu + \frac{1}{8}\left[\gamma^\mu, \gamma^\nu\right]\right]\psi(x) \tag{9.45}$$

$$\delta\bar{\psi}(x) = \bar{\psi}(x)\left[x^\mu\overleftarrow{\partial}^\nu - \frac{1}{8}\left[\gamma^\mu, \gamma^\nu\right]\right]\omega_{\mu\nu} \tag{9.46}$$

We could also consider finite, rather than infinitesimal Lorentz transformations. They are a symmetry of the Dirac equation if the matrix S satisfies

$$S^{-1}\gamma^\mu S \Lambda^{-1\nu}{}_\mu = \gamma^\nu$$

and this should be solved by $S = 1 + s + \ldots = 1 + \frac{1}{8}\left[\gamma^\mu, \gamma^\nu\right]\omega_{\mu\nu} + \ldots$.

9.4 Spin of the Dirac Field

For an infinitesimal spatial rotation in the 1-2 plane (or, about the 3-axis), the only non-zero components of $\omega_{\mu\nu}$ are ω_{12} and $\omega_{21} = -\omega_{12}$ and the transformation of the Dirac field (9.45) in this case is

$$\delta\psi(x) = \omega_{12}\left[x^1\partial^2 - x^2\partial^1 + \frac{1}{2}\gamma^1\gamma^2\right]\psi(x) \tag{9.47}$$

$$= \omega_{12}\left[(\vec{x}\times\vec{\nabla})^3 + \frac{1}{2}\gamma^1\gamma^2\right]\psi(x) \tag{9.48}$$

$$= i\omega_{12}\begin{bmatrix}\left[\vec{x}\times\left(\frac{1}{i}\vec{\nabla}\right) + \frac{1}{2}\vec{\sigma}\right]^3 & 0 \\ 0 & \left[\vec{x}\times\left(\frac{1}{i}\vec{\nabla}\right) + \frac{1}{2}\vec{\sigma}\right]^3\end{bmatrix}\psi(x) \tag{9.49}$$

where we have used the explicit form of the Dirac matrices given in equation (9.19) and σ^3 is a 2×2 Pauli matrix. Similarly, for the conjugate field,

$$\delta\bar{\psi}(x) = -i\omega_{12}\,\bar{\psi}(x)\begin{bmatrix}\left[\vec{x}\times\left(\frac{1}{i}\overleftarrow{\nabla}\right) + \frac{1}{2}\vec{\sigma}\right]^3 & 0 \\ 0 & \left[\vec{x}\times\left(\frac{1}{i}\overleftarrow{\nabla}\right) + \frac{1}{2}\vec{\sigma}\right]^3\end{bmatrix} \tag{9.50}$$

Indeed, for an infinitesimal rotation about an axis in the direction of the vector $\vec{\theta}$ by angle $|\vec{\theta}|$, the transformation is

$$\delta\psi(x) = i\begin{bmatrix}\vec{\theta}\cdot\left(\vec{x}\times\frac{1}{i}\vec{\nabla} + \frac{1}{2}\vec{\sigma}\right) & 0 \\ 0 & \vec{\theta}\cdot\left(\vec{x}\times\frac{1}{i}\vec{\nabla} + \frac{1}{2}\vec{\sigma}\right)\end{bmatrix}\psi(x)$$

$$\delta\bar{\psi}(x) = -i\bar{\psi}(x)\begin{bmatrix}\vec{\theta}\cdot\left(\vec{x}\times\frac{1}{i}\overleftarrow{\nabla} + \frac{1}{2}\vec{\sigma}\right) & 0 \\ 0 & \vec{\theta}\cdot\left(\vec{x}\times\frac{1}{i}\overleftarrow{\nabla} + \frac{1}{2}\vec{\sigma}\right)\end{bmatrix}$$

that is, a rotation is a combination of a rotation of the spatial coordinates, implemented by the angular momentum operator $\vec{L} = \vec{x}\times\vec{p} = \vec{x}\times\frac{1}{i}\vec{\nabla}$ and a rotation of the Dirac field spin, implemented by spin operator $\vec{S} = \frac{1}{2}\vec{\sigma}$ which is made from Pauli matrices. We note that it is only this simultaneous rotation of the coordinates and the spins which is a symmetry of the Dirac fields, unlike in the non-relativistic theories which we studied where rotations of the spatial and spin degrees of freedom were both symmetries by themselves. The coupling of rotation and spin symmetries that is contained in the Dirac equation is what gives rise to the spin-orbit coupling in physics, terms in the Hamiltonian of the form $\vec{L}\cdot\vec{S}$ which correct the non-relativistic limit.

Then, in particular, under infinitesimal rotations, the electron density and the mass operator transform like a scalar fields,

$$\delta\left(\psi^\dagger(x)\psi(x)\right) = \left(\vec{\theta}\cdot\vec{x}\times\vec{\nabla}\right)\left(\psi^\dagger(x)\psi(x)\right)$$

$$\delta\left(\bar{\psi}(x)\psi(x)\right) = \left(\vec{\theta}\cdot\vec{x}\times\vec{\nabla}\right)\left(\bar{\psi}(x)\psi(x)\right)$$

For a Lorentz transformation with infinitesimal velocity $\vec{v}$, the only non-zero components of $\omega_{\mu\nu}$ are $\omega_{0a} = v_a$ and $\omega_{a0} = -v_a$ and

$$\delta\psi(x) = \left[x^0 \vec{v} \cdot \vec{\nabla} - \vec{v} \cdot \vec{x}\partial^0 + \frac{1}{2}\gamma^0 \vec{v} \cdot \vec{\gamma} \right] \psi(x) \tag{9.51}$$

$$\delta\bar{\psi}(x) = \bar{\psi}(x) \left[x^0 \vec{v} \cdot \overleftarrow{\vec{\nabla}} - \vec{v} \cdot \vec{x}\overleftarrow{\partial}^0 - \frac{1}{2}\gamma^0 \vec{v} \cdot \vec{\gamma} \right] \tag{9.52}$$

$$\delta\psi^\dagger(x) = \psi^\dagger(x) \left[x^0 \vec{v} \cdot \overleftarrow{\vec{\nabla}} - \vec{v} \cdot \vec{x}\overleftarrow{\partial}^0 + \frac{1}{2}\gamma^0 \vec{v} \cdot \vec{\gamma} \right] \tag{9.53}$$

The mass operator transforms like a scalar field,

$$\delta\left(\bar{\psi}(x)\psi(x)\right) = (x^0\vec{v} \cdot \vec{\nabla} - \vec{v} \cdot \vec{x}\partial^0)\left(\bar{\psi}(x)\psi(x)\right)$$

However, the density is not a scalar, but has the transformation law

$$\delta\left(\psi^\dagger(x)\psi(x)\right) = (x^0\vec{v} \cdot \vec{\nabla} - \vec{v} \cdot \vec{x}\partial^0)\left(\psi^\dagger(x)\psi(x)\right) + \vec{v} \cdot \bar{\psi}(x)\vec{\gamma}\psi(x) \tag{9.54}$$

Of course, it transforms like the time-component of a vector field, consistent with the fact that it is the time-component of

$$\mathbf{J}^\mu(x) = \left(\psi^\dagger(x)\psi(x), -\bar{\psi}(x)\vec{\gamma}\psi(x)\right) = -\bar{\psi}(x)\gamma^\mu\psi(x) \tag{9.55}$$

Indeed, we could examine the full Lorentz transformation law and see that $\mathbf{J}^\mu(x)$ transforms like a vector field.

9.5   Phase Symmetry and the Conservation of Charge

In the above, we have examined the transformation law for the Dirac field density and we found that it transforms like the time-component of a vector field (9.55). We will see shortly that this vector field obeys the continuity equation, $\partial_\mu \mathbf{J}^\mu(x) = 0$, which we would expect for a "conserved current". When Dirac fermions are coupled to photons $\mathbf{J}^\mu(x)$ where e is the unit of electric charge will be identified with the electric charge and current densities. In non-relativistic terminology, $e\mathbf{J}^0(x)$ is the "charge density" and $e\vec{\mathbf{J}}(x)$ is the "current density". In relativistic physics, we simply call $\mathbf{J}^\mu(x)$ a "current" or a "conserved current".

The Dirac equation can be derived from an action

$$S = \int dx \mathcal{L}(x)$$

where the Lagrangian density is

$$\mathcal{L}(x) = -i\bar{\psi}(x)\left[\frac{1}{2}\overrightarrow{\partial\!\!\!/} - \frac{1}{2}\overleftarrow{\partial\!\!\!/} + m\right]\psi(x)$$

Here we have defined the action as the integral of the Lagrangian density. It should be kept in mind, of course, that the Dirac field describes fermions and therefore the Lagrangian density depends on anti-commuting classical fields $\psi(x)$ and $\bar{\psi}(x)$. The "$-i$" in front and the symmetrization of the derivative operators are present to make the Lagrangian density real. Moreover, the leading terms, up to total derivatives are

$$\mathcal{L}(x) = i\psi^\dagger(x)\frac{\partial}{\partial x^0}\psi(x) + \dots \tag{9.56}$$

which is compatible with the equal-time anti-commutation relations that we have used for the Dirac field. The Dirac equation is easily recovered from this Lagrangian density using the Euler–Lagrange equations. The Euler–Lagrange equation has the form

$$\frac{\partial\mathcal{L}}{\partial\bar{\psi}_a(x)} - \partial_\mu\frac{\partial\mathcal{L}}{\partial(\partial_\mu\bar{\psi}_a(x))} = 0 \tag{9.57}$$

$$\frac{\partial\mathcal{L}}{\partial\psi_a(x)} - \partial_\mu\frac{\partial\mathcal{L}}{\partial(\partial_\mu\psi_a(x))} = 0 \tag{9.58}$$

These lead to the Dirac equation and its conjugate

$$\left[\partial\!\!\!/ + m\right]\psi = 0, \quad \bar{\psi}\left[-\overleftarrow{\partial\!\!\!/} + m\right] = 0$$

respectively.

9.5.1 Conserved Number Current

We may wonder whether there is a concept of conserved particle number in the relativistic Dirac field theory. The particle number of the non-relativistic theory was $\mathcal{N} = \psi^\dagger(x)\psi(x)$ which we can write as $\mathcal{N} = -\bar{\psi}(x)\gamma^0\psi(x)$ which is the time component of a four-vector. The space part of that four-vector would be $\mathbf{J}^a(x) = -\bar{\psi}(x)\gamma^a\psi(x)$ and we can write a four-vector current density as

$$\mathbf{J}^\mu(x) = -\bar{\psi}(x)\gamma^\mu\psi(x) \tag{9.59}$$

We can confirm that this current is conserved by simply using the equation of motion, the Dirac equation. We form the current and we assume that the Dirac spinor satisfies the Dirac equation. Then,

$$-\partial_\mu\mathbf{J}^\mu(x) = \partial_\mu\left(\bar{\psi}(x)\gamma^\mu\psi(x)\right) = \bar{\psi}\left[\overrightarrow{\partial\!\!\!/} + \overleftarrow{\partial\!\!\!/}\right]\psi(x) = \bar{\psi}(x)[-m + m]\psi(x) = 0$$

This continuity equation for the current implies the conservation of charge, which we have so far called the number operator,

$$\frac{d}{dt}\mathcal{N} = \frac{d}{dt}\int d^3x\,\psi^\dagger(x)\psi(x) = \int d^3x\,\vec{\nabla}\cdot\bar{\psi}(x)\vec{\gamma}\psi(x) = \oint d^2\sigma\hat{n}\cdot\bar{\psi}(x)\vec{\gamma}\psi(x)$$

where we have used Gauss's theorem to write the last term as a surface integral of the normal component of the current density at the infinite boundary of three dimensional space. If our boundary conditions are such that the quantum expectation value of the current density goes to zero sufficiently rapidly there, the number operator is conserved.

9.5.2 Relativistic Noether's Theorem for the Dirac Equation

Let us assume that we have identified a transformation of the fields

$$\psi(x) \to \tilde{\psi}(x) = \psi(x) + \delta\psi(x) \tag{9.60}$$

$$\bar{\psi}(x) \to \tilde{\bar{\psi}}(x) = \bar{\psi}(x) + \delta\bar{\psi}(x) \tag{9.61}$$

which is a symmetry of the theory, that is, that the Lagrangian density varies by a quantity which, without use of the equations of motion, can be written as the time-derivative of a density and the divergence of a current density, which we can assemble into the four-divergence of a four-vector,

$$\delta\mathcal{L}(x) = \partial_\mu R^\mu(x) \tag{9.62}$$

Then, by our previous proof of Noether's theorem, we know that it involves a Noether charge density and a spatial current density which we can assemble to make a four-current density

$$\mathcal{J}^\mu(x) = \delta\psi_a(x)\frac{\partial\mathcal{L}(x)}{\partial(\partial_\mu\psi_a(x))} + \delta\bar{\psi}_a(x)\frac{\partial\mathcal{L}(x)}{\partial(\partial_\mu\bar{\psi}_a(x))} - R^\mu(x) \tag{9.63}$$

Noether's theorem tells us that this current density must obey the continuity equation which, in relativistic notation, is

$$\partial_\mu\mathcal{J}^\mu(x) = 0 \tag{9.64}$$

For the case of phase symmetry, the infinitesimal transformation is

$$\delta\psi(x) = -i\psi(x) \tag{9.65}$$

$$\delta\bar{\psi}(x) = i\bar{\psi}(x) \tag{9.66}$$

Then, examining the Lagrangian density, we find that, in this case, $\delta\mathcal{L}(x) = 0$ and therefore $R^\mu(x) = 0$. Plugging this and equations (9.65) and (9.66) into equation (9.63) we see that the Noether current is given by

$$\mathbf{J}^\mu(x) = -\bar{\psi}(x)\gamma^\mu\psi(x) \tag{9.67}$$

which is the current which we have identified in equation (9.59). We have already confirmed that it is conserved, as Noether's theorem tells us it must be.

9.5.3   Alternative Proof of Noether's Theorem

The proof of Noether's theorem that we outlined in the context of non-relativistic field theory in previous chapters is completely general and, as we have already noted, it applies irregardless of whether the field theory has relativistic symmetry or not. In this subsection, we note that there is an alternative proof of Noether's theorem which is sometimes convenient as it gives us a different (and sometimes faster) way to find the Noether current.

Let us assume that we have identified a transformation of the fields

$$\psi(x) \rightarrow \tilde{\psi}(x) + \epsilon\psi(x) \tag{9.68}$$

$$\bar{\psi}(x) \rightarrow \tilde{\bar{\psi}}(x) = \bar{\psi}(x) + \epsilon\delta\bar{\psi}(x) \tag{9.69}$$

which is a symmetry of the theory, that is, that the Lagrangian density varies by a quantity which, without use of the equations of motion, can be written as the four-divergence of a vector,

$$\delta\mathcal{L}(x) = \partial_\mu(\epsilon R^\mu(x)) \tag{9.70}$$

Here, we have explicitly included the infinitesimal parameter of the transformation, ϵ which, since the variation is only to linear order, just occur to linear order in $\delta\mathcal{L}(x)$.

Now, we ask the question, what if ϵ is not just a constant but it depends on the coordinates? The transformation

$$\psi(x) \rightarrow \tilde{\psi}(x) + \epsilon(x)\delta\psi(x) \tag{9.71}$$

$$\bar{\psi}(x) \rightarrow \tilde{\bar{\psi}}(x) = \bar{\psi}(x) + \epsilon(x)\delta\bar{\psi}(x) \tag{9.72}$$

is no longer a symmetry. But, since it becomes a symmetry when $\epsilon(x)$ is put to a constant, we know that the variation of the Lagrangian density must have the form

$$\delta\mathcal{L}(x) = \partial_\mu(\epsilon(x)R^\mu(x)) + \partial_\mu\epsilon(x)\mathcal{J}^\mu(x) \tag{9.73}$$

The first term is what we would get if $\epsilon(x)$ were a constant, independent of x. The second term is the correction to the expression which must occur when $\epsilon(x)$ depends on x. We have been assuming that the Lagrangian density depends only on the fields and the first derivatives of the fields. In that situation, the variation of the Lagrangian density can only produce terms with first order derivatives of $\epsilon(x)$. These are the last terms in equation (9.73).

Then, we can rewrite equation (9.73) as

$$\delta\mathcal{L}(x) = \partial_\mu(\epsilon(x)R^\mu(x)) + \partial_\mu[\epsilon(x)\mathcal{J}^\mu(x)] - \epsilon(x)\partial_\mu\mathcal{J}^\mu(x) \tag{9.74}$$

where $\mathcal{J}^\mu(x)$ is composed of the fields and their derivatives. Again, we emphasize that, when $\epsilon(x)$ becomes a constant, equation (9.73) becomes equation (9.70) and the transformation becomes a symmetry.

Now, we assume that, rather than becoming a constant, $\epsilon(x)$ is a function which vanishes at the asymptotic boundaries of space–time, so that (9.71) is a variation that we might take to derive the equation of motion. Then, at this point, let us assume that the equations of motion are obeyed. When the equations of motion are obeyed, the variation of the Lagrangian density must be the divergence of some quantity. To see this, consider

$$
\begin{aligned}
\delta\mathcal{L}(x) = {}& \epsilon(x)\delta\psi(x)\frac{\partial\mathcal{L}(x)}{\partial\psi(x)} + \epsilon(x)\delta\bar{\psi}(x)\frac{\partial\mathcal{L}(x)}{\partial\bar{\psi}(x)} \\
& + \partial_\mu[\epsilon(x)\delta\psi(x)]\frac{\partial\mathcal{L}(x)}{\partial(\partial_\mu\psi(x))} \\
& + \partial_\mu[\epsilon(x)\bar{\psi}(x)]\frac{\partial\mathcal{L}(x)}{\partial(\partial_\mu\bar{\psi}(x))} \\
= {}& \partial_\mu\left[\epsilon(x)\delta\psi(x)\frac{\partial\mathcal{L}(x)}{\partial(\partial_\mu\psi(x))} + \epsilon(x)\delta\bar{\psi}(x)\frac{\partial\mathcal{L}(x)}{\partial(\partial_\mu\bar{\psi}(x))}\right] \\
& + \epsilon(x)\delta\psi(x)\left[\frac{\partial\mathcal{L}(x)}{\partial\psi(x)} - \partial_\mu\frac{\partial\mathcal{L}(x)}{\partial(\partial_\mu\psi(x))}\right] \\
& + \epsilon(x)\delta\bar{\psi}(x)\left[\frac{\partial\mathcal{L}(x)}{\partial\bar{\psi}(x)} - \partial_\mu\frac{\partial\mathcal{L}(x)}{\partial(\partial_\mu\bar{\psi}(x))}\right] \\
= {}& \partial_\mu\left[\epsilon(x)\delta\psi(x)\frac{\partial\mathcal{L}(x)}{\partial(\partial_\mu\psi(x))} + \epsilon(x)\delta\bar{\psi}(x)\frac{\partial\mathcal{L}(x)}{\partial(\partial_\mu\bar{\psi}(x))}\right] \qquad (9.75)
\end{aligned}
$$

where we have used the Euler–Lagrange equation. This is a total divergence for any function $\epsilon(x)$. In order for equation (9.74) to have this form, the quantity $\mathcal{J}^\mu(x)$ which we identify in equation (9.73) must obey the conservation law

$$
\partial_\mu\mathcal{J}^\mu(x) = 0 \qquad (9.76)
$$

This is Noether's theorem and $\mathcal{J}^\mu(x)$ is the conserved Noether current density.

We can apply this latter version of Noether's theorem to phase symmetry. Recall that the infinitesimal transformation of the fields is

$$
\delta\psi(x) = -i\epsilon(x)\psi(x) \qquad (9.77)
$$
$$
\delta\bar{\psi}(x) = i\epsilon(x)\bar{\psi}(x) \qquad (9.78)
$$

Then, the variation of the Lagrangian density is

$$
\delta\mathcal{L}(x) = -\partial_\mu\epsilon(x)\bar{\psi}(x)\gamma^\mu\psi(x) \qquad (9.79)
$$

From this equation, we could identify the Noether current as $\mathbf{J}^\mu(x) = -\bar{\psi}(x)\gamma^\mu\psi(x)$ which is identical to what we obtained form our first pass at Noether's theorem.

One might wonder whether the currents that are obtained from the two versions of Noether's theorem are indeed identical. To see this, let us take the last line of equation (9.75),

$$\delta\mathcal{L}(x) = \partial_\mu\left[\epsilon(x)\delta\psi(x)\frac{\partial\mathcal{L}(x)}{\partial(\partial_\mu\psi(x))} + \epsilon(x)\delta\bar\psi(x)\frac{\partial\mathcal{L}(x)}{\partial(\partial_\mu\bar\psi(x))}\right] \tag{9.80}$$

$$\sim \partial_\mu\epsilon(x)\left[\delta\psi(x)\frac{\partial\mathcal{L}(x)}{\partial(\partial_\mu\psi(x))} + \delta\bar\psi(x)\frac{\partial\mathcal{L}(x)}{\partial(\partial_\mu\bar\psi(x))}\right] + \dots \tag{9.81}$$

$$= \partial_\mu\epsilon(x)\ \mathcal{J}^\mu(x) + \dots \tag{9.82}$$

where we have recorded all of the possible dependence of the right-hand-side on derivatives of $\epsilon(x)$. We see that, indeed

$$\mathcal{J}^\mu(x) = \delta\psi(x)\frac{\partial\mathcal{L}(x)}{\partial(\partial_\mu\psi(x))} + \delta\bar\psi(x)\frac{\partial\mathcal{L}(x)}{\partial(\partial_\mu\bar\psi(x))} \tag{9.83}$$

which is identical to our previous identification of the Noether current.

9.6   Spacetime Symmetry

9.6.1   Translation Invariance and the Stress Tensor

In the previous sections we have presented a detailed discussion of the Dirac equation and the quantum field theory of the field which obeys the Dirac equation. We observed there that this Dirac field theory can be derived from an action which is a functional of anti-commuting four-component fields $\psi(x)$ and $\bar\psi(x)$ and where the Lagrangian density is

$$\mathcal{L} = -i\bar\psi(x)\left[\slashed\partial + m\right]\psi(x) \tag{9.84}$$

Now, consider a space–time translation $x^\mu \to x^\mu + \epsilon^\mu$, where ϵ^μ is an infinitesimal constant four-vector. The infinitesimal transformation of the Dirac field is

$$\delta\psi(x) = -\epsilon^\mu\partial_\mu\psi(x), \ \ \delta\bar\psi(x) = -\epsilon^\mu\partial_\mu\bar\psi(x)$$

By inspection, and without use of the equation of motion, we see that, when the Dirac field transforms this way, the Lagrangian density transforms as

$$\delta\mathcal{L}(x) = \partial_\mu\left[-\epsilon^\mu\mathcal{L}(x)\right]$$

The fact that the Lagrangian density varies by the divergence of some quantity means that the infinitesimal translation is a symmetry of the theory. Of course, we already knew that this should be the case since, in the previous chapter, we have already seen that the equation of motion has this symmetry. Here, we shall use this observation to construct the Noether current corresponding to the symmetry.

To find the Noether current, we use Noether's theorem which tells us that the Noether current is

$$\mathcal{J}_\epsilon^\mu(x) = -\epsilon^\nu \partial_\nu \psi(x) \frac{\partial \mathcal{L}(x)}{\partial(\partial_\mu \psi(x))} - \epsilon^\nu \partial_\nu \bar{\psi}(x) \frac{\partial \mathcal{L}(x)}{\partial(\partial_\mu \bar{\psi}(x))} + \epsilon^\mu \mathcal{L}(x)$$

$$= \epsilon_\nu \mathbf{T}_0^{\mu\nu}(x)$$

where we give the quantity which appears in the Noether current a special symbol and name, the stress tensor

$$\mathbf{T}_0^{\mu\nu}(x) = \frac{i}{2} \bar{\psi}(x) \left(\gamma^\mu \partial^\nu - \gamma^\mu \overleftarrow{\partial}^\nu \right) \psi(x) \tag{9.85}$$

This version of the stress tensor is sometimes called the "canonical stress tensor". We will eventually "improve" it and the improved one will be called the "stress tensor". Noether's theorem tells us that

$$\partial_\mu \mathbf{T}_0^{\mu\nu}(x) = 0 \tag{9.86}$$

The conserved Noether charges that are the spatial integrals of the Noether charge density simply correspond to the energy and momentum four-vector for the Dirac field,

$$P^\mu = \int d^3x \, \mathbf{T}_0^{0\mu}(x) \tag{9.87}$$

with time and space components being the Hamiltonian and the linear momentum,

$$P^0 = i \int d^3x \, \bar{\psi}(x) \left[\vec{\gamma} \cdot \vec{\nabla} + m \right] \psi(x)$$

$$P^a = -i \int d^3x \, \psi^\dagger(x) \nabla^a \psi(x)$$

In both of the equations above, we have eliminated the time derivatives from the integrand by using the Dirac equation.

We can easily confirm the continuity equation (9.86) explicitly using the Dirac equation. Moreover, using the Dirac equation $(\slashed{\partial} + m)\psi(x) = 0$, which implies that $0 = (-\slashed{\partial} + m)(\slashed{\partial} + m)\psi(x) = (-\partial^2 + m^2)\psi(x)$ and also $\bar{\psi}(x)(-\overleftarrow{\partial}^2 + m^2) = 0$, it is easy to check that the canonical stress tensor $\mathbf{T}_0^{\mu\nu}(x)$ also obeys another continuity equation for derivatives by its second spacetime index,

$$\partial_\nu \mathbf{T}_0^{\mu\nu}(x) = \frac{i}{2} \bar{\psi}(x) \left(\gamma^\mu \partial^2 - \gamma^\mu \overleftarrow{\partial}^2 \right) \psi(x) = 0$$

This implies that both the symmetric and anti-symmetric parts of the canonical stress tensor,

$$\mathbf{T}^{\mu\nu}(x) \equiv \frac{1}{2}\left[\mathbf{T}_0^{\mu\nu}(x) + \mathbf{T}_0^{\mu\nu}(x)\right] \tag{9.88}$$

$$\mathbf{T}_A^{\mu\nu}(x) \equiv \frac{1}{2}\left[\mathbf{T}_0^{\mu\nu}(x) - \mathbf{T}_0^{\mu\nu}(x)\right] \tag{9.89}$$

obey continuity equations,

$$\partial_\mu \mathbf{T}^{\mu\nu}(x) = 0 \tag{9.90}$$

$$\partial_\mu \mathbf{T}_A^{\mu\nu}(x) = 0 \tag{9.91}$$

We will use this fact at some point in the following.

Alternative Derivation of the Noether Current
It is instructive to examine our alternative proof of Noether's theorem in this context. We can also consider an infinitesimal translation, as we have just done, but assuming that the transformation parameter depends on the coordinates, $\epsilon^\mu(x)$, and

$$\delta\psi(x) = -\epsilon^\mu(x)\partial_\mu\psi(x), \quad \delta\bar\psi(x) = -\epsilon^\mu(x)\partial_\mu\bar\psi(x)$$

The variation of the Lagrangian now can be written in the form

$$\delta\mathcal{L}(x) = \partial_\mu\left[\epsilon^\mu(x)\mathcal{L}(x)\right] - \partial_\mu\epsilon_\nu(x)\frac{i}{2}\bar\psi(x)\left(\gamma^\mu\overrightarrow{\partial}^\nu - \gamma^\mu\overleftarrow{\partial}^\nu\right)\psi(x)$$

This equation makes no assumptions about the nature of $\epsilon^\mu(x)$, other than that it is infinitesimal. If we allow it to go to a constant four-vector, we recover the fact that the Lagrangian density transforms as a total derivative. Our alternative proof of Noether's theorem tells us that, when the field equations are obeyed, the coefficient of $\partial_\mu\epsilon_\nu(x)$ must be the conserved Noether current, that is,

$$\partial_\mu\left[\frac{i}{2}\bar\psi(x)\left(\gamma^\mu\overrightarrow{\partial}^\nu - \gamma^\mu\overleftarrow{\partial}^\nu\right)\psi(x)\right] = 0$$

This is just the conservation law for the Noether current that we obtained and confirmed above in our first derivation.

9.6.2 Lorentz Transformations

Now, consider an infinitesimal Lorentz transformation, $x^\mu \to x^\mu + \omega^\mu_\nu x^\nu$ where the transformation of the Dirac field is given in equations (9.45) and (9.46), which we copy here

$$\delta\psi(x) = \omega_{\mu\nu}\left[x^{\mu}\partial^{\nu} + \frac{1}{8}\left[\gamma^{\mu}, \gamma^{\nu}\right]\right]\psi(x)$$

$$\delta\bar{\psi}(x) = \bar{\psi}(x)\left[x^{\mu}\overleftarrow{\partial}^{\nu} - \frac{1}{8}\left[\gamma^{\mu}, \gamma^{\nu}\right]\right]\omega_{\mu\nu}$$

Under this transformation, the Lagrangian density varies as

$$\delta\mathcal{L}(x) = \omega_{\mu\nu}x^{\mu}\partial^{\nu}\mathcal{L}(x) = \partial^{\nu}\left[\omega_{\mu\nu}x^{\mu}\mathcal{L}(x)\right]$$

The fact that the Lagrangian density varies by a total derivative term confirms that the Lorentz transformation is a symmetry of the theory. Of course, we already knew that this should be the case, since we have already demonstrated that the equation of motion has this symmetry.

To find the Noether current, let us assume that the transformation parameter depends on the coordinates, $\omega_{\mu\nu}(x)$. The variation of the Lagrangian now has the form

$$\delta\mathcal{L}(x) = \partial^{\nu}\left[\omega_{\mu\nu}(x)x^{\mu}\mathcal{L}(x)\right]$$
$$- \partial_{\lambda}\omega_{\rho\sigma}\frac{i}{2}\bar{\psi}(x)\left(\gamma^{\lambda}\left[x^{\rho}\partial^{\sigma} + \frac{1}{8}[\gamma^{\rho}, \gamma^{\sigma}]\right] - \left[\overleftarrow{\partial}^{\sigma}x^{\rho} - \frac{1}{8}[\gamma^{\rho}, \gamma^{\sigma}]\right]\gamma^{\lambda}\right)\psi(x)$$

Now, if $\omega_{\mu\nu}(x)$ vanishes sufficiently rapidly at the boundaries of the integral so that boundary terms can be ignored, and if we assume that the equations of motion are obeyed, the variation of the action must vanish for any variation, in particular, with any profile of $\omega_{\rho\sigma}(x)$. Then, we must have

$$\partial_{\lambda}M^{\lambda\rho\sigma}(x) = 0 \tag{9.92}$$

where the Noether current is given by

$$M^{\lambda\rho\sigma}(x) = T_0^{\lambda\sigma}(x)x^{\rho} - T_0^{\lambda\rho}(x)x^{\sigma} + \frac{i}{16}\bar{\psi}(x)\left\{\gamma^{\lambda}, [\gamma^{\rho}, \gamma^{\sigma}]\right\}\psi(x) \tag{9.93}$$

and where $T_0^{\mu\nu}(x)$ is the canonical stress tensor given in equation (9.85) and which was associated with translations. We could also find this current by using the more conventional Noether theorem.

Noether's theorem tells us that $M^{\lambda\rho\sigma}(x)$ must obey equation (9.92). However, we could also take the divergence of the expression for $M^{\lambda\rho\sigma}(x)$ in equation (9.93). Doing the derivatives explicitly, and remembering that $\partial_{\mu}T_0^{\mu\nu}(x) = 0$, we get the identity

$$T_A^{\sigma\rho}(x) = T_0^{\sigma\rho}(x) - T_0^{\rho\sigma}(x) = \partial_{\lambda}\left[\frac{i}{16}\bar{\psi}(x)\left\{\gamma^{\lambda}, [\gamma^{\rho}, \gamma^{\sigma}]\right\}\psi(x)\right] \tag{9.94}$$

This is an equation for the anti-symmetric part of $\mathbf{T}_0^{\mu\nu}(x)$, which states that it is given by the four-divergence of $\frac{i}{16}\bar{\psi}(x)\left\{\gamma^\lambda, [\gamma^\rho, \gamma^\sigma]\right\}\psi(x)$. The latter quantity is clearly anti-symmetric in the indices ρ and σ. We can use the algebraic properties of the gamma-matrices to see that the combination $\left\{\gamma^\lambda, [\gamma^\rho, \gamma^\sigma]\right\}$ is equal to zero unless the three indices λ, ρ, σ are all different. Therefore $\left\{\gamma^\lambda, [\gamma^\rho, \gamma^\sigma]\right\} = -4i\epsilon^{\lambda\rho\sigma\nu}\gamma^5\gamma_\nu$ where $\epsilon^{\lambda\rho\sigma\nu}$ is the totally anti-symmetric tensor with $\epsilon^{0123} = 1$ and $\gamma^5 = -i\gamma^0\gamma^1\gamma^2\gamma^3$.

Then, recalling that we can write $\mathbf{T}_0^{\mu\nu}(x)$ as its symmetric part $\mathbf{T}^{\mu\nu}(x)$ plus its anti-symmetric part which we have found above,

$$\mathbf{T}_0^{\rho\sigma}(x) = \mathbf{T}^{\rho\sigma}(x) + \epsilon^{\rho\sigma\lambda\nu}\partial_\lambda\left[\frac{1}{4}\bar{\psi}(x)\gamma^5\gamma_\nu\psi(x)\right] \tag{9.95}$$

Moreover

$$M_0^{\lambda\rho\sigma}(x) = M^{\lambda\rho\sigma}(x) + \partial_\gamma\left[\frac{1}{4}\left(\epsilon^{\lambda\sigma\gamma\nu}x^\rho - \epsilon^{\gamma\lambda\rho\nu}x^\sigma\right)\bar{\psi}(x)\gamma^5\gamma_\nu\psi(x)\right]. \tag{9.96}$$

$$M^{\lambda\rho\sigma}(x) = \mathbf{T}^{\lambda\sigma}(x)x^\rho - \mathbf{T}^{\lambda\rho}(x)x^\sigma \tag{9.97}$$

Now, we see that, for the improved stress tensor, we could simply use the symmetric part of the canonical stress tensor $\mathbf{T}_0^{\mu\nu}(x)$,

$$\mathbf{T}^{\mu\nu}(x) = \frac{i}{4}\bar{\psi}(x)[\gamma^\mu\partial^\nu + \gamma^\nu\partial^\mu - \gamma^\mu\overleftarrow{\partial}^\nu - \gamma^\nu\overleftarrow{\partial}^\mu]\psi(x) \tag{9.98}$$

$$\mathbf{T}^{\mu\nu}(x) = \mathbf{T}^{\nu\mu}(x), \quad \partial_\lambda\mathbf{T}^{\nu\mu}(x) = 0 \tag{9.99}$$

This is the improved stress tensor for the Dirac field. Since we will use it everywhere from now on, we will simply call it the "stress tensor".

The spatial integral of the stress tensor is the four-momentum. It is unaffected by the replacement of $\mathbf{T}_0^{\mu\nu}(x)$ by $\mathbf{T}^{\mu\nu}(x)$ when the fields fall off fast enough at spatial infinity, since

$$P^\mu = \int d^3x\,\mathbf{T}_0^{0\mu}(x) = \int d^3x\,\mathbf{T}^{0\mu}(x) + \int d^3x\,\nabla_a\epsilon^{a0\mu c}\bar{\psi}(x)\gamma^5\gamma^c\psi(x)$$

where Gauss's theorem could be used to write $\int d^3x\,\nabla_a\epsilon^{a0\mu c}\bar{\psi}(x)\gamma^5\gamma^c\psi(x)$ as a surface integral of the normal component of $\epsilon^{a0\mu c}\bar{\psi}(x)\gamma^5\gamma^c\psi(x)$ on the sphere at infinity. We shall assume that the integrand vanishes sufficiently rapidly that the surface integral is zero. Then

$$P^\mu = \int d^3x\,\mathbf{T}^{0\mu}(x)$$

In addition

$$M^{\rho\sigma} = \int d^3x\,M^{0\rho\sigma}(x)$$

9.6.3   Stress Tensor and Killing Vectors

There are good reasons why it is convenient to have a symmetric stress tensor. By modifying $\mathbf{T}^{\rho\sigma}(x)$ to make it symmetric, we will be able to unify the generator of translations and Lorentz transformations.

An infinitesimal general coordinate transformation is

$$x^{\mu} \rightarrow x^{\mu} + f^{\mu}(x)$$

If we recall that a symmetry of Minkowski space is a special case of a general coordinate transformation which is generated by a Killing vector $\hat{f}^{\mu}(x)$, that is, a vector field which satisfies the Killing equation of Minkowski space,

$$\partial_{\mu}\hat{f}_{\nu}(x) + \partial_{\nu}\hat{f}_{\mu}(x) = 0$$

We might make a candidate for a conserved current by contracting the stress tensor with the vector field which generates a general coordinate transformation, $\mathbf{T}_{f}^{\mu}(x) \equiv \mathbf{T}^{\mu}{}_{\nu}(x) f^{\nu}(x)$. Then, to have a conservation law, we need

$$\partial_{\mu}\mathbf{T}_{f}^{\mu}(x) = 0$$

Using a bit of simple algebra, we see that this conservation law indeed exists when $\mathbf{T}^{\mu\nu}(x)$ is conserved, i.e. $\partial_{\mu}\mathbf{T}^{\mu\nu}(x) = 0$ and also where $\mathbf{T}^{\mu\nu}(x)$ is symmetric and where the vector field $f^{\mu}(x)$ obeys the Killing equation, $\partial_{\mu}\hat{f}_{\nu}(x) + \partial_{\nu}\hat{f}_{\mu}(x) = 0$. Thus, we conclude that the Noether current for any symmetry of Minkowski space is given by

$$\mathbf{T}_{f}^{\mu}(x) \equiv \mathbf{T}^{\mu}{}_{\nu}(x)\hat{f}^{\nu}(x) \tag{9.100}$$

where $\hat{f}^{\nu}(x)$ is the Killing vector corresponding to the symmetry.

We might ask whether the stress tensor can also be used in the case of a search for conformal symmetry, that is, if equation (9.100) yields a conserved charge if $\hat{f}^{\mu}(x)$ satisfies the conformal Killing equation, rather than just the Killing equation. Remember that the conformal Killing equation is

$$\partial_{\mu}\hat{f}_{\nu}(x) + \partial_{\nu}\hat{f}_{\mu}(x) - \frac{\eta_{\mu\nu}}{2}\partial_{\lambda}\hat{f}^{\lambda}(x) = 0$$

By simply testing the conservation law, we see that the current defined in equation (9.100) would be conserved if $\hat{f}^{\mu}(x)$ satisfied the conformal Killing equation and if the stress energy tensor obeyed

$$\partial_{\mu}\mathbf{T}^{\mu\nu}(x) = 0, \quad \mathbf{T}^{\mu\nu}(x) = \mathbf{T}^{\nu\mu}(x), \quad \mathbf{T}^{\mu}{}_{\mu}(x) = 0$$

The first of two of these condition, conservation and symmetry, are needed for the usual space–time symmetries. The third, that the stress tensor be traceless, is the additional condition that is needed for conformal invariance. It is straightforward to check that the stress tensor of our Dirac theory obeys this condition when the fermion mass is equal to zero, that is, it is a conformal field theory when $m = 0$.

Photons

10

We must now turn from our treatment of relativistic fermions to a bosonic degree
of freedom, the massless vector field. The photon, the particle that is responsible
for electromagnetic interactions, is an example of such a field. It, together with the
Dirac field which we studied in the previous chapter, comprise the fields of quantum
electrodynamics which will an example of an interacting quantum field theory that
will be of interest to us.

The classical electric and magnetic fields are familiar to us from classical electrodynamics. In fact, these classical fields play an important role in nature and they
are described by a successful classical field theory whose classical field equations
are Maxwell's equations. That classical field theory contains the electromagnetic
field, whose physical manifestation is familiar to us as the electric field $\vec{E}(x)$ and
the magnetic field $\vec{B}(x)$. These are both spatial three-vector fields. They appear in
abundance in the physical world.

We wish to study how these arise from a quantum theory. To do so, we begin by
remembering that the low energy states of weakly interacting quantum Fermi and
Bose gases are very different. Fermions have a Fermi surface, particles and holes
and we have made use liberal of these concepts to construct quantum field theories
of non-relativistic fermions and their relativistic analog, the Dirac theory in the
previous chapter. Bose gases, on the other hand, tend to occupy the lowest possible
energy states. An important feature that we encountered in our discussion of the
non-relativistic Bose field was that the field operator of the quantum field theory
could be decomposed into a "classical part" and a "quantum part". In that case of
the non-relativistic Bose field, we had $\psi(\vec{x}, t) = \eta(\vec{x}, t) + \tilde{\psi}(\vec{x}, t)$ where $\psi(\vec{x}, t)$ and
$\tilde{\psi}(\vec{x}, t)$ were quantized fields obeying equal time commutation relations and $\eta(\vec{x}, t)$
was a classical field. Moreover, when the interactions were very weak, we showed
that, to a good approximation, the classical part of the Bose field obeyed a field

© The Author(s), under exclusive license to Springer Nature Singapore Pte Ltd. 2023   181
G. W. Semenoff, *Quantum Field Theory*, Graduate Texts in Physics,
https://doi.org/10.1007/978-981-99-5410-0_10

equation which was identical to that of the quantum field theory with classical fields taking the place of the quantum fields.

For a quantum theory of electromagnetic fields, we can invert the logic of that discussion. We know the field equations that classical electric and magnetic fields must obey, they are Maxwell's equations of classical electrodynamics. What is more (perhaps a posteriori) we know that electrodynamics is weakly coupled. The strength of interactions is governed by the fine structure constant

$$\alpha \equiv \frac{e^2}{4\pi\hbar c\epsilon_0} \approx \frac{1}{137.035999...} \tag{10.1}$$

where e is the charge of a typical elementary particle such as the electron and ϵ_0 is the electric permittivity.

If classical fields which obey Maxwell's equations are the classical parts of quantum fields, and if the quantum field theory is weakly coupled, we might expect that the quantum field theory is a theory whose field equation is obtained by simply replacing the classical electric and magnetic fields in Maxwell's equations by their corresponding quantum field operators. This expectation will turn out to be correct. In fact, Maxwell's equations as the field equations for the quantum field theory of electromagnetism turns out to be the only known fully consistent formulation of a quantum mechanical theory of electromagnetic fields in four space–time dimensions. This is due to the combined result of the requirements of spacetime symmetries and the requirement of mathematical consistency of the quantum field theory, the criterion of renormalizability, which we shall learn about in later chapters.

10.1 Relativistic Classical Electrodynamics

Classical electrodynamics is governed by Maxwell's equations which are partial differential equations for the electric and magnetic fields, $\vec{E}$ and $\vec{B}$, respectively,

$$\vec{\nabla} \cdot \vec{E}(\vec{x}, t) = \rho(\vec{x}, t) \tag{10.2}$$

$$-\dot{\vec{E}}(\vec{x}, t) + \vec{\nabla} \times \vec{B}(\vec{x}, t) = \vec{\mathbf{J}}(\vec{x}, t) \tag{10.3}$$

$$\vec{\nabla} \cdot \vec{B}(\vec{x}, t) = 0 \tag{10.4}$$

$$\dot{\vec{B}}(\vec{x}, t) + \vec{\nabla} \times \vec{E}(\vec{x}, t) = 0 \tag{10.5}$$

We are working in a system of units where the constants ϵ_0 and μ_0 that sometimes appear in these equations are set equal to one. (This requires that the speed of light be set equal to one.) We are quoting Maxwell's equations with sources, a charge density $\rho(\vec{x}, t)$ and a current density $\vec{\mathbf{J}}(\vec{x}, t)$ which we will leave unspecified for now. Maxwell's equations are internally consistent only when the charge and current densities satisfy the continuity equation

$$\frac{\partial}{\partial t}\rho(\vec{x}, t) + \vec{\nabla} \cdot \vec{\mathbf{J}}(\vec{x}, t) = 0 \tag{10.6}$$

To put Maxwell's equations into relativistic notation, we identify the electric and magnetic fields with the components of an anti-symmetric two-index tensor field, $F^{\mu\nu}(x)$ as

$$E_i(\vec{x}, t) \equiv F^{0i} \tag{10.7}$$

$$\epsilon^{ijk} B_k(\vec{x}, t) \equiv F^{ij}(x) \tag{10.8}$$

and the charge and current densities as a four-current

$$\left(\rho(\vec{x}, t), \vec{\mathbf{J}}(\vec{x}, t)\right) = \left(\mathbf{J}^0(\vec{x}, t), \vec{\mathbf{J}}(\vec{x}, t)\right) \equiv \mathbf{J}^\mu(x) \tag{10.9}$$

and the continuity equation is

$$\partial_\mu \mathbf{J}^\mu(x) = 0 \tag{10.10}$$

With this notation, Maxwell's equations become

$$\partial_\nu F^{\mu\nu}(x) = \mathbf{J}^\mu(x) \tag{10.11}$$

$$\partial_\mu F_{\nu\lambda}(x) + \partial_\nu F_{\lambda\mu}(x) + \partial_\lambda F_{\mu\nu}(x) = 0 \tag{10.12}$$

These are the relativistic form of Maxwell's equations. They are the field equation of classical electrodynamics. We note that, since $F^{\mu\nu}(x)$ is an anti-symmetric tensor, $\partial_\mu \partial_\nu F^{\mu\nu}(x) = 0$ and equation (10.11) implies $\partial_\mu \mathbf{J}^\mu(x) = 0$. The current must be conserved if equations (10.11) and (10.12) are to be internally consistent.

In equations (10.11) and (10.12) we have presented the field equations of classical electrodynamics and they are also the equations that we expect that the quantized electromagnetic fields must obey. However, we still do not have a guide to determining the operator nature of the fields. The operator nature of the fields is defined by the commutation relations which we have yet to find. Our strategy for finding commutation relations will be to construct a Lagrangian density from which the field equations can be derived by using a variational principle. Then we will deduce the commutation relations by examining the time derivative terms in the Lagrangian density.

Finding a Lagrangian density requires an important preliminary step which involves identifying the appropriate dynamical variable. This is the variable which should be varied when we use the variational principle to find the field equations. This variable turns out to be the four-vector potential field $A_\mu(x)$. This field is normally introduced in order to solve equation (10.12). Consider the ansatz

$$F_{\mu\nu}(x) = \partial_\mu A_\nu(x) - \partial_\nu A_\mu(x) \tag{10.13}$$

By plugging it into equation (10.12) we can see that equation is satisfied identically for any $A_\mu(x)$. Then, equation (10.11) becomes an equation for $A_\mu(x)$, it is

$$\left(-\partial^2 \delta^\mu_\nu + \partial^\mu \partial_\nu\right) A^\nu(x) = \mathbf{J}^\mu(x) \tag{10.14}$$

In classical electrodynamics, this is a partial differential wave-equation which we must solve in order to find the four-vector potential $A_\mu(x)$ which we then plug into formula (10.13) to find the field strengths $\vec{E}(x)$ and $\vec{B}(x)$.

Now, we are ready to write down an action and a Lagrangian density from which equation (10.11) or (10.14) can be derived. The action is, as usual the space–time volume integral of the Lagrangian density

$$S[A] = \int dx \mathcal{L}(x)$$

where the Lagrangian density is

$$\mathcal{L}(x) = -\frac{1}{4} F_{\mu\nu}(x) F^{\mu\nu}(x) + A_\mu(x) J^\mu(x) \qquad (10.15)$$

In the Lagrangian density, $A_\mu(x)$ is the basic dynamical variable, that is, the variable that should be varied in order to find the field equations or test for symmetries. The field strength tensor $F_{\mu\nu}(x)$ is assumed to be constructed from $A_\mu(x)$ as in equation (10.13). It is easy to check that the Euler–Lagrange equations, applied to the field $A_\mu(x)$,

$$\partial_\mu \frac{\partial \mathcal{L}}{\partial(\partial_\mu A_\nu(x))} - \frac{\partial \mathcal{L}}{\partial A_\nu(x)} = 0$$

reproduce the field equation (10.11) or (10.14).

Now, to implement the next step, which finds the commutation relations, we examine the time derivative terms in the Lagrangian density. They are

$$\mathcal{L}(x) = \frac{1}{2} \left(\frac{\partial}{\partial t} \vec{A}(x) \right)^2 - \frac{\partial}{\partial t} \vec{A}(x) \cdot \vec{\nabla} A_0(x) + \dots$$

Here, we see the first major complication in the quantization of the photon. The Lagrangian density does not contain the time derivative of the temporal component of the vector potential field $A_0(x)$. The momentum conjugate to the time component of the gauge field vanishes. Whenever the relationship between the generalized velocities, $\frac{\partial}{\partial t} A_\mu(x)$, and the generalized momenta, $\Pi^\mu(x) = \frac{\partial \mathcal{L}}{\partial(\frac{\partial}{\partial t} A_\mu(x))}$, cannot be solved for the velocities, the dynamical system has constraints. Electrodynamics is a dynamical system with constraints.

We might have reasonably expected that this description of electrodynamics is as a constrained system. After all, in four dimensions, we expect that the photon has two physical polarizations. However, the vector potential field contains four functions, it would seem to be too many to describe the photon. Indeed, this is so and the manifestation of this fact is the constraint that we find when we begin quantizing the theory.

There are systematic ways to study constrained systems which could be, and which routinely are exploited at this point. However, we will choose a technically simpler (albeit perhaps logically not quite as clear) approach which exploits the gauge invariance of the theory, which we will now review.

Gauge invariance of Maxwell's theory is the fact that, if we make the replacement

$$A_\mu(x) \to \tilde{A}_\mu(x) = A_\mu(x) + \partial_\mu \chi(x) \tag{10.16}$$

the field strength tensor

$$
\begin{aligned}
\tilde{F}_{\mu\nu}(x) &= \partial_\mu \tilde{A}_\nu(x) - \partial_\nu \tilde{A}_\mu(x) \\
&= \partial_\mu A_\nu(x) - \partial_\nu A_\mu(x) + (\partial_\mu \partial_\nu - \partial_\nu \partial_\mu)\chi(x) \\
&= \partial_\mu A_\nu(x) - \partial_\nu A_\mu(x) \\
&= F_{\mu\nu}(x)
\end{aligned}
$$

is left unchanged. Then, Maxwell's equations (10.11) or (10.14) are also left unchanged. The Lagrangian density changes by a total derivative,

$$\mathcal{L} \to \mathcal{L} + \partial_\mu \left(\chi(x) \mathbf{J}^\mu(x) \right)$$

(This makes use of conservation of the current $\partial_\mu \mathbf{J}^\mu(x) = 0$.)

Gauge invariance is not a real symmetry as the noun "invariance" would imply. Instead it reflects the redundancy of the set of four dynamical variables that appear in $A_\mu(x)$. We can make use of this redundancy to enforce a constraint on $A_\mu(x)$ called a gauge condition. For various reasons, mainly to do with technical simplicity of calculations that we shall do later on, we are mostly interested in Lorentz invariant gauge conditions such as

$$\partial_\mu A^\mu(x) = 0 \tag{10.17}$$

which can always be imposed by exploiting the gauge invariance. To see this, observe that, if a generic $A_\mu(x)$ did not obey this condition, we could find a new field $\tilde{A}_\mu(x) = A_\mu(x) + \partial_\mu \chi(x)$ which does

$$\partial_\mu \tilde{A}^\mu(x) = \partial_\mu A^\mu(x) + \partial^2 \chi = 0$$

if

$$-\partial^2 \chi(x) = \partial_\mu A^\mu(x)$$

The latter wave equation should always have a solution for $\chi(x)$, always allowing us to fix our relativistic gauge condition (10.17).

At this point, we should note that we have the freedom to gauge transform and to fix a gauge condition such as the relativistic one that we have chosen in equation (10.17) exists only if the physical quantities that we will eventually consider in the quantum theory are invariant under gauge transforms. That is, if the observables that our quantization of the photon will eventually study are quantities such as the field strength tensor $F_{\mu\nu}(x)$, rather than, for example, the basic dynamical variable which is the four-vector potential $A_\mu(x)$. We will be reminded of this later on in our quantization of the photon.

We can thus, alternatively, present Maxwell's equations as a wave equation and a constraint,

$$-\partial^2 A^\mu(x) = J^\mu(x) \tag{10.18}$$

$$\partial_\mu A^\mu(x) = 0 \tag{10.19}$$

It is easy to see that the constraint is consistent with the time evolution which is governed by the wave equation (10.18). To see this, we operate ∂_μ on both sides of (10.18). We use conservation of the current, $\partial_\mu J^\mu(x) = 0$ to find

$$-\partial^2 \left(\partial_\mu A^\mu(x) \right) = 0 \tag{10.20}$$

We see that, even when sources are present, the constraint obeys the free wave equation (10.20). The solutions of this free wave equation are such that, if at an initial time x_i^0, both $\partial_\mu A^\mu(x)\big|_{x^0=x_i^0} = 0$ and $\frac{\partial}{\partial x^0}\partial_\mu A^\mu(x)\big|_{x^0=x_i^0} = 0$, then $\partial_\mu A^\mu(x) = 0$ at all times thereafter.[1] Thus, once we set $\partial_\mu A^\mu(x) = 0$ and $\frac{\partial}{\partial x^0}\partial_\mu A^\mu(x) = 0$ at some time, it will be zero at all times. This is the sense in which the constraint is consistent with the time evolution in the classical field theory.

---

[1] To the general solution of this initial value problem explicitly, let us define a advanced Green function

$$-\partial^2 g_A(x, y) = \delta^{(4)}(x, y) , \quad g_A(x, y) = 0 \text{ if } x^0 > y_0$$

and consider the quantity

$$\partial_\nu A^\nu(x) = \int_{y_1^0}^{y_2^0} dy^0 \int d^3y \left(\partial_\nu A^\nu(y) \right) \delta^{(4)}(y, x) , \quad y_1^0 < x^0 < y_2^0$$

$$= \int_{y_1^0}^{y_2^0} dy^0 \int d^3y \left(\partial_\nu A^\nu(y) \right) \left(-\partial_y^2 g_A(y, x) \right) = \int_{y_1^0}^{y_2^0} dy^0 \int d^3y \partial_\mu \left(\partial_\nu A^\nu(y) \right) \left(\overleftarrow{\partial}_y^\mu - \overrightarrow{\partial}_y^\mu \right) g_A(y, x)$$

$$= \int d^3y \left(\partial_\nu A^\nu(y) \right) \left(\overleftarrow{\partial}_y^0 - \overrightarrow{\partial}_y^0 \right) g_A(y, x) \Big|_{y_1^0}^{y_2^0} = \int d^3y \left(\partial_\nu A^\nu(y) \right) \left(\overleftarrow{\partial}_y^0 - \overrightarrow{\partial}_y^0 \right) g_A(y, x) \Big|_{y^0=y_1^0}$$

where we have used Gauss' theorem to write the volume integral of the four divergence as a surface integral over the boundaries of spacetime, we have assumed that the integrand goes to zero at the spatial boundaries sufficiently rapidly that the spatial boundary integrals can be dropped and we have used the equation $\partial^2 \left(\partial_\nu A^\nu(y) \right) = 0$. Then, assuming the advanced Green function is known (as it is, of course), the final expression evaluates $\partial_\nu A^\nu(x)$ for all $x^0 > y_1^0$ if $\partial_\nu A^\nu(y)$ and $\partial^0 \left(\partial_\nu A^\nu(y) \right)$ are known at $y^0 = y_1^0$. In particular, if $\partial_\nu A^\nu(y_1^0, \vec{y}) = 0$ and $\partial^0 \left(\partial_\nu A^\nu(y) \right)_{y^0=y_1^0} = 0$ then $\partial_\nu A^\nu(x) = 0$ forever thereafter.

10.2 Quantization

We have argued that, at the classical level, the field equation and constraint in equations (10.18) and (10.19) are equivalent to Maxwell's equations for the electric and magnetic fields. Our strategy for going to the quantum theory from the classical theory will be to study the quantum field theory that is defined by the wave equation (10.18) and then, on the set of all solutions of that wave equation, we shall choose a subset of those solutions where the gauge condition (10.19) is satisfied. In practice, we will do this by considering the quantum states that we find in the Heisenberg picture of the quantum field theory and then choosing a subset of those states where the gauge condition (10.19) is satisfied. That subset of states will be called "physical states".

The wave equation (10.18) can be derived from the Euler–Lagrange equations applied to the Lagrangian density

$$\mathcal{L}(x) = -\frac{1}{2}\partial_\mu A_\nu(x)\partial^\mu A^\nu(x) + A_\mu(\vec{x})\mathbf{J}^\mu(x) \tag{10.21}$$

The time derivative terms in this Lagrangian density (10.21) are given by

$$\mathcal{L}(x) = \frac{1}{2}\frac{\partial}{\partial t}A_\mu(x)\eta^{\mu\nu}\frac{\partial}{\partial t}A_\nu(x) + \dots \tag{10.22}$$

From these time-derivative terms, we can identify the generalized momenta

$$\Pi^\mu(x) = \frac{\partial \mathcal{L}}{\partial(\frac{\partial}{\partial t}A_\mu(x))} = \frac{\partial}{\partial t}A_\mu(x)$$

The Poisson bracket would be

$$\left\{A_\mu(\vec{x},t), \Pi_\nu(\vec{y},t)\right\}_{PB} = \eta_{\mu\nu}\delta(\vec{x}-\vec{y})$$
$$\left\{A_\mu(\vec{x},t), A_\nu(\vec{y},t)\right\}_{PB} = 0, \quad \left\{\Pi_\mu(\vec{x},t), \Pi_\nu(\vec{y},t)\right\}_{PB} = 0$$

or

$$\left\{A_\mu(\vec{x},t), \frac{\partial}{\partial t}A_\nu(\vec{y},t)\right\}_{PB} = \eta_{\mu\nu}\delta(\vec{x}-\vec{y})$$
$$\left\{A_\mu(\vec{x},t), A_\nu(\vec{y},t)\right\}_{PB} = 0, \quad \left\{\frac{\partial}{\partial t}A_\mu(\vec{x},t), \frac{\partial}{\partial t}A_\nu(\vec{y},t)\right\}_{PB} = 0$$

In the quantum field theory, these Poisson brackets are replaced by the equal-time commutation relations

$$\left[A_\mu(\vec{x},t), \frac{\partial}{\partial t}A_\nu(\vec{y},t)\right] = i\eta_{\mu\nu}\delta(\vec{x}-\vec{y}) \tag{10.23}$$

$$\left[A_\mu(\vec{x},t), A_\nu(\vec{y},t)\right] = 0, \quad \left[\frac{\partial}{\partial t}A_\mu(\vec{x},t), \frac{\partial}{\partial t}A_\nu(\vec{y},t)\right] = 0 \tag{10.24}$$

(We have set $\hbar = 1$, otherwise it would be a factor on the right-hand-side of equation (10.23).) Here, we have used the assumption that the photons will be bosons so that we have used commutation rather than anti-commutation relations.

This leaves us with the quantum field theory specified by the field equation (10.18) that is obtained by applying the Euler–Lagrange equation to the Lagrangian density (10.21) and with the equal-time commutation relations in equations (10.23) and (10.24).

Let us first examine the wave equation (10.18) with the sources set to zero, $\mathbf{J}^\mu(x) = 0$,

$$-\partial^2 A^\mu(x) = 0 \qquad (10.25)$$

This equation is solved by plane waves

$$\exp\left(i\vec{k}\cdot\vec{x} - i|\vec{k}|x^0\right) , \quad \exp\left(i\vec{k}\cdot\vec{x} + i|\vec{k}|x^0\right)$$

with positive and negative energies given by $k^0 = |\vec{k}|$ and $k^0 = -|\vec{k}|$, respectively. The general solution of the free wave equation is a superposition of these plane waves,

$$A_\mu(x) = \int \frac{d^3k}{\sqrt{(2\pi)^3 2|\vec{k}|}} \left[e^{ik_\nu x^\nu} a_\mu(\vec{k}) + e^{-ik_\nu x^\nu} a_\mu^\dagger(\vec{k}) \right] \qquad (10.26)$$

where, inside the integral, the temporal component of the wave-vector is fixed so that $k^\mu = (|\vec{k}|, \vec{k})$. Notice that we have used the negative energy states differently from our previous use of them for non-relativistic fermions or for solutions of the Dirac equation where they were associated with holes or anti-particles. Here, in equation (10.26) we have used them in order to make the field Hermitian, so that $\left(A_\mu(x)\right)^\dagger = A_\mu(x)$, which is the quantum mechanical mirror of the fact that the four-vector potential $A_\mu(x)$ is real, $\left(A_\mu(x)\right)^* = A_\mu(x)$. Of course, this must be so at the classical level. At the quantum level, it could be said that the photon is its own antiparticle.

This solution (10.26) of the wave equation obeys the equal-time commutation relations (10.23) and (10.24) if the creation and annihilation operators obey the commutation relations

$$\left[a_\mu(\vec{k}), a_\nu^\dagger(\vec{\ell}) \right] = \eta_{\mu\nu} \delta(\vec{k} - \vec{\ell}) \qquad (10.27)$$

$$\left[a_\mu(\vec{k}), a_\nu(\vec{\ell}) \right] = 0, \quad \left[a_\mu^\dagger(\vec{k}), a_\nu^\dagger(\vec{\ell}) \right] = 0 \qquad (10.28)$$

The wave-functions of the positive and negative energy states are

$$\phi_k^+(x) = \frac{e^{ik_\nu x^\nu}}{\sqrt{(2\pi)^3 2|\vec{k}|}}, \quad \phi_k^-(x) = \frac{e^{-ik_\nu x^\nu}}{\sqrt{(2\pi)^3 2|\vec{k}|}}, \quad k_\mu = (-|\vec{k}|, \vec{k}) \qquad (10.29)$$

respectively, and they are orthonormal in the sense that

$$\int d^3x \phi_k^{+*}(x) \left(i\frac{\overrightarrow{\partial}}{\partial x^0} - i\frac{\overleftarrow{\partial}}{\partial x^0} \right) \phi_{k'}^{+}(x) = \delta^3(\vec{k} - \vec{k}') \tag{10.30}$$

$$\int d^3x \phi_k^{-*}(x) \left(i\frac{\overrightarrow{\partial}}{\partial x^0} - i\frac{\overleftarrow{\partial}}{\partial x^0} \right) \phi_{k'}^{-}(x) = -\delta^3(\vec{k} - \vec{k}') \tag{10.31}$$

$$\int d^3x \phi_k^{+*}(x) \left(i\frac{\overrightarrow{\partial}}{\partial x^0} - i\frac{\overleftarrow{\partial}}{\partial x^0} \right) \phi_{k'}^{-}(x) = 0 \tag{10.32}$$

$$\int d^3x \phi_k^{-*}(x) \left(i\frac{\overrightarrow{\partial}}{\partial x^0} - i\frac{\overleftarrow{\partial}}{\partial x^0} \right) \phi_{k'}^{+}(x) = 0 \tag{10.33}$$

The annihilation and creation operators are projected out of the field operator by the integrals

$$a_\mu(\vec{k}) = \int d^3x \phi_k^{+*}(x) \left(i\frac{\overrightarrow{\partial}}{\partial x^0} - i\frac{\overleftarrow{\partial}}{\partial x^0} \right) A_\mu(x) \tag{10.34}$$

$$a_\mu^\dagger(\vec{k}) = -\int d^3x \phi_k^{-*}(x) \left(i\frac{\overrightarrow{\partial}}{\partial x^0} - i\frac{\overleftarrow{\partial}}{\partial x^0} \right) A_\mu(x) \tag{10.35}$$

The operators $a_\mu(\vec{k})$ and $a_\mu^\dagger(\vec{k})$ have the properties of annihilation and creation operators and we can construct a Fock space by beginning with a vacuum state, $|\mathcal{O}>$ which is annihilated by all of the annihilation operators and is unit normalized,

$$a_\mu(\vec{k})|\mathcal{O}> = 0, \quad <\mathcal{O}|a_\mu^\dagger(\vec{k}) = 0, \quad \forall \vec{k}, \mu, \quad <\mathcal{O}|\mathcal{O}> = 1$$

We then create a basis of multi-photon states from this vacuum state by operating on it with creation operators

$$\left\{ |\mathcal{O}>, \; a_\mu^\dagger(\vec{k})|\mathcal{O}>, \; \ldots, \; \frac{1}{\sqrt{n!}} a_{\mu_1}^\dagger(\vec{k}_1) a_{\mu_2}^\dagger(\vec{k}_2) \ldots a_{\mu_n}^\dagger(\vec{k}_n)|\mathcal{O}>, \; \ldots \right\}$$

These states span the Fock space. A generic state in this Fock space is a superposition of the basis states,

$$|\chi> = \sum_{n=0}^\infty \frac{1}{\sqrt{n!}} \int d^3k_1 \ldots d^3k_n \chi^{\mu_1 \cdots \mu_n}(\vec{k}_1, \ldots, \vec{k}_n) a_{\mu_1}^\dagger(\vec{k}_1) \ldots a_{\mu_n}^\dagger(\vec{k}_n)|\mathcal{O}> \tag{10.36}$$

where $\chi^{\mu_1 \cdots \mu_n}(\vec{k}_1, \ldots, \vec{k}_n)$ is a completely symmetric function under permutations of the labels $\mu_i, k_i \leftrightarrow \mu_j, k_j$.

We can use the commutation relations for the creation and annihilation operators to compute the norm of any state. For example,

$$
\begin{aligned}
||\ |\chi> || &=< \chi|\chi > \\
&= \sum_{n=0}^{\infty} \int d^3 k_1 \ldots d^3 k_n \sum_{\mu_1 \ldots \mu_n} \chi^{*\mu_1 \ldots \mu_n}(\vec{k}_1, \ldots, \vec{k}_n) \chi_{\mu_1 \ldots \mu_n}(\vec{k}_1, \ldots, \vec{k}_n)
\end{aligned}
$$

$$(10.37)$$

10.2.1 Negative Normed States

Now, we come to one of the problems which plagues this and any Lorentz covariant quantization of the photon. The inner product of the Fock space is not positive. There are states where the expressions for $< \chi|\chi >$ given in equation (10.37) are negative. Let us look at this for a simple example. Consider the one-photon state

$$
|\chi >= \int d^3 k \chi^{\mu}(\vec{k}) a_{\mu}^{\dagger}(\vec{k})|\mathcal{O} >
$$

Its norm is

$$
< \chi|\chi >= \int d^3 k \chi_{\mu}^{*}(\vec{k}) \chi^{\mu}(\vec{k})
$$

If $\chi_{\mu}(k)$ is (on the average) a time-like vector, $\chi_{\mu}^{*}(\vec{k}) \chi^{\mu}(\vec{k}) < 0$ (all we would need is that $\chi^{\mu}(\vec{k}) = (\xi^0(\vec{k}), \vec{0})$ for example), the state $|\chi >$ has negative norm, $< \chi|\chi > < 0$. This problem persists with multi-photon states. The Fock space that we have constructed has an indefinite metric. These states with negative norm are not physically acceptable in a reasonable quantum mechanical theory, as they would imply the nonsensical result of negative probabilities as the answers to some physical questions.

10.2.2 Physical State Condition

The negative normed states would seem to be a disaster for our quantization of the photon. However, we are not finished with this quantization yet. We still have to enforce the constraint

$$
\partial_{\mu} A^{\mu}(x) = 0
$$

that augmented the field equation.

We will impose this gauge fixing constraint as a physical state condition. This condition chooses a subspace of the Fock space as the space of "physical states". These are states where the gauge condition $\partial_{\mu} A^{\mu}(x) = 0$ is obeyed. If we attempt to

impose the condition in its strongest form, that a physical state, which we denote as
|phys >, must obey

$$\partial_\mu A^\mu(x)|\text{phys} >= 0$$

we discover that this is too many constraints. There would be no physical states at
all, besides the zero vector in the Hilbert space.

We need to impose a weaker condition. The weaker condition is that the matrix
element of the constraint between any two physical states vanishes,

$$< \text{phys}'|\partial_\mu A^\mu(x)|\text{phys} > \; = 0 \tag{10.38}$$

so that, in the physical subspace of the Fock space, the expectation value of the
physical state condition is always obeyed.

We can do this as follows. We have already observed that, the continuity equation
for the current in equation (10.20) implies that the constraint satisfies the free wave
equation

$$-\partial^2 \left(\partial_\mu A^\mu(x) \right) = 0 \tag{10.39}$$

independently of the existence of currents (and interactions which will be introduced
through the currents). We only need that the current is conserved. A general solution
of the free wave equation is

$$\partial_\mu A^\mu(x) = \int \frac{d^3k}{\sqrt{(2\pi)^3 2|\vec{k}|}} \left[e^{ik_\mu x^\mu} \xi(\vec{k}) + e^{-ik_\mu x^\mu} \xi^\dagger(\vec{k}) \right], \; k_\mu = (-|\vec{k}|, \vec{k}) \tag{10.40}$$

$$= \partial_\mu A^{\mu(+)}(x) + \partial_\mu A^{\mu(-)}(x) \tag{10.41}$$

where we have made the decomposition into positive and negative frequency parts

$$\partial_\mu A^{\mu(+)}(x) = \int \frac{d^3k}{\sqrt{(2\pi)^3 2|\vec{k}|}} e^{ik_\nu x^\nu} \xi(\vec{k}) \tag{10.42}$$

$$\partial_\mu A^{\mu(-)}(x) = \int \frac{d^3k}{\sqrt{(2\pi)^3 2|\vec{k}|}} e^{-ik_\nu x^\nu} \xi^\dagger(\vec{k}) \tag{10.43}$$

Moreover, $\left(\partial_\mu A^{\mu(+)}(x) \right)^\dagger = \partial_\mu A^{\mu(-)}(x)$ and $\left(\partial_\mu A^{\mu(-)}(x) \right)^\dagger = \partial_\mu A^{\mu(+)}(x)$.

The entity $\xi(\vec{k})$ is an operator. In free field theory, as we discuss below, it is simply
$ik_\mu a^\mu(\vec{k})$. However, in a field theory where the photon is interacting with other fields,
we might not know much more about $\xi(\vec{k})$, other than the fact that we must use it to
impose the physical state condition

$$\partial_\mu A^{\mu(+)}(x) \; |\text{phys} >= 0 \; \text{ or } \; \xi(\vec{k})|\text{phys} >= 0 \; \forall \vec{k} \tag{10.44}$$

Then, taking the conjugate of this physical state condition yields

$$< \text{phys}| \, \partial_\mu A^{\mu(-)}(x) \; = 0 \;\; \text{or} \;\; < \text{phys}|\xi^\dagger(\vec{k}) = 0 \; \forall \vec{k} \qquad (10.45)$$

and we see that for any two physical states

$$< \text{phys}'|\partial_\mu A^\mu(x)|\text{phys} >=< \text{phys}'| \left(\partial_\mu A^{\mu(-)}(x) + \partial_\mu A^{\mu(+)}(x) \right) |\text{phys} >= 0 \qquad (10.46)$$

This is the best that we can do for the constraint. Its matrix elements between any physical states will be zero.

The fact that $\xi(\vec{k})$ and $\xi^\dagger(\vec{k})$ defined in equations (10.42) and (10.43) are time-independent operators, or, equivalently, the fact that the gauge constraint satisfies the free wave equation (10.39) is important. It guarantees that the time evolution of the system maps physical states to physical states. If we begin with a physical state at an initial time, the state of the system at all later times will be a physical state. Time independence of $\xi(\vec{k})$ and $\xi^\dagger(\vec{k})$ follows from the field equation and conservation of the current, $\partial_\mu J^\mu(x) = 0$, and it generalizes to any version of electrodynamics where the current is conserved.

For the free photon field (that is, when $\mathbf{J}^\mu(x) = 0$), the four-vector potential is decomposed into creation and annihilation operators, $a_\mu^\dagger(\vec{k})$ and $a_\mu(\vec{k})$. In that case $\xi(\vec{k}) = ik^\mu a_\mu(\vec{k})$, $\xi^\dagger(\vec{k}) = -ik^\mu a_\mu^\dagger(\vec{k})$ and the physical state condition is written in terms of creation and annihilation operators as

$$k^\mu a_\mu(\vec{k})|\text{phys} >= 0, \quad < \text{phys}| \, k^\mu a_\mu^\dagger(\vec{k}) = 0, \;\; \forall \vec{k} \qquad (10.47)$$

We remind the reader that, since $k_\mu k^\mu = 0$, and the photon energy is k^0, $k_\mu = (-|\vec{k}|, \vec{k})$ and $k^\mu = (|\vec{k}|, \vec{k})$.

The vacuum state $|\mathcal{O} >$ obeys $a_\mu(\vec{k})|\mathcal{O} >= 0 \, \forall \vec{k}, \mu$ and therefore it trivially obeys

$$\partial_\mu A^{(+)\mu}(x)|\mathcal{O} >= \int \frac{d^3 k}{\sqrt{(2\pi)^3 2|\vec{k}|}} e^{ik \cdot x - i|\vec{k}|x^0} ik^\mu a_\mu(k)|\mathcal{O} >= 0$$

The vacuum is therefore a physical state. Moreover, it has positive norm, $< \mathcal{O}|\mathcal{O} >= 1$.

As another example, consider the one-photon state, which is

$$|\chi >= \int d^3 k \chi^\mu(\vec{k}) a_\mu^\dagger(\vec{k})|\mathcal{O} >$$

If we require that this one-photon state is a physical state, that is, we impose

$$\partial_\mu A^{(+)\mu}(x)|\chi >= 0$$

we get the condition

$$k_\mu \chi^\mu(\vec{k}) = 0$$

We can use this equation to solve for the time-component of $\chi^\mu(k)$ to get

$$\chi^0(\vec{k}) = \frac{1}{|\vec{k}|}\vec{k}\cdot\vec{\chi}(\vec{k})$$

With this condition on $\chi(\vec{k})$, the physical state with one free photon has the norm,

$$< \chi|\chi > = \int d^3k\ \chi^{*\nu}(\vec{k})\chi_\nu(\vec{k}) = \int d^3k \vec{\chi}_a^*(\vec{k})\left(\delta^{ab} - \frac{k^a k^b}{\vec{k}^2}\right)\chi^b(\vec{k}) \geq 0$$

which is non-negative since the quadratic form

$$P_{ab} = \left(\delta_{ab} - \frac{k_a k_b}{\vec{k}^2}\right) \tag{10.48}$$

has the non-negative eigenvalues $(1, 1, 0)$. Thus we see that, amongst one-photon states, we have solved the problem of negative norm by restricting ourselves to physical states, whose norms are greater than equal to zero. We still have the possibility that states can have zero norm. This is still problematic. We will find a way to deal with this issue shortly.

First, we must ask the question as to whether all of the physical states, not just single photon states have a non-negative metric. To see how this works, let us consider a generic n-photon state,

$$|\chi_n > \equiv \frac{1}{\sqrt{n!}}\int d^3k_1\ldots d^3k_n \sum_{\mu_1\ldots\mu_n} \chi^{\mu_1\cdots\mu_n}(\vec{k}_1,\ldots,\vec{k}_n)a^\dagger_{\mu_1}(\vec{k}_1)\ldots a^\dagger_{\mu_n}(\vec{k}_n)|\mathcal{O} >$$

where the object $\chi^{\mu_1\cdots\mu_n}(\vec{k}_1,\ldots,\vec{k}_n)$ is completely symmetric,

$$\chi^{\mu_1\cdots\mu_n}(\vec{k}_1,\ldots,\vec{k}_n) = \chi^{\mu_{P(1)}\cdots\mu_{P(n)}}(\vec{k}_{P(1)},\ldots,\vec{k}_{P(n)})$$

for any permutation $\{P(1), ..., P(n)\}$ of $\{1, ..., n\}$. We can use the definition of the vacuum and the commutators of the creation and annihilation operators to show that this state has the norm given by

$$< \chi_n|\chi_n > = \int d^3k_1\ldots d^3k_n\chi^{\mu_1\cdots\mu_n}(\vec{k}_1,\ldots,\vec{k}_n)\chi^*_{\mu_1\cdots\mu_n}(\vec{k}_1,\ldots,\vec{k}_n)$$

It is possible for the right-hand-side of the above equation to be negative. For example, if the only nonzero component of the function were $\chi^{11\cdots10}(\vec{k}_1,\ldots,\vec{k}_n)$, then $\chi_{11\ldots10}(\vec{k}_1, \ldots, \vec{k}_n) = -\chi^{11\ldots10}(\vec{k}_1, \ldots, \vec{k}_n)$, and the right-hand-side is negative. Multi-photon states can therefore have negative norms.

When we apply the physical state condition, which for free field theory is simply $k^\mu a_\mu(\vec{k})|\chi_n> = 0$, and we use the fact that $\chi^{\mu_1 \cdots \mu_n}(\vec{k}_1, \ldots, \vec{k}_n)$ is completely symmetric, we learn that the state $|\chi_n>$ is a physical state if and only if it obeys the conditions

$$k_{1\mu_1} \chi^{\mu_1 \cdots \mu_n}(\vec{k}_1, \ldots, \vec{k}_n) = 0$$

We can easily solve such a condition. With $k_1^0 = |\vec{k}_1| = \sqrt{\vec{k}_1^2}$ we obtain

$$\chi^{0\mu_2 \cdots \mu_n}(\vec{k}_1, \ldots, \vec{k}_n) = \frac{1}{|\vec{k}_1|} k_{1a_1} \chi^{a_1\mu_2 \cdots \mu_n}(\vec{k}_1, \ldots, \vec{k}_n)$$

and, using this and symmetry of $\chi^{\mu_1 \cdots \mu_n}(\vec{k}_1, \ldots, \vec{k}_n)$ we can see that

$$\chi^{0a_2 \cdots a_n}(\vec{k}_1, \ldots, \vec{k}_n) = \frac{1}{|\vec{k}_1|} k_{1a_1} \chi^{a_1 a_2 a_3 \cdots a_n}(\vec{k}_1, \ldots, \vec{k}_n)$$

$$\chi^{00a_3 \cdots a_n}(\vec{k}_1, \ldots, \vec{k}_n) = \frac{1}{|\vec{k}_1||\vec{k}_2|} k_{1a_1} k_{2a_2} \chi^{a_1 a_2 a_3 \cdots a_n}(\vec{k}_1, \ldots, \vec{k}_n)$$

$$\chi^{000a_4 \cdots a_n}(\vec{k}_1, \ldots, \vec{k}_n) = \frac{1}{|\vec{k}_1||\vec{k}_2||\vec{k}_3|} k_{1a_1} k_{2a_2} k_{3a_3} \zeta^{a_1 a_2 a_3 \cdots a_n}(\vec{k}_1, \ldots, \vec{k}_n)$$

and so on. These equations determine all of the components of $\chi^{\mu_1 \cdots \mu_n}(\vec{k}_1, \ldots, \vec{k}_n)$ where any of $\mu_1 \ldots \mu_n$ are 0 in terms of $\chi^{a_1 a_2 a_3 \cdots a_n}(\vec{k}_1, \ldots, \vec{k}_n)$. Using these, it is easy to prove that, if $|\chi_n>$ is a physical state, its norm is

$$< \chi|\chi > = \int d^3k_1 \ldots d^3k_n \chi^{\mu_1 \cdots \mu_n}(\vec{k}_1, \ldots, \vec{k}_n) \chi^*_{\mu_1 \cdots \mu_n}(\vec{k}_1, \ldots, \vec{k}_n)$$

$$= \int d^3k_1 \ldots d^3k_n \chi^{a_1 \cdots a_n}(\vec{k}_1, \ldots, \vec{k}_n) P_{a_1 b_1} \ldots P_{a_n b_n} \chi^{b_1 \cdots b_n *}(\vec{k}_1, \ldots, \vec{k}_n)$$

where the intermediate matrices are the P_{ab} defined in equation (10.48). Since, as was noted there, this is a non-negative matrix in that it has non-negative eigenvalues: (1, 1, 0) the right-hand-side of the equation above is non-negative. This fact is sufficient to prove that any physical state has non-negative norm.

10.2.3 Null States and the Equivalence Relation

We have seen that the physical state condition eliminates the negative normed states, but the zero normed states remain. In order to remove the zero norm states, we must further restrict our notion of physical states. A physical state which has zero norm is called a null state. In order to eliminate the null states, we shall add to our physical state condition the further statement that a physical state is an equivalence class of states, all of which satisfy the physical state condition $k^\mu a_\mu(k)|\text{phys} >= 0$ and which are related by the equivalence relation

$$|\text{phys} > \sim |\text{phys}' > \quad \text{if and only if} \quad || \, |\text{phys} > -|\text{phys}' > || = 0 \qquad (10.49)$$

In words, two physical states are in the same equivalence class if and only if their difference, as vectors in Fock space, is a null state.

The reader might recall that an equivalence relation, $\sim$, imposed on a set $S = \{a, b, c, \ldots\}$ must have the properties that it is decidable: For any $a, b \in S$, either $a \sim b$ or $a \nsim b$; it is reflexive: $a \sim a$, $\forall a \in S$; it is symmetric $a \sim b$ implies $b \sim a$ for all $a, b \in S$; and it is transitive $a \sim b$ and $b \sim c$ implies $a \sim c$. It is easy to check that our correspondence in (10.49) is indeed an equivalence relation. The beautiful property that an equivalence relation has is that it sections a set into distinct equivalence classes. Any two objects are either equivalent to each other, in which case they are in the same class, or they are not equivalent to each other, in which case they are in distinct equivalence classes. Every object in the set is in some equivalence class.

We can immediately see that our equivalence relation solves the problem of zero norm states. Consider a physical state $|\text{phys} >$. Since there are no states of negative norm, it must have either zero norm, or it must have positive norm. Let us assume that it has zero norm. Then, with the notation that 0 is the zero vector in the Hilbert space,

$$|| \, |\text{phys} > -0 \, || = || \, |\text{phys} > \, || = 0$$

Any zero-norm state $|\text{phys} >$ is in the same equivalence class as the zero vector, 0, in Fock space.[2] Conversely, if it is in the same equivalence class as the zero vector, it has zero norm. This implies that, if a physical state is not in the equivalence class of the zero vector, it must have positive norm. Therefore, all of the other equivalence classes have positive norm. The equivalence relation has given us a positive inner product on equivalence classes. We will call the physical states with zero norm null states, so that

$$\left(\partial_\mu A^\mu(x)\right)^{(+)} |\text{null} >= 0, \quad < \text{null}|\text{null} >= 0 \qquad (10.50)$$
$$|\text{null} > \sim 0 \qquad (10.51)$$

and the equivalence relation for physical states can written

$$|\text{phys} > \sim |\text{phys} > +|\text{null} > \qquad (10.52)$$

---

[2] The reader should not confuse this zero vector, 0, which exists in every vector space, with the vacuum $|\mathcal{O} >$ which is one of the basis states of the vector space that we are discussing. The vacuum has unit norm, $< \mathcal{O}|\mathcal{O} >= 1$ whereas the zero vector has zero norm $||0|| = 0$.

for any null state |null >.

As much as our equivalence relation solves the problem of zero norm states, there is more structure that one would like the decomposition of the physical states into equivalence classes to possess. One obvious requirement is that the sectioning of the physical states into equivalence classes is compatible with the inner product. This will be so if, for any physical state |phys > and any other physical state |phys' >, their inner product does not depend on which vectors we choose from their respective equivalence classes, that is, if

$$< \text{phys}|\text{phys}' > = (< \text{phys}| + < \text{null}|) \left(|\text{phys}' > + |\text{null}' >\right) \qquad (10.53)$$

where |null > and |null' > are any two null states. In particular, this would mean that any physical state in the same equivalence class has the same norm. Equation (10.53) will hold for all cases if, for any null state |null > and for any physical state |phys > (including any null state)

$$< \text{phys}|\text{null} > = 0 , \quad < \text{null}|\text{phys} > = 0 \qquad (10.54)$$

In the following we will see that this indeed the case for the free photon. The upshot of the argument will be the fact that any null state must have the form

$$|\text{null} > = \int d^3 k f(\vec{k}) \xi^\dagger(k) |\text{phys} > \qquad (10.55)$$

where $\xi(k)$ is the operator defined in equation (10.42) and for some function $f(\vec{k})$ and some physical state, |phys >, which could itself be null. Then, remembering the physical state condition in the language of equation (10.43),

$$\xi(k)|\text{phys} > = 0 \qquad (10.56)$$

we see that equation (10.54) follows immediately.

All of this is very nice, but we still need to be wary of another restriction of this approach (or any other) to quantizing the photon. As we observed in our initial discussion of the imposition of a gauge condition in the paragraph following equation (10.17), this formulation of the quantization of the photon will make sense only if we restrict our attention to gauge invariant observables. Here, we again see that this must be so if we consider the expectation value of an operator in a physical state,

$$< \text{phys}|\mathcal{O}(x)|\text{phys} >$$

This expectation value will not depend on which representative of the equivalence class of the physical state if

$$< \text{null}|\mathcal{O}(x)|\text{phys} > = 0 , \quad < \text{phys}|\mathcal{O}(x)|\text{null} > = 0 \qquad (10.57)$$

for any null state |null > and for any physical state |phys >. Given that null states have the form given in equation (10.55), we see that equation (10.57) will hold when

$$\left[\xi(\vec{k}), \mathcal{O}(x)\right] = 0 , \quad \left[\xi^\dagger(\vec{k}), \mathcal{O}(x)\right] = 0 , \quad \forall \vec{k} \qquad (10.58)$$

If we imagine that the operator $\mathcal{O}(x)$ is made from the vector potential $A_\mu(x)$, the commutators in the above equation generate a gauge transform

$$A_\mu(x) \rightarrow A_\mu(x) + \partial_\mu \chi(x) , \quad \partial^2 \chi(x) = 0$$

which is a residual symmetry of the gauge condition $\partial_\mu A^\mu(x) = 0$. This requirement will hold if $\mathcal{O}(x)$ is constructed from $A_\mu(x)$ in such a way that it is gauge invariant, that is, if as we already know that we must, we restrict our attention to gauge invariant operators. Then, we have established that the expectation value of a gauge invariant operator in any physical state will not depend on which state we choose as long as the states are all in the same equivalence class.

We can construct all of the states of the free photon. For this purpose, let us consider the following set of four linearly independent four-vectors:

$$\epsilon_s^\mu(\vec{k}), \; s = 1, 2 , \; k^\mu , \; \delta_0^\mu \qquad (10.59)$$

The two polarization vectors $\epsilon_s^0(\vec{k})$ have the properties

$$\epsilon_s^0(\vec{k}) = 0 , \; k_a \epsilon_s^a(\vec{k}) = \vec{k} \cdot \vec{\epsilon}_s(\vec{k}) = 0 , \; s = 1, 2 \qquad (10.60)$$

$$\epsilon_s^\mu(\vec{k}) \eta_{\mu\nu} \epsilon_{s'}^\nu(\vec{k}) = \delta_{ss'} \qquad (10.61)$$

$$\sum_{s=1}^{2} \epsilon_s^\mu(\vec{k}) \epsilon_s^\nu(\vec{k}) = \begin{cases} 0 \text{ if either } \mu \text{ or } \nu = 0 \\ \delta^{ab} - \frac{k^a k^b}{\vec{k}^2} \text{ otherwise} \end{cases} \qquad (10.62)$$

It is clear that $k_\mu \epsilon_s^\mu(\vec{k}) = 0$.

With this notation, we can form the creation operators

$$a_s^\dagger(\vec{k}) = \epsilon_s^\mu(\vec{k}) a_\mu^\dagger(\vec{k}) , \; a_N^\dagger(\vec{k}) = k^\mu a_\mu^\dagger(\vec{k}) , \; a^{0\dagger}(\vec{k})$$

By repeatedly operating with these operators on the vacuum state,

$$a^{0\dagger}(\vec{k}_1)...a^{0\dagger}(\vec{k}_m) \, a_N^\dagger(\vec{\ell}_1)...a_N^\dagger(\vec{\ell}_n) \, a_{s_1}^\dagger(\vec{p}_1)...a_{s_r}^\dagger(\vec{p}_r)|\mathcal{O} >$$

we can generate a basis for the full Fock space of the theory containing both the physical and unphysical states.

Then, we can begin with the creation operators $a_{s'}^\dagger(\vec{k}')$ with $s = 1, 2$. These operators will turn out to be the operators which create the two physical polarizations of the photon. It is easy to see that they obey the commutation relations

$$\left[a_s(\vec{k}), a_{s'}(\vec{k}') \right] = 0 = \left[a_s^\dagger(\vec{k}), a_{s'}^\dagger(\vec{k}') \right]$$

$$\left[a_s(\vec{k}), a_{s'}^\dagger(\vec{k}') \right] = \delta_{ss'}\delta^3(\vec{k} - \vec{k}')$$

We can use them to create a basis of multi-photon states,

$$|O>, a_s^\dagger(\vec{k})|O>, \ldots, \frac{1}{\sqrt{n!}}a_{s_1}^\dagger(\vec{k}_1) \ldots a_{s_n}^\dagger(\vec{k}_n)|O>, \ldots \qquad (10.63)$$

These states span a subspace of the full Fock space. Generic states in their span are

$$|\chi> = \sum_{n=0}^{\infty} \frac{1}{\sqrt{n!}} \int d^3k_1 \ldots d^3k_n \chi^{s_1 \ldots s_n}(\vec{k}_1, \ldots, \vec{k}_n) a_{s_1}^\dagger(\vec{k}_1) \ldots a_{s_n}^\dagger(\vec{k}_n)|O> \quad (10.64)$$

These states are physical states in that they obey the physical state condition

$$\partial_\mu A^{\mu(+)}(x)|\chi> = 0 \qquad (10.65)$$

The norm of these states cannot be negative, that is, $|| \, |\chi> \, || \geq 0$. Moreover, we can easily check that, for these states, the norm is positive in that and the equality is satisfied only by the zero vector, $|| \, |\chi> \, || = 0$ if an only if $|\chi> = 0$.

The physical state equivalence relation states that two states are equivalent, $|\chi> \, \sim \, |\tilde{\chi}>$ if $|| \, |\chi> - |\tilde{\chi}> \, || = 0$. Indeed, for two states $|\chi>$ and $|\tilde{\chi}>$ of the sort that are defined in equation (10.64), $|| \, |\chi> - |\tilde{\chi}> \, || = 0$ if and only if they are identical, $|\chi> = |\tilde{\chi}>$. Thus, each of the states in the span of the basis in equation (10.63) is a physical state in a unique equivalence class.

This means that each distinct state of type $|\chi>$ is in its own equivalence class. Now let us examine the states that are obtained by operating $a_N^\dagger(\vec{k})$, at least once and perhaps repeatedly, as well as operating $a_s^\dagger(\vec{q})$ some number of times on the vacuum. The generic state that we would obtain is

$$\sum_{m=1}^{\infty}\sum_{n=0}^{\infty} \frac{1}{\sqrt{m!n!}} \int d\vec{k}_1 \ldots d\vec{k}_m d\vec{q}_1 \ldots \vec{q}_n \chi_N^{s_1 \ldots s_n}(\vec{k}_1, \ldots, \vec{k}_m; \vec{q}_1, \ldots, \vec{q}_n) \cdot$$
$$\cdot a_N^\dagger(\vec{k}_1) \ldots a_N^\dagger(\vec{k}_m) a_{s_1}^\dagger(\vec{q}_1) \ldots a_{s_n}^\dagger(\vec{q}_n)|O> \qquad (10.66)$$

Using the commutation relations

$$\left[a_N(\vec{k}), a_N^\dagger(\vec{\ell}) \right] = 0 \, , \quad \left[a_N(\vec{k}), a_s^\dagger(\vec{\ell}) \right] = 0$$

and the physical state condition

$$a_N(\vec{k}) \, |\text{phys} > \, = 0$$

we can see that the generic state given in equation (10.66) is a physical state and it is also a null state. We can then form a state that is in the equivalence class of the physical state $|\chi >$ as

$$|\chi_N > \, = \, |\chi > \tag{10.67}$$

$$+ \sum_{m=1}^{\infty} \sum_{n=0}^{\infty} \frac{1}{\sqrt{m!n!}} \int d\vec{k}_1 \ldots d\vec{k}_m d\vec{q}_1 \ldots \vec{q}_n \chi_N^{s_1 \ldots s_n}(\vec{k}_1, \ldots, \vec{k}_m; \vec{q}_1, \ldots, \vec{q}_n) \cdot$$

$$\cdot a_N^\dagger(\vec{k}_1) \ldots a_N^\dagger(\vec{k}_m) a_{s_1}^\dagger(\vec{q}_1) \ldots a_{s_n}^\dagger(\vec{q}_n) |\mathcal{O} > \tag{10.68}$$

We still need to show that the generic null state that we have added to $|\chi >$ covers all possible null states. This would have the implication that, if we fix $|\chi >$ and we sweep across all states of the form $|\chi_N >$, we find the complete equivalence class of physical states that are equivalent to $|\chi >$. We will show that this is indeed the case, as all or the vectors that are not in exclusively this subspace of the Fock space are not physical states. This set of states then indeed has the desired properties which we summarize as

$$\partial_\mu A^{\mu(+)}(x)|\chi > \, = 0 \, , \quad \partial_\mu A^{\mu(+)}(x)|\chi_N > \, = 0 \tag{10.69}$$

$$|| \, |\chi > \, || = 0 \iff |\chi > \, = 0 \tag{10.70}$$

$$|\chi > \sim |\chi_N > \tag{10.71}$$

$$< \chi | \chi' > \, = \, < \chi_N | \chi'_{N'} > \tag{10.72}$$

$$< \chi | \mathcal{O}(x) | \chi' > \, = \, < \chi_N | \mathcal{O}(x) | \chi'_{N'} > \tag{10.73}$$

for gauge invariant operators $\mathcal{O}(x)$.

Now, by our construction of the array of physical states of the types $\{|\chi >, |\chi >_N\}$ we know that any other state in the Fock space must be obtained by operating at least one of the remaining creation operators $\alpha^{0\dagger}(k)$. These remaining states therefore have the general form

$$|v > \, = \sum_{m=1}^{\infty} \sum_{n,\bar{n}=0}^{\infty} \frac{1}{\sqrt{m!n!\bar{n}!}} \int d^3k_1 \ldots d^3k_m d^3q_1 \ldots d^3q_n d^3\ell_1 \ldots d^3\ell_{\bar{n}} \cdot$$

$$\cdot v^{s_1 \ldots s_{\bar{n}}}(\vec{k}_1, \ldots, \vec{k}_m; \vec{q}_1, \ldots, \vec{q}_n; \ell_1, \ldots, \ell_{\bar{n}}) \cdot$$

$$\cdot a^{0\dagger}(\vec{k}_1) \ldots a^{0\dagger}(\vec{k}_m) \, a_N^\dagger(\vec{q}_1) \ldots a_N^\dagger(\vec{q}_n) a_{s_1}^\dagger(\vec{\ell}_1) \ldots a_{s_n}^\dagger(\vec{\ell}_{\bar{n}}) |\mathcal{O} > \tag{10.74}$$

There is the question as to whether any of the states of the form $|v >$, whose only characteristic is that they all contain at least one operation of $a^{0\dagger}(\vec{k})$, can be physical

states. To explore this question, consider what we get by operating the physical state
condition on the state $|\upsilon>$. We get

$$k_\mu a^\mu(\vec{k})|\upsilon>= \sum_{m=2}^{\infty} \sum_{n,\bar{n}=0}^{\infty} \frac{n|\vec{k}|}{\sqrt{m!n!\bar{n}!}} \int d^3k_2 \ldots d^3k_m d^3q_1 \ldots d^3q_n d^3\ell_1 \ldots d^3\ell_{\bar{n}}.$$

$$\cdot \upsilon^{s_1 \ldots s_{\bar{n}}}(\vec{k}, \vec{k}_2, ..., \vec{k}_m; \vec{q}_1, ..., \vec{q}_n; \ell_1, ..., \ell_{\bar{n}}).$$

$$\cdot a^{0\dagger}(\vec{k}_2) \ldots a^{0\dagger}(\vec{k}_m) \, a_N^\dagger(\vec{q}_1) \ldots a_N^\dagger(\vec{q}_n) a_{s_1}^\dagger(\vec{\ell}_{\bar{1}}) \ldots a_{s_n}^\dagger(\vec{\ell}_{\bar{n}})|\mathcal{O}> \qquad (10.75)$$

Our question now asks whether there is a non-zero solution to the equation
$\ell_\mu a^\mu(\ell)|\upsilon>= 0$. Since the vectors

$$a^{0\dagger}(\vec{k}_2) \ldots a^{0\dagger}(\vec{k}_m) \, a_N^\dagger(\vec{q}_1) \ldots a_N^\dagger(\vec{q}_n) a_{s_1}^\dagger(\vec{\ell}_{\bar{1}}) \ldots a_{s_n}^\dagger(\vec{\ell}_{\bar{n}})|\mathcal{O}>$$

are linearly independent for each distinct choice of

$$\{\vec{k}_2, \ldots, \vec{k}_m; \vec{q}_1, \ldots, \vec{q}_n; \vec{\ell}_{\bar{1}}, \ldots, \vec{\ell}_{\bar{n}}\}$$

and

$$\{s_1, \ldots, s_{\bar{n}}\}$$

the right-hand-side of equation (10.75) can only vanish, for all $\vec{k}$ if all of the coefficients vanish, that is, if

$$\upsilon^{s_1 \ldots s_{\bar{n}}}(\vec{k}_1, \vec{k}_2, ..., \vec{k}_m; \vec{q}_1, \ldots, \vec{q}_n; \ell_1, ..., \ell_{\bar{n}}) = 0$$

As a result, we conclude that no states of the form $|\upsilon>$ can be physical states.

We have found that any state in the Fock space is a linear combination of a physical
state and an unphysical state,

$$c_1|\chi_N> + c_2|\upsilon>$$

What is more, this combination is a physical state if and only if $c_2 \neq 0$. Thus we
reach the conclusion that all of the physical states are contained in the set of states
of type $|\chi_N>$. The have also demonstrated that the equivalence classes of these
states are in one-to-one components of states of the type $|\chi>$. These contain the
multi-photon states that are made from the two physical polarizations of the photon.

10.3 Space–Time Symmetries of the Photon

Once we can obtain the field equations from an action principle using the Lagrangian density and the Euler–Lagrange equations, we can identify symmetries of the theory by examining the Lagrangian density and we can use Noether's theorem to identify the conserved currents which correspond to the symmetries of the theory.

In the present case, the Maxwell equations (10.11) and (10.14) are obtained by applying the Euler–Lagrange equations to the Lagrangian density in equation (10.15). If we set the source, $\mathbf{J}^{\mu}(x)$ to zero, the source-free Maxwell theory is invariant under space–time translations. If we consider an infinitesimal translation

$$\delta_{\mu} A_{\lambda}(x) = -\partial_{\mu} A_{\lambda}(x)$$

(where we have omitted the infinitesimal parameter) we find that the Maxwell Lagrangian density varies as

$$\delta_{\mu} \mathcal{L} = -\partial_{\nu} \left(\delta_{\mu}^{\nu} \mathcal{L}(x) \right)$$

which is a total divergence. The space–time translation is therefore a symmetry of the source-free Maxwell theory of the photon on Minkowski space–time.

Since the translation is a symmetry, Noether's theorem tells us that it is associated with a conservation law. To find the conserved quantity, we form the Noether current density as

$$\mathcal{J}_{\mu}^{\nu}(x) = -\delta_{\mu} A_{\lambda}(x) \frac{\partial \mathcal{L}}{\partial(\partial_{\nu} A_{\lambda}(x))} + \delta_{\mu}^{\nu} \mathcal{L}(x)$$

$$= \partial_{\mu} A_{\lambda}(x) F^{\nu\lambda}(x) - \delta_{\mu}^{\nu} \frac{1}{4} F_{\rho\sigma}(x) F^{\rho\sigma}(x)$$

From this Noether current, we can deduce the stress tensor

$$\mathbf{T}_{0}^{\mu\nu}(x) = F^{\mu\lambda}(x) \partial^{\nu} A_{\lambda}(x) - \frac{1}{4} \eta^{\mu\nu} F_{\rho\sigma}(x) F^{\rho\sigma}(x)$$

We know from Noether's theorem that this stress tensor must be conserved,

$$\partial_{\mu} \mathbf{T}_{0}^{\mu\nu}(x) = 0$$

and it is easy to check that this is so by using Maxwell's equations.

The stress tensor $\mathbf{T}_{0}^{\mu\nu}(x)$ is not gauge invariant. Its physical interpretation is therefore problematic. Remember that our quantization of the photon is guaranteed to make sense only when we only compute expectation values of gauge invariant operators. Aside from not being gauge invariant, this stress tensor is not symmetric and it cannot be used to form the conserved current corresponding to Lorentz transformations.

The situation with gauge invariance can be improved by realizing that, as well as a space–time transformation, we are free to do gauge transformations. The gauge field transforms under a coordinate transformation as a vector field

$$\delta_f A_\lambda(x) = -f^\rho(x)\partial_\rho A_\lambda(x) - \partial_\lambda f^\rho(x)A_\rho(x)$$

and the right-hand-side of this equation is not gauge invariant. We can fix this by augmenting this coordinate transformation with a gauge transformation

$$\tilde{\delta}_f A_\lambda(x) = \partial_\lambda \left(f^\rho(x)A_\rho(x)\right)$$

so that the total transformation is, to linear order, the sum of the two transformations,

$$\hat{\delta}_f A_\lambda(x) \equiv \delta_f A_\lambda(x) + \tilde{\delta}_f A_\lambda(x) = -f^\rho(x)F_{\lambda\rho}(x)$$

This transformation is also a symmetry of the Maxwell theory. It is also gauge invariant. Note that, this is an infinitesimal general coordinate transformation. If we now specialize to a transformation that we expect to be a symmetry, where $f^\rho(x)$ is a Killing vector, and if we use this symmetry and apply Noether's theorem, we find the stress tensor

$$\Theta^{\mu\nu}(x) = F^{\mu\lambda}(x)F^\nu{}_\lambda(x) - \frac{1}{4}\eta^{\mu\nu}F_{\rho\sigma}(x)F^{\rho\sigma}(x) \qquad (10.76)$$

The resulting tensor, $\Theta^{\mu\nu}(x)$, is

1. Gauge invariant
2. Conserved, $\partial_\mu\Theta^{\mu\nu}(x) = 0$
3. Symmetric, $\Theta^{\mu\nu}(x) = \Theta^{\nu\mu}(x)$
4. Traceless, $\Theta^\mu{}_\mu(x) = 0$

These properties allow us to find a Noether current corresponding to any of the space–time symmetries of the theory by contracting the tensor with a conformal Killing vector,

$$\mathcal{J}^\mu_f(x) = \Theta^{\mu\nu}(x)f_\nu(x)$$

This current satisfies

$$\partial_\mu\mathcal{J}^\mu_f(x) = 0$$

if $f_\mu(x)$ satisfies the conformal Killing equation

$$\partial_\mu f_\nu(x) + \partial_\nu f_\mu(x) - \frac{1}{2}\eta_{\mu\nu}\partial_\lambda f^\lambda(x) = 0$$

In the absence of sources, classical electrodynamics has not only translation and Lorentz invariance, it is conformally invariant.

The energy density is the time-time component of the stress tensor

$$\Theta^{00}(x) = \frac{1}{2}\left(\vec{E}^2(\vec{x}, t) + \vec{B}^2(\vec{x}, t)\right)$$

and the momentum density is the Poynting vector

$$\Theta^{0i} = \left(\vec{E}(\vec{x}, t) \times \vec{B}(\vec{x}, t)\right)^i$$

These can be used to study the energy and momentum that are stored in a classical configuration of electric and magnetic fields. Using

$$E^a(x) = \int \frac{d^3k}{\sqrt{(2\pi)^3 2|\vec{k}|}}\left[e^{ik_\mu x^\mu}\left(ik^a a^0(\vec{k}) - i|\vec{k}|a^a(\vec{k})\right)\right.$$
$$\left. + e^{-ik_\mu x^\mu}\left(-i\vec{k}a^0(\vec{k}) + i|\vec{k}|\vec{a}(\vec{k})\right)\right]$$
$$= \int \frac{d^3k}{\sqrt{(2\pi)^3 2|\vec{k}|}}\left[e^{ik_\mu x^\mu}i|\vec{k}|P^a_b a^b(\vec{k}) - e^{-ik_\mu x^\mu}i|\vec{k}|P^a_b a^{\dagger b}(\vec{k})\right] + \text{constraint}$$

we can see that the Hamiltonian and linear momentum operators are just what one would expect

$$H = \int d^3k |\vec{k}|\, a_s^\dagger(k)a_s(k) \; , \; \vec{P} = \int d^3k\, \vec{k}\, a_s^\dagger(k)a_s(k)$$

where we have dropped terms which would vanish when we take matrix elements in physical states. These are just what we would expect for the energies and momenta of multi-photon states.

10.4 Massive Photon

Maxwell's equations are the field equations which describe a massless photon. The elementary excitation of the quantum field theory is a particle which travels at the speed of light and it has two physical polarizations. We have just found that the quantum description of the theory, at least if we want to retain manifest covariance, is complicated by the appearance of negative normed unphysical states, and even once these are eliminated using the physical state condition, null states remain and require a further sectioning of the physical states into equivalence classes. Given this level of complication, which is partially due to gauge invariance, it might be useful to ask, the question "what if the photon had a small mass". Such a theory would not be gauge invariant. The photon would be a massive spin-one particle we would expect to have three physical polarizations, the states helicities $(+1, 0, -1)$. Of course, it is clear that to fit known physics, we can never completely rule out the existence of a photon mass. The current experimental bounds on the mass of the photon are exceedingly small, but they can never tell us that it is precisely zero. If the limit as the mass becomes small is a smooth one, we should be able to reproduce all of the behaviour of the massless photon if the mass is small enough. For example, the extra photon polarization, the one with helicity 0, should decouple from observable phenomena. In this section, we will examine the quantization of a massive photon which is still coupled to a conserved current.

To introduce a photon mass, we would modify the first of Maxwell's equations (10.11) so that it has the form

$$\partial_\nu F^{\mu\nu}(x) + m^2 A^\mu(x) = \mathbf{J}^\mu(x) \tag{10.77}$$

Now, with a photon mass, the Maxwell equation has an explicit dependence on the vector field $A^\mu(x)$. The Maxwell equation is no longer gauge invariant. This means that the vector potential is less redundant than it was for a strictly massless photon. However, we will not be able to completely avoid constraints. The photon has spin one. A massive photon therefore must have three spin states. The four components of $A_\mu(x)$ are still too many variables to describe it. There must still be a constraint.

In fact, we can easily find out what the constraint is. To do so, we operate ∂_μ on the field equation (10.77). We remember that, $F^{\mu\nu}(x)$ is anti-symmetric so that $\partial_\mu \partial_\nu F^{\mu\nu}(x) = 0$. We shall also assume that the current is conserved, $\partial_\mu \mathbf{J}^\mu(x) = 0$, Then we obtain

$$\partial_\mu A^\mu(x) = 0 \tag{10.78}$$

as a consequence of the field equation. This is the same equation as we imposed using gauge invariance of the massless photon that we discussed in the previous section. As we did there, we will treat this equation as an equation of constraint.

Once $A_\mu(x)$ satisfies this constraint, its field equation (10.77) and constraint (10.78) are summarized as

$$(-\partial^2 + m^2)A^\nu(x) = \mathbf{J}^\nu(x) \tag{10.79}$$
$$\partial_\mu A^\mu(x) = 0 \tag{10.80}$$

Every solution of the field equation (10.77) is a solution of the pair of equations (10.79) and (10.80) and vice versa. The field equation (10.79) contains the relativistic wave operator $-\partial^2 + m^2$, and it will describe a massive particle where m (in units where $\hbar = 1$ and $c = 1$) is the mass.

As with the massless photon, we will begin by quantizing the theory which has the field equation (10.79), (with $\mathbf{J}^\mu(x) = 0$) irrespective of the constraint (10.80). For this, we need to identify the commutation relations. The identification of the commutation relations follows the same logic as it did for the massless photon and the result is the same as it was there,

$$\left[A_\mu(\vec{x}, t), \frac{\partial}{\partial t} A_\nu(\vec{y}, t)\right] = i\eta_{\mu\nu}\delta(\vec{x} - \vec{y}) \tag{10.81}$$

$$\left[A_\mu(\vec{x}, t), A_\nu(\vec{y}, t)\right] = 0, \quad \left[\frac{\partial}{\partial t} A_\mu(\vec{x}, t), \frac{\partial}{\partial t} A_\nu(\vec{y}, t)\right] = 0 \tag{10.82}$$

The solution of the field equation is

$$A_\mu(x) = \int \frac{d^3k}{\sqrt{(2\pi)^3 2E(\vec{k})}} \left[e^{ik_\nu x^\nu} a_\mu(\vec{k}) + e^{-ik_\nu x^\nu} a_\mu^\dagger(\vec{k})\right] \tag{10.83}$$

$$k^\mu = (E(\vec{k}), \vec{k}) \,, \quad E(\vec{k}) = \sqrt{\vec{k}^2 + m^2}$$

This solution of the wave equation obeys the equal-time commutation relations (10.81) and (10.82) if the creation and annihilation operators obey the commutation relations

$$\left[a_\mu(\vec{k}), a_\nu^\dagger(\vec{\ell})\right] = \eta_{\mu\nu}\delta(\vec{k} - \vec{\ell}) \tag{10.84}$$

$$\left[a_\mu(\vec{k}), a_\nu(\vec{\ell})\right] = 0, \quad \left[a_\mu^\dagger(\vec{k}), a_\nu^\dagger(\vec{\ell})\right] = 0 \tag{10.85}$$

The construction of the quantum states begins with the vacuum state $|\mathcal{O}>$ which obeys

$$a_\mu(\vec{k})|\mathcal{O}>= 0 \,, \quad < \mathcal{O}|a_\mu^\dagger(\vec{k}) = 0 \,, \quad \forall \vec{k}, \mu \,, \quad < \mathcal{O}|\mathcal{O}>= 1$$

The quantum states are elements of the Fock space which is found by taking superpositions of the basis vectors

$$|\mathcal{O}> , \; a_\mu^\dagger(\vec{k})|\mathcal{O}> , \dots, \frac{1}{\sqrt{n!}} a_{\mu_1}^\dagger(\vec{k}_1) a_{\mu_2}^\dagger(\vec{k}_2) \dots a_{\mu_n}^\dagger(\vec{k}_n)|\mathcal{O}> , \dots$$

This basis contains states which have positive, negative and zero norms. We are again faced with the task of imposing a physical state condition in order to eliminate the negative normed states.

If we take the four-divergence of the field equation (10.79) and use the fact that the current is conserved, we find the equation

$$(-\partial^2 + m^2)\left[\partial_\mu A^\mu(x)\right] = 0$$

and $\partial_\mu A^\mu(x)$ can be decomposed into positive and negative frequency parts. The only difference between this equation and the one for the massless photon is the replacement of the wave operator $-\partial^2$ with $-\partial^2 + m^2$. As it did there, this equation tells us that $\partial_\mu A^\mu(x)$ can be set equal to zero if it and its time derivative are set equal to zero at some time, they are zero at all other times. Thus, in the classical field theory, the physical state condition can be imposed on the initial data. Secondly, since the above equation implies that $\partial_\mu A^\mu(x)$ can be decomposed into positive and negative frequency parts. Then, as we did for the massless photon, when we quantize the theory, we can impose the physical state condition

$$\partial_\mu A^{\mu(+)}(x)|\text{phys} >= 0$$

to isolate a subspace of the Fock space which we call the space of physical states.

As before, if we require that

$$|\zeta > = \frac{1}{\sqrt{n!}} \int d^3k_1 \ldots d^3k_n \zeta^{\mu_1 \cdots \mu_n}(\vec{k}_1, \ldots, \vec{k}_n) a^\dagger_{\mu_1}(\vec{k}_1) \ldots a^\dagger_{\mu_n}(\vec{k}_n)|\mathcal{O} >$$

is a physical state, the physical state condition again tells us that polarization tensor satisfies

$$k_{1\mu_1} \zeta^{\mu_1 \cdots \mu_n}(\vec{k}_1, \ldots, \vec{k}_n) = 0$$

and we can use this equation to eliminate the time-components of the coefficient functions $\zeta^{\mu_1 \cdots \mu_n}(\vec{k}_1, \ldots, \vec{k}_n)$. Then, combining this and the fact that $\zeta^{\mu_1 \cdots \mu_n}(\vec{k}_1, \ldots, \vec{k}_n)$ is a symmetric function, the inner product of physical states,

$$< \zeta|\zeta > = \int d^3k_1 \ldots d^3k_n \zeta^{\mu_1 \cdots \mu_n}(\vec{k}_1, \ldots, \vec{k}_n) \zeta^*_{\mu_1 \cdots \mu_n}(\vec{k}_1, \ldots, \vec{k}_n)$$

can be written as

$$< \zeta|\zeta > = \int d^3k_1 \ldots d^3k_n \zeta^{a_1 \cdots a_n}(\vec{k}_1, \ldots, \vec{k}_n) \tilde{P}_{a_1 b_1} \ldots \tilde{P}_{a_n b_n} \zeta^{b_1 \cdots b_n *}(\vec{k}_1, \ldots, \vec{k}_n)$$

where,

$$\tilde{P}_{ab} = \left(\delta_{ab} - \frac{k_a k_b}{k^2 + m^2} \right)$$

is now a positive definite matrix with eigenvalues $(1, 1, m^2/(m^2 + \vec{k}^2))$. Thus we see that, for a massive photon, once the physical state condition has been imposed, the Hilbert space metric is positive definite and there is no need for the further equivalence relation. The cost of this simplicity is that the photon has an extra mode, it is a massive spin-1 particle and it has three instead of two polarization states. The extra mode should decouple when $m \to 0$ whence the states which contain it become null states and they need to be eliminated by imposing the equivalence relation.

10.5   Quantum Electrodynamics

Now that we have discussed the quantum field theories of the Dirac field and the photon, we are ready to put the two together. The resulting theory is called quantum electrodynamics. It describes electrons and positrons, described by the Dirac theory interacting with photons. Since the active components of practically every substance and physical system outside of the physics laboratory are made from electrons and photons, this is an extremely important theory for physics. It is the theory which describes the spectral lines which occur in the emission of photons by excited atoms, for example, as well as other properties of such systems.

The way to couple the photon to the Dirac field is to identify the current $\mathbf{J}^\mu$ which occurs in Maxwell's equations with the conserved Noether current corresponding to the phase symmetry of the Dirac theory,

$$\mathbf{J}^\mu(x) = -e\bar{\psi}(x)\gamma^\mu\psi(x)$$

The coefficient e on the right-hand-side is related to the electric charge of the electron. Also, we saw that, in the Maxwell action the current was coupled by adding the term $A_\mu(x)\mathbf{J}^\mu(x)$ to the Lagrangian density. We could therefore add the coupling term

$$A_\mu(x)\mathbf{J}^\mu(x) = -e\bar{\psi}(x)\slashed{A}(x)\psi(x)$$

to the sum of the Dirac and Maxwell Lagrangian densities to get

$$\mathcal{L}(x) = -i\bar{\psi}(x)\left[\tfrac{1}{2}\slashed{\partial} - \tfrac{1}{2}\overleftarrow{\slashed{\partial}} - ie\slashed{A}(x) + m\right]\psi(x) - \frac{1}{4}F_{\mu\nu}(x)F^{\mu\nu}(x) \qquad (10.86)$$

This is the Lagrangian density of quantum electrodynamics. It describes electrons which carry charge e interacting with massless photons. The field theory which it describes has gauge invariance. The Lagrangian density is left unchanged by the substitution

$$A_\mu(x) \to \tilde{A}_\mu(x) = A_\mu(x) + \partial_\mu\chi(x) \qquad (10.87)$$

$$\psi(x) \to \tilde{\psi}(x) = e^{ie\chi(x)}\psi(x) \; , \;\; \bar{\psi}(x) \to \tilde{\bar{\psi}}(x) = \bar{\psi}(x)e^{-ie\chi(x)} \qquad (10.88)$$

Note that the phase symmetry of the Dirac field theory has been promoted to a local invariance, that is, one which has a space and time-dependent parameter, $\chi(x)$.

Also, as a consequence, the Dirac field itself is not gauge invariant. The derivative of the Dirac field now appears as the "covariant derivative" $D_\mu\psi(x) = [\partial_\mu - ieA_\mu(x)]\psi(x)$. Under the gauge transformation $D_\mu\psi(x) \to e^{ie\chi(x)}D_\mu\psi(x)$, so that the covariant derivative $D_\mu\psi(x)$ transforms in the same way that $\psi(x)$ does.

Applying the Euler–Lagrange equation to (10.86) yields the coupled Dirac and Maxwell equations

$$\left[\slashed{\partial} - ie\slashed{A}(x) + m\right]\psi(x) = 0$$
$$\partial_\mu F^{\mu\nu}(x) = e\bar{\psi}(x)\gamma^\nu\psi(x)$$

where $F_{\mu\nu}(x)$ is now assumed to be constructed from $A_\mu(x)$ as $F_{\mu\nu}(x) = \partial_\mu A_\nu(x) - \partial_\nu A_\mu(x)$. These field equations are no longer linear equations. They are coupled nonlinear partial differential equations and it is not known how to solve them exactly, except in a few very special cases. The main approach to solving them is perturbative. In that approach, one begins by setting $e = 0$ and solving the free field theory of the Dirac field and the photon field, which we already know how to do. Then, we use perturbation theory to correct these solutions for the presence of the coupling which is of order e. The corrections to free field theory are governed by a dimensionless parameter which is called the fine structure constant which turns out to be small for quantum electrodynamics. This makes the perturbation theory exceedingly accurate. In the following chapters we will spend a large amount of time in learning how to do perturbation theory.

However, as we have already discussed, to quantize the theory of the photon, we must begin by fixing a gauge, that is, by using the gauge invariance of the Lagrangian density to impose a gauge condition on the photon field. The condition that we will choose will be the Lorentz invariant $\partial_\mu A^\mu(x) = 0$. Then, the field equations and gauge constraint are

$$\left[\slashed{\partial} - ie\slashed{A}(x) + m\right]\psi(x) = 0 \tag{10.89}$$
$$\partial^2 A^\nu(x) = e\bar{\psi}(x)\gamma^\nu\psi(x) \tag{10.90}$$
$$\partial_\mu A^\mu(x) = 0 \tag{10.91}$$

The strategy that we will take is similar to what we did with the quantization of the free photon. We will first understand the theory described by the field equations (10.89) and (10.90). Then, we will impose the gauge condition (10.91) as a constraint on the physical states.

The field equations (10.89) and (10.90) can be obtained from the Lagrangian density

$$\mathcal{L} = -i\bar{\psi}(x)\left[\tfrac{1}{2}\overrightarrow{\slashed{\partial}} - \tfrac{1}{2}\overleftarrow{\slashed{\partial}} - ie\slashed{A} + m\right]\psi(x) - \frac{1}{2}\partial_\mu A_\nu(x)\partial^\mu A^\nu(x)$$

As we did for the free Dirac field and for the free photon, we can examine the time derivative terms in this Lagrangian density and deduce the equal-time commutation and anti-commutation relations. They are

$$\left\{\psi_a(x), \psi_b^\dagger(y)\right\}_{x^0=y^0} = \delta_{ab}\delta(\vec{x} - \vec{y}) \tag{10.92}$$

$$\{\psi_a(x), \psi_b r(y)\}_{x^0=y^0} = 0 \tag{10.93}$$

$$\left\{\psi_a^\dagger(x), \psi_b^\dagger(y)\right\}_{x^0=y^0} = 0 \tag{10.94}$$

$$\left[A_\mu(x), \partial_0 A_\nu(y)\right]_{x^0=y^0} = i\eta_{\mu\nu}\delta(\vec{x} - \vec{y}) \tag{10.95}$$

$$\left[A_\mu(x), A_\nu(y)\right]_{x^0=y^0} = 0 \tag{10.96}$$

$$\left[\partial_0 A_\mu(x), \partial_0 A_\nu(y)\right]_{x^0=y^0} = 0 \tag{10.97}$$

It is this quantum field theory, the one contained in the equations of motion (10.89) and (10.90), the constraint (10.91) and the equal time commutation relations (10.92)–(10.97) that we must solve. We will proceed to do this in later chapters using perturbation theory.

10.5.1 C, P and T

For completeness, and without detailed discussion or proof, we will discuss the discrete symmetries of parity P, charge conjugation C and time reversal T. Parity is the coordinate transformation

$$P: \ (x^0, \vec{x}) \to (x'^0, \vec{x}') = (x^0, -\vec{x})$$

Unlike the spacetime symmetries that we have studied so far, parity cannot be an infinitesimal transformation. Under parity, the photon and Dirac fields transform as

$$P: \ (A'^0(x), \vec{A}'(x)) = (A^0(x'), -\vec{A}(x')) \tag{10.98}$$

$$P: \ \psi'(x) = \gamma^0\psi(x'), \ \bar{\psi}'(x) = -\bar{\psi}(x')\gamma^0 \tag{10.99}$$

and it is easy to check that, with these transformations of the fields, the Lagrangian density of electrodynamics transforms like a scalar

$$P: \ \mathcal{L}(x) \to \mathcal{L}(x')$$

so that, after a change of integration variable, the action $\mathbf{S} = \int dx \mathcal{L}(x)$ is invariant. It is also easy to confirm that the field equations are invariant.

Another symmetry which is related to particle-hole symmetry is called charge conjugation which we shall denote by C. Under this transformation

$$C: \quad A^{\mu}(x) \rightarrow -A^{\mu}(x) \tag{10.100}$$

$$C: \quad \psi(x) \rightarrow C\bar{\psi}^{t}(x) \tag{10.101}$$

$$C: \quad \bar{\psi}(x) \rightarrow \psi^{t}(x)C^{-1} \tag{10.102}$$

$$C^{-1}\gamma_{\mu}C = -\gamma_{\mu}^{t} \tag{10.103}$$

where the superscript t indicates transpose. Here, the specific form of the matrix C depends on the representation of the Dirac matrices. The Lagrangian density of electrodynamics is invariant under C. The effect of the transformation is to interchange particles and anti-particles and, at the same time, multiply electric and magnetic fields by a minus sign.

Finally, there is time reversal invariance. Its transformation is somewhat more complicated than P or C. The first term in the Dirac equation $i\partial_0\psi + \dots$ is not invariant under $x^0 \rightarrow x^0$ as is. Also, the time evolution that is generated by the Hamiltonian, $< f|e^{iHt}|i>$ is time asymmetric because the Hamiltonian itself has a strictly non-negative spectrum. In fact, what is needed is an additional conjugation, so that

$$T: \quad < \psi_1|e^{iHt}|\psi_2 > \rightarrow < \psi_1|e^{-iHt}|\psi_2 >^* \tag{10.104}$$

Such a transformation operators on the photon and Dirac fields as

$$T: \quad (A_0(x^0, \vec{x}), \vec{A}(x^0, \vec{x}) \rightarrow (A_0(-x^0, \vec{x}), -\vec{A}(-x^0, \vec{x}) \tag{10.105}$$

$$T: \quad \psi(x^0, \vec{x}) \rightarrow T\psi(-x^0, \vec{x}) \tag{10.106}$$

$$T: \quad \bar{\psi}(x^0, \vec{x}) \rightarrow \psi^{\dagger}(-x^0, \vec{x})T^{-1}\gamma^0 \tag{10.107}$$

$$T^{-1}\gamma^{\mu}T = (\gamma^{\mu})^{\dagger} , \quad T^{-1}(\text{c number})T = (\text{c number})^* \tag{10.108}$$

The last, conjugation of c-numbers operation takes care of the fact that time reversal also must map quantum matrix elements onto their conjugates. Again, the form of T depends on the representation of the Dirac matrices.

There are a few facts about relativistic quantum field theories which we will state here without proof. One is the the CPT theorem. If we combine the transformations C, P and T in this way, it is possible to show that all local Lorentz invariant operators (that is all Lorentz invariant products of fields and their derivatives all evaluated at the same spacetime point) and only local Lorentz invariant operators have this symmetry. This means that any quantum field theory whose Lagrangian density contains only local Lorentz invariant terms will also have CPT symmetry and vice-versa. This has the implication that, if theories which describe nature are Lorentz invariant, then CPT must also be a symmetry of nature.

A second fact is the spin-statistics theorem which CPT symmetry is used to prove. It states that, in any Lorentz invariant quantum field theory, integer spin fields must be bosons whereas half-integer spin fields must be fermions.

Functional Methods

11

We have now understood some aspects of quantum field theory with scalar fields as well as the theories of the Dirac field and the photon field when they do not interact with other fields. In order to proceed further, it will be convenient to, once again, repackage the information that is contained in either the Lagrangian density or the field equations and equal-time commutation relations of a quantum field theory in a third form, the set of all correlation functions of the quantum field theory. An example of a correlation function for scalar fields would be

$$< \mathcal{O}|\phi(x_1)\phi(x_2)\dots\phi(x_k)|\mathcal{O} >$$

or for Dirac fields would be

$$< \mathcal{O}|\psi_{a_1}(x_1)\dots\psi_{a_n}(x_n)\psi^\dagger_{b_1}(y_1)\dots\psi^\dagger_{b_n}(y_n)|\mathcal{O} >$$

The set of all possible correlation functions in a quantum field theory can be regarded as containing information which is equivalent to our other descriptions of the theory, either by field equations and equal-time commutation relations or by a Lagrangian density and action. It is also of course much closer to the solution of the theory in that the correlation functions can be used to find the expectation values of observables in the vacuum state.

Of course, a given quantum field theory has an infinite number of correlation functions and our description of the theory in terms of them would not be useful or even viable without a compact means for presenting them. This compact description of correlation functions can be found in functional methods.

In this chapter, we will introduce functional methods. This includes functional differentiation and functional integration. Then, we will give a more complete definition of what we mean by a correlation function. We will use functional methods

© The Author(s), under exclusive license to Springer Nature Singapore Pte Ltd. 2023
G. W. Semenoff, *Quantum Field Theory*, Graduate Texts in Physics,
https://doi.org/10.1007/978-981-99-5410-0_11

to search for expressions which encode correlation functions of the quantum field theory of the relativistic real scalar field as an example. This will involve studying generating functionals from which correlation functions can be found by taking functional derivatives and finding that the formal expressions for the correlation functions themselves are most elegantly presented as functional integrals.

11.1   Functional Derivative

In order to use functional methods, we must learn how to take functional derivatives and functional integrals. We will begin with functional derivatives which is the easier of the two. In order to understand the concept of functional derivative, consider, as a simple illustration, a real-number valued function $\phi(x)$ of one real variable x and a functional, $Z[\phi]$ of that function. Remember that a functional is a mathematical object into which we put a function, in this case $\phi(x)$. The output of the functional is a number, the value of the functional when it is evaluated on that function. An alternative way to define what we mean by a functional is to begin with a discrete, complete, infinite set of square integrable functions,

$$\{f_1(x), f_2(x), ...\}$$

We can assume that these functions are orthonormal

$$\int dx f_m(x) f_n(x) = \delta_{mn}$$

and that they obey a completeness relation

$$\sum_{n=1}^{\infty} f_n(x) f_n(y) = \delta(x - y)$$

We will assume that any of the functions of interest to us can be expanded in the basis of square integrable functions as

$$\phi(x) = \sum_n c_n f_n(x) \tag{11.1}$$

For a specific function $\phi(x)$, the infinite sequence of coefficients c_n are found by doing the integrals

$$c_n = \int dx f_n(x)\phi(x) \tag{11.2}$$

The function $\phi(x)$ is completely specified by the coefficients in this expansion. In principle, if we know the infinite-component vector $(c_1, c_2, ...)$ we can reconstruct the function $\phi(x)$ using equation (11.2). Alternatively, if we know the function $\phi(x)$

we can do the integrals in (11.2) to find the coefficients c_n and construct the vector $(c_1, c_2, ...)$.

If we plug the expansion of $\phi(x)$ into the functional, $Z[\phi]$, the functional becomes an ordinary function of the components of the vector $(c_1, c_2, ...)$,

$$Z[\phi] = Z(c_1, c_2, ...) \tag{11.3}$$

We can define the functional derivative of $Z[\phi]$ by $\phi(x)$ in terms of its ordinary derivative by each of the coefficients c_n,

$$\frac{\delta Z[\phi]}{\delta \phi(x)} \equiv \sum_n f_n(x) \frac{\partial}{\partial c_n} Z(c_1, c_2, ...) \tag{11.4}$$

This defines a functional derivative in terms of an infinite number of ordinary derivatives.

In particular, we can use this definition of functional derivative to generalize the Liebnitz rules for ordinary derivatives to similar rules for functional derivatives. For example, it is easy to see that the product and chain rules for differentiation follow,

$$\frac{\delta}{\delta \phi(x)} ((Z_1[\phi])(Z_2[\phi])) = \left(\frac{\delta}{\delta \phi(x)} Z_1[J] \right) Z_2[\phi] + Z_1[\phi] \left(\frac{\delta}{\delta \phi(x)} Z_2[\phi] \right) \tag{11.5}$$

$$\frac{\delta}{\delta \phi(x)} f(Z[\phi]) = \frac{\partial f(Z[\phi])}{\partial Z[\phi]} \frac{\delta Z[\phi]}{\delta \phi(x)} \tag{11.6}$$

Another useful identity is

$$\frac{\delta \phi(y)}{\delta \phi(x)} = \delta(x - y) \tag{11.7}$$

As a specific example, consider a function $K(y_1, y_2, ..., y_n)$ which is a completely symmetric function of its n arguments and form the functional

$$K[\phi] = \int dy_1 dy_2 \dots dy_n \phi(y_1) \phi(y_2) \dots \phi(y_n) K(y_1, y_2, ..., y_n) \tag{11.8}$$

If we use the expansion of the function in equation (11.1), we can present the functional as the function

$$K(c_1, c_2, ...) = \sum_{m_1,...,m_n=1}^{\infty} c_{m_1} \dots c_{m_n} K_{m_1...m_n} \tag{11.9}$$

where

$$K_{m_1...m_n} = \int dy_1 \dots dy_n f_{m_1}(y_1) \dots f_{m_n}(y_n) K(y_1, \dots, y_n)$$

We can easily derive the rule for taking a functional derivative of $K[\phi]$,

$$
\frac{\delta}{\delta\phi(x)} K[\phi]
$$

$$
= \frac{\delta}{\delta\phi(x)} \int dy_1 dy_2 \dots dy_n \phi(y_1)\phi(y_2)\dots\phi(y_n) K(y_1, y_2, ..., y_n)
$$

$$
= n \int dy_2 \dots dy_n \phi(y_2)\dots\phi(y_n) K(x, y_2, ..., y_n) \tag{11.10}
$$

These identities are all that we will ever really need for a functional derivative. However, it is useful to note that the functional derivative is very similar to the variational derivative that we used when we studied the derivation of the Euler-Lagrange equations of motion from an action principle. We consider two functions which differ by an infinitesimal amount, $\phi(x)$, and $\phi(x) + \delta\phi(x)$. Then we expand the functional $Z[\phi + \delta\phi]$ to linear order in $\delta\phi(x)$,

$$
Z[\phi + \delta\phi] = Z[\phi] + \int dx \, \delta\phi(x) \frac{\delta Z[\phi]}{\delta\phi(x)} + \dots \tag{11.11}
$$

The coefficient of $\delta\phi(x)$ in the linear term, which we have already written as $\frac{\delta Z[\phi]}{\delta\phi(x)}$, is the functional derivative of $Z[\phi]$ by $\phi(x)$. When we are finding this coefficient, we are allowed to assume that $\delta\phi(x)$ has support in a compact region and goes to zero on the boundaries of the system so that we can integrate by parts and ignore the surface terms that this would generate, as many times as is needed to get the functional into the form in equation (11.11). The similarity with the variational derivative makes it easy to demonstrate the following equations

$$
\frac{\delta}{\delta\phi(x)} e^{i \int dy\phi(y)f(y)} = if(x) \, e^{i \int dy\phi(y)f(y)} \tag{11.12}
$$

$$
\frac{\delta}{\delta\phi(x)} e^{-\frac{1}{2}\int dydy'\phi(y)\Delta(y,y')\phi(y')}
$$

$$
= - \int dy' \Delta(x, y')\phi(y') \, e^{-\frac{1}{2}\int dydy'\phi(y)\Delta(y,y')\phi(y')} \tag{11.13}
$$

$$
\frac{\delta}{\delta\phi(x_1)} \frac{\delta}{\delta\phi(x_2)} e^{-\frac{1}{2}\int dydy'\phi(y)\Delta(y,y')\phi(y')} = \left(-\Delta(x_1, x_2)+\right.
$$

$$
+ \left. \int dy'dy'' \Delta(x_1, y')\Delta(x_2, y'')\phi(y')\phi(y'')\right) e^{-\frac{1}{2}\int dydy'\phi(y)\Delta(y,y')\phi(y')} \tag{11.14}
$$

These equations will be useful in the following.

11.2   Functional Integral

We can define a functional integral using similar ideas as those which we used for functional derivatives. In a functional integral, there will be an integration measure which indicates an integral over a set of functions. There will also be an integrand which will be a functional depending on the functions over which we are integrating. If we consider the example of a functional $F[\phi]$ of a single real function of one real variable, $\phi(x)$, we wish to define the expression

$$\int [d\phi(x)] \, F[\phi]$$

As we did for functional derivatives, we can assume that we could expand the function $\phi(x)$ in an infinite series of square integrable functions,

$$\phi(x) = \sum_n c_n f_n(x)$$

where the coefficients in the series are determined by doing the integrals

$$c_n = \int dx f_n(x)\phi(x)$$

Then, any functional of $\phi(x)$ has an equivalent description as a function of the infinite array of coefficients, $\{c_1, c_2, \ldots\}$,

$$F[\phi] = F(c_1, c_2, \ldots)$$

We define the functional integral as

$$\int [d\phi(x)] \, F[\phi] \equiv \int_{-\infty}^{\infty} dc_1 \int_{-\infty}^{\infty} dc_2 \ldots F(c_1, c_2, \ldots) \qquad (11.15)$$

where each component is integrated over the entire real line $-\infty < c_i < \infty$.[1]

Using this definition, we can find some examples of functional integrals. For example, the functional delta function can be gotten from an integral over functional plane waves. If we take a function

$$P(x) = \sum_n p_n f_n(x)$$

---

[1] In principle, one could consider definite integrals over other intervals, or even indefinite integrals. However, the only functional integrals which we will use are definite integrals over the entire real line as we are defining them here.

and we form the integral

$$\int [d\phi(x)] e^{i \int dy \phi(y) P(y)} = \int dc_1 dc_2 \dots e^{i \sum_n c_n p_n}$$

$$= \prod_{n=1}^{\infty} 2\pi \delta(p_n) \equiv \delta(P(x)/2\pi)$$

We can also do a Gaussian integral. Consider the integral

$$\mathcal{I}_G = \int [d\phi(x)] \, e^{-\frac{1}{2} \int dy dz \phi(y) K(y,z)\phi(z)}$$

If we introduce the expansion of the integration variable, wherever they appear in terms of the complete set of square integrable functions to get

$$\mathcal{I}_G = \int dc_1 dc_2 \dots e^{-\frac{1}{2} \sum_{mn} c_m K_{mn} c_n}$$

where we define the quadratic form which appears in the exponent of the resulting Gaussian as

$$K_{mn} = \int dx \int dy f_m(x) K(x, y) f_n(y)$$

It is easy to see that $K_{mn} = K_{nm}$, that is, it is a symmetric matrix. A fact from linear algebra is that a symmetric matrix can be diagonalized by an orthogonal transformation. That is, there exists a real matrix O such that $OO^t = I$ where O^t is its transpose[2] and I is the unit matrix and such that

$$[OKO^t]_{mn} = k_m \delta_{mn}$$

Here, k_m are the eigenvalues of the matrix K_{mn}.

What is more, we can change the integration variable by rotating the vector over which we are integrating, $c_n \to \tilde{c}_n = O_{nm} c_m$, Since $OO^t = 1$ implies that $|\det O| = 1$, the Jacobian for this change of variables is equal to one. The result is the Gaussian integral (after relabelling $\tilde{c}_n$ as c_n),

$$\prod_{m=1}^{\infty} \int dc_m \exp\left[-\frac{1}{2}c_m^2 k_m\right] = \sqrt{\prod_{m=1}^{\infty} \frac{2\pi}{k_m}} = \left[\det(K/2\pi)\right]^{-\frac{1}{2}}$$

---

[2] Such a matrix is called an orthogonal matrix.

The integrals over each of the c_i are finite only when the eigenvalues of K are all positive. The quadratic form $K(y, z)$ is called positive if all of its eigenvalues are positive. We have also made use of the fact that the determinant of a matrix is equal to a product over its eigenvalues. The integral of a Gaussian is thus proportional to the inverse of the square root of the determinant of the quadratic form in the exponent.

Then, we can also get a similar integration formula for the offset Gaussian

$$\int [d\phi(x)] e^{-\frac{1}{2}\int dy dz \phi(y) K(y,z)\phi(z) + i\int dy J(y)\phi(y)}$$

$$= \det^{-\frac{1}{2}}(K/2\pi) e^{-\frac{1}{2}\int dy dz J(y) K^{-1}(y,z) J(z)} \tag{11.16}$$

where $K^{-1}(y, z)$ is the inverse of the quadratic form $K(y, z)$,

$$\int dy K(x, y) K^{-1}(y, z) = \delta(x - z) = \int dy K^{-1}(x, y) K(y, z)$$

Here, we are assuming that this inverse exists, as it should when the eigenvalues of K_{mn} are all non-zero. We can also take functional derivatives of equation (11.16), we can find the correlation functions of the integration variable. For example, we use the equation for functional derivatives of an exponential

$$\phi(x_1)\phi(x_2) e^{i\int dy J(y)\phi(y)} = \frac{1}{i}\frac{\delta}{\delta J(x_1)}\frac{1}{i}\frac{\delta}{\delta J(x_2)} e^{i\int dy J(y)\phi(y)}$$

to show that

$$\int [d\phi(x)] e^{-\frac{1}{2}\int dy dx \phi(y) K(y,z)\phi(z)} \phi(x_1)\phi(x_2)$$

$$= \int [d\phi(x)] e^{-\frac{1}{2}\int dy dz \phi(y) K(y,z)\phi(z)} \frac{1}{i}\frac{\delta}{\delta J(x_1)}\frac{1}{i}\frac{\delta}{\delta J(x_2)} e^{i\int dy J(y)\phi(y)} \Bigg|_{J=0}$$

$$= \frac{1}{i}\frac{\delta}{\delta J(x_1)}\frac{1}{i}\frac{\delta}{\delta J(x_2)} \int [d\phi(x)] e^{-\frac{1}{2}\int dy dz \phi(y) K(y,z)\phi(z) + i\int dy J(y)\phi(y)} \Bigg|_{J=0}$$

$$= \frac{1}{i}\frac{\delta}{\delta J(x_1)}\frac{1}{i}\frac{\delta}{\delta J(x_2)} \det^{-\frac{1}{2}}(K/2\pi) e^{-\frac{1}{2}\int dy dz J(y) K^{-1}(y,z) J(z)} \Bigg|_{J=0}$$

$$= \det^{-\frac{1}{2}}(K/2\pi) K^{-1}(x_1, x_2). \tag{11.17}$$

and, by a straightforward generalization,

$$
\int [d\phi(x)]e^{-\frac{1}{2}\int dydx\phi(y)K(y,z)\phi(z)}\phi(x_1)\phi(x_2)\phi(x_3)\phi(x_4)
$$

$$
= \det^{-\frac{1}{2}}(K/2\pi)\{K^{-1}(x_1,x_2)K^{-1}(x_3,x_4)
$$

$$
+K^{-1}(x_1,x_3)K^{-1}(x_2,x_4) + K^{-1}(x_1,x_4)K^{-1}(x_2,x_3)\} \quad (11.18)
$$

and by a further straightforward generalization

$$
\int [d\phi(x)]e^{-\frac{1}{2}\int dydx\phi(y)K(y,z)\phi(z)}\phi(x_1)\ldots\phi(x_n) =
$$

$$
= \det^{-\frac{1}{2}}(K/2\pi)\sum_{\substack{\text{pairings}}}\prod_{\substack{\text{pairs}\\<ab>}}K^{-1}(x_a,x_b) \quad (11.19)
$$

The right-hand-side has a sum over all possible pairings of the coordinates x_a, x_b. Each distinct pairing is counted exactly once and it involves a product over all of the pairs of $K^{-1}(x_a, x_b)$, one for each pair. This integration formula will be very useful in the following sections and chapters.

11.3   Generating Functional for Free Scalar Fields

A generating functional is a functional of a source field which we can use to find correlation functions of quantum fields by taking functional derivatives by the source. In our example of the free scalar field, we would like to find a functional $Z[J]$ of a source field $J(x)$ so that the correlation functions are given by taking functional derivatives by $J(x)$ and then evaluating at $J(x) = 0$,

$$
< \mathcal{O}|T\phi(x_1)\ldots\phi(x_n)|\mathcal{O} >= \frac{1}{i}\frac{\delta}{\delta J(x_1)}\cdots\frac{1}{i}\frac{\delta}{\delta J(x_n)}Z[J]\Big|_{J=0} \quad (11.20)
$$

where the $n = 0$ term or the right-hand-side is 1. Here, the n point correlation function of the scalar field is obtained by taking n functional derivatives of the generating functional and then putting the argument, $J(x)$ to zero. Another way to write the same information that is contained in equation (11.20) is

$$
Z[J] = \sum_{n=0}^{\infty}\frac{i^n}{n!}\int dx_1\ldots dx_n J(x_1)\ldots J(x_n) < \mathcal{O}|T\phi(x_1)\ldots\phi(x_n)|\mathcal{O} >
$$

$$
(11.21)
$$

By plugging the expression for $Z[J]$ on the right-hand-side of the above equation (11.21) into the right-hand-side of equation (11.20) and using the rules for functional differentiation, specifically the example in equation (11.10), and then setting $J = 0$, we see that $Z[J]$ is indeed the generating functional. We can write it in a shorthand notation by formally summing the series on its right-hand-side to form the expression with a time-ordered exponential

$$Z[J] = < \mathcal{O}|T e^{i \int dx J(x)\phi(x)}|\mathcal{O} > \qquad (11.22)$$

We emphasize that this is a formal expression. It can always to be understood to be defined by its Taylor expansion in powers of the exponent.

If we know the explicit form of the generating functional, then, in principle, we know all of the time-ordered correlation functions of the quantum field $\phi(x)$. This can be regarded as an exact solution of the quantum field theory.

We will now proceed to find an explicit formula for the generating functional. Consider the functional derivative,

$$\frac{1}{i} \frac{\delta}{\delta J(y)} Z[J] =$$

$$\sum_{n=1}^{\infty} \frac{i^n}{n!} \frac{1}{i} \frac{\delta}{\delta J(y)} \int dx_1 \dots dx_n J(x_1) \dots J(x_n) < \mathcal{O}|T\phi(x_1)\dots\phi(x_n)|\mathcal{O} >$$

$$= \sum_{n=1}^{\infty} \frac{i^n n}{n! i} \int dx_1 \dots dx_{n-1} J(x_1) \dots J(x_{n-1}) < \mathcal{O}|T \phi(y) \phi(x_1) \dots \phi(x_{n-1})|\mathcal{O} >$$

$$(11.23)$$

where, in the last term, we have used the fact that the n point function is symmetric in its arguments. Then, we operate the wave-operator on the quantity in equation (11.23)

$$\left(-\partial_y^2 + m^2\right) \frac{1}{i} \frac{\delta}{\delta J(y)} Z[J]$$

$$= \sum_{n=1}^{\infty} \frac{i^{n-1}}{(n-1)!} \int dx_1 \dots dx_{n-1} J(x_1) \dots J(x_{n-1}) \cdot$$

$$\cdot (-\partial_y^2 + m^2) < \mathcal{O}|T \phi(y) \phi(x_1) \dots \phi(x_{n-1})|\mathcal{O} >$$

Then, we use the fact that the time derivatives in the expression

$$(\partial_y^2 - m^2) < \mathcal{O}|T \phi(y) \phi(x_1) \dots \phi(x_{n-1})|\mathcal{O} >$$

produce time delta-functions when they operate on the time-ordering theta-functions, and also the fact that the field operator satisfies the equation of motion $(\partial_y^2 -$

$m^2)\phi(y) = 0$ to get

$$(-\partial_y^2 + m^2) < \mathcal{O}|T \phi(y) \phi(x_1)\ldots\phi(x_{n-1})|\mathcal{O} >$$

$$= -i \sum_{k=1}^{n-1} \delta(y - y_k) < \mathcal{O}|T\phi(x_1)\ldots\phi(x_{k-1})\phi(x_{k+1})\ldots\phi(x_{n-1})|\mathcal{O} >$$

We plug this into the above equations to get

$$\left(-\partial_y^2 + m^2\right) \frac{1}{i} \frac{\delta}{\delta J(y)} Z[J]$$

$$= \sum_{n=0}^{\infty} \frac{i^n}{n!} \int dx_1 \ldots dx_n J(x_1)\ldots J(x_n)\cdot$$

$$\cdot \sum_{k=0}^{n} (-i)\delta(y - x_k) < \mathcal{O}|T \phi(x_1)\ldots\phi(x_{k-1})\phi(x_{k+1})\ldots\phi(x_n)|\mathcal{O} >$$

$$= J(y) \sum_{n=0}^{\infty} \frac{i^n}{n!} \int dx_1 \ldots dx_n J(x_1)\ldots J(x_n) < \mathcal{O}|T \phi(x_1)\ldots\phi(x_n)|\mathcal{O} >$$

$$= J(y) Z[J]$$

The final result is the functional differential equation

$$\left(-\partial^2 + m^2\right) \frac{1}{i} \frac{\delta}{\delta J(y)} Z[J] = J(y) Z[J] \tag{11.24}$$

This is a combination of a functional differential equation and an ordinary partial differential equation. It is straightforward to find a solution. To this end, we divide each side of (11.24) by $Z[J]$ and we make use of a Green function, $G(x, y)$, for the wave operator. The Green function obeys the equation

$$\left(-\partial^2 + m^2\right) G(x, y) = \delta^{(4)}(x - y) \tag{11.25}$$

Of course, the Green function is ambiguous and the ambiguity can only be fixed by specifying some boundary conditions for it. We will fix which Green function we are using shortly.

Then we can rewrite the functional differential equation (11.24) in the simple form

$$\frac{\delta}{\delta J(y)} \ln Z[J] = \int dx \, iG(y, z) J(z) \tag{11.26}$$

From this equation we conclude that $\ln Z[J]$ must be a quadratic functional of J. The functional anti-derivative of equation (11.26) is

$$\ln Z[J] = \text{constant} + \frac{i}{2} \int dxdy \, J(x)G(x, y)J(y) \tag{11.27}$$

We can use the boundary condition $Z[0] = 1$, to fix the constant in equation (11.27) whence we have

$$Z[J] = \exp\left(\frac{i}{2} \int dx\,dy\, J(x) G(x, y) J(y)\right) \tag{11.28}$$

The only thing left to do is to determine the Green function. We do this by requiring that the generating functional gives the correct result for the 2 point function,

$$\frac{1}{i}\frac{\delta}{\delta J(x_1)}\frac{1}{i}\frac{\delta}{\delta J(x_2)} Z[J]\bigg|_{J=0} = -i G(x, y) = \Delta(x, y) \tag{11.29}$$

where $\Delta(x, y)$ is the propagator, that is, the 2 point function defined in equation (7.30) and written more explicitly in equation (7.47). Equation (11.29) determines the Green function as being proportional to the 2 point function. We already know by independent means that the 2 point function is proportional to a Green function with just the correct coefficient to solve equation (11.29).

Our final result is then the expression for the generating functional of correlation functions of the non-interacting scalar field theory,

$$Z[J] = \exp\left(-\frac{1}{2} \int dx\,dy\, J(x) \Delta(x, y) J(y)\right) \tag{11.30}$$

This formula can then be used to find all of the correlation functions of the non-interacting scalar field. It is possible to check by independent means that the result is in fact correct. This formula represents a good example of how we can find a compact summary of all of the correlation functions of a quantum field theory.

There is a weak point in the logic of our discussion above. The diligent reader might notice that our derivation, in equations (11.23)–(11.24), of the functional differential equation (11.24) and its solution (11.28) depends only on the nature of time ordering, as well as the field equation and the equal-time commutation relations. Up to the solution of the functional differential equation, $|\mathcal{O}>$ could have been any state. The first place that we input the fact that $|\mathcal{O}>$ is the vacuum state was when we determined the Green function by asserting that the generating functional should generate the correct 2 point function. Up to that point, it could have been any state.

There is a more general way to know that we are on the right track. This is to recall our discussion in Sect. 7.9 where we examined the consequences of translation invariance in time. The fact that the multi-point functions are computed in the vacuum state had a consequence. That consequence was analyticity of the correlation functions in the second and fourth quadrants of the complex $x^0 - y^0$-plane, for the difference of any two of the time variables in the correlation function. This would not be so if the expectation values were taken in any other state. Indeed, it is easy to check that the propagator $\Delta(x, y)$ has the correct analyticity property. Moreover, in the free field theory that we have discussed here, the n-point functions are made from simple products of the two-point functions and they therefore all have the correct analyticity property. The analyticity must also be there in the interaction field theory.

Indeed, it must persist in perturbation theory and we will see it when we do perturbative computations as the possibility of performing a Wick rotation of Feynman integrals. We will return to this issue later.

11.3.1 Wick's Theorem for Scalar Fields

Equation (11.30) is our explicit solution for the generating functional. From it, we can deduce the general n point correlation function by taking functional derivatives. It is clear that all of the correlation functions with an odd number of fields vanish. When there are an even number of fields, that is, when n is an even number, the result is

$$< \mathcal{O}|T\phi(x_1)\ldots\phi(x_n)|\mathcal{O} >= \sum_{\substack{\text{pairings}}} \prod_{\substack{\text{pairs} \\ <ab>}} \Delta(x_a, x_b) \qquad (11.31)$$

The sum on the right-hand-side of (11.31) is a sum over all of the possible pairings of the coordinate labels $(1, 2, ..., n)$. There are $\frac{n!}{2^{\frac{n}{2}}\frac{n}{2}!}$ distinct pairings. In this sum, each distinct pairing is counted once. The product of the functions $\Delta(x_a, x_b)$ is to be taken over all of the $\frac{n}{2}$ pairs in each pairing.

Equation (11.31) is usually called Wick's theorem. As an example of how it works, consider the 4 point correlation function

$$< \mathcal{O}|T\phi(x_1)\phi(x_2)\phi(x_3)\phi(x_4)|\mathcal{O} > \qquad (11.32)$$
$$= \frac{1}{i}\frac{\delta}{\delta J(x_1)}\frac{1}{i}\frac{\delta}{\delta J(x_2)}\frac{1}{i}\frac{\delta}{\delta J(x_3)}\frac{1}{i}\frac{\delta}{\delta J(x_4)}Z[J]\Big|_{J=0}$$
$$= \Delta(x_1, x_2)\Delta(x_3, x_4) + \Delta(x_1, x_3)\Delta(x_2, x_4) + \Delta(x_1, x_4)\Delta(x_3, x_2) \qquad (11.33)$$

We see that, in this case, there are three distinct pairings, $[(1, 2), (3, 4)], [(1, 3)(2, 4)]$ and $[(1, 4), (2, 3)]$ of the four indices $\{1, 2, 3, 4\}$. (Indeed, using our formula which counts the number of pairings, results in $\frac{4!}{2^2 \times 2!} = 3$.) The result is a sum over these three pairings of the products of the 2 point functions, one term in the sum for each pairing, and the individual terms containing a product over 2 point functions with the pairs of indices in the pairings.

Wick's theorem will be useful when we study interacting quantum field theories using perturbation theory.

11.3.2 Generating Functional as a Functional Integral

In this section, we will convert the expression that we have already found for the generating functional in equation (11.30) to an equivalent expression which uses a functional integral. The appropriate expression is

$$Z[J] = \frac{\int [d\phi(x)] e^{i \int dy \left[-\frac{1}{2} \partial_\mu \phi(y) \partial^\mu \phi(y) - \frac{m^2 - i\epsilon}{2} \phi^2(x) + \phi(y) J(y) \right]}}{\int [d\phi(x)] e^{i \int dy \left[-\frac{1}{2} \partial_\mu \phi(y) \partial^\mu \phi(y) - \frac{m^2 - i\epsilon}{2} \phi^2(x) \right]}}. \tag{11.34}$$

$$= \frac{\int [d\phi(x)] e^{iS + i \int dy \phi(y) J(y)}}{\int [d\phi(x)] e^{iS}} \tag{11.35}$$

$$S = \int dy \mathcal{L}(x) \tag{11.36}$$

$$\mathcal{L}(x) = -\frac{1}{2} \partial_\mu \phi(y) \partial^\mu \phi(y) - \frac{m^2}{2} \phi^2(x) \tag{11.37}$$

Note that the integrand in equation (11.34) contains a convergence factor

$$\exp\left(-\frac{\epsilon}{2} \int dx \phi^2(x) \right)$$

where ϵ is an infinitesimal positive real number. This factor helps the the integral, whose integrand otherwise contains only oscillating factors, converge. It also appears in the quadratic form as $(\partial^2 - m^2 + i\epsilon)$ and, by doing so, it gives this quadratic form a unique inverse. When we do the Gaussian integral, we must invert this quadratic form. If the $i\epsilon$ were not there, we would still need some more information in order to decide which Green function to use as the inverse. The presence of the $i\epsilon$ does this for us. It chooses the time-ordered Green function. It is thus what contains the information that the result of the functional integral

$$\frac{1}{i} \frac{\delta}{\delta J(x_1)} \cdots \frac{1}{i} \frac{\delta}{\delta J(x_n)} Z[J] \bigg|_{J=0} = \frac{\int [d\phi(x)] e^{iS} \phi(x_1) \phi(x_2) \dots \phi(x_n)}{\int [d\phi(x)] e^{iS}} \tag{11.38}$$

will produce the time-ordered 2-point function when $n = 2$ and, generally, the time-ordered n point function.

We can confirm that equations (11.34) and (11.30) are identical. We do this by doing the functional integration explicitly. This is possible because it has an offset Gaussian integrand and in the previous section we learned how to do such and integral, the result was given in equation (11.16). Using that formula, we see that the Gaussian integral reproduces equation (11.30) with $i\Delta(x, y)$ given by the inverse of the quadratic form $(-\partial^2 + m^2 - i\epsilon)$ which is indeed proportional to the time-ordered 2 point function (see equation (7.47)). The $i\epsilon$ factor is often suppressed when the functional integral is written down, but it must always be assumed to be there.

Alternatively, we can show that the functional integral formula (11.34) obeys the functional differential equation (11.24). To show that it obeys the functional differential equation, consider

$$(-\partial^2 + m^2)\frac{1}{i}\frac{\delta}{\delta J(x)}Z[J]$$

$$= \frac{\int [d\phi(x)]e^{i\int dy\left[-\frac{1}{2}\partial_\mu\phi(y)\partial^\mu\phi(y)-\frac{m^2-i\epsilon}{2}\phi^2(y)+\phi(y)J(y)\right]}(-\partial^2+m^2)\phi(x)}{\int [d\phi(x)]e^{i\int dy\left[-\frac{1}{2}\partial_\mu\phi(y)\partial^\mu\phi(y)-\frac{m^2-i\epsilon}{2}\phi^2(y)\right]}}$$

$$= \frac{\int [d\phi(x)]e^{i\int dy[\phi(y)J(y)]}\frac{i\delta}{\delta\phi(x)}e^{i\int dy\left[-\frac{1}{2}\partial_\mu\phi(y)\partial^\mu\phi(y)-\frac{m^2-i\epsilon}{2}\phi^2(y)\right]}}{\int [d\phi(x)]e^{i\int dy\left[-\frac{1}{2}\partial_\mu\phi(y)\partial^\mu\phi(y)-\frac{m^2-i\epsilon}{2}\phi^2(y)\right]}}$$

$$= \frac{\int [d\phi(x)]\frac{i\delta}{\delta\phi(x)}\left\{e^{i\int dy[\phi(y)J(y)]}e^{i\int dy\left[-\frac{1}{2}\partial_\mu\phi(y)\partial^\mu\phi(y)-\frac{m^2-i\epsilon}{2}\phi^2(y)\right]}\right\}}{\int [\phi(x)]e^{i\int dy\left[-\frac{1}{2}\partial_\mu\phi(y)\partial^\mu\phi(y)-\frac{m^2-i\epsilon}{2}\phi^2(y)\right]}}$$

$$- \frac{\int [d\phi(x)]\frac{i\delta}{\delta\phi(x)}e^{i\int dy[\phi(y)J(y)]}e^{i\int dy\left[-\frac{1}{2}\partial_\mu\phi(y)\partial^\mu\phi(y)-\frac{m^2-i\epsilon}{2}\phi^2(y)\right]}}{\int [d\phi(x)]e^{i\int dy\left[-\frac{1}{2}\partial_\mu\phi(y)\partial^\mu\phi(y)-\frac{m^2-i\epsilon}{2}\phi^2(y)\right]}}$$

$$= J(x)\frac{\int [d\phi(x)]e^{i\int dy\left[-\frac{1}{2}\partial_\mu\phi(y)\partial^\mu\phi(y)-\frac{m^2-i\epsilon}{2}\phi^2(y)+\phi(y)J(y)\right]}}{\int [d\phi(x)]e^{i\int dy\left[-\frac{1}{2}\partial_\mu\phi(y)\partial^\mu\phi(y)-\frac{m^2-i\epsilon}{2}\phi^2(y)\right]}}$$

$$= J(x)\,Z[J] \tag{11.39}$$

where we have dropped the total functional derivative term.[3] We recover equation (11.24), the functional differential equation that the generating functional must obey. The functional integral representation of the generating functional is therefore a solution of this equation.

---

[3] We assume that the damping factor in the integrand, $e^{-\int dx\frac{\epsilon}{2}\phi^2(x)} = e^{-\frac{\epsilon}{2}\sum_n c_n^2}$, (where $c_n = \int dx\phi_n(x)\phi(x)$ with $\phi_n(x)$ a complete set of square integrable functions), makes the integrand go to zero when any of the $c_n \to \pm\infty$ which we could regard as the boundaries of the space of functions over which we are integrating. This allows us to drop terms which are total functional derivatives when the integrand contains this factor.

Wick's theorem, equation (11.31) for scalar fields written in terms of the functional integrals is

$$\frac{\int [d\phi(x)]e^{i\int dy\left[-\frac{1}{2}\partial_\mu\phi(y)\partial^\mu\phi(y)-\frac{m^2-i\epsilon}{2}\phi^2(y)+\phi(y)J(y)\right]}\phi(x_1)\phi(x_2)\ldots\phi(x_n)}{\int [d\phi(x)]e^{i\int dy\left[-\frac{1}{2}\partial_\mu\phi(y)\partial^\mu\phi(y)-\frac{m^2-i\epsilon}{2}\phi^2(y)\right]}}$$

$$= \sum_{\substack{\text{pairings}}} \prod_{\substack{\text{pairs}\\<ab>}} \Delta(x_a, x_b) \tag{11.40}$$

We will often use Wick's theorem in this form.

11.4   The Interacting Real Scalar Field

So far, we have studied the application of functional methods to the non-interacting scalar field. It is appropriate to ask what we can say about theories with interactions. Let us consider the example of the real scalar field with self-interactions governed by the action and Lagrangian density, field equation and commutation relations which we summarize here for the convenience of the reader:

$$\mathbf{S} = \int dx\mathcal{L}(x)$$

$$\mathcal{L}(x) = -\frac{1}{2}\partial_\mu\phi(x)\partial^\mu\phi(x) - \frac{m^2}{2}\phi^2(x) - \frac{\lambda}{4!}\phi^4(x)$$

$$(-\partial^2 + m^2)\phi(x) = -\frac{\lambda}{3!}\phi^3(x)$$

$$\left[\phi(x), \frac{\partial}{\partial y^0}\phi(y)\right]\delta(x^0 - y^0) = \delta(x - y)$$

$$[\phi(x), \phi(y)]\delta(x^0 - y^0) = 0, \quad \left[\frac{\partial}{\partial x^0}\phi(x), \frac{\partial}{\partial y^0}\phi(y)\right]\delta(x^0 - y^0) = 0$$

The generating functional for time-ordered correlation functions in this interaction theory is defined as

$$Z[J] \equiv <\mathcal{O}|Te^{i\int dx\phi(x)J(x)}|\mathcal{O}> \tag{11.41}$$

Here, we are using the same notation for the fields $\phi(x)$, the vacuum state $|\mathcal{O}>$ and the generating functional $Z[J]$ as we did in the non-interacting theory. To be clear, these are not the same objects that they were there and our inability to solve this theory explicitly means that we have much less information about them. Our goal here is to find the functional integral representation of $Z[J]$. In fact, from the free field theory, where the integrand in the functional integral contained the exponential of the action, $e^{i\mathbf{S}}$, a good guess might be that we obtain the generating functional of

the interacting theory by simply replacing the action of the noninteracting theory by
the action of the self-interacting theory. That is we replace equation (11.34) by

$$Z[J] = \frac{\int [d\phi(x)]e^{iS+i\int dy\phi(y)J(y)}}{\int [d\phi(x)]e^{iS}} \qquad (11.42)$$

$$S = \int dy\mathcal{L}(x) \qquad (11.43)$$

$$\mathcal{L}(x) = -\frac{1}{2}\partial_\mu\phi(y)\partial^\mu\phi(y) - \frac{m^2}{2}\phi^2(x) - \frac{\lambda}{4!}\phi^4(x) \qquad (11.44)$$

One reason that we might think this is the right answer is that it gives us the
identity

$$0 = \frac{\int [d\phi(x)]\frac{1}{i}\frac{\delta}{\delta\phi(x)}\left\{e^{iS[\phi]}\right\}}{\int [d\phi(x)]e^{iS[\phi]}}$$

$$= \frac{\int [d\phi(x)]e^{iS[\phi]}\left[(\partial^2 - m^2)\phi(x) - \frac{\lambda}{3!}\phi^3(x)\right]}{\int [d\phi(x)]e^{iS[\phi]}}$$

$$=< \mathcal{O}|\left\{(\partial^2 - m^2)\phi(x) - \frac{\lambda}{3!}\phi^3(x)\right\}|\mathcal{O} >$$

which tells us that the expectation value of the field equation is obeyed. This gives
us some confidence that the guess is correct. In fact, we can show that it is correct.
For this, we generalize the argument of the previous section where we derived a
functional differential equation that $Z[J]$ must satisfy to the interacting field theory.
It is easy to see that most of the steps are the same with the additional complication
that there is a term with the interaction term from the equation of motion inserted
into the time-ordered products. This term can be dealt with in a simple way and the
functional differential equation that we obtain is

$$\left\{(-\partial^2 + m^2)\frac{1}{i}\frac{\delta}{\delta J(y)} + \frac{\lambda}{3!}\left(\frac{1}{i}\frac{\delta}{\delta J(y)}\right)^3\right\} Z[J] = J(y) Z[J] \qquad (11.45)$$

This is a linear functional differential equation that we will not be able to solve
explicitly. However, we can find a formal solution. We use the fact that we could

easily solve it in the limit where λ is put to zero. We modify that solution by operating on it with a functional differential operator to get

$$Z[J] = \left. \frac{e^{-i \int dx \frac{\lambda}{4!}\left(\frac{1}{i}\frac{\delta}{\delta J(x)}\right)^4} e^{-\frac{1}{2}\int dydz J(y)\Delta(y,z)J(z)}}{e^{-i \int dx \frac{\lambda}{4!}\left(\frac{1}{i}\frac{\delta}{\delta J(x)}\right)^4} e^{-\frac{1}{2}\int dydz J(y)\Delta(y,z)J(z)}} \right|_{J=0} \tag{11.46}$$

where $\Delta(y, z)$ is the free-field 2 point function which is defined in equation (7.47) and which we re-copy here:

$$\Delta(x, y) = -i \int \frac{dk}{(2\pi)^4} \frac{e^{ik_\mu (x-y)^\mu}}{k_\mu k^\mu + m^2 - i\epsilon}$$

We have also used the fact that $Z[0] = 1$ to determine the denominator of equation (11.46).

The functional differential operator in both the numerator and denominator of equation (11.46) can be defined by a Taylor expansion in powers of λ, so that

$$e^{-i \int dx \frac{\lambda}{4!}\left(\frac{1}{i}\frac{\delta}{\delta J(x)}\right)^4}$$

$$\equiv \sum_{n=0}^{\infty} \frac{(-i)^n}{n!} \frac{\lambda^n}{(4!)^n} \int dy_1 \dots dy_n \left(\frac{1}{i}\frac{\delta}{\delta J(y_1)}\right)^4 \dots \left(\frac{1}{i}\frac{\delta}{\delta J(y_n)}\right)^4$$

This formula gives us natural way to Taylor expand the generating functional in powers of the coupling constant λ and therefore it is a natural approach to perturbation theory. We will make use of this expansion when we discuss perturbation theory in later chapters.

We can easily see that equation (11.46) satisfies the functional differential equation (11.45). To see this, we plug it into the equation to get

$$\left\{ (-\partial^2 + m^2) \frac{1}{i}\frac{\delta}{\delta J(y)} + \frac{\lambda}{3!}\left(\frac{1}{i}\frac{\delta}{\delta J(y)}\right)^3 - J(y) \right\} \cdot$$

$$\cdot e^{-i \int dx \frac{\lambda}{4!}\left(\frac{1}{i}\frac{\delta}{\delta J(x)}\right)^4} e^{-\frac{1}{2}\int dwdz J(w)\Delta(w,z)J(z)}$$

Then we note that the functional derivative operators commute with each other, so we can write the above formula as

$$e^{-i \int dx \frac{\lambda}{4!}\left(\frac{1}{i}\frac{\delta}{\delta J(x)}\right)^4} \left\{ (-\partial^2 + m^2) \frac{1}{i}\frac{\delta}{\delta J(y)} \right\} e^{-\frac{1}{2}\int dwdz J(w)\Delta(w,z)J(z)}$$

$$+ e^{-i \int dx \frac{\lambda}{4!}\left(\frac{1}{i}\frac{\delta}{\delta J(x)}\right)^4} \left\{ \frac{\lambda}{3!}\left(\frac{1}{i}\frac{\delta}{\delta J(y)}\right)^3 \right\} e^{-\frac{1}{2}\int dwdz J(w)\Delta(w,z)J(z)}$$

$$- J(y) e^{-i \int dx \frac{\lambda}{4!}\left(\frac{1}{i}\frac{\delta}{\delta J(x)}\right)^4} e^{-\frac{1}{2}\int dwdz J(w)\Delta(w,z)J(z)}$$

which becomes

$$\left[e^{-i \int dx \frac{\lambda}{4!} \left(\frac{1}{i} \frac{\delta}{\delta J(x)} \right)^4} , J(y) \right] e^{-\frac{1}{2} \int dw dz J(w) \Delta(w,z) J(z)}$$

$$+ e^{-i \int dx \frac{\lambda}{4!} \left(\frac{1}{i} \frac{\delta}{\delta J(x)} \right)^4} \left\{ \frac{\lambda}{3!} \left(\frac{1}{i} \frac{\delta}{\delta J(y)} \right)^3 \right\} e^{-\frac{1}{2} \int dw dz J(w) \Delta(w,z) J(z)}$$

Then, using the Taylor expansion of $e^{-i \int dx \frac{\lambda}{4!} \left(\frac{1}{i} \frac{\delta}{\delta J(x)} \right)^4}$ we can easily show that

$$\left[e^{-i \int dx \frac{\lambda}{4!} \left(\frac{1}{i} \frac{\delta}{\delta J(x)} \right)^4} , J(y) \right] = -e^{-i \int dx \frac{\lambda}{4!} \left(\frac{1}{i} \frac{\delta}{\delta J(x)} \right)^4} \frac{\lambda}{3!} \left(\frac{1}{i} \frac{\delta}{\delta J(y)} \right)^3$$

and conclude that the generating functional given in equation (11.46) indeed solves the functional differential equation (11.45).

By the same token, it is now easy to show that equation (11.42) is the correct functional integral representation of the generating functional for the interacting field theory. One simply takes the ratio

$$Z[J] = \frac{e^{-i \int dx \frac{\lambda}{4!} \left(\frac{1}{i} \frac{\delta}{\delta J(x)} \right)^4} \int [d\phi(x)] e^{i \int dy \left[-\frac{1}{2} \partial_\mu \phi(y) \partial^\mu \phi(y) - \frac{m^2 - i\epsilon}{2} \phi^2(y) + \phi(y) J(y) \right]}}{e^{-i \int dx \frac{\lambda}{4!} \left(\frac{1}{i} \frac{\delta}{\delta J(x)} \right)^4} \int [d\phi(x)] e^{i \int dy \left[-\frac{1}{2} \partial_\mu \phi(y) \partial^\mu \phi(y) - \frac{m^2 - i\epsilon}{2} \phi^2(y) + \phi(y) J(y) \right]}} \Bigg|_{J=0}$$

and, using

$$e^{-i \int dx \frac{\lambda}{4!} \left(\frac{1}{i} \frac{\delta}{\delta J(x)} \right)^4} e^{i \int dy \phi(y) J(y)} = e^{-i \int dx \frac{\lambda}{4!} \phi^4(x)} e^{i \int dy \phi(y) J(y)}$$

returns equation (11.42). Here, we have normalized the generating functional so that it goes to one when J is set to zero.

More Functional Integrals

12

In the last chapter, we found a functional integral representation for the correlation functions of a real scalar field. It was presented as a ratio of functional integrals

$$Z[J] = \frac{\int [d\phi] e^{iS[\phi] + i \int dx \phi(x) J(x)}}{\int [d\phi] e^{iS[\phi]}}$$

All that is needed to write this functional integral is knowledge of the basic field degrees of freedom, so that we know what the integration variables are, and the action of the quantum field theory. This was so even for the interacting field theory where the action and Lagrangian density are, for example,

$$S[\phi] = \int dx \, \mathcal{L}(x)$$

$$\mathcal{L}(x) = -\frac{1}{2} \partial_\mu \phi(x) \partial^\mu \phi(x) - \frac{m^2}{2} \phi^2(x) - \frac{\lambda}{4!} \phi^4(x)$$

Since we know how to do Gaussian integrals with polynomials in the integrand, and they can be summarized as Wick's theorem, we are ready to begin perturbation theory. We write the integrands in powers of the coupling constant λ and evaluate the terms that are generated by this expansion. We will put this off the details until the next chapter.

We might immediately generalize the functional integral formula to other interesting quantum field theories where we know the basic degrees of freedom and the action. We would then write a functional integral over the fields and put e^{iS} in the integrand. For the simple field theories that we shall study in the following, this is essentially the correct approach. However there remain some subtleties which require

© The Author(s), under exclusive license to Springer Nature Singapore Pte Ltd. 2023   229
G. W. Semenoff, *Quantum Field Theory*, Graduate Texts in Physics,
https://doi.org/10.1007/978-981-99-5410-0_12

some attention. For the photon field, there is the complication of gauge invariance and the need to "fix a gauge". For the Dirac field, the integration variables in the functional integral should be anti-commuting functions. We will examine these issues by studying the functional integral representation for correlations of free photons and of free Dirac fields first. Later on, we will put them together to study quantum electrodynamics.

12.1   Functional Integrals for the Photon Field

Let us begin by studying the free photon field. Recall that we had begun with Maxwell's equations written in relativistic form as the gauge invariant wave equation for the four-vector potential field

$$-\partial^2 A_\mu(x) + \partial_\mu \partial_\nu A^\nu(x) = 0$$

We then used the gauge invariance to fix a relativistic gauge condition. The result was a presentation of the field theory described by the field equation above as the field equation, constraint and the equal-time commutation relations for the photon field $A_\mu(x)$,

$$-\partial^2 A_\mu(x) = 0 \tag{12.1}$$

$$\left[A_\mu(x), \frac{\partial}{\partial y^0} A_\nu(y)\right]_{x^0=y^0} = i\eta_{\mu\nu}\delta(\vec{x} - \vec{y}) \tag{12.2}$$

$$\left[A_\mu(x), A_\nu(y)\right]_{x^0=y^0} = 0, \quad \left[\frac{\partial}{\partial x^0} A_\mu(x), \frac{\partial}{\partial y^0} A_\nu(y)\right]_{x^0=y^0} = 0 \tag{12.3}$$

and, the additional gauge constraint

$$\partial_\mu A^\mu(x) = 0 \tag{12.4}$$

The field equation and the commutation relations in equations (12.1)–(12.3) could be obtained from the action

$$S[A] = \int dx \, \mathcal{L}(x)$$

$$\mathcal{L}(x) = -\frac{1}{2}\partial_\mu A_\nu(x)\partial^\mu A^\nu(x)$$

and the constraint is compatible with the time evolution in the sense that the field equation implies that $-\partial^2\left(\partial_\mu A^\mu(x)\right) = 0$ and therefore, if $\partial_\mu A^\mu(x)$ and its time derivative were set to zero at an initial time, they are forever zero.

Also, recall that we found a solution of the field equation (12.1) and the commutation relations (12.2) and (12.3) as a quantum field theory and then we imposed the constraint (12.4) as a physical state condition,

$$\partial_\mu A^{(+)\mu}(x)|\text{phys} >= 0$$

that is, we used this condition to separate out a subset of the states of the quantum field theory and then we imposed an equivalence relation on that subset and called the equivalence classes physical states.

As in our discussion of the real scalar field, it will be very useful to think of the information that is contained in the quantum field theory as being encoded in its correlation functions, that is in the quantities

$$G_{\mu_1 \ldots \mu_n}(x_1 \ldots x_n) \equiv \; <\mathcal{O}|T A_{\mu_1}(x_1) A_{\mu_2}(x_2) \ldots A_{\mu_n}(x_n)|\mathcal{O}> \qquad (12.5)$$

Moreover, for the free photon, we expect that such multi-point functions are entirely determined once the 2 point function

$$\Delta_{\mu\nu}(x, y) \equiv \; <\mathcal{O}|T A_\mu(x) A_\nu(y)|\mathcal{O}> \qquad (12.6)$$

is known. We can build up a multi-point function from 2 point functions using Wick's theorem, which applies to free fields.

However, there is an immediate problem which is already apparent in the 2 point function. Even though the vacuum $|\mathcal{O}>$ is a physical state, the field operator $A_\nu(y)$ creates both physical and unphysical states when it operates on the vacuum. An immediate consequence is that the constraint is not obeyed, for example,

$$<\mathcal{O}|T \partial_\mu A^\mu(x) A_\nu(y)|\mathcal{O}> \neq 0$$

We will ignore this fact for now. We can take the position that only the expectation values of gauge invariant operators make physical sense anyway, and the full correlation functions in (12.6) are simply an intermediate construction which we shall need in order to later form the correlations of gauge invariant operators. For example, we can evaluate the photon two-point function because it is technically expedient but at the end of the day we confine ourselves to using the final result to evaluate objects like

$$<\mathcal{O}|T F_{\mu\nu}(x) F_{\rho\sigma}(y)|\mathcal{O}>$$

Then, we rely on the fact that the gauge invariant operators commute with the physical state condition and as a consequence, when they act on physical states, they create only physical states. This will turn out to be the best that we can do.

If we operate the wave operator on the 2 point function and use the equal-time commutation relations, we discover that

$$-\partial^2 \Delta_{\mu\nu}(x, y) = -i \eta_{\mu\nu} \delta(x - y)$$

which reveals that the 2 point function for free fields is proportional to a Green function. It must be the time-ordered Green function and it is easy to find as a Fourier integral

$$\Delta_{\mu\nu}(x, y) = \int \frac{dk}{(2\pi)^4} e^{ik_\lambda (x-y)^\lambda} \frac{-i\eta_{\mu\nu}}{k^2 - i\epsilon} \qquad (12.7)$$

where the replacement of k^2 by $k^2 - i\epsilon$ in the denominator implements the time-ordered boundary conditions.

Now, we can ask whether the correlation functions can be obtained by taking functional derivatives of a generating functional which we would formally define as

$$Z[J] = \sum_{n=0}^{\infty} \frac{i^n}{n!} \int dx_1 \ldots dx_n G_{\mu_1 \ldots \mu_n}(x_1, \ldots, x_n) J^{\mu_1}(x_1) \ldots J^{\mu_n}(x_n) \quad (12.8)$$

It is easy to see that this generating functional is completely determined by the 2 point function,

$$Z[J] = e^{-\frac{1}{2} \int dx dy J^{\mu}(x) \Delta_{\mu\nu}(x,y) J^{\nu}(y)} \quad (12.9)$$

To see this, as in the case of the relativistic scalar field in the previous chapter, we can use the field equation and the commutation relations to derive a functional differential equation for the generating functional. The result is

$$-\partial^2 \frac{1}{i} \frac{\delta}{\delta J^{\mu}(y)} Z[J] = J_{\mu}(y) Z[J] \quad (12.10)$$

which, using the normalization $Z[0] = 1$, has the solution

$$Z[J] = \exp\left(-\frac{1}{2} \int dx dy J^{\mu}(x) \Delta_{\mu\nu}(x, y) J^{\nu}(y)\right) \quad (12.11)$$

All of the higher point correlation functions are determined by this formula, they are simply given by the generalization of Wick's theorem to the photon field,

$$< \mathcal{O}|T A_{\mu_1}(x_1) \ldots A_{\mu_n}(x_n)|\mathcal{O}> = \sum_{\substack{pairings \\ <ab>}} \prod_{pairs} \Delta_{\mu_a \mu_b}(x_a, x_b) \quad (12.12)$$

We can also see that the generating functional in equation (12.11) is the result of doing the Gaussian functional integral

$$Z[J] = \frac{\int [dA_{\mu}] e^{i \int dy \left[-\frac{1}{2} \partial_{\mu} A_{\nu}(y) \partial^{\mu} A^{\nu}(y) + \frac{i\epsilon}{2} A_{\nu}(y) A^{\nu}(y) + A_{\mu}(y) J_{\mu}(y)\right]}}{\int [dA] e^{i \int dy \left[-\frac{1}{2} \partial_{\mu} A_{\nu}(y) \partial^{\mu} A^{\nu}(y) + \frac{i\epsilon}{2} A_{\nu}(y) A^{\nu}(y)\right]}} \quad (12.13)$$

We have included the "$i\epsilon$" terms in the action so that (after one integration by parts) the quadratic terms in the exponent is $-\int dy \frac{1}{2} A_{\mu}(x)(-\partial^2 - i\epsilon) A^{\mu}(x)$ so that, when we find the Green function for the operator $(-\partial^2 - i\epsilon)$, it will be the time-ordered one. We will generally omit the infinitesimal $i\epsilon$-term with the understanding that, whenever we invert a quadratic form which is a differential operator, we take the inverse to contain the time-ordered Green function.

Alternatively, we can show that the functional integral formula (12.13) obeys the functional differential equation (12.10). This, together with the input that the Green function that is used to invert the wave-operator is the time-ordered one, would also establish that it is the correct generating functional.

We conclude that equation (12.13) is the appropriate representation of the generating functional of the quantum field theory that is governed by the action

$$S[A] = \int dy \left[-\frac{1}{2} \partial_\mu A_\nu(y) \partial^\mu A^\nu(y) \right]$$

If we use the fact that this theory should only be used to compute gauge invariant correlation functions, that is, that its correct usage is to compute[1]

$$< \mathcal{O} | \mathcal{T} \text{ gauge invariant operators } | \mathcal{O} >=$$

$$\frac{\int [dA_\mu] e^{i \int dy \left[-\frac{1}{2} \partial_\mu A_\nu(y) \partial^\mu A^\nu(y) \right]} \text{gauge invariant functions}}{\int [dA] e^{i \int dy \left[-\frac{1}{2} \partial_\mu A_\nu(y) \partial^\mu A^\nu(y) \right]}} \tag{12.14}$$

we can transform the integral to a more symmetric form. If we do the change of the functional integration variable

$$A_\mu(x) \rightarrow \tilde{A}_\mu(x) = A_\mu(x) + \partial_\mu \chi(x)$$

---

[1] Equivalently, we could constrain the current by which we take functional derivatives to be a conserved current, that is, so that it obeys $\partial_\mu J^\mu(x) = 0$. This would mean that functional derivatives of it are constrained to produce only correlations of the transverse components of the vector potential field, $t^{\mu\nu} A_\nu(x)$, where the transverse and longitudinal projection operators are defined by

$$t^{\mu\nu}(x, y) \equiv \left(\eta_{\mu\nu} - \frac{\partial_\mu \partial_\nu}{\partial^2} \right) \delta(x - y) \equiv \int \frac{dk}{(2\pi)^4} e^{ikx} \left(\eta_{\mu\nu} - \frac{k_\mu k_\nu}{k^2} \right)$$

$$\ell^{\mu\nu}(x, y) \equiv \left(\frac{\partial_\mu \partial_\nu}{\partial^2} \right) \delta(x - y) \equiv \int \frac{dk}{(2\pi)^4} e^{ikx} \left(\frac{k_\mu k_\nu}{k^2} \right)$$

obey

$$t^{\mu\nu}(x, y) + \ell^{\mu\nu}(x, y) = \eta^{\mu\nu} \delta(x - y)$$

$$\int dy \, t^{\mu\nu}(x, y) t_\nu{}^\lambda(y, z) = t^{\mu\lambda}(x, z), \quad \int dy \, \ell^{\mu\nu}(x, y) \ell_\nu{}^\lambda(y, z) = \ell^{\mu\lambda}(x, z)$$

$$\int dy \, t^{\mu\nu}(x, y) \ell_\nu{}^\lambda(y, z) = 0 = \int dy \, \ell^{\mu\nu}(x, y) t_\nu{}^\lambda(y, z)$$

Then, the transverse part of the vector potential is gauge invariant,

$$t^{\mu\nu} A_\nu(x) = - \int dy \int \frac{dk}{(2\pi)^4} \frac{e^{ik(x-y)}}{k^2} \partial^\mu F_{\mu\nu}(y)$$

the exponent in the integrand in the numerator changes as

$$\int dy \left[-\frac{1}{2}\partial_\mu A_\nu(y)\partial^\mu A^\nu(y) \right] \rightarrow$$

$$\int dy \left[-\frac{1}{2}\partial_\mu A_\nu(y)\partial^\mu A^\nu(y) + \partial^2 \chi(y)\partial_\mu A^\mu(y) - \frac{1}{2}\partial^2 \chi(y)\partial^2 \chi(y) \right]$$

where we have integrated the second and third terms in the last line by parts and dropped the surface terms. Then, since $\chi(x)$ was introduced by a simple change of the integration variable, and the result of the functional integral cannot not depend on the field $\chi(x)$, we can do the Gaussian integral over it (and divide by an infinite constant which will cancel with a similar factor from the denominator of equation (12.14)). We obtain

$$< \mathcal{O}|T \text{ gauge invariant operators } |\mathcal{O} >=$$

$$\frac{\int [dA_\mu]e^{i\int dy\left[-\frac{1}{4}F_{\mu\nu}(y)F^{\mu\nu}(y)\right]} \text{ gauge invariant functions}}{\int [dA]e^{i\int dy\left[-\frac{1}{4}F_{\mu\nu}(y)F^{\mu\nu}(y)\right]}} \qquad (12.15)$$

We now have a gauge invariant form of the functional integral representation of the generating functional where we have the action

$$S[A] = \int dy \left[-\frac{1}{4}F_{\mu\nu}(y)F^{\mu\nu}(y) \right]$$

from which we derive the gauge invariant form of Maxwell's equations.

Later, when we do computations, it will actually be the gauge-fixed functional integral (12.14) which will be of most use to us. In fact, it is useful to derive a slightly more general version of it. To do that, we begin with the gauge invariant integral in equation (12.15) and we make use of the Faddeev-Popov substitution where we insert the identity

$$1 = \int [d\chi]\delta(\partial_\mu A^\mu(x) - \partial^2 \chi(x) - f(x))|\det(-\partial^2)|$$

into the functional integrals in both the numerator and the denominator if equation (12.15). In this expression, when the functional delta-function is used to evaluate the integral over $\chi(x)$ it produces a Jacobian which cancels the factor $|\det(-\partial^2)|$. Then the numerator in equation (12.15) becomes

$$\int [dA][d\chi]\, \delta(\partial_\mu A^\mu(x) - \partial^2 \chi(x) - f(x))\, |\det(-\partial^2)|\cdot$$

$$\cdot e^{i\int dy\left[-\frac{1}{4}F_{\mu\nu}(y)F^{\mu\nu}(y)\right]} \text{ gauge invariant functions}$$

Upon doing a gauge transformation $A_\mu(x) \to A_\mu(x) + \partial_\mu \chi(x)$, assuming that the integration measure $[dA_\mu]$ is invariant under this transformation, and using the gauge invariance of the exponent, $\int dy \left[-\frac{1}{4} F_{\mu\nu}(y) F^{\mu\nu}(y) \right]$, and the other terms in the integrand, the expression becomes

$$\left(\int [d\chi] \right) \int [dA] \, \delta(\partial_\mu A^\mu(x) - f(x)) \, | \det(-\partial^2)| \cdot$$

$$\cdot \, e^{i \int dy \left[-\frac{1}{4} F_{\mu\nu}(y) F^{\mu\nu}(y) \right]} \text{ gauge invariant functions}$$

where the infinite factor of the volume of the set of gauge transformations $\left(\int [d\chi] \right)$ has been extracted. This factor cancels between the numerator and the denominator when we take the ratio in equation (12.15).

The expression in the equation above cannot depend on the function $f(x)$. We will now make use of this fact. We multiply the equation by $\exp \left(-i \frac{\xi}{2} \int dx f(x)^2 \right)$ and then we do the functional integral over $f(x)$ and use the functional delta function to evaluate the integral. We do this for both the numerator and the denominator in equation (12.15) so that the extra factors that we produce are identical in both places and cancel in the ratio. We obtain the gauge fixed functional integral

$$< \mathcal{O} | \mathcal{T} \text{ gauge invariant operators } | \mathcal{O} > =$$

$$\frac{\int [dA] \, e^{i S_{gf}[A]} \text{gauge invariant functions}}{\int [dA] \, e^{i S_{gf}[A]}} \tag{12.16}$$

$$\mathbf{S}_{gf}[A] = \int dy \left[-\frac{1}{2} \partial_\mu A_\nu(y) \partial^\mu A^\nu(y) + \frac{1-\xi}{2} \partial_\mu A^\mu(x) \partial_\nu A^\nu(x) \right] \tag{12.17}$$

Finally, we could summarize the above result in the form of a generating functional,

$$Z[J] = \frac{\int [dA] \, e^{i S_{gf}[A] + i \int dy A_\mu(y) J^\mu(y)}}{\int [dA] \, e^{i S_{gf}[A]}} \tag{12.18}$$

with which we can find correlation functions by taking functional derivatives by $J^\mu(x)$. We note that, if we choose the gauge parameter to be $\xi = 1$, the expression for $Z[J]$ that we have found in equation (12.18) is identical to the one that we started with in equation (12.13). For other values of the gauge parameter, they are not equivalent, and indeed, in spite of the notation, the $Z[J]$ in equation (12.18) is not identical to the $Z[J]$ in equation (12.13). However they are guaranteed to give identical correlation functions of gauge invariant operators.

The 2 point function is

$$
\Delta_{\mu\nu}(x, y) = -i \int \frac{dk}{(2\pi)^4} e^{ik_\mu(x^\mu - y^\mu)} \left[\frac{\eta_{\mu\nu}}{k_\nu k^\nu - i\epsilon} - \left(1 - \frac{1}{\xi}\right) \frac{k_\mu k_\nu}{(k_\nu k^\nu - i\epsilon)^2} \right]
$$

$$(12.19)$$

and it depends on the gauge fixing parameter, ξ. Computations of gauge invariant correlation functions should obtain results which are independent of ξ. As a result, the parameter ξ can be chosen for our convenience. Some convenient choices are

1. $\xi = 1$ the "Feynman gauge", which is used for most perturbative computations
2. $\xi = \infty$ the "Landau gauge", which is sometimes convenient when demonstrating gauge invariance is important.
3. $\xi = \frac{1}{3}$ the "Fried-Yennie gauge", where the electron self-energy will not require an infrared regularization.

12.2   Functional Methods for Fermions

In order to find functional integral representations of the generating functionals for correlations of fermions, we shall need a generalization of the definitions of functional derivative and functional integral. We recall that, even at the level of classical field theory, it was convenient to describe fermions using anti-commuting functions. This was an essential part of the use of the Lagrangian density, the Euler-Lagrange equations and Noether's theorem when they are applied to field theories describing fermions. It will turn out that anti-commuting functions are essential for the generating functional and the functional integral representation of a quantum field theory of fermions. For this, we have to generalize our functional calculus to anti-commuting functions. Let us begin with some of the properties of anti-commuting numbers. Consider two numbers, η_1 and η_2 which have the property

$$
\eta_1 \eta_2 = -\eta_2 \eta_1, \quad \eta_1^2 = 0, \quad \eta_2^2 = 0
$$

Given these properties, functions of η_1 and η_2, $g(\eta_1, \eta_2)$ can only have a very simple form. Their Taylor expansion in the coordinates η_1 and η_2 can only have four terms

$$
g(\eta_1, \eta_2) = g_0 + g_1 \eta_1 + g_2 \eta_2 + g_{12} \eta_1 \eta_2
$$

where (g_0, g_1, g_2, g_{12}) are four real numbers (or complex numbers if we were considering complex-valued functions). The derivative of the function is defined as

$$
\frac{\partial}{\partial \eta_1} g(\eta_1, \eta_2) \equiv g_1 + g_{12} \eta_2
$$

This is just what we would expect a derivative to do. Moreover

$$\frac{\partial}{\partial \eta_2} g(\eta_1, \eta_2) \equiv g_2 - g_{12}\eta_1$$

and

$$\frac{\partial}{\partial \eta_2} \frac{\partial}{\partial \eta_1} g(\eta_1, \eta_2) = g_{12}, \quad \frac{\partial}{\partial \eta_1} \frac{\partial}{\partial \eta_2} g(\eta_1, \eta_2) = -g_{12}$$

which contains all of the information that we need to know about derivatives by anti-commuting numbers, including the fact that they anti-commute,

$$\frac{\partial}{\partial \eta_1} \frac{\partial}{\partial \eta_2} = -\frac{\partial}{\partial \eta_2} \frac{\partial}{\partial \eta_1},$$
$$\frac{\partial}{\partial \eta_1} \frac{\partial}{\partial \eta_1} = 0, \quad \frac{\partial}{\partial \eta_2} \frac{\partial}{\partial \eta_2} = 0$$

and

$$\frac{\partial}{\partial \eta_1} \eta_2 = -\eta_2 \frac{\partial}{\partial \eta_1},$$
$$\frac{\partial}{\partial \eta_2} \eta_1 = -\eta_1 \frac{\partial}{\partial \eta_2}$$
$$\frac{\partial}{\partial \eta_1} (\eta_1 g(\eta_1, \eta_2)) = g(\eta_1, \eta_2) - \eta_1 \frac{\partial}{\partial \eta_1} g(\eta_1, \eta_2),$$
$$\frac{\partial}{\partial \eta_2} (\eta_2 g(\eta_1, \eta_2)) = g(\eta_1, \eta_2) - \eta_2 \frac{\partial}{\partial \eta_2} g(\eta_1, \eta_2)$$

Besides derivatives, we need to know how to integrate. It turns out that the mathematically consistent way of defining an integral is to simply say that it does exactly the same thing as a derivative, that is

$$\int d\eta_1 g(\eta_1, \eta_2) \equiv \frac{\partial}{\partial \eta_1} g(\eta_1, \eta_2)$$
$$\int d\eta_2 g(\eta_1, \eta_2) \equiv \frac{\partial}{\partial \eta_2} g(\eta_1, \eta_2),$$
$$\int d\eta_2 d\eta_1 g(\eta_1, \eta_2) \equiv \frac{\partial}{\partial \eta_2} \frac{\partial}{\partial \eta_1} g(\eta_1, \eta_2)$$

This turns out to be a consistent calculus for anti-commuting numbers. For example, the definition of integral has the consequence

$$\int d\eta_1 \frac{\partial}{\partial \eta_1} g(\eta_1, \eta_2) = 0$$
$$\int d\eta_2 d\eta_1 \frac{\partial}{\partial \eta_1} g(\eta_1, \eta_2) = 0$$

so that we can integrate by parts without surface terms,

$$\int d\eta_2 d\eta_1 g(\eta_1, \eta_2)\frac{\partial}{\partial \eta_1}g'(\eta_1, \eta_2) = -\int d\eta_2 d\eta_1 \frac{\partial}{\partial \eta_1}g(-\eta_1, -\eta_2)g'(\eta_1, \eta_2)$$

Also, the integration is translation invariant in that

$$\int d\eta_1 g(\eta_1, \eta_2) = \int d\eta_1 g(\eta_1 + \zeta, \eta_2)$$

where ζ is a third anti-commuting variable. Also, for a change of variables as

$$\int d\eta_1 d\eta_2 g(\eta_1, \eta_2) = \int d\eta_1 d\eta_2 \frac{1}{\det m}\, g(m_{11}\eta_1 + m_{12}\eta_2, m_{21}\eta_1 + m_{22}\eta_2)$$

so that is we consider a linear transformation of the integration variables

$$\begin{bmatrix} \eta_1' \\ \eta_2' \end{bmatrix} = \begin{bmatrix} m_{11} & m_{12} \\ m_{21} & m_{22} \end{bmatrix}\begin{bmatrix} \eta_1 \\ \eta_2 \end{bmatrix}$$

the change of the integration measure would be

$$\int d\eta_1' d\eta_2'... = \int d\eta_1 d\eta_2 \frac{1}{\det m}...$$

where the Jacobian appears in the inverse of its usual position for integration with commuting variables.

We will need to be able to do Gaussian integrals. Of most interest will be integral of the form

$$\int d\eta\, d\bar{\eta}\, e^{\bar{\eta} D\eta} = D$$

or its generalization to higher dimensions numbers of variables. It is easy to confirm that

$$\int d\eta_2 d\bar{\eta}_2 d\eta_1 d\bar{\eta}_1 e^{\sum_{i,j}\bar{\eta}_i D_{ij}\eta_j} = D_{11}D_{22} - D_{12}D_{21} = \det D$$

More generally with two sets of n anti-commuting variables, $\eta_1, ..., \eta_n$ and $\bar{\eta}_1, ..., \bar{\eta}_n$ and the integral

$$\int d\eta_n d\bar{\eta}_n \dots d\eta_1 d\bar{\eta}_1 \exp\left(\sum_{jk} \bar{\eta}_j D_{jk} \eta_k\right)$$

$$= (-1)^{n(n-1)/2} \int d\eta_n \dots d\eta_1 d\bar{\eta}_n \dots d\bar{\eta}_1 \exp\left(\sum_{jk} \bar{\eta}_j D_{jk} \eta_k\right)$$

$$= (-1)^{n(n-1)/2} \int d\eta_n \dots d\eta_1 d\bar{\eta}_n \dots d\bar{\eta}_1 \exp\left(\sum_{k} \bar{\eta}_1 D_{1k} \eta_k + \sum_{j\neq 1,k} \bar{\eta}_j D_{jk} \eta_k\right)$$

$$= (-1)^{n(n-1)/2} \int d\eta_n \dots d\eta_1 d\bar{\eta}_n \dots d\bar{\eta}_2 d\bar{\eta}_1 \sum_{\ell=1}^{n} \cdot$$

$$\cdot \left\{ (1 + \bar{\eta}_1 D_{1\ell} \eta_\ell) \exp\left(\sum_{j\neq 1,k} \bar{\eta}_j D_{jk} \eta_k\right)\right\}$$

$$= (-1)^{n(n-1)/2} \int d\eta_n \dots d\eta_1 d\bar{\eta}_n \dots d\bar{\eta}_2 \sum_{\ell=1}^{n} \cdot$$

$$\cdot \left\{ D_{1\ell} \eta_\ell \exp\left(\sum_{j\neq 1,k\neq \ell} \bar{\eta}_j D_{jk} \eta_k\right)\right\}$$

$$= \sum_{\ell=1}^{n} \left\{ (-1)^{\ell-1} D_{1\ell} \ (-1)^{(n-1)(n-2)/2} \cdot \right.$$

$$\left. \cdot \int d\eta_n \dots d\eta_{\ell+1} d\eta_{\ell-1} \dots d\eta_1 d\bar{\eta}_n \dots d\bar{\eta}_2 \exp\left(\sum_{j\neq 1,k\neq \ell} \bar{\eta}_j D_{jk} \eta_k\right)\right\}$$

The expression above is precisely the expansion of the determinant of the matrix D using minors and co-factors[2] which is normally used to calculate a determinant and we must conclude that, for any n,

$$\int d\eta_n d\bar{\eta}_n \dots d\eta_1 d\bar{\eta}_1 \exp\left(\sum_{jk} \bar{\eta}_j D_{jk} \eta_k\right) = \det D \qquad (12.20)$$

Now that we have introduced derivatives and integrals which use anti-commuting numbers, we must generalize what we have learned to a discussion of functional derivatives and integrals for anti-commuting functions. As was the case with functional derivatives by ordinary, commuting functions, which could be defined in terms of ordinary derivatives by ordinary, commuting variables, functional derivatives by

---

[2] The co-factor here is $D_{1\ell}$ and the minor is the determinant of the matrix that would be obtained by removing the first row and the ℓ'th column of D.

anti-commuting functions can be defined as derivatives by anti-commuting variables. Let us consider an anti-commuting function $\eta(x)$. Here, x denotes the coordinates of the space on which the function is defined. An anti-commuting function is one which obeys

$$\eta(x)\eta(y) = -\eta(y)\eta(x)$$

for any arguments x and y. In particular, this implies that $\eta^2(x) = 0$.

We can define an anti-commuting function in terms of anti-commuting numbers. Consider a complete orthonormal set of normalized square-integrable functions, $\{f_1(x), f_2(x), ...\}$

$$\int dx f_m(x) f_n(x) = \delta_{mn}, \quad \sum_n f_n(x) f_n(y) = \delta(x - y)$$

Note that these are ordinary commuting, real-number-valued functions or ordinary real variables. We can expand $\eta(x)$ in a series of these functions

$$\eta(x) = \sum_{n=1}^{\infty} \eta_n f_n(x), \quad \eta_n = \int dx f_n(x)\eta(x)$$

where the coefficients, $\{\eta_1, \eta_2, ...\}$ are anti-commuting numbers,

$$\eta_m \eta_n = -\eta_n \eta_m, \quad \eta_n^2 = 0, \; \forall m, n$$

A functional $Z[\eta]$ of $\eta(x)$, can always be written as an ordinary function of the infinite set anti-commuting numbers $Z(\eta_1, \eta_2, ...)$ by plugging the expansion of $\eta(x)$ into the $Z[\eta]$.

Then the functional derivative by $\eta(x)$ can be defined as

$$\frac{\delta Z[\eta]}{\delta \eta(x)} \equiv \sum_{n=1}^{\infty} f_n(x) \frac{\partial}{\partial \eta_n} Z(\eta_1, \eta_2, ...)$$

The derivatives by the anti-commuting variables themselves must be anti-commuting which leads to similar formulas for the functional derivatives

$$\frac{\delta}{\delta \eta(x)} \frac{\delta}{\delta \eta(y)} = -\frac{\delta}{\delta \eta(y)} \frac{\delta}{\delta \eta(x)}, \quad \left(\frac{\delta}{\delta \eta(x)}\right)^2 = 0$$

$$\frac{\delta}{\delta \eta(x)} \eta(y) Z[\eta] = \delta(x - y) Z[\eta] - \eta(y) \frac{\delta}{\delta \eta(x)} Z[\eta]$$

In addition, we can define the functional integral

$$\int [d\eta] Z[\eta] \equiv \int d\eta_1 d\eta_2 ... Z(\eta_1, \eta_2, ...) = \frac{\partial}{\partial \eta_1} \frac{\partial}{\partial \eta_2} ... Z(\eta_1, \eta_2, ...)$$

We will be particularly interested in Gaussian integrals which are of a form similar to those which we studied above. For this purpose, we introduce a second, independent anti-commuting function $\bar{\eta}(x)$. Then, we consider the Gaussian

$$\int [d\eta d\bar{\eta}] \exp\left(\int dy dz \bar{\eta}(y) D(y,z) \eta(z)\right)$$

$$\equiv \int d\eta_1 d\bar{\eta}_1 d\eta_2 d\bar{\eta}_2 \ldots \exp\left(\sum_{m,n} \bar{\eta}_m D_{mn} \eta_n\right)$$

where we have defined $[d\eta d\bar{\eta}] \equiv d\eta_1 d\bar{\eta}_1 d\eta_2 d\bar{\eta}_2 \ldots$ and

$$D_{mn} = \int dy dz f_m(y) D(y,z) f_n(z)$$

Then, the infinite-dimensional generalization of our Gaussian integral formula (12.20) gives a formal definition of the Gaussian functional integral

$$\int [d\eta d\bar{\eta}] \exp\left(\int dy dz \bar{\eta}(y) D(y,z) \eta(z)\right) = \det D$$

where $\det D$ is defined as the determinant of the infinite matrix D_{mn}. Note that, unlike the case of a bosonic Gaussian integral, the determinant appears with a positive power.

We will also be particularly interested in an off-set Gaussian integral of the form

$$\int [d\eta d\bar{\eta}] \exp\left(\int dy dz \bar{\eta}(y) D(y,z) \eta(z) + \int dy (\bar{\xi}(y) \eta(y) + \bar{\eta}(y) \xi(y))\right)$$

This integral can be done by a change of variables,

$$\eta(x) \rightarrow \eta(x) - \int dy D^{-1}(x,y) \xi(y)$$

$$\bar{\eta}(x) \rightarrow \bar{\eta}(x) - \int dy \bar{\xi}(y) D^{-1}(x,y)$$

The functional integration measure is invariant under such a change. It also requires that the inverse of the quadratic form exists so that

$$\int dy D(x,y) D^{-1}(y,z) = \delta(x-z), \quad \int dy D^{-1}(x,y) D(y,z) = \delta(x-z)$$

Thus, plugging in this change of variables and doing the Gaussian integral yields

$$
\int [d\eta d\bar{\eta}] \exp\left(\int dy dz \bar{\eta}(y) D(y,z) \eta(z) + \int dy (\bar{\xi}(y)\eta(y) + \bar{\eta}(y)\xi(y)) \right)
$$

$$
= \det D \, \exp\left(-\int dx dy \bar{\xi}(x) D^{-1}(x,y)\xi(y) \right)
$$

$$
(12.21)
$$

We shall make extensive use of the above formula in the following sections.

12.3   Generating Functionals for Non-relativistic Fermions

While we are on the subject of propagators, let us revisit the non-relativistic theory and discuss the computation of correlation functions in that theory. The non relativistic theory is defined by the field equation and anti-commutation relations

$$
\left(i\hbar \frac{\partial}{\partial t} + \frac{\hbar^2 \vec{\nabla}^2}{2m} + \mu \right) \psi_\sigma(\vec{x}, t) = 0
\qquad (12.22)
$$

$$
\{\psi_\sigma(\vec{x}, t), \psi^{\dagger \rho}(\vec{y}, t)\} = \delta_\sigma^\rho \delta(\vec{x} - \vec{y})
$$
$$
\{\psi_\sigma(\vec{x}, t), \psi_\rho(\vec{y}, t)\} = 0, \quad \{\psi^{\dagger \sigma}(\vec{x}, t), \psi^{\dagger \rho}(\vec{y}, t)\} = 0
$$

Consider the ground state $|\mathcal{O}>$ and the correlation functions

$$
g_\sigma{}^\rho(\vec{x}, t; \vec{x}', t') \equiv \qquad\qquad\qquad\qquad\qquad (12.23)
$$
$$
< \mathcal{O}|\theta(t - t')\psi_\sigma(\vec{x}, t)\psi^{\dagger \rho}(\vec{x}', t') - \theta(t' - t)\psi^{\dagger \rho}(\vec{x}', t')\psi_\sigma(\vec{x}, t)|\mathcal{O} >
$$
$$
\equiv < \mathcal{O}|T\psi_\sigma(\vec{x}, t)\psi^{\dagger \rho}(\vec{x}', t')|\mathcal{O} > \qquad (12.24)
$$

Often important for applications are the retarded and advanced correlation functions

$$
g_{R\sigma}{}^\rho(\vec{x}, t; \vec{x}', t') \equiv < \mathcal{O}|\theta(t - t')\{\psi_\sigma(\vec{x}, t), \, \psi^{\dagger \rho}(\vec{x}', t')\}|\mathcal{O} > \qquad (12.25)
$$
$$
g_{A\sigma}{}^\rho(\vec{x}, t; \vec{x}', t') \equiv - < \mathcal{O}|\theta(t' - t)\{\psi_\sigma(\vec{x}, t), \, \psi^{\dagger \rho}(\vec{x}', t')\}|\mathcal{O} > \qquad (12.26)
$$

We can get an explicit expression for the 2 point correlation function. To do this, we can substitute the expressions for the operators in terms of creation and annihilation operators into the expression (12.24) and evaluate the resulting Green function explicitly. Recall that the expression for the quantum field in terms of creation and annihilation operators is

$$\psi_\sigma(\vec{x}, t) = \int_{|\vec{k}|>k_F} \frac{d^3k}{(2\pi)^{\frac{3}{2}}} e^{i\vec{k}\cdot\vec{x} - i\left(\frac{\hbar^2 \vec{k}^2}{2m} - \mu\right)t/\hbar} \alpha_\sigma(\vec{k})$$

$$+ \int_{|\vec{k}|\leq k_F} \frac{d^3k}{(2\pi)^{\frac{3}{2}}} e^{-i\vec{k}\cdot\vec{x} - i\left(\frac{\hbar^2 \vec{k}^2}{2m} - \mu\right)t/\hbar} \beta_\sigma^\dagger(\vec{k})$$

where the creation and annihilation operators anti-commute, with the non-vanishing anti-commutators being

$$\left\{\alpha_\sigma(\vec{k}), \alpha^{\rho\dagger}(\vec{k}')\right\} = \delta_\sigma^{\ \rho}\delta(\vec{k} - \vec{k}') \tag{12.27}$$

$$\left\{\beta^\sigma(\vec{k}), \beta_\rho^\dagger(\vec{k}')\right\} = \delta^\sigma_{\ \rho}\delta(\vec{k} - \vec{k}') \tag{12.28}$$

Then we obtain

$$g_\sigma^{\ \rho}(\vec{x}, t : \vec{x}', t') = \theta(t - t')\delta_\sigma^\rho \int_{k>k_F} \frac{d^3k}{(2\pi)^3} e^{i\vec{k}\cdot(\vec{x}-\vec{x}') - i\left(\frac{\hbar^2 \vec{k}^2}{2m} - \mu\right)(t-t')/\hbar}$$

$$+ \theta(t' - t)\delta_\sigma^\rho \int_{k<k_F} \frac{d^3k}{(2\pi)^3} e^{i\vec{k}\cdot(\vec{x}-\vec{x}') + i\left(\frac{\hbar^2 \vec{k}^2}{2m} - \mu\right)(t-t')/\hbar} \tag{12.29}$$

Here, k_F is the Fermi wave-number which gives the Fermi energy, $\epsilon_F = \frac{\hbar^2 k_F^2}{2m}$ and, in the absence of interactions or temperature, the Fermi energy is equal to the chemical potential $\epsilon_F = \mu$. We note that the dependence on the indices σ and ρ is trivial. To streamline the notation, we define

$$g_\sigma^{\ \rho}(\vec{x}, t : \vec{x}', t') \equiv \delta_\sigma^{\ \rho}\, g(x - x')$$

where we are now denoting $(\vec{x}, t) \equiv x$. We will also denote the spacetime volume integral $\int dt d^3x \equiv \int dx$. Now we can use the contour integral representation of the theta-function to introduce a frequency integral and we obtain the expression

$$g(x - x') = \int \frac{d\omega d^3k}{(2\pi)^4} \frac{ie^{i\vec{k}\cdot(\vec{x}-\vec{x}') - i\omega(t-t')}}{\omega - \left(\frac{\hbar^2 \vec{k}^2}{2m} - \mu\right)/\hbar + i\varepsilon\,\text{sign}(k - k_F)}$$

We can also easily find the retarded and advanced green functions,

$$g_R(x - x') = \int \frac{d\omega d^3k}{(2\pi)^4} \frac{ie^{i\vec{k}\cdot(\vec{x}-\vec{x}') - i\omega(t-t')}}{\omega - \left(\frac{\hbar^2 \vec{k}^2}{2m} - \mu\right)/\hbar + i\varepsilon}$$

$$g_A(x - x') = \int \frac{d\omega d^3k}{(2\pi)^4} \frac{ie^{i\vec{k}\cdot(\vec{x}-\vec{x}') - i\omega(t-t')}}{\omega - \left(\frac{\hbar^2 \vec{k}^2}{2m} - \mu\right)/\hbar - i\varepsilon}$$

The above expressions obey the equation for a Green function. For example,

$$\left(i\hbar \frac{\partial}{\partial t} + \frac{\hbar^2 \vec{\nabla}^2}{2m} + \mu \right) g(x-y) = i\hbar\, \delta(x-y)$$

It is easy to use the field equation and the anti-commutation relations to confirm that $g(x-y)$ should indeed obey this equation.

Now, we shall consider the following generating functional for the correlation functions of the fermion fields,

$$Z[\eta, \eta^\dagger] \equiv$$
$$< \mathcal{O}|T \sum_{m=0}^{\infty} \frac{1}{m!} \left(i \int dx\, \eta^{\dagger \rho}(x)\psi_\rho(x) \right)^m \sum_{n=0}^{\infty} \frac{1}{n!} \left(i \int dy\, \psi^{\dagger\sigma}(y)\eta_\sigma(y) \right)^n |\mathcal{O} >$$

(12.30)

where the time ordering means that the individual terms in the product are ordered according to the values of their time arguments. As with the scalar and photon field which we studied earlier, the ordering is such that the times are always decreasing as we read the operator product from left to right. Under the time-ordering symbol, the source functions, the functional derivatives by the source functions, and the fields anti-commute with each other. We have defined the product as above with the source functions paired with the operators. We can make a formal summation of the series to write the functional in the more compact notation,

$$Z[\eta, \eta^\dagger] =< \mathcal{O}|T e^{i \int dx\, [\eta^{\dagger\rho}(x)\psi_\rho(x)+\psi^{\dagger\sigma}(x)\eta_\sigma(x)]}|\mathcal{O} >$$

(12.31)

It is easy to derive a simple functional differential equation that this generating functional should satisfy and to find the solution of that equation. The generating functional is the exponential of a quadratic in the sources,

$$Z[\eta, \eta^\dagger] = \exp\left(- \int dx dy\, \eta^{\sigma\dagger}(x)\, g(x,y)\, \eta_\sigma(y) \right)$$

(12.32)

Functional derivatives of $Z[\eta, \eta^\dagger]$ by the anti-commuting functions $\eta(x)$ and $\eta^\dagger(x)$ yield the time-ordered correlation functions of non-relativistic fermions

$$< \mathcal{O}|\psi_{\sigma_1}(x_1)\ldots\psi_{\sigma_k}(x_k)\psi^{\dagger\rho_1}(y_1)\ldots\psi^{\dagger\rho_k}(y_k)|\mathcal{O} >$$
$$= \frac{\delta}{\delta\eta^{\dagger\sigma_1}(x_1)} \cdots \frac{\delta}{\delta\eta^{\dagger\sigma_k}(x_k)} \frac{\delta}{\delta\eta_{\rho_1}(x_1)} \cdots \frac{\delta}{\delta\eta_{\rho_k}(x_k)} Z[\eta, \eta^\dagger] \bigg|_{\eta=0=\eta^\dagger}$$

These correlation functions vanish unless the number of ψ's and $\psi^\dagger$'s are equal. As we shall learn later on, this is a consequence of symmetry. We have then also anticipated that the factors of $\frac{1}{i}$ and $-\frac{1}{i}$ that would accompany the individual functional derivatives also cancel.

We would now like to present the generating functional as a functional integral. The functional integral is

$$Z[\eta, \eta^\dagger] = \frac{\int [d\psi d\psi^\dagger] \, e^{\frac{i}{\hbar}S + i \int dx[\eta^{\dagger\sigma}(x)\psi_\sigma(x) + \psi^{\dagger\sigma}(x)\eta_\sigma(x)]}}{\int [d\psi d\psi^\dagger] \, e^{\frac{i}{\hbar}S}} \tag{12.33}$$

where the action is

$$S = \int dx \mathcal{L}(x)$$

and the Lagrangian density is

$$\mathcal{L}(x) = \psi^{\dagger\sigma}(x) \left(\frac{i\hbar}{2} \frac{\partial}{\partial t} - \frac{i\hbar}{2} \overleftarrow{\frac{\partial}{\partial t}} \right) \psi_\sigma(x) - \frac{\hbar^2}{2m} \vec{\nabla} \psi^{\dagger\sigma}(x) \cdot \vec{\nabla} \psi_\sigma(x) + \mu \psi^{\dagger\sigma}(x) \psi_\sigma(x)$$

Inside the functional integration, and the action and Lagrangian density written above, $\psi_\sigma(x)$ and $\psi^{\dagger\sigma}(x)$ are anti-commuting functions. Even though we use the same notation for them as for the quantum fields, they are not quantum fields, when they appear inside a functional integral they are just anti-commuting functions. The integration is over the space of all such functions.

We need to know very little about the precise definition of this integration in order to show that equation (12.33) is equivalent to (12.32). All that we need to know is that the integration measure is invariant under translations in function space. That allows us to do the transformation of variables in the integral in the numerator,

$$\psi_\sigma(x) = \tilde{\psi}_\sigma(x) - \frac{1}{i} \int dy g(x, y) \eta_\sigma(y)$$

$$\psi^{\dagger\sigma}(x) = \tilde{\psi}^{\dagger\sigma}(x) - \frac{1}{i} \int dy \, \eta^{\dagger\sigma}(y) g(y, x)$$

where we use the fact that $\frac{1}{i} g(x - y)$ is the Green function for the non-relativistic wave operator. Then, we also need the property that integrations by parts within the exponent are allowed and they do not generate any residual surface terms. Our transformation of variables completes the square in the exponent. The functional integral in the numerator becomes

$$\int [d\psi d\psi^\dagger] \, e^{\frac{i}{\hbar}S + i \int dx[\eta^{\dagger\sigma}(x)\psi_\sigma(x) + \hbar\psi^{\dagger\sigma}(x)\eta_\sigma(x)]}$$

$$= \int [d\psi d\psi^\dagger] \, e^{\frac{i}{\hbar}S} \, e^{-\int dxdy \, \eta^{\dagger\sigma}(x) g(x, y) \eta_\sigma(y)}$$

and the second factor in the integrand, which does not depend on the integration variables can be factored out. It is identical to the functional in equation (12.32). The integral is then identical to the one in the denominator of equation (12.33) and it cancels. Therefore (12.33) and (12.32) are identical.

There are a few more interesting functional integral formulae which are implied by the above development. If we use the functional integral for a generating functional, taking functional derivatives of it by the sources and then setting the sources equal to zero yields the following integral for time-ordered correlation functions:

$$
< \mathcal{O}|T\psi_{\rho_1}(x_1)\ldots\psi_{\rho_r}(x_r)\psi^{\dagger\sigma_1}(y_1)\ldots\psi^{\dagger\sigma_s}(y_s)|\mathcal{O} >
$$

$$
= \frac{\int [d\psi d\psi^\dagger]e^{\frac{i}{\hbar}S}\,\psi_{\rho_1}(x_1)\ldots\psi_{\rho_r}(x_r)\psi^{\dagger\sigma_1}(y_1)\ldots\psi^{\dagger\sigma_s}(y_s)}{\int [d\psi d\psi^\dagger]\,e^{\frac{i}{\hbar}S}} \tag{12.34}
$$

This has the implication that all correlation functions can be found by simply integrating the classical fields with the appropriate functional integration measure. Here, for the case of free field theory, this is a Gaussian integral (the exponent is quadratic in the fields) and the factorization into 2 point functions is also a property of a Gaussian functional integral.

$$
< \mathcal{O}|T\psi_{\rho_1}(x_1)\ldots\psi_{\rho_r}(x_r)\psi^{\dagger\sigma_1}(y_1)\ldots\psi^{\dagger\sigma_s}(y_s)|\mathcal{O} > \tag{12.35}
$$

$$
= \sum_{\text{pairings}} (-1)^{\deg} \prod_{\text{pairs }(\rho_i x_i,\sigma_i y_j)} \delta_{\rho_i}^{\sigma_j} g(x_i - y_j) \tag{12.36}
$$

where deg is the number of neighbours which need to be interchanged in order to put the fields in the pairing adjacent to each other.

12.3.1 Interacting Non-relativistic Fermions

Now, what about the interacting non-relativistic field theory, where the Lagrangian density has an interaction term,

$$
\mathcal{L}(x) = \psi^{\dagger\sigma}(x)\left(\frac{i\hbar}{2}\frac{\partial}{\partial t} - \frac{i\hbar}{2}\frac{\overleftarrow{\partial}}{\partial t}\right)\psi_\sigma(x) - \frac{\hbar^2}{2m}\vec{\nabla}\psi^{\dagger\sigma}(x)\cdot\vec{\nabla}\psi_\sigma(x)
$$

$$
+ \mu\psi^{\dagger\sigma}(x)\psi_\sigma(x) - \frac{\lambda}{2}\left(\psi^{\dagger\sigma}(x)\psi_\sigma(x)\right)^2
$$

The action is the space-time integral of the Lagrangian density,

$$
\mathbf{S} = \int dx \mathcal{L}(x)
$$

and the equation of motion is

$$\frac{\delta}{\delta \psi^{\dagger \sigma}(x)} S = \left(i\hbar \frac{\partial}{\partial t} + \frac{\hbar^2 \vec{\nabla}^2}{2m} + \mu \right) \psi_\sigma(x) - \lambda \psi^{\dagger \rho}(x) \psi_\rho(x) \psi_\sigma(x) = 0$$

where we have observed that the field equation can be derived by setting a functional derivative of the action to zero. The natural conjecture is that the correlation functions of this interacting theory is described by the functional integral

$$Z[\eta, \eta^\dagger] = \frac{\int [d\psi d\psi^\dagger] \, e^{\frac{i}{\hbar} S + i \int dx [\eta^{\dagger \sigma}(x) \psi_\sigma(x) + \psi^{\dagger \sigma}(x) \eta_\sigma(x)]}}{\int [d\psi d\psi^\dagger] \, e^{\frac{i}{\hbar} S}} \tag{12.37}$$

where we simply use the action of the interacting field theory. This guess turns out to be correct. The functional integral in equation (12.37) represents the full interacting quantum field theory when we insert the action which is the space-time integral of $\mathcal{L}(x)$ for the interacting field theory. In the tutorial we shall basically give a proof of this statement by finding a functional differential equation for the interacting theory and showing that the functional integral is a solution of that equation.

12.4   The Dirac Field

In this section, we shall include Dirac fields in correlations functions. Let us begin with the non-interacting Dirac fields which we denote by $\psi(x)$ and $\bar{\psi}(x)$ and we shall examine the quantitites

$$< \mathcal{O}|\mathcal{T} \psi_{a_1}(x_1) \dots \psi_{a_n}(x_n) \bar{\psi}_{b_1}(y_1) \dots \bar{\psi}_{b_n}(y_n)|\mathcal{O} >$$

with the same anti-symmetric time ordering that we used for non-relativistic fermions. The non-interacting fields obey the Dirac equation

$$\left[\partial\!\!\!/ + m \right] \psi(x) = 0, \quad \bar{\psi}(x) \left[-\overleftarrow{\partial\!\!\!/} + m \right] = 0$$

and the equal time anti-commutation relations

$$\{\psi_a(x), \psi_b^\dagger(y)\}_{x^0 = y^0} = \delta_{ab} \delta(\vec{x} - \vec{y})$$

$$\{\psi_a(x), \psi_b(y)\}_{x^0 = y^0} = 0, \quad \{\psi_a^\dagger(x), \psi_b^\dagger(y)\}_{x^0 = y^0} = 0$$

The generating functional for correlation functions of Dirac fields is

$$Z[\eta, \bar{\eta}] = < \mathcal{O}|\mathcal{T} \exp \left(i \int dx \, [\bar{\eta}(x)\psi(x) + \bar{\psi}(x)\eta(x)] \right) |\mathcal{O} > \tag{12.38}$$

and correlation functions are obtained from it by functional derivatives by the anti-commuting functions $\bar{\eta}(x)$ and $\eta(x)$,

$$
< \mathcal{O}|T\psi_{a_1}(x_1)\ldots\psi_{a_n}(x_n)\bar{\psi}_{b_1}(y_1)\ldots\bar{\psi}_{b_n}(y_n)|\mathcal{O} >
$$

$$
= \frac{\delta}{\delta\bar{\eta}_{a_1}(x_1)}\cdots\frac{\delta}{\delta\bar{\eta}_{a_n}(x_n)}\frac{\delta}{\delta\eta_{b_1}(y_1)}\cdots\frac{\delta}{\delta\eta_{b_n}(y_n)}Z[\eta,\bar{\eta}]\Big|_{\eta=0=\bar{\eta}}
$$

In the following we will get an explicit form for this generating functional and we will discuss its properties.

12.4.1 2 Point Function for the Dirac Field

The time-ordered 2 point correlation function of the Dirac theory is

$$
g_{ab}(\vec{x},t;\vec{x}',t') \equiv < \mathcal{O}|T\psi_a(\vec{x},t)\bar{\psi}_b(\vec{x}',t')|\mathcal{O} >
$$
$$
= < \mathcal{O}|\theta(t-t')\psi_a(\vec{x},t)\bar{\psi}_b(\vec{x}',t') - \theta(t'-t)\bar{\psi}_b(\vec{x}',t')\psi_a(\vec{x},t)|\mathcal{O} > \quad (12.39)
$$

We shall use the same notation, $g(x, y)$ for the 2 point function in the Dirac theory as we used for non-relativistic fermions. This should not cause confusion as we shall never discuss both theories in the same context. The 2 point function is proportional to a Green function for the Dirac wave operator. To see this, we operator the Dirac wave operator on the time-ordered function to obtain

$$
[\slashed{\partial} + m]g(x, y) =
$$
$$
[\slashed{\partial} + m]_{ac} < \mathcal{O}|\theta(x^0 - y^0)\psi_c(x)\bar{\psi}_b(y) - \theta(y^0 - x^0)\bar{\psi}_b(y)\psi_c(x)|\mathcal{O} >
$$
$$
= \gamma^0\delta(x^0 - y^0) < \mathcal{O}|\{\psi_a(x), \bar{\psi}_b(y)\}|\mathcal{O} > + < \mathcal{O}|T[\slashed{\partial} + m]_{ac}\psi_c(x), \bar{\psi}_b(y)|\mathcal{O} >
$$
$$
= (\gamma^0)^2_{ab}\delta(x - y) = -\delta_{ab}\delta(x - y)
$$

where we have used the field equation, the anti-commutation relation and the observation that the derivative of a Heavyside function is a delta function $\frac{d}{dx}\theta(x) = \delta(x)$ We see that $g(x, y')$ is proportional to a Green function for the Dirac wave equation

$$
[\slashed{\partial} + m]g(x, y) = -\delta(x - y) \quad (12.40)
$$

where we have suppressed the Dirac spinor indices.

A quick way to fin this propagator would be to note that the algebra of Dirac matrices can be used to show that

$$
(\slashed{\partial} + m)(-\slashed{\partial} + m) = -\partial^2 + m^2
$$

and the propagator can thus be written as $-(-\slashed{\partial} + m)$ operating on the time-ordered Green function that was also used in the propagator for the scalar field. This allows us to write

$$
\begin{aligned}
g(x, y) &= -(-\slashed{\partial} + m) \int \frac{dp}{(2\pi)^4} e^{ip(x-y)} \frac{1}{p^2 + m^2 - i\epsilon} \\
&= \int \frac{dp}{(2\pi)^4} e^{ip(x-y)} \frac{i\slashed{p} - m}{p^2 + m^2 - i\epsilon}
\end{aligned}
\tag{12.41}
$$

where we have anticipated the position of $i\epsilon$ in analogy with the scalar field propagator. Note that, consistent with this expression,

$$
-[i\slashed{p} + k]^{-1} = \frac{i\slashed{p} - m}{p^2 + m^2 - i\epsilon}
$$

To confirm that the $i\epsilon$ is in the right place, we could also I find the propagator explicitly. We plug our solution of the Dirac equation into the defining equation for the 2 point function. The explicit solution was

$$
\psi(x) = \int \frac{d^3k}{(2\pi)^{\frac{3}{2}}} \left[e^{ik_\mu x^\mu} \left\{ \psi_{++} a_+(\vec{k}) + \psi_{+-} a_-(\vec{k}) \right\} + \tag{12.42}
$$

$$
e^{-ik_\mu x^\mu} \left\{ \psi_{-+} b_+^\dagger(\vec{k}) + \psi_{--} b_-^\dagger(\vec{k}) \right\} \right] \tag{12.43}
$$

where

$$
k^\mu = (E, \vec{k}), \quad E = \sqrt{\vec{k}^2 + m^2}
$$

where the first $\pm$ in the subscript on the wave-functions denote positive and negative energy solutions whereas the second $\pm$ denote positive and negative helicity. We found these spinors explicitly when we discussed the non-interacting Dirac theory,

$$
\psi_{++}(\vec{k}) = \frac{1}{\sqrt{2}} \begin{bmatrix} i\sqrt{1 - \frac{|\vec{k}|}{E}} u_+ \\ \sqrt{1 + \frac{|\vec{k}|}{E}} u_+ \end{bmatrix} \qquad \psi_{+-}(\vec{k}) = \frac{1}{\sqrt{2}} \begin{bmatrix} i\sqrt{1 + \frac{|\vec{k}|}{E}} u_- \\ \sqrt{1 - \frac{|\vec{k}|}{E}} u_- \end{bmatrix}
$$

$$
\psi_{-+}(\vec{k}) = \frac{1}{\sqrt{2}} \begin{bmatrix} i\sqrt{1 - \frac{|\vec{k}|}{E}} u_+ \\ -\sqrt{1 + \frac{|\vec{k}|}{E}} u_+ \end{bmatrix} \qquad \psi_{--}(\vec{k}) = \frac{1}{\sqrt{2}} \begin{bmatrix} i\sqrt{1 + \frac{|\vec{k}|}{E}} u_- \\ -\sqrt{1 - \frac{|\vec{k}|}{E}} u_- \end{bmatrix}
$$

Then, we can see explicitly that

$$
\psi_{++} \bar{\psi}_{++} + \psi_{+-} \bar{\psi}_{+-} = \frac{i\slashed{k} - m}{-2iE}
$$

$$
\psi_{-+} \bar{\psi}_{-+} + \psi_{--} \bar{\psi}_{--} = \frac{i\slashed{k} + m}{-2iE}
$$

and

$$< \mathcal{O}|\psi(x)\bar{\psi}(y)|\mathcal{O}> = \int \frac{d^3k}{(2\pi)^3} e^{ik_\mu(x^\mu - y^\mu)} \frac{i\not{k} - m}{-2iE}$$

$$< \mathcal{O}|\bar{\psi}(y)\psi(x)|\mathcal{O}> = \int \frac{d^3k}{(2\pi)^3} e^{-ik_\mu(x^\mu - y^\mu)} \frac{i\not{k} + m}{-2iE}$$

We can represent the Heavyside step function as a contour integral,

$$\theta(x^0 - y^0) < \mathcal{O}|\psi(x)\bar{\psi}(y)|\mathcal{O}> = \int \frac{dk}{(2\pi)^4} e^{ik_\mu \cdot (x-y)^\mu} \frac{i\gamma^0 E - i\vec{\gamma} \cdot \vec{k} + m}{2E(k^0 + E - i\epsilon)}$$

$$\theta(y^0 - x^0) < \mathcal{O}|\bar{\psi}(y)\psi(x)|\mathcal{O}> = \int \frac{dk}{(2\pi)^4} e^{ik_\mu(x-y)^\mu} \frac{i\gamma^0 E - i\vec{\gamma} \cdot \vec{k} + m}{2E(-k^0 + E - i\epsilon)}$$

so that

$$g(x, y) = < \mathcal{O}|T\psi(x)\bar{\psi}(y)|\mathcal{O}> = \int \frac{dk}{(2\pi)^4} e^{ik_\mu(x-y)^\mu} \frac{i\not{k} - m}{k_\mu k^\mu + m^2 - i\epsilon} \quad (12.44)$$

It can easily be checked that this expression satisfies the equation for a Green function for the Dirac operator.

In the discussion above, we have computed $< \mathcal{O}|T\psi(x)\bar{\psi}(y)|\mathcal{O}>$ for the Dirac field theory. Here, we also note that the other possible 2 point correlation functions $< \mathcal{O}|T\psi(x)\psi(y)|\mathcal{O}>$ and $< \mathcal{O}|T\bar{\psi}(x)\bar{\psi}(y)|\mathcal{O}>$ must both vanish. This is due to the phase symmetry of the theory. In fact, phase symmetry implies that any correlation function must vanish unless it has the same number of ψ's and $\bar{\psi}$'s. We will see this explicitly when we derive the generating functional in the next section.

12.4.2 Generating Functional for the Dirac Field

Now that we have found the 2 point correlation function for the Dirac field theory in the previous section, we are ready to find the full generating functional for all of the time-ordered correlation functions.

$$Z[\eta, \bar{\eta}] = < \mathcal{O}|T \exp\left(i \int dx \, [\bar{\eta}(x)\psi(x) + \bar{\psi}(x)\eta(x)]\right)|\mathcal{O}> \quad (12.45)$$

As we did for the photon, we can easily derive a functional differential equation that the generating functional in equation (12.45) must satisfy. Then, we can solve the equation with the boundary condition that

$$Z[\eta = 0, \bar{\eta} = 0] = 1 \quad (12.46)$$

The functional differential equation for $Z[\eta, \bar{\eta}]$ is obtained by operating the Dirac wave operator on the functional derivative,

$$\left[\overrightarrow{\partial} + m\right] \frac{\delta}{\delta\bar{\eta}(x)} Z[\eta, \bar{\eta}] =$$

$$\gamma^0 \partial_0 < \mathcal{O}|T e^{i \int_{x^0}^{\infty} dy^0 \int d^3y \, [\bar{\eta}(y)\psi(y) + \bar{\psi}(y)\eta(y)]} i\psi(x) e^{i \int_{-\infty}^{x^0} dy^0 \int d^3y \, [\bar{\eta}(y)\psi(y) + \bar{\psi}(y)\eta(y)]}|\mathcal{O} >$$

$$- \gamma^0 < \mathcal{O}|T e^{i \int_{x^0}^{\infty} dy^0 \int d^3y \, [\bar{\eta}(y)\psi(y) + \bar{\psi}(y)\eta(y)]} i\dot{\psi}(x) e^{i \int_{-\infty}^{x^0} dy^0 \int d^3y \, [\bar{\eta}(y)\psi(y) + \bar{\psi}(y)\eta(y)]}|\mathcal{O} >$$

$$= \gamma^0 < \mathcal{O}|T e^{i \int_{x^0}^{\infty} dy^0 \int d^3y \, [\bar{\eta}(y)\psi(y) + \bar{\psi}(y)\eta(y)]} \left[\psi(x), \int d^3y [\bar{\eta}(y)\psi(y) + \bar{\psi}(y)\eta(y)]\right]_{y^0 = x^0} \cdot$$

$$\cdot \, e^{i \int_{-\infty}^{x^0} dy^0 \int d^3y \, [\bar{\eta}(y)\psi(y) + \bar{\psi}(y)\eta(y)]}|\mathcal{O} >$$

$$= \eta(x) Z[\eta, \bar{\eta}] \tag{12.47}$$

where we have used the equation of motion and the equal-time anti-commutation relations of the free Dirac field. The upshot of the above is the functional differential equation

$$\left[\overrightarrow{\partial} + m\right] \frac{\delta}{\delta\bar{\eta}(x)} Z[\eta, \bar{\eta}] = \eta(x) Z[\eta, \bar{\eta}] \tag{12.48}$$

This equation, together with its complex conjugate and the boundary condition (12.46) has the solution

$$Z[\eta, \bar{\eta}] = \exp\left(-\int dx dy \, \bar{\eta}(x) g(x, y) \eta(y)\right) \tag{12.49}$$

which is our explicit solution for the generating functional for correlation functions of the Dirac field theory.

We now see explicitly that a non-vanishing multi-point correlation function must have the same number of ψ's as $\bar{\psi}$'s, since taking n functional derivatives by $\bar{\eta}$ will generate the exponential in (12.49) times an n'th order monomial in η. It is only n further functional derivatives by η which will give a non-zero result when η and $\bar{\eta}$ are both set equal to zero. We also see that taking repeated functional derivatives of equation (12.49) will an equation for a multi-point correlation function

$$< \mathcal{O}|T\psi_{a_1}(x_1) \dots \psi_{a_n}(x_n)\bar{\psi}_{b_1}(y_1) \dots \bar{\psi}_{b_n}(y_n)|\mathcal{O} > \tag{12.50}$$

$$= \sum_{\substack{\text{pairings} \\ <ij>}} (-1)^{\#\text{perm}} \prod_{\text{pairs}} g_{a_i b_j}(x_i - y_j)$$

where the pairings are of x's with y's and the integer #perm is the number of permutations of neighbours which is required to put the operators in the product in the order of the pairings. For example, the 4 point function is given by

$$< \mathcal{O}|T\psi_{a_1}(x_1)\psi_{a_2}(x_2)\bar{\psi}_{b_1}(y_1)\bar{\psi}_{b_2}(y_2)|\mathcal{O} > = - g_{a_1 b_1}(x_1 - y_1)g_{a_2 b_2}(x_2 - y_2)$$

$$+ g_{a_1 b_2}(x_1 - y_2)g_{a_2 b_1}(x_2 - y_1) \tag{12.51}$$

The relative minus sign of the two terms comes from the difference of the number of interchanges of neighbours that would be needed to put the operators in the appropriate order in each term.

12.4.3 Functional Integral for the Dirac Field

We have developed a functional integral formulation of the non-relativistic quantum field theory. It is straightforward to generalize this development to the Dirac field. The result uses the Dirac action

$$\mathbf{S}[\psi, \bar{\psi}] = -i \int dx \bar{\psi}(x) \left[\tfrac{1}{2} \overrightarrow{\partial\!\!\!/} - \tfrac{1}{2} \overleftarrow{\partial\!\!\!/} + m \right] \psi(x) \tag{12.52}$$

The development is identical to that of the non-relativistic theory with $\psi^\dagger$ and $\eta^\dagger$ replaced by $\bar{\psi}$ and $\bar{\eta}$ and the non-relativistic action is replaced by the Dirac action. The functional integral representation of the generating function for time-ordered correlation functions is

$$Z[\eta, \bar{\eta}] = \frac{\displaystyle\int [d\psi d\bar{\psi}] e^{i\mathbf{S}[\psi, \bar{\psi}] + i \int dx [\bar{\eta}(x)\psi(x) + \bar{\psi}(x)\eta(x)]}}{\displaystyle\int [d\psi d\psi^\dagger] e^{i\mathbf{S}[\psi, \bar{\psi}]}} \tag{12.53}$$

or

$$Z[\eta, \bar{\eta}] = \exp\left(-\int dx \, \bar{\eta}(x) g(x, y) \eta(y) \right) \tag{12.54}$$

We can find this by translating the variables in the functional integral, as we did for the non-relativistic field theory. Now, $g(x, y)$ is a matrix. For brevity of notation, we have suppressed the indices.

In this free field theory, all correlation functions factorize into products of 2 point correlation functions. As in the non-relativistic case, phase symmetry implies that a correlation function will vanish unless it contains the same number of ψ's and $\bar{\psi}$'s. The general formula is

$$< \mathcal{O} | T \psi_{a_1}(x_1) \ldots \psi_{a_r}(x_r) \bar{\psi}_{b_1}(y_1) \ldots \bar{\psi}_{b_r}(y_r) | \mathcal{O} >$$

$$= \frac{\delta}{\delta \bar{\eta}_{a_1}(x_1)} \cdots \frac{\delta}{\delta \bar{\eta}_{a_r}(x_r)} \frac{\delta}{\delta \eta_{b_1}(y_1)} \cdots \frac{\delta}{\delta \eta_{b_r}(y_s)} Z[\eta, \bar{\eta}] \Big|_{\eta = \bar{\eta} = 0}$$

$$= \frac{\displaystyle\int [d\psi d\bar{\psi}] e^{i\mathbf{S}[\psi, \bar{\psi}]} \psi_{a_1}(x_1) \ldots \psi_{a_r}(x_r) \bar{\psi}_{b_1}(y_1) \ldots \bar{\psi}_{b_r}(y_r)}{\displaystyle\int [d\psi d\bar{\psi}] e^{i\mathbf{S}[\psi, \bar{\psi}]}}$$

$$= \sum_P (-1)^\# \prod_{i=1}^r g_{a_i b_{P(i)}}(x_i - y_{P(i)}) \tag{12.55}$$

In the last formula, the sum is over permutations and # is the number of exchanges of neighbours needed to put the operators in each pair next to each other.

12.5   Functional Quantum Electrodynamics

As we have already discussed in Sect. 10.5, the quantum field theory which is composed of the coupled Maxwell and Dirac theories is called quantum electrodynamics. The Dirac field is usually associated with the electron and positron but of course, they could represents any charged relativistic spin-$\frac{1}{2}$ particle, so we will generally refer to it as the "Dirac field". Classical electrodynamics coupled to the relativistic Dirac field theory has the action and Lagrangian density

$$\mathbf{S} = \int dx \mathcal{L}(x)$$

$$\mathcal{L}(x) = -i\bar{\psi}(x)\left[\slashed{\partial} - ie\slashed{A}(x) + m\right]\psi(x) - \frac{1}{4}F_{\mu\nu}(x)F^{\mu\nu}(x) \qquad (12.56)$$

In this field theory, we must take the field strength tensor to be given in terms of the four-vector potential field by the equation

$$F_{\mu\nu}(x) = \partial_\mu A_\nu(x) - \partial_\nu A_\mu(x)$$

and the basic dynamical variable is the four-vector potential $A_\mu(x)$. The field equations that follow from the application of the Euler-Lagrange equations to the Lagrangian density are those of the coupled Maxwell-Dirac theory

$$\partial_\mu F^{\mu\nu}(x) = e\bar{\psi}(x)\gamma^\nu\psi(x) \qquad (12.57)$$

$$\left[\slashed{\partial} - ie\slashed{A}(x) + m\right]\psi(x) = 0 \qquad (12.58)$$

This field theory is invariant under the gauge transformation

$$A_\mu(x) \to A_\mu(x) + \partial_\mu\chi(x), \ \psi(x) \to e^{ie\chi(x)}\psi(x), \ \bar{\psi}(x) \to \bar{\psi}(x)e^{-ie\chi(x)}\bar{\psi}(x)$$

As in the case of the free photon field which we studied in earlier chapters, we shall need to fix a gauge. We will choose the covariant gauge condition

$$\partial_\mu A^\mu(x) = 0$$

This constraint is used to simplify the Maxwell equation and the Lagrangian so that we can quantize the theory. With this constraint the field equations become

$$-\partial^2 A_\mu(x) = -e\bar{\psi}(x)\gamma_\mu\psi(x) \qquad (12.59)$$

$$\left[\partial\!\!\!/ - ie A\!\!\!/(x) + m\right]\psi(x) = 0 \tag{12.60}$$

These equations can be obtained from the gauge-fixed action with the Lagrangian density

$$\mathbf{S}_{\text{gf}}[A, \psi, \bar{\psi}] = \int dx \mathcal{L}_{\text{gf}}(x) \tag{12.61}$$

$$\mathcal{L}_{\text{gf}}(x) = -i\bar{\psi}(x)\left[\partial\!\!\!/ + m\right]\psi(x) - \frac{1}{2}\partial_\mu A_\nu(x)\partial^\mu A^\nu(x) - eA_\mu(x)\bar{\psi}(x)\gamma^\mu\psi(x) \tag{12.62}$$

which is then compatible with the equal time commutation and anti-commutation relations, the non-vanishing of which are

$$\left[A_\mu(x), A_\nu(y)\right]\delta(x^0 - y^0) = i\eta_{\mu\nu}\delta(x - y) \tag{12.63}$$

$$\left\{\psi_a(x), \psi_b^\dagger(y)\right\}\delta(x^0 - y^0) = \delta_{ab}\delta(x - y) \tag{12.64}$$

The idea is then to study the field theory defined by the action and Lagrangian density (12.61) and (12.62) or equivalently the field equations and commutators (12.59), (12.60), (12.63) and (12.64) and to impose the constraint on the solutions of the theory as a physical state condition. Since the field equations imply

$$-\partial^2\left(\partial_\mu A^\mu(x)\right) = 0$$

so that, even in the interacting field theory, $\partial_\mu A^\mu(x)$ can be decomposed into positive and negative frequency parts, the physical state condition is as it was for the free photon,

$$\partial_\mu A^{(+)\mu}(x)|\text{phys} >= 0 \tag{12.65}$$

together with an equivalence relation which eliminates null states.

However, as we saw with the free photon, there is another way to deal with the gauge constraint. We begin by ignoring it altogether and simply studying the quantum field theory specified by (12.61) and (12.62) or (12.59), (12.60), (12.63) and (12.64). That field theory has correlation functions which do not obey the gauge constraint and which include unphysical negative normed states. However, if we confine our attention to the correlation functions of gauge invariant operators, we can show that this quantum field theory is equivalent to the fully gauge invariant Maxwell theory and also to one with the gauge fixing condition consistently imposed as a constraint.

A generating functional for time-ordered correlation functions in the quantum field theory described by (12.59), (12.60), (12.63) and (12.64) is given by

$$Z[J, \eta, \bar{\eta}] =< \mathcal{O}|\mathcal{T}\exp\left(i\int dx\left[A_\mu(x)J^\mu(x) + \bar{\eta}(x)\psi(x) + \bar{\psi}(x)\eta(x)\right]\right)|\mathcal{O}> \tag{12.66}$$

It is straightforward to use the field equations and the commutation and anti-commutation relations, for the field operators to show that this generating functional must obey the functional differential equations

$$\left[\not{\partial} + m - ie\gamma^\mu \frac{1}{i} \frac{\delta}{\delta J^\mu(x)} \right] \frac{1}{i} \frac{\delta}{\delta \bar{\eta}(x)} Z[J, \eta, \bar{\eta}] = \eta(x) Z[J, \eta, \bar{\eta}] \qquad (12.67)$$

$$Z[J, \eta, \bar{\eta}] \frac{1}{i} \frac{\overleftarrow{\delta}}{\delta \eta(x)} \left[-\overleftarrow{\not{\partial}} + m - ie\gamma^\mu \frac{1}{i} \frac{\overleftarrow{\delta}}{\delta J^\mu(x)} \right] = Z[J, \eta, \bar{\eta}] \bar{\eta}(x) \qquad (12.68)$$

$$\left[-\partial^2 \frac{1}{i} \frac{\delta}{\delta J^\mu(x)} + e \frac{\delta}{\delta \eta(x)} \gamma_\mu \frac{\delta}{\delta \bar{\eta}(x)} \right] Z[J, \eta, \bar{\eta}] = J_\mu(x) Z[J, \eta, \bar{\eta}] \qquad (12.69)$$

These equations for the generating functional are solved by the functional integral

$$Z[J, \eta, \bar{\eta}] = \frac{\displaystyle\int [dA d\psi d\bar{\psi}] e^{iS_{\mathrm{gf}}[A, \psi, \bar{\psi}] + i \int dx [A_\mu(x) J^\mu(x) + \bar{\eta}(x)\psi(x) + \bar{\psi}(x)\eta(x)]}}{\displaystyle\int [dA d\psi d\bar{\psi}] e^{iS_{\mathrm{gf}}[A, \psi, \bar{\psi}]}}$$

$$(12.70)$$

where the gauge-fixed action $S_{\mathrm{gf}}[A, \psi, \bar{\psi}]$ is given in equation (12.61).

Now, we could to do what we did for the free photon. By transforming the integration variables in the functional integral in equation (12.70) we could unfix the relativistic gauge to find a fully gauge invariant functional integral. We could also re-fix the gauge to get the gauge fixed functional integral with a gauge parameter. In a generic covariant gauge,

$$Z_\xi[J, \eta, \bar{\eta}] = \frac{\displaystyle\int [dA d\psi d\bar{\psi}] e^{iS_\xi[A, \psi, \bar{\psi}] + i \int dx [A_\mu(x) J^\mu(x) + \bar{\eta}(x)\psi(x) + \bar{\psi}(x)\eta(x)]}}{\displaystyle\int [dA d\psi d\bar{\psi}] e^{iS_\xi[A_\nu, \psi, \bar{\psi}]}}$$

$$(12.71)$$

$$\mathbf{S} = \int dx \, \mathcal{L}(x)$$

$$\mathcal{L}_\xi(x) = -i\bar{\psi}(x) \left[\not{\partial} - ie\not{A}(x) + m \right] \psi(x) - \frac{1}{4} F_{\mu\nu}(x) F^{\mu\nu}(x) - \frac{\xi}{2} (\partial_\mu A^\mu(x))^2$$

$$(12.72)$$

The functional integral in equation (12.70) is identical to the integral in equation (12.72) in the Feynman gauge, $\xi = 1$.

However, we caution the reader that the gauge unfixing and fixing that we describe above involves gauge transformations and such transformations are not symmetries of the coupling of the functional integration variables in the functional integrals to the sources, $J^\mu(x)$, $\eta(x)$ and $\bar\eta(x)$ in (12.70). As a consequence, the generating functionals in different gauges are not actually equal, nor are the correlation functions which are generated by taking functional derivatives of them. However, what is guaranteed is the fact that, if the generating functionals are used to compute the correlation functions of gauge invariant operators, then those correlation functions are indeed equal in different gauges. They can be computed using (12.70) or equivalently the fully gauge invariant functional integral or the gauge fixed functional integral with the gauge fixing parameter ξ given in equations (12.71) and (12.72).

The Weakly Coupled Real Scalar Field

13

In a previous chapter, we showed that the generating functional for correlations in the theory of a single self-interacting real scalar field is given by the ratio of functional integrals

$$Z[J] = \frac{\int [d\phi] e^{iS[\phi] + i \int dx \phi(x) J(x)}}{\int [d\phi] e^{iS[\phi]}} \qquad (13.1)$$

where the action and Lagrangian density are

$$\mathbf{S}[\phi] = \int dx \, \mathcal{L}(x) \qquad (13.2)$$

$$\mathcal{L}(x) = -\frac{1}{2} \partial_\mu \phi(x) \partial^\mu \phi(x) - \frac{m^2}{2} \phi^2(x) - \frac{\lambda}{4!} \phi^4(x) \qquad (13.3)$$

The natural objects which are computed by the functional integral are the time-ordered correlation functions

$$\Gamma(x_1, x_2, \ldots, x_n) \equiv \, < \mathcal{O}|T\phi(x_1)\phi(x_2)\ldots\phi(x_n)|\mathcal{O} >$$

which can be gotten by taking functional derivatives of the generating functional (13.1),

$$\Gamma(x_1, x_2, \ldots, x_n) = \frac{1}{i^n} \frac{\delta}{\delta J(x_1)} \frac{\delta}{\delta J(x_2)} \cdots \frac{\delta}{\delta J(x_n)} Z[J] \Big|_{J=0}$$

© The Author(s), under exclusive license to Springer Nature Singapore Pte Ltd. 2023   257
G. W. Semenoff, *Quantum Field Theory*, Graduate Texts in Physics,
https://doi.org/10.1007/978-981-99-5410-0_13

and which can therefore also be written as the functional integral

$$\Gamma(x_1, x_2, \ldots, x_n) = \frac{\int [d\phi] e^{iS[\phi]} \phi(x_1) \phi(x_2) \ldots \phi(x_n)}{\int [d\phi] e^{iS[\phi]}} \qquad (13.4)$$

In this way, the correlation functions of the quantum field theory is expressed as a certain average over classical field configurations in the classical field theory where the weight in the average of a given classical field configuration, $\phi(x)$, is the exponential of the action, $e^{iS[\phi]}$, evaluated on that field configuration. This suggests an analogy between quantum field theory and classical statistical mechanics which indeed exists and which can be very useful from both a conceptual and practical point of view. However, a thorough discussion of this analogy is beyond the scope of this book and we will not pursue it here. For now, we will concentrate, for the most part, on the computation of objects like (13.4) and the interpretation of the results.

Generally, we know much less about the interacting field theory than we know about free field theory. It is thus far not possible to find an explicit exact solution of the interacting quantum field theory whose Lagrangian density is written above when the interaction term $-\frac{\lambda}{4!} \phi^4(x)$ is present. For example, it is not known how to do the functional integrals in equation (13.1) or equation (13.4) explicitly. Nor is it possible to find an exact solution of the field equations and commutation relations.

What we do know how to solve is the free field theory which results when we put the coupling constant λ to zero. We found a complete solution in Sect. 7.5. We have also found the explicit generating functional as well as all of the correlation functions in Sect. 11.3. The basis of the quantum states of the free field theory were constructed by operating creation operators on the vacuum state, for example,

$$\frac{1}{\sqrt{n!}} a^\dagger(\vec{k}_1) \ldots a^\dagger(\vec{k}_n) |\mathcal{O}> \qquad (13.5)$$

which is interpreted as a state containing n identical particles, in this case bosons. The particles have momenta $\vec{k}_1, \ldots, \vec{k}_n$ and energies $k_1^0 = \sqrt{\vec{k}_1^2 + m^2}, \ldots, k_n^0 = \sqrt{\vec{k}_n^2 + m^2}$. The states (13.5) are eigenstates of the total energy and momentum, P^μ, with eigenvalues

$$p^\mu = k_1^\mu + \ldots + k_n^\mu$$

The energies and momenta of these non-interacting particles are separately conserved and the total energy and momentum is just a sum of these quantities for each particle.

When we introduce interactions, we expect that some aspects of the field theory do not change very much. For example, there should still be a vacuum state which we can define as being the lowest energy state of the theory. Here, we are assuming that the interaction does not destabilize the system. For example, in our Lagrangian density, (13.3) the coupling constant λ should be positive. Otherwise, the classical energy of this theory would be unbounded from below and we could expect this to result in an instability of the quantized theory. We will assume that $\lambda > 0$. We will still denote the vacuum state of the interacting theory by $|\mathcal{O}>$.

The interacting field theory has a conserved energy and momentum,

$$P^\mu = \int d^3x \; T^{0\mu}(x) \tag{13.6}$$

and we must be able to organize the quantum states of the interacting theory into eigenstates of the four-momentum, so that

$$P^\mu \mid p >= p^\mu \mid p >$$

We could also reasonably expect that the energy and momentum of the vacuum vanishes, $P^\mu|\mathcal{O}> \; = 0$ and then, in fact

$$-P_\mu P^\mu \; |\mathcal{O}>= 0$$

The vacuum is an eigenvector of $-P^2$ with eigenvalue $-p^2 = 0$. If there are no massless particles in the theory, the vacuum is an isolated discrete eigenstate of $-P^2$. We will put off the important but remarkably subtle case of massless particles for a later discussion. For now, we will assume that all of the particles are massive.

In free field theory, any one-particle state is also a discrete eigenvector of $-P^2$

$$-P_\mu P^\mu \; |\text{one particle state} >= m^2 \; |\text{one particle state} >$$

where m is the mass of the particle. On the other hand, two-particle states have a continuum spectrum with $-P^2 \geq 4m^2$, three particle state with $-P^2 \geq 9m^2$ and so on.

We can expect that some of this structure survives when we turn on an interaction, as long as that interaction is sufficiently weak. For example, we could still define what we mean by a "one-particle state" as a state which is a discrete eigenstate of $-P^2$ with eigenvalue m^2. We would interpret m as the mass of the particle. This is consistent with the spectral theorem that we discussed in Sect. 7.7 and where we showed that the poles in the variable $-p^2$ in the momentum space 2 point functions are associated with discrete eigenstates of $-P^2$. In general, we should therefore examine the 2 point functions of all possible operators and identify those which have poles as those are indicative of the single particle states. However, when interactions are weak and we use perturbation theory, we expect that the only single particle states are as they were in free field theory, those which give rise to the pole in the 2 point function of the scalar field itself. (When $\lambda > 0$ the interaction is repulsive there should be no bound states.)

Multi-particle states on the other hand should reflect the fact that the particles occupying those states interact with each other. The multi-particle states can still be written as eigenstates of P^μ and $-P^2$ with continuum spectra but now the the eigenvalues of P^μ cannot be simple sums of single particle energies and momenta but they should also contain an interaction energy. When interactions are repulsive, we would expect that this interaction energy is positive.

We could still ask the question as to why we need an approach to perturbation theory as elaborate as the computation of the functional integral (13.4) in an expansion in the the parameter λ. For example, as a simple alternative, we could recall from time-independent perturbation theory the formula for the shift at first order of the energy of a state. If $|E>$ is an eigenstate of the non-interacting Hamiltonian with eigenvalue E, that is, if

$$H_0|E> = E|E>$$

then when we turn on a perturbation, $H_0 \rightarrow H_0 + H_I$, the shift in the energy of that state is given by the formula

$$\delta E = < E|H_I|E >$$

In the above formula, the matrix element on the right-hand-side should be computed in the unperturbed theory.

To apply this first order perturbation theory to the quantum field theory, we could separate the Hamiltonian of our interacting scalar field theory into free field and interaction parts as

$$H = \int d^3x \, \mathbf{T}^{00}(0, \vec{x}) = H_0 + H_I \tag{13.7}$$

$$H_0 = \int d^3k \sqrt{\vec{k}^2 + m^2} \, a^\dagger(\vec{k})a(\vec{k}) \tag{13.8}$$

$$H_I = \frac{\lambda}{4!} \int d^3x \, \phi^4(x) \tag{13.9}$$

Then, for example, we can use time-independent perturbation theory to compute the correction to the energy splitting between the vacuum state and the one-particle state,

$$\sqrt{\vec{k}^2 + m^2} \rightarrow \sqrt{\vec{k}^2 + m^2} + \delta E(\vec{k})$$

where we compute $\delta E(\vec{k})$ as follows:

$$\delta E(\vec{k})\delta(\vec{k} - \vec{k}') = < \mathcal{O}|a(\vec{k})H_I a^\dagger(\vec{k}')|\mathcal{O} > - < \mathcal{O}|H_I|\mathcal{O} > \delta(\vec{k} - \vec{k}') \tag{13.10}$$

$$\delta E(\vec{k}) = \frac{1}{\sqrt{\vec{k}^2 + m^2}} \frac{\lambda}{4} \int \frac{d^3\ell}{(2\pi)^3 2\sqrt{\vec{\ell}^2 + m^2}} \tag{13.11}$$

To go from equation (13.10) to equation (13.11) we have substituted the expansion of the free field in terms of creation and annihilation operators into the interaction Hamiltonian and computed the matrix element.

The result is highly problematic. The wave-vector integral on the right-hand-side of equation (13.11) diverges. The energy of the one-particle state relative to the vacuum would appear to be infinite. The difference between the energies of states, especially the two lowest lying states in the spectrum, should be observable and, obviously, finite. To address this difficulty, first of all, we need a more sensible mathematical definition of the correction. For this, the only thing that we can do is to modify it so that the diverging integral no longer diverges, but is finite. The obvious way to do this is to impose a momentum cutoff. This replaces equation (13.11) by

$$\delta E(\vec{k}) = \frac{1}{\sqrt{\vec{k}^2 + m^2}} \frac{\lambda}{4} \int \frac{d^3\ell}{(2\pi)^3 2\sqrt{\vec{\ell}^2 + m^2}} \theta(\Lambda^2 - \vec{\ell}^2) \qquad (13.12)$$

Now the shift is finite and we could imagine that λ is small enough that $\delta E(\vec{k})$ is actually a small correction to $E(\vec{k})$, as would be needed for the perturbation expansion to be accurate. Then, we can observe that the energy and its correction can be written as

$$E(\vec{k}) + \delta E(\vec{k}) = \sqrt{\vec{k}^2 + m^2} + \left[\frac{\lambda}{2} \int \frac{d^3\ell\, \theta(\Lambda^2 - \vec{\ell}^2)}{(2\pi)^3 2\sqrt{\vec{\ell}^2 + m^2}}\right] \frac{\partial}{\partial m^2} \sqrt{\vec{k}^2 + m^2}$$

$$(13.13)$$

Equation (13.13) suggests an interpretation of the correction. It has the form of a Taylor expansion and it suggests that the entire correction can be accounted for by a shift in the mass of the particle,

$$m^2 \rightarrow m^2 + \frac{\lambda}{2} \int \frac{d^3\ell\, \theta(\Lambda^2 - \vec{\ell}^2)}{(2\pi)^3 2\sqrt{\vec{\ell}^2 + m^2}} \qquad (13.14)$$

In the interacting theory, the mass squared of the particle is corrected from m^2 to M^2 where

$$M^2 = m^2 + \left[\frac{\lambda}{2} \int \frac{d^3\ell}{(2\pi)^3 2\sqrt{\vec{\ell}^2 + m^2}} \theta(\Lambda^2 - \vec{\ell}^2)\right]$$

Given this information, it is most elegant to re-organize the problem in the following way. We can, to first order in λ, re-write the above equation as

$$m^2 = M^2 - \left[\frac{\lambda}{2} \int \frac{d^3\ell}{(2\pi)^3 2\sqrt{\vec{\ell}^2 + M^2}} \theta(\Lambda^2 - \vec{\ell}^2)\right]$$

Then, we could plug this expression back into the Lagrangian density which would become

$$\mathcal{L}(x) = -\frac{1}{2}\partial_\mu\phi(x)\partial^\mu\phi(x) - \frac{M^2}{2}\phi^2(x) - \frac{\lambda}{4!}\phi^4(x) - z_M\frac{M^2}{2}\phi^2(x) \qquad (13.15)$$

$$z_M = -\frac{\lambda}{2M^2} \int \frac{d^3\ell}{(2\pi)^3 2\sqrt{\vec{\ell}^2 + M^2}} \theta(\Lambda^2 - \vec{\ell}^2) \qquad (13.16)$$

and revamp our perturbation theory problem by taking the free and interacting Hamiltonians as

$$H = \int d^3x \, \mathbf{T}^{00}(0, \vec{x}) \; = \; H_0 \; + \; H_I \tag{13.17}$$

$$H_0 = \int d^3k \, \sqrt{\vec{k}^2 + M^2} \, a^\dagger(\vec{k}) a(\vec{k}) \tag{13.18}$$

$$H_I \;=\; \int d^3x \left[\frac{\lambda}{4!}\phi^4(x) + z_M \frac{M^2}{2}\phi^2(x) \right] \tag{13.19}$$

Now, rather than the parameter m, which is called the "bare mass", the Lagrangian density and the Hamiltonian contain the mass of the one-particle state of the interacting field theory, M, and another interaction which appears as the the term $\int d^3x \, z_M \frac{M^2}{2}\phi^2(x)$ with the parameter, z_M. z_M is now to be determined so that the correction to the mass cancels exactly. That criterion gives the value of z_M in equation (13.16). We have seen in our explicit computation that this will work to the leading order in λ, and that this way of organizing the perturbative computation is equivalent to our starting point. The term with the parameter z_M that is added to the Lagrangian density and the interaction Hamiltonian is called a "counterterm".

We introduced a cutoff Λ in order to define the quantum field theory. Indeed, at least in perturbation theory, there will not be a way to define what we mean by the quantum field theory without such a cutoff, so we will have to assume that, in one form or another, it is always there.[1] However, what we have managed to do is to confine the cutoff dependence to the counterterm by requiring that M is the mass of the one-particle state.

One might wonder whether such a process generalizes to other quantities or to higher orders in perturbation theory. The simple computations of time-independent perturbation theory, an example of which we have used above, rapidly become extremely cumbersome at higher orders. What is more, they obscure the Lorentz invariance of the theory. It is desirable to take an approach which retains as many of the symmetries of the quantum field theory as possible, in this case, Poincaré symmetry. The approach that begins with the functional integral (13.4) is ideal for this purpose. It will turn out that it is possible to define the quantum field theory using a cutoff which is Lorentz invariant and then it is possible to maintain manifest Lorentz invariance in computations. Not only does this give a definition of the quantum field theory which is Lorentz invariant, but it simplifies many computations. Beyond that, it will be possible to absorb all of the cutoff dependence into counterterms. In a renormalizable quantum field theory, only a finite number of different counterterms are needed, although each one generically gets contributions to any order in λ. We will develop this approach in the following section.

---

[1] We could have been more careful and chosen a Lorentz invariant way to introduce a cutoff.

13.1 Counterterms

Before we begin our detailed study of the interacting scalar field theory, we must recognize the fact that what we are doing here is constructing a mathematical model of a (perhaps hypothetical) physical system. We are describing the physical system by the quantum field theory which is defined by the functional integral that we have written in equation (13.4). That equation has a two explicit parameters, m^2 and λ. In addition to these, there is a third parameter, the coefficient of the kinetic term $-\frac{1}{2}\partial_\mu\phi(x)\partial^\mu\phi(x)$ which has been set to one in equation (13.4). By re-scaling the integration variable ϕ in the functional integrals, we can always set this coefficient to one. However, we do not have a way to know whether the scalar field with that normalization is the correct amplitude of whatever physical quantity we might associate with it. For that reason, we should include the scale factor by replacing $-\frac{1}{2}\partial_\mu\phi(x)\partial^\mu\phi(x)$ in the Lagrangian density by $-\frac{Z}{2}\partial_\mu\phi(x)\partial^\mu\phi(x)$. Then, it is also convenient to introduce similar scale factors for the remaining two terms in the Lagrangian density so that it has the form

$$\mathcal{L}(x) = -\frac{Z}{2}\partial_\mu\phi(x)\partial^\mu\phi(x) - \frac{Z_{m^2}m^2}{2}\phi^2(x) - \frac{\lambda Z_\lambda}{4!}\phi^4(x) \qquad (13.20)$$

Our strategy will be to specify m^2 and λ as physical constants describing the system, for example the mass of a particle and the strength of the interaction in a particular circumstance. The remaining quantities, sometimes called "renormalization constants", Z, Z_m and Z_λ are then the three adjustable parameters of the model. These parameters are to be adjusted so that the physical quantities computed with the quantum field theory model, energy levels, scattering amplitudes, et cetera, match those of the actual physical system which is being modelled.

Part of the modelling that we have described above is renormalization. It is a general fact, which we have already encountered, and which we will confirm by explicit calculation shortly, that perturbative calculations encounter infinities—ultraviolet divergences—coming from the high energy and momentum regime in Feynman integrals. An important role for the renormalization constants Z, Z_m and Z_λ is to cancel these ultraviolet divergences. In a renormalizable quantum field theory, such as the one described by the Lagrangian density (13.20), the infinities can be canceled, order by order in perturbation theory by adjusting the renormalization constants. The renormalization constants become functions of the high energy cutoff in such a way that all of the n point functions, when expressed as functions of the finite parameters m and λ as well as spacetime coordinates, remain finite as the cutoff is taken to be very large. A general proof of this statement is straightforward, but it is also beyond the scope of this introductory monograph and we will not give it here. We will, however, observe that the criterion of renormalizability greatly constrains the number and the types of terms that we can write in the Lagrangian density. For our self-interacting real scalar field with $\phi \to -\phi$ symmetry and in four space–time dimensions, all of the allowed terms are already contained in equation (13.20).

The appearance of ultraviolet divergences was at one time thought to be a pathology of relativistic quantum field theory and it cast doubt on its validity as a possible

description of nature. With a clearer understanding of renormalization and particularly the renormalization group this evolved to a view of the divergences as being
natural and even necessary.

Sometimes $Z_{m^2}m^2$ and $Z_\lambda\lambda$ are referred to as the "bare mass" and the "bare
coupling", respectively, and m^2 as the "renormalized mass" or "physical mass" and
λ the "renormalized coupling". Also $Z^{\frac{1}{2}}\phi(x)$ is sometimes called the "bare field"
and $\phi(x)$ the "renormalized field".

The general rules which determine the structure of a Lagrangian density for a
relativistic field theory are surprisingly simple. At the same time they are quite constraining. First of all, there is a constraint of locality. All of the terms in the Lagrangian
density should be monomials in the field and derivatives of the field, evaluated at
the same spacetime point. Such terms are sometimes called "local operators". This
restriction is our way of maintaining control over some of the causal properties of
the quantum field theory, like the absence of action at a distance. This limits our
consideration to theories where the action is a single spacetime volume integral of a
Lagrangian density and the Lagrangian density is a linear combination of terms such
as

$$\phi^2(x),\ \phi^4(x),\ ...,\ \partial_\mu\phi(x)\partial^\mu\phi(x),\ \partial_\mu\phi(x)\partial^\mu\phi(x)\phi^2(x),\ (\partial_\mu\phi(x)\partial^\mu\phi(x))^2,\ ...$$

Then, beyond locality, there are symmetry considerations. For example, if the field
theory is to be translation invariant and Lorentz invariant the local operators should
be chosen so that they have these symmetries. This includes internal symmetries
like the $\phi \to -\phi$ symmetry that we have imposed on the scalar field theory. Finally,
there is the constraint of renormalizability. Which local operators are allowed in the
Lagrangian density depends on their scaling dimensions.

Let us examine the scaling dimensions of local operators in a scalar field theory
in D spacetime dimensions. With our natural units where $\hbar = 1$ and $c = 1$, the
action should be dimensionless. The action is the spacetime volume integral of the
Lagrangian density. The Lagrangian density must have dimension of inverse distance
to the power of the number of spacetime dimensions, D,

$$\dim[\mathcal{L}] = D$$

Then, this implies that

$$\dim[\partial_\mu\phi\partial^\mu\phi] = D$$

and

$$\dim[\phi] = \frac{D-2}{2}$$

Then

$$\dim[\phi^k] = k(D-2)/2$$

and the dimension of any local operator is equal to the number of derivatives it
contains plus $\frac{D-2}{2}$ times the number of fields that it contains.

We say that a local operator $O(x)$ is a "relevant operator" if $\dim[O] < D$. It is a "marginal operator" if $\dim[O] = D$ and it is an "irrelevant operator" if $\dim[O] > D$.

A quantum field theory is renormalizable if its Lagrangian density contains all of the relevant and marginal local operators which are consistent with the symmetries, in this case Poincare symmetry and the internal symmetry. A cursory examination of our example Lagrangian density in equation (13.20) indeed indicates that it contains all of the relevant and marginal operators that are consistent with the symmetries. It therefore describes a renormalizable quantum field theory.

For perturbation theory, it is useful to write the renormalization constants as a power series in the coupling that governs the interaction strength. We will take that parameter to be λ, the renormalized coupling. Then

$$Z = 1 + z, \quad z = \sum_{n=1}^{\infty} z^{(n)}, \quad z^{(n)} \sim \lambda^n \tag{13.21}$$

$$Z_{m^2} = 1 + z_{m^2}, \quad z_{m^2} = \sum_{n=1}^{\infty} z_{m^2}^{(n)}, \quad z_{m^2}^{(n)} \sim \lambda^n \tag{13.22}$$

$$Z_\lambda = 1 + z_\lambda, \quad z_\lambda = \sum_{n=1}^{\infty} z_\lambda^{(n)}, \quad z_\lambda^{(n)} \sim \lambda^n \tag{13.23}$$

The zeroth order term in each of the Z's is fixed by the free field theory which emerges when we set λ to zero. In that limit, m is the mass of the scalar field and we will assume that the coefficient in front of the derivative term in the Lagrangian density reverts to one. This fixes the normalization of the free field commutation relations.

Then, it is useful to put the terms in the Lagrangian density with the constants z, z_{m^2}, z_λ with the interactions, so that the action is written as a non-interacting part and an interaction action as

$$S[\phi] = S_0[\phi] + S_I[\phi]$$

$$S_0[\phi] = \int dx \mathcal{L}_0(x), \quad S_I[\phi] = \int dx \mathcal{L}_I(x)$$

$$\mathcal{L}_0(x) = -\frac{1}{2} \partial_\mu \phi(x) \partial^\mu \phi(x) - \frac{m^2}{2} \phi^2(x)$$

$$\mathcal{L}_I(x) = -\frac{\lambda}{4!} \phi^4(x) - \frac{z}{2} \partial_\mu \phi(x) \partial^\mu \phi(x) - \frac{m^2}{2} z_{m^2} \phi^2(x) - \frac{\lambda}{4!} z_\lambda \phi^4(x)$$

The terms in the interaction action with coefficients z, z_{m^2}, z_λ are called "counterterms". They will be important in that they are the quantitites which are to be adjusted, order by order in perturbation theory, so that the mass m and the coupling constant λ are indeed the parameters of the physical theory. Since they contain higher orders in λ, it is useful to consider the interaction action itself as a sum of terms of different orders in λ,

$$S_I = \sum_{\ell=1}^{\infty} S_I^{(\ell)} \tag{13.24}$$

where $S_I^{(\ell)}$ is the order λ^ℓ term in the interaction action. Explicitly,

$$S_I^{(1)}[\phi] = \int dx \left\{ -\frac{\lambda}{4!}\phi^4(x) - \frac{z^{(1)}}{2}\partial_\mu\phi(x)\partial^\mu\phi(x) - \frac{m^2}{2}z_{m^2}^{(1)}\phi^2(x) \right\} \tag{13.25}$$

$$S_I^{(\ell>1)}[\phi] = \int dx \left\{ -\frac{z^{(\ell)}}{2}\partial_\mu\phi(x)\partial^\mu\phi(x) - \frac{m^2}{2}z_{m^2}^{(\ell)}\phi^2(x) - \frac{\lambda}{4!}z_\lambda^{(\ell-1)}\phi^4(x) \right\} \tag{13.26}$$

The exponential of the action which appears inside the generating functional then has an expansion in λ of the form

$$e^{iS[\phi]} = e^{iS_0[\phi]} \sum_{\ell_1,\ell_2,\dots=0}^{\infty} \frac{\left(iS_I^{(1)}[\phi]\right)^{\ell_1}}{\ell_1!} \frac{\left(iS_I^{(2)}[\phi]\right)^{\ell_2}}{\ell_2!} \dots$$

We are now ready to perform the perturbative expansion of either the generating functional or correlations functions which are obtained from it,

$$\Gamma(y_1, \dots, y_n) = <\mathcal{O}|T\phi(y_1)\dots\phi(y_n)|\mathcal{O}> \tag{13.27}$$

$$= \frac{\int [d\phi]e^{iS_0[\phi]}\phi(y_1)\dots\phi(y_n)\sum_{\ell_1,\ell_2,\dots=0}^{\infty}\frac{\left(iS_I^{(1)}[\phi]\right)^{\ell_1}}{\ell_1!}\frac{\left(iS_I^{(2)}[\phi]\right)^{\ell_2}}{\ell_2!}\dots}{\int [d\phi]e^{iS_0[\phi]}\sum_{\ell_1,\ell_2,\dots=0}^{\infty}\frac{\left(iS_I^{(1)}[\phi]\right)^{\ell_1}}{\ell_1!}\frac{\left(iS_I^{(2)}[\phi]\right)^{\ell_2}}{\ell_2!}\dots} \tag{13.28}$$

In order to compute the order k contribution, which we shall denote by $\Gamma^{(k)}(y_1, \dots, y_n)$, we assemble all of the terms so that the total order adds up to k. (Note that this should entail a further expansion of the denominator. We will not do this expansion explicitly here as we will soon find a way to avoid considering the denominator at all.) What we obtain is an assembly of Gaussian integrals which we we can evaluate explicitly using techniques which we developed in Sect. 11.2 and Wick's theorem for scalar fields which we developed in Sect. 11.3.1.

Then, the Gaussian integrals can be evaluated using Wick's theorem. If n is even,

$$\frac{\int [d\phi]e^{iS_0[phi]}\phi(x_1)\dots\phi(x_n)}{\int [d\phi]e^{iS_0[\phi]}} = \sum_{\text{pairings}} \prod_{\text{pairs } (ij)} \Delta(x_i, x_j) \tag{13.29}$$

The above integral vanishes if n is odd.

We note that, because of the $\phi \to -\phi$ symmetry that we have imposed, all correlation functions with an odd number of fields vanish,

$$\Gamma(y_1, ..., y_{2n+1}) = 0$$

In making this statement we are assuming that the symmetry is not spontaneously broken. We will make this assumption for the remainder of this chapter.

13.2 Computation of the 2 Point Function

As an example of a perturbative computation, let us consider the expansion of the 2 point function, $\Gamma(y_1, y_2)$ for which we use the special notation

$$D(y_1, y_2) \equiv \Gamma(y_1, y_2)$$

and

$$\begin{aligned} D(y_1, y_2) &\equiv\; < \mathcal{O}|T\phi(y_1)\phi(y_2)|\mathcal{O} > \\ &= D^{(0)}(y_1, y_2) + D^{(1)}(y_1, y_2) + D^{(2)}(y_1, y_2) + \dots \end{aligned}$$

where we use $D^{(k)}(y_1, y_2)$ to denote the contributions to the 2 point function of order λ^k. To find these contributions, we expand the functional integral representation of the 2 point function, equation (13.28) with $n = 2$, in powers of the interaction and then we use Wick's theorem, equation (13.29), to evaluate the resulting Gaussian integrals.

We begin by carefully expanding equation (13.28) with $n = 2$ to each of the lower orders in λ. The leading order contribution is just the free field theory result,

$$D^{(0)}(y_1, y_2) = \frac{\int [d\phi] e^{i S_0[\phi]} \phi(y_1)\phi(y_2)}{\int [d\phi] e^{i S_0[\phi]}}$$

This leading order approximation is free field theory, the 2 point function at this order is the propagator, as it should be,

$$D^{(0)}(y_1, y_2) = \Delta(y_1, y_2) \tag{13.30}$$

The first order correction is given by

$$D^{(1)}(y_1, y_2) = D_N^{(1)}(y_1, y_2) + D_D^{(1)}(y_1, y_2)$$

where we have separated it into two parts, the first coming from computing the leading corrections to the numerator in the ratio of functional integrals,

$$D_N^{(1)}(y_1, y_2) = \frac{\int [d\phi] e^{iS_0[\phi]} \phi(y_1)\phi(y_2) iS_I^{(1)}[\phi]}{\int [d\phi] e^{iS_0[\phi]}} \tag{13.31}$$

and the second coming from the expansion of the denominator in the ratio of functional integrals,

$$D_D^{(1)}(y_1, y_2) = -\frac{\int [d\phi] e^{iS_0[\phi]} \phi(y_1)\phi(y_2)}{\int [d\phi] e^{iS_0[\phi]}} \frac{\int [d\phi] e^{iS_0[\phi]} iS^{(1)}[\phi]}{\int [d\phi] e^{iS_0[\phi]}} \tag{13.32}$$

Let us begin by studying the contribution in equation (13.31) which is

$$i \int dx \frac{\int [d\phi] e^{iS_0[\phi]} \phi(y_1)\phi(y_2) \left[-\frac{z^{(1)}}{2} \partial_\mu \phi(x) \partial^\mu \phi(x) - \frac{m^2 z_m^{(1)}}{2} \phi^2(x) - \frac{\lambda}{4!} \phi^4(x) \right]}{\int [d\phi] e^{iS_0[\phi]}}$$

$$\tag{13.33}$$

Use of Wick's theorem in the above equation is easily seen to lead to the expression

$$D_N^{(1)}(y_1, y_2) = -i \int dx \, \Delta(y_1, x) \left(-z^{(1)} \partial^2 + z_{m^2}^{(1)} m^2 \right) \Delta(x, y_2)$$

$$- i \frac{\lambda}{2} \int dx \, \Delta(y_1, x) \Delta(x, x) \Delta(x, y_2)$$

$$- i \Delta(y_1, y_2) \int dx \left[\frac{\lambda}{8} \Delta(x, x) \Delta(x, x) + \frac{1}{2} \lim_{x' \to x} \left(-z^{(1)} \partial^2 \right. \right.$$

$$\left. \left. + m^2 z_{m^2}^{(1)} \right) \Delta(x, x') \right] \tag{13.34}$$

Then, we can study the contribution of equation (13.32). The use of Wick's theorem in equation (13.32) gives the contribution

$$D_D^{(1)}(y_1, y_2) = i \Delta(y_1, y_2) \int dx \left[\frac{\lambda}{8} \Delta(x, x) \Delta(x, x) + \frac{1}{2} \lim_{x' \to x} \left(-z^{(1)} \partial^2 \right. \right.$$

$$\left. \left. + m^2 z_{m^2}^{(1)} \right) \Delta(x, x') \right] \tag{13.35}$$

When we add the two terms together to form $D^{(1)}(y_1, y_2) = D_N^{(1)}(y_1, y_2) + D_D^{(1)}(y_1, y_2)$, the contribution $D_D^{(1)}(y_1, y_2)$ in equation (13.35) cancels the terms in the last line of equation (13.34) for $D_N^{(1)}(y_1, y_2)$. This is an example of a general rule, a corollary of the linked cluster theorem which we shall prove in a later Sect. 14.4.1, that the contribution from the denominator serves to cancel a particular class of contributions from the numerator.

Then, the result is simply the first line of equation (13.34), that is,

$$D^{(1)}(y_1, y_2) = -i \int dx\, \Delta(y_1, x) \left(-z^{(1)}\partial^2 + z_{m^2}^{(1)} m^2\right) \Delta(x, y_2)$$
$$ - i \int dx\, \frac{\lambda}{2} \Delta(y_1, x) \Delta(x, x) \Delta(x, y_2) \tag{13.36}$$

For future reference, we shall write the above result, equation (13.36), in the form

$$D^{(1)}(y_1, y_2) = \int dw_1 dw_2\, \Delta(x, w_1) \Pi^{(1)}(w_1, w_2) \Delta(w_2, y) \tag{13.37}$$

$$\Pi^{(1)}(w_1, w_2) = -i \int dw \left(-z^{(1)}\partial^2 + z_{m^2}^{(1)} m^2\right) \Delta(w_1 - w_2)$$
$$ - i\frac{\lambda}{2} \Delta(w_1, w_1)\delta(w_1 - w_2) \tag{13.38}$$

Once we have a better view of the organization of the perturbative expansion of the 2 point function, we will make extensive use of the quantity $\Pi(w_1, w_2)$ which we will call the "self-energy" of the scalar field. We have computed $\Pi^{(1)}(w_1, w_2)$ which is the order λ contribution to $\Pi(w_1, w_2)$.

We note that the second term on the right-hand-side of equations (13.36) and (13.38) contain an ill defined divergent quantity $\Delta(x, x)$. If we use the fourier transform of the propagator, we can see that it is given by the integral

$$\Delta(x, x) = -i \int \frac{dk}{(2\pi)^4} \frac{1}{k^2 + m^2 - i\epsilon} \tag{13.39}$$

The divergence of this integral comes from the large k regime where the volume integral $\int dk/k^2$ is infinite. For this reason it is sometimes called an "ultraviolet divergence" and the integral is said to be "ultraviolet divergent". In fact, if we first integrate the right-hand-side over the variable k^0, holding $\vec{k}$ fixed, we can see that we obtain the same integral that we found for the mass shift of our time-independent perturbation theory computation in equation (13.14). Here that quantity appears in a manifestly relativistic form. However, it still needs a cutoff or some other way of defining the otherwise infinite integral. We will return to this important point later.

Now we can clearly see the role of and the need for the counterterm contributions on the right-hand-sides of equations (13.36) and (13.38). The counterterms can be adjusted in such a way that they also have divergent parts which then cancel the divergence coming from the last term in those equations, rendering the overall expression finite.[2] Aside from canceling the divergence, the counterterms should be determined

---

[2] This is very closely related to the discussion around equation (13.13) and (13.14) as the correction being a divergent shift of the mass of the scalar field.

by some criterion. We will formulate concrete criteria once we have developed more formalism.

The next order, $\sim \lambda^2$, contribution to the perturbative computation of the 2 point function is given by

$$
D^{(2)}(y_1, y_2) = \frac{\int [d\phi] e^{iS_0[\phi]} \phi(y_1)\phi(y_2)\frac{1}{2!}\left(iS_I^{(1)}[\phi]\right)^2}{\int [d\phi] e^{iS_0[\phi]}}
$$

$$
- \frac{\int [d\phi] e^{iS_0[\phi]} \phi(y_1)\phi(y_2) iS_I^{(1)}[\phi]}{\int [d\phi] e^{iS_0[\phi]}} \frac{\int [d\phi] e^{iS_0[\phi]} iS_I^{(1)}[\phi]}{\int [d\phi] e^{iS_0[\phi]}}
$$

$$
- \frac{\int [d\phi] e^{iS_0[\phi]} \phi(y_1)\phi(y_2)}{\int [d\phi] e^{iS_0[\phi]}} \frac{\int [d\phi] e^{iS_0[\phi]} \frac{1}{2!}\left(iS_I^{(1)}[\phi]\right)^2}{\int [d\phi] e^{iS_0[\phi]}}
$$

$$
+ \frac{\int [d\phi] e^{iS_0[\phi]} \phi(y_1)\phi(y_2)}{\int [d\phi] e^{iS_0[\phi]}} \left(\frac{\int [d\phi] e^{iS_0[\phi]} iS_I^{(1)}[\phi]}{\int [d\phi] e^{iS_0[\phi]}}\right)^2
$$

$$
+ \frac{\int [d\phi] e^{iS_0[\phi]} \phi(y_1)\phi(y_2) iS_I^{(2)}[\phi]}{\int [d\phi] e^{iS_0[\phi]}}
$$

$$
- \frac{\int [d\phi] e^{iS_0[\phi]} \phi(y_1)\phi(y_2)}{\int [d\phi] e^{iS_0[\phi]}} \frac{\int [d\phi] e^{iS_0[\phi]} iS_I^{(2)}[\phi]}{\int [d\phi] e^{iS_0[\phi]}} \tag{13.40}
$$

We see that the complexity of the perturbative calculation increases rapidly with the order. Before we set out to evaluate these contributions, we will pause to develop a more elegant approach. The first step will be the use of Feynman diagrams which we outline in the next section.

13.3   Feynman Diagrams

The terms in the perturbative expansion have an elegant diagrammatic representation where the individual contributions are called Feynman diagrams. This representation can help us to visualize the perturbation expansion and it can also help us to organize it. In the diagrams, a 2 point function $\Delta(x_1, x_2)$ is represented as a solid line joining the points x_1 and x_2. The diagram also has vertices which are localized at points and are each associated with a factor of λ. Such a vertex emits four lines. The counterterms

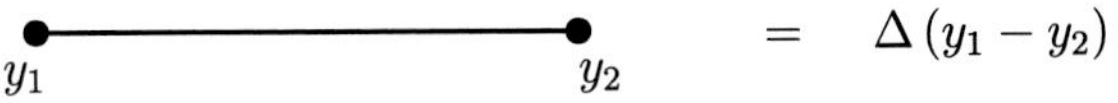

$$= \quad \Delta\,(y_1 - y_2)$$

Fig. 13.1 The solid line in a Feynman diagram represents a propagator $\Delta(y_1, y_2)$

also are treated as vertices, the ones with z_{m^2} and z emit two lines and z_λ emits four lines. The line, as a component of a Feynman diagram, is depicted in Fig. 13.1. The vertices are depicted in Fig. 13.2.

In order to evaluate the order k contribution to the n point function

$$< \mathcal{O}|T\phi(y_1)\ldots\phi(y_n)|\mathcal{O} >$$

the use of Wick's theorem is equivalent to the following diagrammatic procedure:

1. Be prepared that there can be a large number of diagrams. The number of diagrams increases rapidly with the order in perturbation theory and with the number of external points $y_1, \ldots, y_n$.
2. Plot the points $y_1, \ldots y_n$ around the boundary of the diagrams. Each of the points $y_1, \ldots y_n$ can be the endpoint of a single line.
3. Draw the internal vertices at points $x_1, \ldots, x_k$ in the centre of each of the diagrams. Each of these vertices of type $\lambda\phi^4$ or $\lambda z_\lambda^{(\ell)}\phi^4$ will be the endpoint of four lines. Each counterterm vertex $z^{(\ell)}\partial_\mu\phi\partial^\mu\phi + z_{m^2}^{(\ell)}m^2\phi^2$ is the endpoint of two lines.
4. Then draw lines between pairs of points, remembering that four lines end at each vertex and two or four at each counterterm vertex. Also one line ends at each external point $y_1, \ldots, y_n$. Each distinct possible way of drawing lines with these criteria correspond to a pairing in Wick's theorem. Summing over all of the possible ways of drawing the lines is equivalent to the sum over the pairings in Wick's theorem. We must count each distinct pairing once. There are often a number of pairings which result in the same diagram and which will therefore make an identical contribution when the diagram is evaluated. We must record this multiplicity as it will become a factor in the contribution of that diagram.

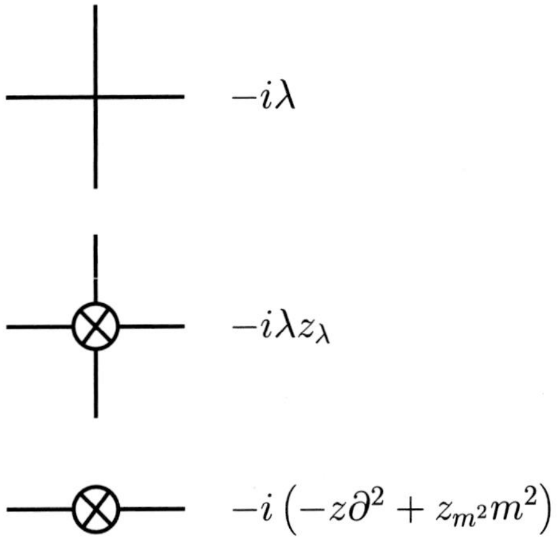

Fig. 13.2 The vertices which are used in a Feynman diagram are depicted. The top figure is the interaction vertex and the second and third figures are counterterm vertices

$$-i\lambda$$

$$-i\lambda z_\lambda$$

$$-i\left(-z\partial^2 + z_{m^2}m^2\right)$$

5. We can drop all diagrams which have sub-diagrams which are not connected to any external points. These are called "vacuum diagrams". The fact that vacuum diagrams cancel is a consequence of a theorem called the linked cluster theorem. We give a proof of this theorem and this important consequence of it later. We have already seen an example of this useful theorem in action when we computed $D^{(1)}(y_1, y_2)$. The role of the contributions to $D_N^{(1)}(y_1, y_2)$ in equation (13.35) was to cancel the contributions which we can see by inspection correspond to vacuum diagrams that arise in equation (13.34). Then, the terms that remain in the contributions to $D^{(1)}(y_1, y_2)$ do not contain any vacuum diagrams.
6. We associate a propagator $\Delta(z_i, z_j)$ with each line in the Feynman diagram. z_i and z_j are the endpoints of the line.
7. If there are k interaction vertices in the diagram, we multiply by the factor $\frac{1}{k!}\left(-i\frac{\lambda}{4!}\right)^k$ and integrate over the space–time point x_i where the vertex is located. We do the same for the counterterm vertices with the appropriate coefficients.
8. We multiply the contribution of each distinct diagram by an integer which is the number of pairings which yield the same diagram. This multiplicity will cancel all or part of the $\frac{1}{k!}$ and the $\frac{1}{4!}$ factors. The leftover factors are integers occurring in the denominator and they are called "symmetry factors". As well as counting the number of Wick pairings which give the same diagram, the symmetry factor can be related to the order of the discrete symmetry group of the diagram itself.

Now, let us follow these rules to re-evaluate the order λ correction to the 2 point function $D^{(1)}(y_1, y_2)$ which we have already found directly from Wick's theorem. We begin by plotting the two external points and the various vertices as in Fig. 13.3 which has the 4 point vertex and Fig. 13.4 which has the 2 point counterterm vertices.

Then, we begin drawing connections between the external points and the vertices in all possible ways. These connections lead to the Feynman diagrams in Figs. 13.5, 13.6, 13.7 and 13.8. These correspond to the computations that we have already done which are summarized in equation (13.34).

Equation (13.34) has four terms. The first term on the right-hand-side of equation (13.34) corresponds to the Feynman diagram in Fig. 13.5. The second term corresponds to the Feynman diagram in Fig. 13.6, the third term to Fig. 13.7 and the fourth term to Fig. 13.8.

This accounts for the contributions in equation (13.34). There are additional contributions contained in equation (13.35). We will not consider Feynman diagrams corresponding to them. Instead we will simply note that their role is to cancel the

$$y_1 \qquad\qquad x \qquad\qquad y_2$$

Fig. 13.3 The beginning of the Feynman diagrams with two external points y_1 and y_2 and a single internal 4 point interaction vertex is depicted. The points are to be paired in all possible ways and the pairs will be inserted as internal lines which correspond to propagators. The pairings lead to the diagrams in Figs. 13.5 and 13.7

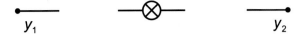

Fig. 13.4 The beginning of the Feynman diagrams with two external points y_1 and y_2 and a single internal 2 point counterterm vertex is depicted. The points are to be paired in all possible ways and the pairs will be inserted as internal lines which correspond to propagators. The pairings lead to the diagrams in Figs. 13.6 and 13.8

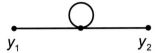

Fig. 13.5 This is a connected Feynman diagram containing the 4 point interaction vertex. It is connected and it accounts for twelve pairings of the points in Fig. 13.3

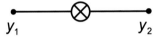

Fig. 13.6 This is a connected diagram which contains the counterterm 2 point vertex. It is connected and it accounts for two pairings of the points in Fig. 13.4

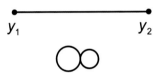

Fig. 13.7 This is diagram contains the 4 point vertex and it has a piece which is not connected to any of the external lines. It is called a vacuum diagram. There are three different pairings of the points in Fig. 13.3 which lead to this diagram

contributions of the Feynman diagrams in Figs. 13.7 and 13.8. That is, they precisely cancel all of the contributions which have vacuum diagrams. We will eventually prove this as part of a larger discussion of the linked cluster theorem. Indeed, the role of all of the contributions due to the denominator in the ratio of functional integrals that we began with is to cancel the Feynman diagrams from the numerator which contain vacuum diagrams. With this theorem, we can ignore all of Feynman diagrams originating from the denominator and simultaneously drop all of the contributions from the numerator which lead to vacuum diagrams.

That leaves us with the Feynman diagrams in Figs. 13.5 and 13.6 which are given by the first and second terms on the right-hand-side of equation (13.36), respectively.

13.4 Simplifications of Feynman Diagrams

There are a number of considerations which can decrease the number of Feynman diagrams which must be computed in order to study a given correlation function. For example, as we have already noted at the end of the previous section, the linked

Fig. 13.8 This diagram contains the counterterm 2 point vertex. There is one pairing of the points in Fig. 13.4 which lead to this diagram. This is diagram has a piece which is not connected to any of the external lines and, like the diagram in Fig. 13.7, it is called a vacuum diagram

cluster theorem allows us to ignore all Feynman diagrams which contain vacuum diagrams as well as all contributions from the denominators.

Another, sometimes considerable simplification of the computation of a multi-point function is to consider only the connected Feynman diagrams which contribute to it. A connected Feynman diagram is a diagram which is a connected graph. Within such a diagram, one can trace a path between any two points, either internal or external that continuously follows propagators. We denote the n point function by $\Gamma(y_1, ..., y_n)$ and the connected n point function by $\Gamma_C(y_1, ..., y_n)$. $\Gamma(y_1, ..., y_n)$ is given by the sum of all Feynman diagrams (excluding vacuum diagrams) with n external points and $\Gamma_C(y_1, ..., y_n)$ is given by the sum of all connected Feynman diagrams with n external points.

There is a formal definition, which does not rely on perturbation theory, of a connected n point function and a simple way of constructing the full n point function from connected n- and lower-point functions and vice-versa. If the generating functional for n point functions is $Z[J]$, that is,

$$\Gamma(y_1, ..., y_n) = \frac{1}{i}\frac{\partial}{\partial J(y_1)} \cdots \frac{1}{i}\frac{\partial}{\partial J(y_n)} Z[J]_{J=0}$$

then the generating functional for connected n point functions is given by the logarithm, $W[J] = \ln Z[J]$, so that

$$\Gamma_C(y_1, ..., y_n) = \frac{1}{i}\frac{\partial}{\partial J(y_1)} \cdots \frac{1}{i}\frac{\partial}{\partial J(y_n)} \ln Z[J]_{J=0}$$

Using this relationship or its inverse, $Z[J] = e^{W[J]}$, we can easily find $\Gamma(y_1, ..., y_n)$ if $\Gamma_C(y_1, ..., y_m)$ for $m = 2, 4, ..., n$ are known. If we focus on connected correlation functions, we only need to consider connected Feynman diagrams. We give a proof of this fact in Sect. 14.4.2. In many cases, this can decrease the total number of diagrams that we need to study.

As a third simplification of the diagrammatic technique, there is another classification of diagrams that is useful to consider. A connected diagram is said to be irreducible if it cannot be rendered disconnected by removing a single internal or external line. (An internal line is a line which connects two vertices in the interior of a diagram. An external line has one or both endpoints on external points.) We denote the irreducible n point correlation function by $\Gamma_I(y_1, ..., y_n)$. We will study the combinatorics of how irreducible correlation functions are related to connected correlation functions in Sect. 14.4.4. We will find a systematic way of obtaining these irreducible functions from a generating functional. It will also give a systematic relationship between $\Gamma_I(y_1, ..., y_n)$ and $\Gamma_C(y_1, ..., y_n)$ which will allow us to reconstruct the connected functions $\Gamma_C(y_1, ..., y_n)$ or the full n point functions $\Gamma(y_1, ..., y_n)$ once the irreducible functions are known.

An important example is the 2 point function. Recall that we denoted the 2 point function as

$$D(y_1, y_2) \equiv \Gamma(y_1, y_2) = \int \frac{dk}{(2\pi)^4} e^{ik(y_1 - y_2)} D(k)$$

By its combinatorial definition, the irreducible 2 point function is the negative of the inverse of the two-point function.

$$\Gamma_I(y_1, y_2) = -\Gamma^{-1}(y_1, y_2) = -i(-\partial^2 + m^2)\delta(y_1, y_2) + \Pi(y_1, y_2) \quad (13.41)$$

Here, the first term on the right hand side is the negative of the inverse of the propagator and the second term, $\Pi(y_1, y_2)$, is the sum of irreducible Feynman diagrams with two external lines. Its Fourier transform is denoted by

$$\Pi(y_1, y_2) = \int \frac{dk}{(2\pi)^4} e^{ik(y_1 - y_2)} \Pi(k)$$

Equation (13.41) implies that the 2 point function is given by the geometric series

$$D(k) = \Delta(k) + \Delta(k)\Pi(k)\Delta(k) + \Delta(k)\Pi(k)\Delta(k)\Pi(k)\Delta(k) + ... \quad (13.42)$$

which can be summed to get

$$D^{-1}(k) = \Delta^{-1}(k) - \Pi(k) \quad (13.43)$$

If we compute an irreducible function, then we only need to consider Feynman diagrams which are irreducible. For example, the full perturbative expression for the 2 point function to order λ^2 is given by the expansions of the functional integrals in equation (13.40). Using Wick's theorem would generate a large number of contributions. But, if we consider only the irreducible diagrams, they are given by

$$\begin{aligned}
\Pi^{(2)}(y_1, y_2) = &-\frac{\lambda^2}{6}\Delta^3(y_1, y_2) - \frac{\lambda^2}{4}\int dx\delta(y_1 - y_2)\Delta^2(y_1, x)\Delta(x, x) \\
&- i\lambda z_\lambda^{(1)}\delta(y_1 - y_2)\Delta(y_1, y_1) \\
&- i\lambda\delta(y_1 - y_2)\int dx\Delta(y_1, x)\left(z^{(1)}\partial^2 - z_{m^2}^{(1)}m^2\right)\Delta(x, y_2) \\
&+ i(z^{(2)}\partial^2 - z_{m^2}^{(2)}m^2)\delta(y_1 - y_2) \quad (13.44)
\end{aligned}$$

Another considerable simplification of the procedure, which is already hinted at in equation (13.42), is to draw the diagrams in momentum space. The diagrams themselves are identical whether in coordinate space or in momentum space. What differs is how the components of the diagrams are labeled and in how they correspond to a mathematical expression.

The momentum space diagrams will be contributions to the Fourier transform of the n point function, connected n point function or the irreducible n point function. These Fourier transforms are defined by the expressions

$$\Gamma(y_1, \ldots, y_n)$$
$$= \int \frac{dk_1}{(2\pi)^4} \cdots \frac{dk_n}{(2\pi)^4} e^{ik(y_1+y_2+\ldots+y_n)} \Gamma(k_1, \ldots, k_n)(2\pi)^4 \delta(k_1 + \ldots + k_n)$$

$$(13.45)$$

$$\Gamma_C(y_1, \ldots, y_n)$$
$$= \int \frac{dk_1}{(2\pi)^4} \cdots \frac{dk_n}{(2\pi)^4} e^{ik(y_1+y_2+\ldots+y_n)} \Gamma_C(k_1, \ldots, k_n)(2\pi)^4 \delta(k_1 + \ldots + k_n)$$

$$(13.46)$$

$$\Gamma_I(y_1, \ldots, y_n)$$
$$= \int \frac{dk_1}{(2\pi)^4} \cdots \frac{dk_n}{(2\pi)^4} e^{ik(y_1+y_2+\ldots+y_n)} \Gamma_I(k_1, \ldots, k_n)(2\pi)^4 \delta(k_1 + \ldots + k_n)$$

$$(13.47)$$

Here, we have anticipated the translation invariance of the n point functions under simultaneous translations of all of the points. This symmetry results in conservation of momentum which results in the Dirac delta functions which occur in the integrands. The functions $\Gamma(k_1, \ldots, k_n)$ etc. do not carry this delta function around with them. However, they only make sense when they conserve energy and momentum. For this reason, they will always be assumed to have momentum arguments which add to zero as four-vectors, $k_1^\mu + \ldots + k_n^\mu = 0$. For example, the 2 point function is

$$\Gamma(k, -k) = D(k)$$

The two momenta add to zero. Of course, since we use the two point function a lot, we have given it a symbol, which in momentum space is $D(k)$.

Also, notice that, like the propagator, $\Delta(k) = \frac{-i}{k^2+m^2-i\epsilon}$, which depends only on k^2, due to Lorentz invariance, the 2 point function and the irreducible 2 point function are also functions of k^2. It is easy to write the 2 point function in terms of the irreducible 2 point function as

$$D(k^2) = \frac{-i}{k^2 + m^2 + i\Pi(k^2)} \tag{13.48}$$

Because of where it occurs in this equation, as a (k^2-dependent) correction to the terms k^2 and m^2, $i\Pi(k^2)$ is sometimes called the "self-energy" of the scalar field.

Then we can reformulate the Feynman rules in momentum space as follows:

1. To compute $\Gamma(k_1, \ldots, k_n)$ or $\Gamma_C(k_1, \ldots, k_n)$ or $\Gamma_I(k_1, \ldots, k_n)$ to a given order in perturbation theory, we draw all of the Feynman diagrams or we draw all of the connected Feynman diagrams or we draw all of the irreducible Feynman diagrams with n external lines and some number of 4 point vertices or four and 2 point counterterm vertices in the interior of the diagram. The number of vertices plus the number of counterterm vertices times the order of each counterterm vertex should add up to the order in perturbation theory that we are interested in.

2. We drop all vacuum diagrams.

3. We give each of the lines, both internal and external, an orientation. The orientation will determine the sign of the momentum that is flowing through the line. We label the external lines by the momenta $k_1, k_2, \ldots, k_n$ which are all oriented so that the four-momenta k_i are flowing into the body of the diagram.

4. We assign momenta to all of the internal lines of the diagram in such a way that it is consistent with the conservation of momenta at all of the internal vertices and internal counterterm vertices. This momentum assignment must also be consistent with the values of the momenta of the external lines. The overall momentum conservation, $k_1^{\mu} + k_2^{\mu} + \ldots + k_n^{\mu} = 0$ will occur automatically due to conservation of momentum at all of the internal vertices.

5. To each internal line we ascribe a factor of $\Delta(\ell)$ where ℓ is the momentum of that internal line.

6. If we are computing $\Gamma(k_1, \ldots, k_n)$ or $\Gamma_C(k_1, \ldots, k_n)$, we must also ascribe a factor of $\Delta(\ell)$, where ℓ is the momentum, to each external line. If we are computing $\Gamma_I(k_1, \ldots, k_n)$, these propagators for the external lines are omitted.

7. To each vertex, we ascribe a factor of $-i\lambda$ if it is a 4 point interaction vertex, $-i\lambda z_{\lambda}^{(k)}$ if it is a 4 point counterterm vertex, $-i[z^{(k)}\ell^2 + z_{m^2}^{(k)}]$ (where ℓ is the momentum of the line into which the counterterm is inserted) for the 2 point counterterm vertices.

8. We then integrate $\int \prod_i \frac{d\ell_i}{(2\pi)^4} \ldots$ for each of the independent momentum of the internal lines. There is an independent four-momentum for each closed loop in the diagram. Their number is sometimes called the number of loops. The perturbation theory in the coupling constant turns out to be closely related to a perturbative expansion in the number of loops in the diagrams. The sum over diagrams contribution to a given n point function where the diagrams have no internal loops at all is called the "tree approximation". The sum of diagrams where each diagram has one internal loop is called the "one-loop approximation".

9. We take into account symmetry factors of the diagrams.

Let us take these rules and revisit the computation of the 2 point function. We need only compute the irreducible part $\Pi(k)$ and then we can use it to construct the full 2 point function using equation (13.48). The first correction is of order λ. The momentum space Feynman diagrams corresponding to these two contributions are depicted in Fig. 13.9. They are equivalent to the expression

$$\Pi^{(1)}(k) = -i\frac{\lambda}{2} \int \frac{d\ell}{(2\pi)^4} \Delta(\ell) - i\left(z^{(1)}k^2 + z_{m^2}^{(1)}m^2\right) \qquad (13.49)$$

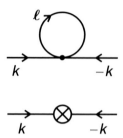

Fig. 13.9 The momentum space Feynman diagrams which contribute to the self-energy $\Pi(k)$ to order λ in perturbation theory are depicted. The first and second diagrams in the figure correspond to the first and second terms on the right-hand-side of equation (13.49), respectively

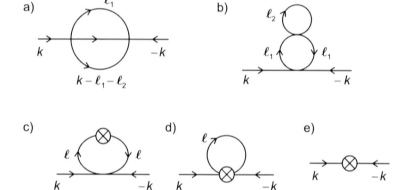

Fig. 13.10 The momentum space Feynman diagrams which contribute to the self-energy $\Pi(k)$ to order λ^2 in perturbation theory are depicted. The labels a, b, c, d, e correspond to the contributions which are similarly labeled in equation (13.50)

At the second order we find

$$
\begin{aligned}
\Pi^{(2)}(k) = &-\frac{\lambda^2}{6}\int\frac{\ell_1}{(2\pi)^4}\frac{\ell_2}{(2\pi)^4}\Delta(\ell_1)\Delta(\ell_2)\Delta(k-\ell_1-\ell_2) && (a) \\
&-\frac{\lambda^2}{4}\int\frac{d\ell_1}{(2\pi)^4}\Delta^2(\ell_1)\int\frac{d\ell_2}{(2\pi)^4}\Delta(\ell_2) && (b) \\
&-\frac{\lambda}{2}\int\frac{d\ell}{(2\pi)^4}\Delta^2(\ell)\left(z^{(1)}\ell^2+z^{(1)}_{m^2}m^2\right) && (c) \\
&-i\lambda z^{(1)}_\lambda\int\frac{d\ell}{(2\pi)^4}\Delta(\ell) && (d) \\
&-i\left(z^{(2)}k^2+z^{(2)}_{m^2}m^2\right) && (e) \quad (13.50)
\end{aligned}
$$

The momentum space Feynman diagrams corresponding to these contribution are depicted in Fig. 13.10 which are labeled a, b, c, d, e, for each of the contributions that are detailed in equation (13.50).

Fig. 13.11 The momentum space Feynman diagram which contributes to the irreducible 4 point function $\Gamma_I(k_1, k_2, k_3, k_4)$ at order λ in perturbation theory is depicted. There is only one diagram. It corresponds to the contribution in equation (13.51)

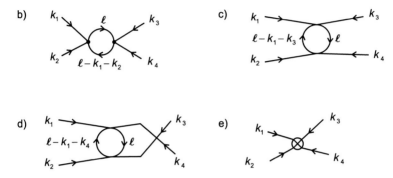

Fig. 13.12 The momentum space Feynman diagrams which contribute to the irreducible 4 point function $\Gamma_I(k_1, k_2, k_3, k_4)$ at order λ^2 in perturbation theory are depicted. The labels b, c, d, e correspond to the contributions which are similarly labeled in equations (13.52)–(13.55)

What is left to do is to evaluate the integrands and to perform the integrations in equations (13.49) and (13.50). We will discuss techniques for doing the these integrals in a later section of this chapter.

Let us consider, as another example, the application of the momentum space Feynman rules to the computation of the irreducible 4 point function $\Gamma_I(k_1, k_2, k_3, k_4)$. We will study its expansion in perturbation theory up to order λ^2.

First of all, if we put λ to zero, the irreducible 4 point function vanishes (as do all connected and irreducible n point functions with $n > 2$). So $\Gamma_I(k_1, k_2, k_3, k_4)$ must be at least of order λ. At order λ it obtains a contribution from only one Feynman diagram which is depicted in Fig. 13.11 and the contribution is given in equation (13.51) below.

The Feynman diagrams which contribute at order λ^2 are depicted in Fig. 13.12 and the contribution to the 4 point function from the diagrams labelled b, c, d, e are in equations (13.52), (13.53), (13.54) and (13.55), respectively.

$$\Gamma_I(k_1, k_2, k_3, k_4)$$

$$= -i\lambda \tag{13.51}$$

$$-\frac{\lambda^2}{2}\int\frac{d\ell}{(2\pi)^4}\Delta(\ell)\Delta(\ell - k_1 - k_2) \tag{13.52}$$

$$-\frac{\lambda^2}{2}\int\frac{d\ell}{(2\pi)^4}\Delta(\ell)\Delta(\ell - k_1 - k_3) \tag{13.53}$$

$$-\frac{\lambda^2}{2}\int\frac{d\ell}{(2\pi)^4}\Delta(\ell)\Delta(\ell - k_1 - k_4) \tag{13.54}$$

$$-i\lambda^2 z_\lambda^{(1)}$$

$$+\ldots \tag{13.55}$$

The ellipses in the equation above denotes contributions of order λ^3 and higher orders. The leading correction in (13.51) is simply proportional to the coupling constant. The next order λ^2 contributions are given in equations (13.52)–(13.55). They contain a one-loop Feynman integral. The contribution in (13.55) is from the only counterterm that contributes at order λ^2.

Equations (13.51)–(13.55) for the irreducible 4 point function and the irreducible 2 point function that is given in equations (13.49) and (13.50) are our computations of the lower order correlation functions up to order λ^2. Both expressions contain Feynman integrals which are yet to be evaluated and counterterms which are yet to be determined. We will do both in the following sections.

13.5   Computation of a One-Loop Feynman Integral

The Feynman integrals corresponding to the one-loop Feynman diagrams in Fig. 13.12 and which contribute to the 4 point function in equations (13.52)–(13.55) are all of the form

$$\mathcal{I} = \int\frac{d\ell}{(2\pi)^4}\Delta(\ell)\Delta(\ell - k) \tag{13.56}$$

or, explicitly, with the propagators inserted

$$\mathcal{I} = \int\frac{d\ell}{(2\pi)^4}\frac{-i}{\ell^2 + m^2 - i\epsilon}\frac{-i}{(k - \ell)^2 + m^2 - i\epsilon} \tag{13.57}$$

for various values of k^μ. We will now discuss how we can do the integral over the loop momentum ℓ^μ in equation (13.57).

This integration has several subtleties, beginning with the fact that the integral itself diverges. The divergence comes from two sources, one is the large ℓ regime. The other is due to the fact that the denominators in the integrand have zeros (or almost zeros, the $i\epsilon$ saves us there, but we must understand how this occurs in more detail).

13.5.1 Dimensional Regularization

In order to make sense of the ultraviolet divergence which occurs in integrals such as those in equation (13.57), we must introduce a regularization. We have already done this in our first pass at perturbative quantum field theory in equation (13.13). In that case we introduced a large momentum cutoff which rendered the integral finite, albeit cutoff-dependent. We could do the same for the integral in equation (13.57).

As intuitive and logically clear as the process of imposing an explicit high momentum cutoff would be, we will choose to employ a different technique. For the quantum field theory that we are interested in here, a very elegant and systematic way to regulate the ultraviolet divergences is to consider the dimension of the spacetime in which we are working to be a parameter. We will say that there are 2ω spacetime dimensions made up of $2\omega - 1$ space dimensions and one time dimension. Then we consider the situation where 2ω is small enough that the Feynman integrals that we encounter converge. We evaluate the integrals and we discover that we can in fact treat 2ω as a parameter which can be varied continuously. Then, we treat the integral as a function of a complex variable, 2ω. Its analytic continuation to a function of the complex variable 2ω has isolated pole singularities on the real axis. The original four space–time dimensional quantum field theory is recovered by taking the limit where $\omega \to 2$. The result of the integral will indeed still diverge in that limit and we will eventually need to find a way to deal with the divergence.

In 2ω space–time dimensions, the Feynman integral from equation (13.57) becomes

$$\mathcal{I} = \mu^{4-2\omega} \int \frac{d^{2\omega}\ell}{(2\pi)^{2\omega}} \frac{-i}{\ell^2 + m^2 - i\epsilon} \frac{-i}{(k-\ell)^2 + m^2 - i\epsilon} \tag{13.58}$$

In order to preserve the scaling dimension of the integral, we have introduced a pre-factor $\mu^{4-2\omega}$ with a parameter μ which has the scaling dimension of a momentum or energy or mass. Doing this will allow us to keep the coupling constant λ dimensionless. Now, we will do the integral for a value of the dimension 2ω where the integral converges and define it everywhere else by analytic continuation in the parameter 2ω. Some formulae which will help us to do the integral are summarized in Sect. 13.8.3.

13.5.2 Wick Rotation

The integral that we are left to do in equation (13.58) is a volume integral over Minkowski space–time. Such an integral has some subtleties. A way of avoiding the subtleties is to now perform a "Wick rotation". The Wick rotation replaces ℓ^0 by $i\ell^0$ and k^0 by ik^0 wherever they appear in the integrand in equation (13.58). The integration measure dk is also replaced by $i\,dk$. The result is

$$\mathcal{I} = -i\mu^{4-2\omega} \int \frac{d^{2\omega}\ell}{(2\pi)^{2\omega}} \frac{1}{\ell^2 + m^2} \frac{1}{(k-\ell)^2 + m^2} \tag{13.59}$$

where, now, $\ell^2 = (\ell^0)^2 + \vec{\ell}^2$ and $k^2 = (k^0)^2 + \vec{k}^2$ are the moduli squared of vectors in 2ω-dimensional Euclidean space. We will undo this Wick rotation later, once we have done the integral.

The reason why we can do the Wick rotation has to do with the places where singularities appear in the propagator in Minkowski space. To see this, consider the partial fractions decomposition of the propagator into two functions of k^0 with simple poles in the complex k^0-plane,

$$
\begin{aligned}
\Delta(k) &= \frac{1}{k^2 + m^2 - i\epsilon} \\
&= \frac{1}{2\sqrt{\vec{k}^2 + m^2}} \left[\frac{1}{\sqrt{\vec{k}^2 + m^2} - i\epsilon + k_0} + \frac{1}{\sqrt{\vec{k}^2 + m^2} - i\epsilon - k_0} \right]
\end{aligned}
\tag{13.60}
$$

The poles occur only in the second and fourth quadrants of the complex k^0-plane. Similarly, $\Delta(k - \ell)$ has singularities only in the second and fourth quadrants of the complex $k^0 - \ell^0$-plane.

We can therefore replace both k^0 and $k^0 - \ell_0$ by $e^{i\phi}k^0$ and $e^{i\phi}(k^0 - \ell_0)$ and dk^0 by $e^{i\phi}dk^0$ without changing the value of the integral if $0 \le \phi \le \pi/2$. The Wick rotation is then accomplished by putting $\phi = \pi/2$ so that $e^{i\phi} = i$.

The result is the expression in equation (13.59). We can undo the Wick rotation by performing the reverse of the above procedure once we have performed the integration. We will discuss some details about Wick rotations in Sect. 13.5.8.

13.5.3 Feynman Parameters

We are left with the integral in equation (13.59) which is now a volume integral over 2ω-dimensional Euclidean space.

The next step in doing this integral will be to combine the denominators by using the Feynman parameter formula, the general case of which is derived in Sect. 13.8.2. The purpose of this step is to better position ourselves to take advantage of the symmetry of the integration under rotations in 2ω-dimensional space. The formula that is appropriate to our case is

$$
\frac{1}{AB} = \int_0^1 d\alpha \frac{1}{[\alpha A + (1 - \alpha)B]^2}
$$

Application of this formula to the integral in equation (13.58) results in the expression

$$
\mathcal{I} = -i\mu^{4-2\omega} \int_0^1 d\alpha \int \frac{d^{2\omega}\ell}{(2\pi)^{2\omega}} \frac{1}{[\alpha(\ell^2 + m^2) + (1 - \alpha)((k - \ell)^2 + m^2)]^2}
$$

$$= -i\mu^{4-2\omega} \int_0^1 d\alpha \int \frac{d^{2\omega}\ell}{(2\pi)^{2\omega}} \frac{1}{[\ell^2 + m^2 + (1-\alpha)k^2 - 2k \cdot \ell(1-\alpha)]^2}$$

Now, we change variables, $\ell \to \ell + (1-\alpha)k$. We get

$$\mathcal{I} = -i\mu^{4-2\omega} \int_0^1 d\alpha \int \frac{d^{2\omega}\ell}{(2\pi)^{2\omega}} \frac{1}{[\ell^2 + m^2 + \alpha(1-\alpha)k^2]^2} \qquad (13.61)$$

The advantage of using Feynman parameters here was to expose the rotation symmetry of the integrand by getting the denominator as a function of ℓ^2. This has been achieved in equation (13.61) at the expense of introducing an extra integration over α.

13.5.4 Integration in 2ω-Dimensions

We can now use use the dimensional regularization formula which is derived in Sect. 13.8.3 to do the integral over the loop momentum ℓ in equation (13.61). The result is

$$\mathcal{I} = -i\mu^{4-2\omega} \frac{\Gamma[2-\omega]}{(4\pi)^\omega \Gamma[2]} \int_0^1 d\alpha [\alpha(1-\alpha)k^2 + m^2]^{\omega-2} \qquad (13.62)$$

The integral formula which we applied here assumes that 2ω is small enough that the integral exists (here, $\omega < 2$). The result (13.62) is a function of the space–time dimension, 2ω, and the integral over the Feynman parameter α remains to be done. Notice that the value of the integral depends on ω in such a way that we can now think of ω as a continuously varying real number. We can also think of it as a complex number. As such, we can analytically continue the expression in equation (13.62) to the entire complex ω-plane. When we do this we see that its only singularities in the complex plane come from singularities of the Euler gamma-function $\Gamma[2 - \omega]$. (The integrand and the remaining integral over α are non-singular for any finite value of a complex number ω.) The function $\Gamma[2 - \omega]$ as a holomorphic function of complex ω has simple poles on the positive real axis located at $\omega = 2, 3, 4, \ldots$, which are dimensions where the original integral would have diverged.[3]

As a function of a complex variable, the Euler gamma function has poles at the negative integers,

$$\Gamma(z - n) \sim \frac{1}{z}, \quad n = 0, 1, 2, \ldots$$

---

[3] Interestingly, this excludes the case of odd spacetime dimensions. Indeed if $2\omega = 5$ this one-loop integral is finite. Of course the scalar field theory is neither finite nor is it renormalizable in five dimensions. The divergences do appear at higher orders of perturbation theory.

and, in particular, in the region of direct interest to us, it has the behaviour

$$\Gamma[2 - \omega] = \frac{1}{2 - \omega} - \gamma + \dots \tag{13.63}$$

where γ is the Euler–Mascheroni constant (see Sect. 13.8.1 and the ellipses are corrections which go to zero as $\omega \to 2$. This defines the singularity of our Feynman integral as $2\omega \to 4$ as a simple pole $\sim \frac{1}{2-\omega}$).

13.5.5 Asymptotic Expansion at $2\omega \sim 4$

We can study the integral (13.62) in the regime where $2\omega \sim 4$. In that region, we use the asymptotic expansion of the Euler gamma function (13.63) to get

$$\mathcal{I} = -\frac{i}{16\pi^2} \frac{1}{2 - \omega} + \frac{i}{16\pi^2} \int\limits_0^1 d\alpha \ln \left[\frac{\alpha(1 - \alpha)k^2 + m^2}{4\pi\mu^2 e^{-\gamma}} \right] \tag{13.64}$$

where we have dropped all of the terms in the expansion which go to zero as $\omega \to 2$.
   The remaining integral over α is elementary and it can easily be done. The result is

$$\begin{aligned}
\mathcal{I} = & -\frac{i}{16\pi^2} \frac{1}{2 - \omega} \\
& + \frac{i}{16\pi^2} \ln \frac{m^2}{4\pi\mu^2 e^{2-\gamma}} + \frac{i}{16\pi^2} \frac{\sqrt{4m^2 + k^2}}{\sqrt{k^2}} \ln \frac{\sqrt{4m^2 + k^2} + \sqrt{k^2}}{2m}
\end{aligned} \tag{13.65}$$

   However, particularly for doing the inverse Wick rotation, it is actually more enlightening to leave the contribution in integral form as it appears in equation (13.64).
   We see in the expressions in equations (13.64) and (13.65) that the singular part, the pole term $\sim \frac{1}{2-\omega}$ has separated into a simple additive contribution. Also, the momentum k^2 in both of these equations is still Euclidean.

13.5.6 Inverse Wick Rotation

In the expression (13.65), we can now go back to Minkowski spacetime by doing an the inverse Wick rotation. That is, we put $k^0 \to -ik^0$ so that k^2 now has the appropriate form for the energy-momentum four vector, $k^2 = \vec{k}^2 - (k^0)^2$.

$$\mathcal{I} = -\frac{i}{16\pi^2} \frac{1}{2 - \omega} + \frac{i}{16\pi^2} \int\limits_0^1 d\alpha \ln \left[\frac{\alpha(1 - \alpha)k^2 + m^2 - i\epsilon}{4\pi\mu^2 e^{-\gamma}} \right] \tag{13.66}$$

Notice that, as well as replacing the Euclidean k^2 by its Minkowski space version, we have re-introduced the "$i\epsilon$" inside the square root and inside the logarithm. Here, we should think of the $i\epsilon$ as recording the direction of the inverse Wick rotation so that the resulting expression is analytic in the first and third quadrants of the complex k^0-plane. This is related to the analyticity of correlations functions in complex time that we have discussed in Sect. 7.9.

The reason why we need to be careful about this is that k^2 can now be negative and therefore the argument of the logarithm can also be negative. This happens when $k^2 < -4m^2$. The presence of $i\epsilon$ helps us to define the logarithm in that regime. If we consider the integral as a function of k^2 in the complex k^2-plane, since $0 \leq \alpha(1-\alpha) \leq 1/4$, the argument of the logarithm is real and positive for all real values in the range $k^2 > -4m^2$. It has a branch point singularity at $k^2 = -4m^2$. Then, it has a branch cut on the negative real k^2-axis on the segment $k^2 \in (-\infty, -4m^2)$.

13.5.7   The Mass Tadpole

We have also encountered the Feynman integral

$$\mathcal{I}_1 = \int \frac{dk}{(2\pi)^4} \frac{-i}{k^2 + m^2 - i\epsilon} \tag{13.67}$$

which occurred as a sub-diagram in both the 2 point and 4 point functions. It is a loop diagram where a single propagator leaves and returns to the same vertex. Such a diagram is called a "tadpole". The integral over k is ultraviolet divergent. If we regulate it using dimensional regularization and then follow the procedure for computing one-loop integrals that we have outlined above, it is found to have the value near four space–time dimensions

$$\mathcal{I}_1 = \frac{\Gamma[1-\omega]}{(4\pi)^\omega} \frac{(\mu^2)^{2-\omega}}{[m^2]^{1-\omega}} = \frac{m^2}{16\pi^2} \Gamma[-1+(2-\omega)] \left[\frac{4\pi\mu^2}{m^2} \right]^{2-\omega} \tag{13.68}$$

Then we use the asymptotic expansion of the gamma function

$$\Gamma[-1+(2-\omega)] = \frac{1}{-1+(2-\omega)} \Gamma[2-\omega]$$

$$= (-1 - (2-\omega) - \ldots) \left(\frac{1}{2-\omega} + \gamma + \ldots \right)$$

$$= -\frac{1}{2-\omega} - (\gamma + 1) + \ldots$$

to obtain

$$\mathcal{I}_1 = -\frac{1}{2-\omega} \frac{m^2}{16\pi^2} + \frac{m^2}{16\pi^2} \ln \left[\frac{m^2}{4\pi\mu^2 e^{1+\gamma}} \right] \tag{13.69}$$

We again see that the divergent part has separated as a simple pole singularity at $\omega \to 2$.

13.5.8 Euclidean Quantum Field Theory

The possibility of doing a Wick rotation in an individual Feynman integral is very closely related to the fact that the time ordered correlation functions in general are analytic in certain regions of the complex time plane. As we discussed in Sect. 7.9, the 2 and higher point functions

$$< \mathcal{O}|T\phi(t_1, \vec{x}_1)\phi(t_2, \vec{x}_2) \ldots \phi(t_n, \vec{x}_n)|\mathcal{O} >$$

as functions of their relative time coordinates $t_i - t_j$, when analytically continued as function of complex variables $t_i - t_j$ must be analytic in in the first and third quadrants of the complex plane for each of these variables. In Sect. 7.9 we saw that this indeed holds true for any quantum field theory with a conserved energy and where the energies of the states of the theory are bounded from below.

The fact that we saw in equation (13.60), that the Fourier transform of the free field theory 2 point function has singularities confined to the second and fourth quadrants of the complex frequency plane is directly related to the fact that the 2 point function itself is analytic in the first and third quadrants of the complex time plane. Indeed, if we consider the full 2 point function, $D(x_1, x_2)$, time translation invariance implies that it is a function only of the difference of the times, $t = x_1^0 - x_2^0$, and analyticity implies that the function of the relative time $D(t, \vec{x}_1, \vec{x}_2)$ is analytic in the first and third quadrants of the complex t-plane. Then, consider the Fourier transform,

$$D(\omega, \vec{x}_1, \vec{x}_2) = \int_{-\infty}^{\infty} dt e^{i\omega t} D(t, \vec{x}_1, \vec{x}_2) \tag{13.70}$$

Taking analyticity into account allows us to deform the t-integration contour in equation (13.70) by rotating it through the first and third quadrants of the complex plane, $t \rightarrow e^{i\varphi}t$ if we do a compensating rotation of the frequency $\omega \rightarrow e^{-i\varphi}\omega$ into the second and fourth quadrants of the complex frequency plane. This must be done in order to keep the exponent in $e^{i\omega t}$ imaginary. The segments at infinity that we must add to the contour in order to do this are assumed to have vanishing contribution. Then we can take the rotation angle to 90°, $\varphi \rightarrow \pi/2$ and the integral in equation (13.70) becomes

$$D(-i\omega, \vec{x}_1, \vec{x}_2) = \int_{-\infty}^{\infty} dt e^{i\omega t} D(it, \vec{x}_1, \vec{x}_2) \tag{13.71}$$

This is the Wick rotated 2 point function.

We could apply a similar argument using analyticity to the Fourier transforms of any of the n point functions to obtain functions of complex frequencies, the Wick rotated n point functions. These would be obtained by replacing k_j^0 in $\Gamma(k_1, \ldots, k_n)$ by $-ik_j^0$. And the inverse Wick rotation would be implemented by the reverse process of replacing k_j^0 by ik_j^0.

One might wonder whether the Wick rotated n point functions could be computed directly by using Wick rotated Feynman rules. That would entail formulating the Feynman rules directly in Euclidean space, the four (or 2ω)-dimensional space where the metric is simply the unit matrix, $\delta_{\mu\nu}$, rather than the Minkowski space $\eta_{\mu\nu}$ and we would avoid the step of having to Wick rotate individual Feynman integrals. If at the end of computations we are actually interested in Minkowski space n point functions these could be found by inverse Wick rotation of the final expressions.

To find Wick rotated Feynman rules, we can define the quantum field theory directly in Euclidean space, by taking the functional integral and doing a Wick rotation of the time argument of the fields there. We simply take functional integral expression for the correlation functions or for the generating functional and put $t \to it$ everywhere so that the Euclidean generating functional is

$$Z_E[J] = \frac{\int [d\phi] e^{-S_E[\phi] + \int dx \phi(x) J(x)}}{\int [d\phi] e^{-S_E[\phi]}} \tag{13.72}$$

where the Euclidean action and Lagrangian density are

$$S_E[\phi] = \int dx \, \mathcal{L}_E(x) \tag{13.73}$$

$$\mathcal{L}_E(x) \frac{1}{2} \delta^{\mu\nu} \partial_\mu \phi(x) \partial_\nu \phi(x) + \frac{m^2}{2} \phi^2(x) + \frac{\lambda}{4!} \phi^4(x) \tag{13.74}$$

The Euclidean space correlation functions are obtained from the generating functional by taking functional derivatives,

$$\Gamma_E(x_1, x_2, \ldots, x_n) = \frac{\delta}{\delta J(x_1)} \frac{\delta}{\delta J(x_2)} \cdots \frac{\delta}{\delta J(x_n)} Z_E[J] \Big|_{J=0}$$

The correlation functions $\Gamma(x_1, \ldots, x_n)$ of the Minkowski space theory would be obtained from the Euclidean correlation functions $\Gamma_E(x_1, \ldots, x_n)$ by the inverse Wick rotation.

We might note that the functional integral in equation (13.74) is putatively better behaved in Euclidean space as the integrand is a damped, rather than oscillating functional of the fields. Of course, this does not help us to do the integral explicitly. If we resort to perturbation theory, Wick's theorem and the diagrammatic expansion can be done directly by studying (13.74) and using the rules for Gaussian functional integration.

13.5.9 The 2 Point and 4 Point Functions

Having done the one-loop Feynman integrals, we now have expressions for the leading corrections to the irreducible 2- and 4- point functions. In summary, the irreducible 2 point function is

$$\Pi(k) = i\frac{\lambda m^2}{32\pi^2}\frac{1}{2-\omega} - i\frac{\lambda m^2}{32\pi^2}\ln\left[\frac{m^2}{4\pi\mu^2 e^{1+\gamma}}\right] - i\left(z^{(1)}k^2 + z^{(1)}_{m^2}m^2\right) + \mathcal{O}(\lambda^2)$$

$$(13.75)$$

and the irreducible 4 point function is

$$\Gamma_I(k_1, k_2, k_3, k_4) = -i\lambda - i\lambda z_\lambda^{(1)}$$

$$+ \frac{i\lambda^2}{32\pi^2}\frac{1}{2-\omega} - \frac{i\lambda^2}{32\pi^2}\int_0^1 d\alpha \ln\left[\frac{\alpha(1-\alpha)(k_1+k_2)^2 + m^2 - i\epsilon}{4\pi\mu^2 e^{-\gamma}}\right]$$

$$+ \frac{i\lambda^2}{32\pi^2}\frac{1}{2-\omega} - \frac{i\lambda^2}{32\pi^2}\int_0^1 d\alpha \ln\left[\frac{\alpha(1-\alpha)(k_1+k_3)^2 + m^2 - i\epsilon}{4\pi\mu^2 e^{-\gamma}}\right]$$

$$+ \frac{i\lambda^2}{32\pi^2}\frac{1}{2-\omega} - \frac{i\lambda^2}{32\pi^2}\int_0^1 d\alpha \ln\left[\frac{\alpha(1-\alpha)(k_1+k_4)^2 + m^2 - i\epsilon}{4\pi\mu^2 e^{-\gamma}}\right]$$

$$+ \mathcal{O}(\lambda^3)$$

$$(13.76)$$

Both expressions contain divergent parts as well as counterterms. In the next section we will discuss the schemes for determining the counterterms.

13.6   Subtraction Schemes

Now, we are in a position to formulate the conditions which determine the counterterms. These are clearly needed in order to find concrete results. There is also clearly some arbitrariness in formulating these conditions. The one thing that is necessary is that the counterterms should be chosen so that the ultraviolet divergences are cancelled. Beyond that we are free to invoke the most convenient possible prescription.

In order to determine the three counterterms, we need three conditions. One scheme that is often used is sometimes called "on-shell subtraction". In this scheme one chooses the counterterms so that the pole in the variable k^2 in the 2 point function is at $k^2 = -m^2$ as it is for the free field 2 point function. Remembering spectral theorem of Sect. 7.7, which tells us that that the position of the pole in the two-point function is equal to the mass squared of the particle, we see that the on-shell subtraction is equivalent to requiring that the parameter m which occurs in the Lagrangian density and in the propagator in the Feynman rules is actually the mass of the particle that is described by the scalar field. A second requirement is that the residue of the pole in the 2 point function remains equal to $-i$. This fixes the normalization of the field operators in a particular way. These two conditions, are implemented by choosing the counterterms so that the irreducible 2 point function obeys

on-shell subtraction:

$$\Pi(k^2 = -m^2) = 0 \tag{13.77}$$

$$\frac{d}{dk^2}\Pi(k^2)_{k^2=-m^2} = 0 \tag{13.78}$$

The two conditions in equations (13.77) and (13.78) must be augmented by third condition. This can be gotten by defining the coupling constant λ to be equal to the irreducible 4 point function at some canonical choice of external momenta. Several choices are commonly used. To be concrete, we shall use the following one,

on-shell subtraction:

$$\lim_{k^2 \to -m^2} \Gamma_I(k, k, -k, -k) = -i\lambda \tag{13.79}$$

As natural as the renormalization conditions (13.77)–(13.79) are, for some purposes, there is an advantage to be gained in using some other prescriptions. Two very elegant and very commonly used alternatives are the minimal subtraction, called "MS", and the modified minimal subtraction scheme, called "$\overline{MS}$". The MS scheme adjusts the counterterms to cancel the singularities of the Feynman integrals only. In that scheme, the counterterms have no finite parts at all.

The $\overline{MS}$ scheme adjusts the counter-terms to cancel the singularities in the Feynman integrals and in addition to cancel the dependence on some constants that commonly accompany Feynman integrals (for example the term $\ln 4\pi e^{-\gamma}$ at one loop order).

Both the MS and the $\overline{MS}$ subtraction schemes have the beauty of simplicity. There is also a slight technical advantage in that the renormalization constants in the end will depend only on λ and they do not depend on the dimensionful parameters m or μ.

We are now in a position to determine the counterterms to the leading order in λ. In the on-shell subtraction scheme

on − shell subtraction :

$$Z = 1 + \mathcal{O}(\lambda^2)$$

$$Z_{m^2} = 1 + \frac{\lambda}{32\pi^2}\frac{1}{2-\omega} - \frac{\lambda}{32\pi^2}\ln\left[\frac{m^2}{4\pi\mu^2 e^{1+\gamma}}\right] + \mathcal{O}(\lambda^2) \tag{13.80}$$

$$\lambda Z_\lambda = \lambda + \lambda z_\lambda^{(1)} + \ldots = \lambda + \frac{3\lambda^2}{32\pi^2}\left[\frac{1}{2-\omega} - \ln\frac{m^2}{4\pi\mu^2 e^{-\gamma}} + \frac{2}{3}\right] + \mathcal{O}(\lambda^3) \tag{13.81}$$

and the irreducible 2 and 4 point functions are then given by on $-$ shell subtraction :

$$\Pi(k) = -i \left(p^2 + m^2 + \mathcal{O}(\lambda^2) \right) \tag{13.82}$$

$$\Gamma_I(k_1, k_2, k_3, k_4) = -i\lambda$$

$$-\frac{i\lambda^2}{32\pi^2} \int_0^1 d\alpha \left\{ \ln \left[\frac{\alpha(1-\alpha)(k_1+k_2)^2 + m^2 - i\epsilon}{m^2} \right] + 2 \right\}$$

$$-\frac{i\lambda^2}{32\pi^2} \int_0^1 d\alpha \ln \left[\frac{\alpha(1-\alpha)(k_1+k_3)^2 + m^2 - i\epsilon}{m^2} \right]$$

$$-\frac{i\lambda^2}{32\pi^2} \int_0^1 d\alpha \ln \left[\frac{\alpha(1-\alpha)(k_1+k_4)^2 + m^2 - i\epsilon}{m^2} \right] + \mathcal{O}(\lambda^3) \tag{13.83}$$

We note that the result is Lorentz invariant and it depends on the momenta in only three combinations, $(k_1 + k_2)^2$, $(k_1 + k_3)^2$ and $(k_1 + k_4)^2$. If we put the four-momenta on their mass shells, $k_i^2 \to m^2$, there are only two independent Lorentz invariant variables. In that case, it is useful to use the Mandelstam variables which are defined by

$$\mathbf{s} = -(k_1 + k_2)^2, \ \mathbf{t} = -(k_1 + k_3)^2, \ \mathbf{u} = -(k_1 + k_4)^2 \tag{13.84}$$

The Mandelstam variables are not independent. They obey

$$\mathbf{s} + \mathbf{t} + \mathbf{u} = 4m^2$$

and the 4 point function in terms of them is

$$\lim_{k_i^2 \to -m^2} \Gamma_I(k_1, k_2, k_3, k_4) = -i\lambda$$

$$-\frac{i\lambda^2}{32\pi^2} \int_0^1 d\alpha \left\{ \ln \left[\frac{-\alpha(1-\alpha)\mathbf{s} + m^2 - i\epsilon}{m^2} \right] + 2 \right\}$$

$$-\frac{i\lambda^2}{32\pi^2} \int_0^1 d\alpha \ln \left[\frac{-\alpha(1-\alpha)\mathbf{t} + m^2 - i\epsilon}{m^2} \right]$$

$$-\frac{i\lambda^2}{32\pi^2} \int_0^1 d\alpha \ln \left[\frac{-\alpha(1-\alpha)\mathbf{u} + m^2 - i\epsilon}{m^2} \right] + \mathcal{O}(\lambda^3)$$

$$\tag{13.85}$$

For completeness, the counterterms in the MS scheme are

$$MS \text{ scheme}:$$

$$Z = 1 + \mathcal{O}(\lambda^2) \tag{13.86}$$

$$Z_{m^2} = 1 + \frac{\lambda}{32\pi^2} \frac{1}{2 - \omega} + \dots$$

$$\lambda Z_\lambda = \lambda + \frac{3\lambda^2}{32\pi^2} \frac{1}{2 - \omega} + \dots \tag{13.87}$$

and the irreducible 2 and 4 point functions are given by

$$MS \text{ scheme}:$$

$$\Pi(k) = -i \left(p^2 + m^2 + \frac{\lambda m^2}{32\pi^2} \ln\left[\frac{m^2}{4\pi\mu^2 e^{1+\gamma}} \right] + \dots \right) \tag{13.88}$$

$$\Gamma_I(k_1, k_2, k_3, k_4) = -i\lambda$$

$$- \frac{i\lambda^2}{32\pi^2} \int_0^1 d\alpha \, \ln\left[\frac{\alpha(1 - \alpha)(k_1 + k_2)^2 + m^2 - i\epsilon}{4\pi\mu^2 e^{-\gamma}} \right]$$

$$- \frac{i\lambda^2}{32\pi^2} \int_0^1 d\alpha \, \ln\left[\frac{\alpha(1 - \alpha)(k_1 + k_3)^2 + m^2 - i\epsilon}{4\pi\mu^2 e^{-\gamma}} \right]$$

$$- \frac{i\lambda^2}{32\pi^2} \int_0^1 d\alpha \, \ln\left[\frac{\alpha(1 - \alpha)(k_1 + k_4)^2 + m^2 - i\epsilon}{4\pi\mu^2 e^{-\gamma}} \right]$$

$$+ \mathcal{O}(\lambda^3) \tag{13.89}$$

In the $\overline{MS}$ scheme we find

$$\overline{MS} \text{ scheme}:$$

$$Z = 1 + \mathcal{O}(\lambda^2) \tag{13.90}$$

$$Z_{m^2} = 1 + \frac{\lambda}{32\pi^2} \frac{1}{2 - \omega} - \frac{\lambda}{32\pi^2} \ln\left[4\pi e^{1+\gamma} \right] + \dots$$

$$\lambda Z_\lambda = \lambda + \frac{3\lambda^2}{32\pi^2} \frac{1}{2 - \omega} - \frac{3\lambda^2}{32\pi^2} \ln[4\pi e^{-\gamma}] + \dots \tag{13.91}$$

and the irreducible 2 and 4 point functions

$$\overline{MS} \text{ scheme}:$$

$$- D^{-1}(p) = -i \left(p^2 + m^2 + \frac{\lambda m^2}{32\pi^2} \ln\left[\frac{m^2}{\mu^2} \right] + \dots \right) \tag{13.92}$$

$$\Gamma_I(k_1, k_2, k_3, k_4) = -i\lambda$$

$$-\frac{i\lambda^2}{32\pi^2} \int_0^1 d\alpha \, \ln\left[\frac{\alpha(1-\alpha)(k_1+k_2)^2 + m^2 - i\epsilon}{\mu^2}\right]$$

$$-\frac{i\lambda^2}{32\pi^2} \int_0^1 d\alpha \, \ln\left[\frac{\alpha(1-\alpha)(k_1+k_3)^2 + m^2 - i\epsilon}{\mu^2}\right]$$

$$-\frac{i\lambda^2}{32\pi^2} \int_0^1 d\alpha \, \ln\left[\frac{\alpha(1-\alpha)(k_1+k_4)^2 + m^2 - i\epsilon}{\mu^2}\right]$$

$$+ \mathcal{O}(\lambda^3) \tag{13.93}$$

There are some notable facts that we have encountered in our computation of 2 and 4 point functions. The 2 point functions are rather trivial, the net result is little more than a change of mass between bare mass and renormalized mass. This is due to the low order of perturbation theory—there are more interesting results at higher orders which we have not explored. The corrections to the renormalization constant Z begin at order λ^2. The 4 point function, on the other hand indeed already has some nontrivial features at the leading order in λ.

We have also learned that the expressions in the three different subtraction schemes appear rather similar but not identical. Of course, the differences arise from the differing prescriptions for how the counterterms are chosen. These differences are not fundamental, in fact they can be accounted for by a simple, finite re-definition of the parameters, λ and m and a finite re-scaling of the ϕ-field. In the next section we will examine this possibility in more detail, in fact we will find a way to exploit it to find information that expands the validity of perturbation theory.

13.7   Renormalization Group

In the computations that we performed in the previous section, we found it convenient when using dimensional regularization to introduce a dimensionful parameter, μ. The purpose was to keep the coupling constant λ dimensionless when we change the spacetime dimension. That is desirable, as the expansion in powers of λ is then an expansion a dimensionless parameter and its accuracy should be controlled by the magnitude of that parameter. But, we then found that the dimensionful parameter μ crept into the renormalized expressions for the 2 and 4 point functions and we expect it to also appear in general n point functions.

We can think of the appearance of μ as being a vestige of the fact that the quantum field theory is only well defined when it has a cutoff which would be an upper bound on the energy or the momentum in Feynman integrals. Such a cutoff would be measured in momentum, or inverse distance units and it should be set very large, approaching infinitely large, compared to any other dimensionful parameters or the

momenta of external lines. Dimensional regularization avoided the explicit introduction of a cutoff, but it still leaves a residue in the form of the finite, and thus far
arbitrary parameter μ.

What is more, now that μ has appeared, the accuracy of our perturbation theory
is not entirely governed by the magnitude of λ. This is easiest to see if we put the
particle mass to zero. The 2 and 4 point functions in the $\overline{MS}$ scheme become

$$\Pi(k) = 0 + \mathcal{O}(\lambda^2) \tag{13.94}$$

$$\Gamma_I(k_1, k_2, k_3, k_4) = -i\lambda +$$
$$+ \frac{i\lambda^2}{32\pi^2} \left[\ln \frac{4(k_1 + k_2)^2 - i\epsilon}{\mu^2} + \ln \frac{4(k_1 + k_3)^2 - i\epsilon}{\mu^2} + \ln \frac{4(k_1 + k_4)^2 - i\epsilon}{\mu^2} \right]$$
$$+ \mathcal{O}(\lambda^3) \tag{13.95}$$

We have retained the $i\epsilon$'s to remind ourselves of where the logarithms are singular.
To avoid this singularity, the rest of our discussion will consider the behaviour of
the 4 point function in what is called the Euclidean regime, where $(k_1 + k_2)^2 > 0$,
$(k_1 + k_3)^2 > 0$ and $(k_1 + k_4)^2 > 0$.

We see that, in the massless limit, $\Gamma_I(k_1, k_2, k_3, k_4)$ does not become scale invariant, as we might have expected in a field theory which has no dimensionful parameters. In any dimensional analysis of the 4 point function, the presence of μ would
have to be taken into account. Renormalization generally violates scale invariance
in this way.

Also, notice that this perturbative result for the 4 point function becomes large
when the k's are either much larger or much smaller than μ. The large logarithms
that occur there threaten the accuracy of perturbation theory. The constant μ defines
a momentum scale where the interaction is weak. For momenta much larger or much
smaller than this scale, the perturbative corrections are not necessarily small and
perturbation theory is not accurate.

We can fix this problem, at least to some extent, by exploiting the fact that, so
far, the parameter μ is arbitrary and it appeared nowhere in the original Lagrangian
of the scalar field theory until we introduced dimensional regularization where we
replaced λ by $\lambda\mu^{4-2\omega}$. That means that, if we had used dimensional regularization
to define the quantum field theory, and then if we had computed the n point function
of the originally normalized fields, $Z^{\frac{1}{2}}\phi$ as a function of the bare mass $m^2 Z_{m^2}$ and
the bare coupling constant $\lambda Z_\lambda \mu^{2\omega-4}$, the result could not depend on the parameter
μ. We can express this fact as a differential equation,

$$Z^{-n/2} \cdot \mu \frac{d}{d\mu} \left[Z^{n/2} \Gamma(p_1, \ldots, p_N) \right] = 0 \tag{13.96}$$

We then remind ourselves that, when written as a function of m and λ, the function
$\Gamma(p_1, \ldots, p_N)$ is finite. It must depend on the scale μ, both explicitly, and implicitly
through the dependence of m and λ on μ. This is the gist of renormalizability—that the
counterterms can be adjusted so that they absorb all of the ultraviolet singularities

and render all of the correlation functions finite as $\omega \to 2$. We can use equation (13.96) to find a differential equation that the n point function must satisfy

$$\left[\mu \frac{\partial}{\partial \mu} + \alpha m \frac{\partial}{\partial m} + \beta \frac{\partial}{\partial \lambda} + \frac{n}{2} \gamma \right] \Gamma(p_1, \ldots, p_n) = 0 \qquad (13.97)$$

$$\alpha(\lambda, m, \mu) = \frac{1}{m} \mu \frac{\partial}{\partial \mu} m \qquad (13.98)$$

$$\beta(\lambda, m, \mu) = \mu \frac{\partial}{\partial \mu} \lambda \qquad (13.99)$$

$$\gamma(\lambda, m, \mu) = \frac{1}{Z} \mu \frac{\partial}{\partial \mu} Z \qquad (13.100)$$

Here, α, β and γ are the renormalization group functions which record how the constants m and λ and the normalization of ϕ depend on μ. This equation tells us that a change of μ can be compensated by a change of m, λ and a rescaling of the correlation function.

We could equally well place a connected correlation function in the above equation. The algebraic relation between n point functions and connected n point functions would lead us to the same equation for a connected n point function. If, on the other hand we are interested in an irreducible n point function, it must be obtained from the correlation function by, amongst other operations, multiplying by a factor of $D^{-1}(p_1) \ldots D^{-1}(p_n)$ which amputates the external legs. This replaces equation (13.97) by the equation

$$\left[\mu \frac{\partial}{\partial \mu} + \alpha m \frac{\partial}{\partial m} + \beta \frac{\partial}{\partial \lambda} - \frac{n}{2} \gamma \right] \Gamma_I(p_1, \ldots, p_n; m, \lambda, \mu) = 0 \qquad (13.101)$$

This is the renormalization group equation for the irreducible n point function.

Since this interacting scalar quantum field theory is renormalizable, all of the n point functions have a finite limit as $\omega \to 2$ when they are expressed as functions of m^2 and λ. Then, since equation (13.101) must hold for all n and for all $p_1, \ldots, p_n$, the renormalization group functions α, β and γ must also be finite in this limit. The renormalization group functions do not have ultraviolet singularities.

Conversely, it is easy to see that equation (13.101) mixes orders in perturbation theory. This fact can be used to fashion an inductive proof of renormalizability. The idea is to assume that the renormalization group functions and the n point functions are finite to some order in perturbation theory and to use the renormalization group equation to prove that they are finite to the next order. We will not elaborate the rather straightforward details of this proof here.

To use the renormalization group equation, we can combine it with dimensional analysis. Dimensional analysis tells us that

$$\Gamma_I(sp_1, \ldots, sp_n; \mu, m) = s^{4-n} \Gamma_I(p_1, \ldots, p_n; m/s, \lambda, \mu/s) \qquad (13.102)$$

We can confirm that the explicit forms of the 2 and 4 point functions given in equations (13.92) and (13.93) indeed scale in this way. Then, we can take a derivative of this equation by s to get

$$s\frac{\partial}{\partial s}\Gamma_I(sp_1,\ldots,sp_n;m,\lambda,\mu) =$$

$$s^{4-n}\left(4-n-\mu\frac{\partial}{\partial\mu}-m\frac{\partial}{\partial m}\right)\Gamma_I(p_1,\ldots,p_n;m/s,\lambda,\mu/s) \qquad (13.103)$$

Then, we can combine equations (13.101) and (13.104) to eliminate the terms with $\mu\frac{\partial}{\partial\mu}$. We get

$$s\frac{d}{ds}\Gamma_I(sp_1,\ldots,sp_n;m,\lambda,\mu) =$$

$$\left(4-n-\frac{n}{2}\gamma+(\alpha-1)m\frac{\partial}{\partial m}+\beta\frac{\partial}{\partial\lambda}\right)\Gamma_I(p_1,\ldots,p_n;m/s,\lambda,\mu/s) \qquad (13.104)$$

This equation tells us how the correlation functions change when we increase or decrease the magnitude of their momentum arguments. It is a type of flow equation. It is particularly simple in the $\overline{MS}$ scheme where the renormalization group functions α, β and γ depend only on the coupling constant λ. This is thought to be the case to all orders in perturbation theory. In that case, if we can solve the flow equations

$$s\frac{d}{ds}\lambda(s) = \beta(\lambda(s)) \qquad (13.105)$$

$$s\frac{d}{ds}m(s) = \alpha(\lambda(s))m(s) \qquad (13.106)$$

$$s\frac{d}{ds}Z(s) = \gamma(\lambda(s)) \qquad (13.107)$$

we can find a solution of the differential equation (13.104). It is solved by

$$\Gamma_I(sp_1,\ldots,sp_n;m,\lambda,\mu) = s^{4-n}Z(s)^{-\frac{n}{2}}\Gamma_I(p_1,\ldots,p_n;m(s)/s,\lambda(s),\mu) \qquad (13.108)$$

From this equation we see that, scaling the momenta in an n point function can be compensated by a flow of the parameters m and λ as well as the normalization $Z^{\frac{1}{2}}$. This flow is called a renormalization group flow.

If we remember that $Z=1+\mathcal{O}(\lambda^2)$, so that $\gamma\sim\mathcal{O}(\lambda^2)$, simply by requiring that our computations of the 2 and 4 point functions, given in the $\overline{MS}$ scheme in equations (13.92) and (13.93), satisfy the renormalization group equations give us the leading orders in the renormalization group functions

$$\alpha = \frac{\lambda}{32\pi^2}+\mathcal{O}(\lambda^2) \qquad (13.109)$$

$$\beta = \frac{3\lambda^2}{16\pi^2} + \mathcal{O}(\lambda^3) \tag{13.110}$$

$$\gamma = \mathcal{O}(\lambda^2) \tag{13.111}$$

The solution of equations (13.105)–(13.107) with the explicit functions in equations (13.109)–(13.111) is given by the flow equations for the mass and the coupling constant,

$$m(s) = m(1) \left(1 - \frac{3}{16\pi^2}\lambda(1)\ln s \right)^{-\frac{1}{6}} + \dots \tag{13.112}$$

$$\lambda(s) = \frac{\lambda(1)}{1 - \frac{3}{16\pi^2}\lambda(1)\ln s} + \dots \tag{13.113}$$

$$\gamma(s) = 0 + \dots \tag{13.114}$$

Equations (13.112) and (13.113) contain the "running mass" and the "running coupling" of the scalar field theory. The ellipses denote higher corrections that would be found by evaluating the renormalization group functions to a higher order of perturbation theory and then solving the flow equations.

We can see now that, since $\lambda(s) \to 0$ as $s \to 0$, by enhancing our perturbative result so that it satisfies the renormalization group flow equation we see that the four-point function in the massless limit behaves as

$$\Gamma_I(k_1, k_2, k_3, k_4) = -i\lambda(s) +$$
$$+ \frac{i\lambda(s)^2}{32\pi^2} \left[\ln \frac{4(k_1+k_2)^2 - i\epsilon}{s^2\mu^2} + \ln \frac{4(k_1+k_3)^2 - i\epsilon}{s^2\mu^2} + \ln \frac{4(k_1+k_4)^2 - i\epsilon}{s^2\mu^2} \right]$$
$$+ \mathcal{O}(\lambda(s)^3) \tag{13.115}$$

where we should now regard the right-hand-side as being independent of s. Indeed, if we Taylor expand it to second order in $\lambda(1)$ the dependence on s cancels. However, it contains higher orders in $\lambda(1)\ln(s)$.

Let us review what we have done here. First of all, we computed the 4 point function to second order in $\lambda(1)$. Then we have used the renormalization group to determine higher orders in $(\lambda(1)\ln s)^n$ with $n > 2$ and we have implicitly summed over n to get the above formula. These higher orders would become important when $\ln s$ has large enough magnitude to compensate for the fact that $\lambda(1)$ is small. This is so when $s \gg 1$ or $s \ll 1$. We should now regard the summed up result (13.115) as being accurate when $\lambda(s)$ is small. We can see from the explicit form of $\lambda(s)$ that this is so when $s < 1$.

Now, when $k_1, \dots, k_4$ are really large or really small, we can control the size of the logarithms in the right-hand-side of (13.115) by also choosing s to be really large or really small. In those cases, we see that $\lambda(s)$ becomes larger or smaller and our perturbation theory is less or more accurate, respectively. Since $\lambda(s)$ gets smaller as s gets smaller this has fixed our problems at small values of $k_1, ..., k_4$. Our renormalization group improved perturbation theory is exceedingly good in that regime. It has, however, exacerbated the problem at large values of $k_1, ..., k_4$ where the coupling runs to larger values. This is unavoidable and, in fact, the perturbation theory that we are using is simply not a good way to study correlation functions in the high energy limit.

What has happened here is sometimes called "dimensional transmutation". The accuracy of perturbation theory is no longer controlled by a dimensionless coupling constant. It is now controlled by the scale parameter μ. It is accurate if the magnitudes of k_i are much less than μ and it is not accurate if they are much greater than μ.

Even though we do not have a way of studying the high energy limit, we can still say some things about it since it depends on the form of the beta function for the coupling constant. We took the mass to zero at the outset of the discussion above to simplify it somewhat. In the high energy limit the scale of the k_i is much larger than the mass and it can be ignored even if it is not zero. Then the behaviour of the running coupling constant is obtained from the beta function by integrating equation (13.105) to get

$$
\ln s = \int_{\lambda(1)}^{\lambda(s)} \frac{d\tilde{\lambda}}{\beta(\tilde{\lambda})} \tag{13.116}
$$

Here, we are still assuming that we are using a mass-independent subtraction scheme MS or $\overline{MS}$ where the beta function depends only on the coupling constant. Even though we are confined to doing perturbative computations and we can therefore only evaluate the beta function at small values of λ, we can still discuss that the integral in equation (13.116) tells us about the general properties of the renormalization group flow of the coupling constant. It is clear from the formula that zeros of the beta function are important. The integrand diverges there and the flow of the coupling stops at those values. The values of the coupling constant, λ^*, where the flow stops are solutions of the equation $\beta(\lambda^*) = 0$ and they are called "fixed points". We already know that there is a fixed point at zero coupling. Since the beta function is positive there, $\lambda(s)$ flows to zero as $s \rightarrow 0$. In this case we call $\lambda^* = 0$ an infrared fixed point.

As s is increased, $\lambda(s)$ flows away from zero. There are four possibilities for asymptotic, large s behaviour in a theory with a single coupling constant such as this scalar field theory.

When the beta function is positive for all positive values of λ and when it increases faster that a linear function at large values of λ,[4] as depicted in Fig. 13.13, so that the integral in equation (13.116) has a finite limit as $\lambda(s) \rightarrow \infty$, the running coupling

---

[4] Beware that our perturbative computation tells us nothing about how the beta function behaves at large λ.

Fig. 13.13 When the beta
function is positive for all
values of λ and when it
increases faster than a linear
function for large λ so that
the integral in equation
(13.116) converges, the
running coupling constant
exhibits a Landau pole. It
goes to infinity at a finite
large value of s

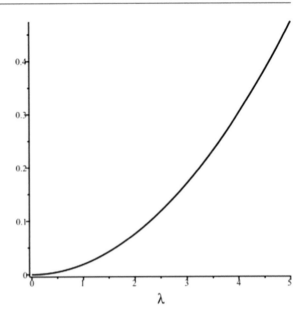

Fig. 13.14 When the beta
function remains positive for
all values of λ but it increases
slower than a linear function
for large λ the running
coupling constant goes to
infinity as $s \to \infty$. In this
case, there is no Landau pole

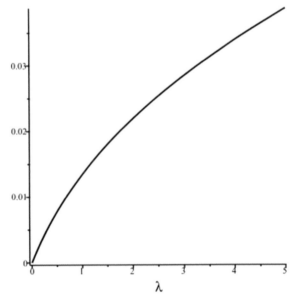

will flow to infinity at some finite of s. This divergence of the running coupling is
called a "Landau pole". It is often regarded as a pathological behaviour, perhaps an
indication that the scalar field theory must have a large but finite cutoff.

On the other hand if, as depicted in Fig. 13.14, the beta function is positive for all
positive values of λ such that the integral on the right-hand-side of equation (13.116)
diverges as $\lambda(s) \to \infty$, that is, $\beta(\lambda)$ grows at most like a linear function of λ for
large λ then there is no Landau pole but $\lambda(s)$ flows to infinity as s goes to infinity.

Fig. 13.15 If the beta function begins positive and then changes sign, the value of $\lambda = \lambda^*$ where it crosses zero is a fixed point to which the coupling runs as $s \to \infty$. There is an infrared fixed point at $\lambda^* = 0$ and an ultraviolet fixed point at $\lambda^* > 0$ where $\beta(\lambda^*) = 0$

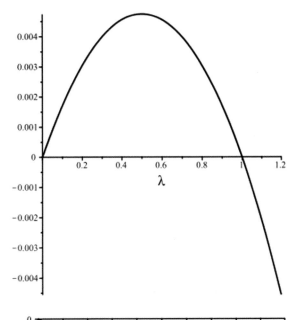

Fig. 13.16 When the beta function begins negative the coupling $\lambda(s)$ flows to smaller values as s increases, approaching zero as $s \to \infty$

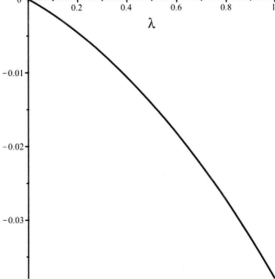

The third possibility is that the beta function starts out positive at small values of λ (as we know it does for the scalar field theory since our perturbative computation is valid there) but it turns around and crosses zero at a finite value which we call $\lambda = \lambda^*$ as depicted in Fig. 13.15. In this case the flow of the coupling goes to the fixed point $\lim_{s \to \infty} \lambda(s) = \lambda^*$. where $\beta(\lambda^*) = 0$ is called an ultraviolet fixed point.

In the fourth possibility, the beta function is negative for small values as depicted in Fig. 13.16. In this case, $\lim_{s \to \infty} \lambda(s) = 0$ and the high energy limit of the theory is

Fig. 13.17 When the beta
function begins negative the
coupling $\lambda(s)$ flows to
smaller values as s increases,
approaching zero as $s \to \infty$

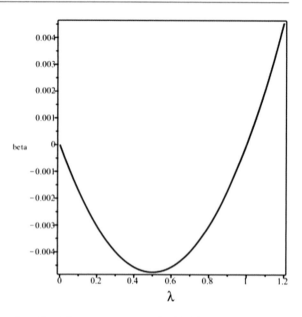

weakly coupled and it flows to free field theory at asymptotically high energies. This
phenomenon is called "asymptotic freedom". We already know that the scalar field
theory in four dimensions does not behave in this way. In fact, most four dimensional
field theories do not. The exception is Yang-Mills theory which is a fascinating case
but it is far beyond the scope of what we will discuss in this treatise. On the other
hand, if at strong coupling such theories have a beta function as in Fig. 13.16, their
massless limit is strongly coupled at low energies. There is another possibility, that
the beta function begins negative and, as depicted in Fig. 13.17, it goes to zero at a
finite value of the coupling.

In this case, the coupling constant still flows to zero at large s, but at small s it
flows toward the infrared fixed point at the value of λ where the beta function is zero.
This situation indeed occurs in the scalar field theory that we are describing if we
take the spacetime dimension to be $2\omega = 4 - \epsilon$ with ϵ a small positive number. The
beta function for the scalar field theory in $2\omega = 4 - \epsilon$ dimensions is

$$\beta(\lambda) = -\epsilon\lambda + \frac{3}{16\pi^2}\lambda^2 \qquad (13.117)$$

This scalar field theory is asymptotically free, $\lim_{s\to\infty} \lambda(s) = 0$ and in the infrared (if
we set the mass to zero, $m = 0$), $\lim_{s\to 0} \lambda(s) = \lambda^* = \frac{16\pi^2}{3}\epsilon$ which is small (and the
field theory is still weakly coupled) if ϵ is small. This infrared fixed point is called the
Wilson-Fisher fixed point. It is important for the application of the scalar field theory
as an emergent quantum field theory which describes classical statistical mechanics
systems in the universality class of the Ising model and some other classical spin
systems. Here the dimension $4 - \epsilon$ is close to four but the perturbation theory in this
case has been solved as a double expansion in λ and in ϵ to quite a high order and
re-summation techniques for asymptotic series are used to extrapolate the results to

three or even two dimensions where they agree rather well with the results of other computation techniques.

13.8   Appendix: Integration Formulae

13.8.1 Euler's Gamma Function

The Euler gamma function occurs often in Feynman integrals and particularly when dimensional regularization is used. It is the generalization of the factorial $n!$ which is defined for non-negative integers n.

It can be defined by its integral representation. To see that, first, consider the integral representation of the factorial,

$$n! = \int\limits_0^\infty dx\, x^n e^{-x}$$

which can be derived as follows:

$$\int_0^\infty dx\, x^n e^{-x} = \lim_{a \to 1} (-1)^n \frac{d^n}{da^n} \int\limits_0^\infty dx\, e^{-ax} = \lim_{a \to 1} (-1)^n \frac{d^n}{da^n} \frac{1}{a} = n!$$

To define the Euler gamma function, we generalize this integral representation to the case where n is any complex number. The resulting integral representation is given in equation (13.119), which we recopy here

$$\Gamma[z] = \int\limits_0^\infty ds\, s^{z-1} e^{-s}$$

Notice the offset of one in the definition $n \to z - 1$.
The Euler gamma function has the property

$$z\Gamma[z] = \Gamma[z + 1]$$

In particular, for $n = 1, 2, 3, \ldots,$

$$\Gamma[n] = (n - 1)!, \quad \Gamma[1] = 1$$

The integral formula also gives

$$\Gamma[1/2] = \sqrt{\pi}$$

When z is a complex number, $\Gamma(z)$ is holomorphic everywhere in the complex plane with the exception of simple poles at $z = 0, -1, -2, \ldots$. In the vicinity of $z = 0$, it has the asymptotic expansion

$$\Gamma(z) = \frac{1}{z} - \gamma + \frac{1}{2}\left(\gamma^2 + \frac{\pi^2}{6}\right)z - \frac{1}{6}\left(\gamma^3 + \frac{\gamma\pi^2}{2} + 2\zeta(3)\right)z^2 + \ldots$$

with the Euler–Mascheroni constant

$$\gamma = 0.57721566490153286060\ldots$$

and the Riemann eta-function is defined as

$$\zeta(s) = \sum_{n=1}^{\infty} \frac{1}{n^s}$$

has the specific value

$$\zeta(3) = 1.2020569031595942854\ldots$$

is sometimes called Apéry's constant.

Another useful asymptotic expansion occurs near any negative integer argument where

$$\Gamma[-n + \epsilon] = \frac{(-1)^n}{n!}\left\{\frac{1}{\epsilon} + \psi(n+1) + \frac{\epsilon}{2}\left[\frac{\pi^2}{3} + \psi^2(n+1) - \psi'(n+1)\right] + \ldots\right\}$$

where the Euler digamma function is

$$\psi(z) = \frac{\Gamma[z]'}{\Gamma[z]}$$

$$\psi(n) = -\gamma + 1 + \frac{1}{2} + \ldots + \frac{1}{n-1}$$

$$\psi'(n) = \frac{\pi^2}{6} - 1 - \frac{1}{2^2} - \ldots - \frac{1}{(n-1)^2}$$

Another common formula for the gamma function is the Euler infinite product,

$$\Gamma[z] = \frac{1}{z}\prod_{n=1}^{\infty}\frac{\left(1 + \frac{1}{n}\right)^z}{1 + \frac{z}{n}}$$

and also the Weierstrass formula

$$\Gamma[z] = \frac{e^{-\gamma z}}{z}\prod_{n=1}^{\infty}\frac{e^{z/n}}{1 + \frac{z}{n}}$$

There are some interesting explicit formulae such as

$$\Gamma[z]\Gamma[1-z] = \frac{\pi}{\sin \pi z}$$

$$\Gamma[z-n] = (-1)^{n-1} \frac{\Gamma[-z]\Gamma[1+z]}{\Gamma[n+1-z]}, \quad n = 1, 2, 3, \ldots$$

$$\Gamma[z]\Gamma[z+1/2] = 2^{1-2z}\sqrt{\pi}\Gamma[2z]$$

13.8.2 Feynman Parameter Formula

Feynman parameters are useful in doing Feynman integrals with multiple denominators. The general formula is

$$\frac{1}{D_1^{\nu_1} \ldots D_n^{\nu_n}} =$$

$$\frac{\Gamma[\nu_1 + \ldots + \nu_n]}{\Gamma[\nu_1] \ldots \Gamma[\nu_n]} \int_0^1 \frac{d\alpha_1 \ldots d\alpha_n \, \delta\left(1 - \sum_{i=1}^n \alpha_i\right) \alpha_1^{\nu_1-1} \ldots \alpha_n^{\nu_n-1}}{[\alpha_1 D_1 + \ldots + \alpha_n D_n]^{\nu_1 + \ldots + \nu_n}} \tag{13.118}$$

We prove this formula as follows. Begin with the Schwinger representation of the left-hand-side, which uses the integral representation of the Euler gamma function[5]

$$\Gamma[\nu] = \int_0^\infty d\alpha \, \alpha^{\nu-1} \, e^{-\alpha} \tag{13.119}$$

to get the formula

$$\frac{1}{D^\nu} = \frac{1}{\Gamma[\nu]} \int_0^\infty d\alpha \alpha^{\nu-1} e^{-\alpha D}$$

In the equation above, the D-dependence on the left-hand-side is recovered by scaling the integration variable α. Applying this formula leads to

$$\frac{1}{D_1^{\nu_1} \ldots D_n^{\nu_n}} = \prod_{i=1}^n \frac{1}{D_i^{\nu_i}} = \prod_{i=1}^n \frac{1}{\Gamma[\nu_i]} \int_0^\infty \frac{d\alpha_i}{\alpha^{1-\nu_i}} e^{-\alpha_i D_i}$$

In this formula, the α_i are called "Schwinger parameters".

---

[5] We discuss this gamma function in more detail in Sect. 13.8.1.

The Schwinger parameterization of a Feynman integral itself can often be quite useful. To go from Schwinger parameters to Feynman parameters, we consider the following identity,

$$1 = \int\limits_0^\infty d\lambda\, \delta \left(\lambda - \sum_{i=1}^n \alpha_i \right)$$

which we insert into the integration to get

$$\frac{1}{D_1^{\nu_1} \dots D_n^{\nu_n}} = \int\limits_0^\infty d\lambda \prod_{i=1}^n \left[\frac{1}{\Gamma[\nu_i]} \int\limits_0^\infty \frac{d\alpha_i}{\alpha_i^{1-\nu_i}} e^{-\alpha_i D_i} \right] \delta \left(\lambda - \sum \alpha_i \right)$$

We have changed the order of integration.
We then change the integration variables as

$$(\lambda, \alpha_1, \dots \alpha_n) \to (\lambda, \lambda\alpha_1, \dots, \lambda\alpha_n)$$

$$d\lambda d\alpha_1 \dots d\alpha_n \delta \left(\lambda - \sum_{i=1}^n \alpha_i \right) \to \lambda^{n-1} d\lambda d\alpha_1 \dots d\alpha_n \delta \left(1 - \sum_{i=1}^n \alpha_i \right)$$

we find

$$\frac{1}{D_1^{\nu_1} \dots D_n^{\nu_n}} = \int\limits_0^\infty \frac{d\lambda}{\lambda^{1-\sum_i \nu_i}} \prod_{i=1}^n \left[\frac{1}{\Gamma[\nu_i]} \int\limits_0^\infty \frac{d\alpha_i}{\alpha^{1-\nu_i}} e^{-\lambda\alpha_i D_i} \right] \delta \left(1 - \sum \alpha_i \right)$$

$$= \frac{\Gamma[\nu_1 + \dots + \nu_n]}{\Gamma[\nu_1] \dots \Gamma[\nu_n]} \int\limits_0^1 \frac{d\alpha_1 \dots d\alpha_n \delta(1 - \sum \alpha_i)\alpha_1^{\nu_1-1} \dots \alpha_n^{\nu_n-1}}{[\alpha_1 D_1 + \dots + \alpha_n D_n]^{\nu_1 + \dots + \nu_n}}$$

where we have integrated over λ to produce an Euler gamma-function which appears in the numerator.

The final result is the Feynman parameter formula

$$\frac{1}{D_1^{\nu_1} \dots D_n^{\nu_n}} = \frac{\Gamma[\nu_1 + \dots + \nu_n]}{\Gamma[\nu_1] \dots \Gamma[\nu_n]} \int\limits_0^1 \frac{d\alpha_1 \dots d\alpha_n \delta(1 - \sum \alpha_i)\alpha_1^{\nu_1-1} \dots \alpha_n^{\nu_n-1}}{[\alpha_1 D_1 + \dots + \alpha_n D_n]^{\nu_1 + \dots + \nu_n}}$$

where the α_i are now called "Feynman parameters". This is a formula which we often use when computing Feynman integrals.

An example of such an application begins with the expression with two denominators,

$$\frac{1}{[(p-k)^2 + m^2][p^2 + m^2]} = \int\limits_0^1 d\alpha \frac{1}{[(p-\alpha k)^2 + \alpha(1-\alpha)k^2 + m^2]^2}$$

This formula is useful since the eventual integration over p can be made more symmetric by translating the integration variable, $p \to p + \alpha k$.

13.8.3 Dimensional Regularization Integral

In this subsection, we will derive the following integral formulae:

In Minkowski space

$$\int \frac{d^{2\omega} p}{(2\pi)^{2\omega}} \frac{1}{[p^2 + m^2 - i\epsilon]^A} = i \frac{\Gamma[A - \omega]}{(4\pi)^\omega \Gamma[A]} \frac{1}{[m^2]^{A-\omega}} \tag{13.120}$$

In Euclidean space

$$\int \frac{d^{2\omega} p}{(2\pi)^{2\omega}} \frac{1}{[p^2 + m^2]^A} = \frac{\Gamma[A - \omega]}{(4\pi)^\omega \Gamma[A]} \frac{1}{[m^2]^{A-\omega}} \tag{13.121}$$

which are very useful in doing loop integrals with dimensional regularization. We can derive these formulas as follows. We will concentrate on the Minkowski space one (13.120) We begin with the integral

$$\int \frac{d^{2\omega-1} p \, dp_0}{(2\pi)^{2\omega-1} \, 2\pi} \frac{1}{[-p_0^2 + \vec{p}^2 + m^2 - i\epsilon]^A}$$

We first perform a "Wick rotation" of the integration of p_0. This begins with the observation that, the integrand is analytic in the first and third quadrants in the complex p_0-plane. (This is closely related to the fact that time-ordered 2 point functions such as $\Delta(x, y)$ are analytic in the second and fourth quadrants of the complex $x^0 - y^0$-plane. See Sect. 7.9 for a thorough discussion of the latter.) We add to this contour the quarter-circles at the boundaries of the first and third quadrants and add and subtract the integration along the imaginary axis. Then, the integral over the closed loop consisting of the real axis, the quarter-circles at the boundaries of the first and third quadrants and the imaginary axis vanishes, as this contour encloses no poles. That remains is an integral from $-\infty$ to ∞ along the imaginary axis. Now, wherever it appears, $p^2 = p_0^2 + p_1^2 + p_2^2 + p_3^2$ and the integration measure gets a factor of i.

$$\int \frac{d^{2\omega-1} p \, dp_0}{(2\pi)^{2\omega-1} \, 2\pi} \frac{1}{[-p_0^2 + \vec{p}^2 + m^2 - i\epsilon]^A} = i \int \frac{d^{2\omega} p}{(2\pi)^{2\omega}} \frac{1}{[p^2 + m^2]^A}$$

Now, we use the Euler gamma function (For more detail in Sect. 13.8.1.) in order to introduce a Schwinger parameter,

$$i \int \frac{d^{2\omega} p}{(2\pi)^{2\omega}} \frac{1}{[p^2 + m^2]^A} = \frac{i}{\Gamma[A]} \int_0^\infty \frac{ds}{s^{1-A}} \int \frac{d^{2\omega} p}{(2\pi)^{2\omega}} e^{-s(p^2+m^2)}$$

$$= \frac{i}{\Gamma[A]} \int_0^\infty \frac{ds}{s^{1-A}} e^{-sm^2} \left(\int \frac{d^2 p}{(2\pi)^2} e^{-sp^2} \right)^\omega$$

$$= \frac{i}{(4\pi)^\omega \Gamma[A]} \int_0^\infty \frac{ds}{s^{1+\omega-A}} e^{-sm^2} = i \frac{\Gamma[\omega - A]}{(4\pi)^\omega \Gamma[A]} \frac{1}{[m^2]^{A-\omega}}$$

where we have used the integral representation of Euler's gamma function (13.119).
Some properties of the gamma function are reviewed in Sect. 13.8.1

By a similar procedure we could obtain the following dimensional regularization
formulae, all given in Minkowski space

$$\int \frac{d^{2\omega} p}{(2\pi)^{2\omega}} \frac{1}{[p^2 + 2pq + m^2 - i\epsilon]^A} = i \frac{\Gamma[A - \omega]}{(4\pi)^\omega \Gamma[A]} \frac{1}{[-q^2 + m^2 - i\epsilon]^{A-\omega}}$$

$$\tag{13.122}$$

$$\int \frac{d^{2\omega} p}{(2\pi)^{2\omega}} \frac{p^\mu}{[p^2 + 2pq + m^2 - i\epsilon]^A} = -i \frac{\Gamma[A - \omega]}{(4\pi)^\omega \Gamma[A]} \frac{q^\mu}{[-q^2 + m^2 - i\epsilon]^{A-\omega}}$$

$$\tag{13.123}$$

$$\int \frac{d^{2\omega} p}{(2\pi)^{2\omega}} \frac{p^\mu p^\nu}{[p^2 + 2pq + m^2 - i\epsilon]^A}$$
$$= i \frac{\Gamma[A - \omega]}{(4\pi)^\omega \Gamma[A]} \frac{1}{[-q^2 + m^2 - i\epsilon]^{A-\omega}} \left[q^\mu q^\nu + \frac{1}{2(A - 1 - \omega)} \eta^{\mu\nu} (-q^2 + m^2 - i\epsilon) \right]$$

$$\tag{13.124}$$

$$\int \frac{d^{2\omega} p}{(2\pi)^{2\omega}} \frac{p^\mu p^\nu p^\lambda}{[p^2 + 2pq + m^2 - i\epsilon]^A}$$
$$= -i \frac{\Gamma[A - \omega]}{(4\pi)^\omega \Gamma[A]} \frac{1}{[-q^2 + m^2 - i\epsilon]^{A-\omega}} \left[4q^\mu q^\nu q^\lambda \right.$$
$$\left. + \frac{1}{2(A - 1 - \omega)} [\eta^{\mu\nu} q^\lambda + \eta^{\nu\lambda} q^\mu + \eta^{\lambda\mu} q^\nu](-q^2 + m^2 - i\epsilon) \right]$$

$$\tag{13.125}$$

More Theory of the Real Scalar Field

14

14.1 The S Matrix

The idea of a scattering experiment is that the experimenter initially prepares an incoming state. This is a state with some particles which are traveling toward each other. At the outset, the particles are well separated and they do not interact with each other. In other words, they travel like free particles. After they are prepared, the particles approach each other, they collide or otherwise interact with each other and then they eventually propagate away from the collision and reach a region where they are again well-separated and they again behave like free particles whence properties of this final state are measured by a particle detector.

We can describe the states of the in-coming particles, at the initial time when they are well separated and do not interact by simply using the quantum states of non-interacting fields. In our case, the example of the self-interacting scalar field, this will be a free scalar field which we call the in-field. This field should have a mass which matches the position of the pole in the scalar field 2 point function. For example, in the on-shell subtraction scheme, we would compute the 2 point function with the renormalization prescription which determines the counterterms so that

$$D(p^2) \sim \frac{-i}{p^2 + m^2 - i\epsilon} + \left[\text{ finite as } p^2 \to -m^2 \right] \qquad (14.1)$$

Then, the in-coming fields are described by the states of the free field theory with the field equation and equal-time commutation relations

© The Author(s), under exclusive license to Springer Nature Singapore Pte Ltd. 2023
G. W. Semenoff, *Quantum Field Theory*, Graduate Texts in Physics,
https://doi.org/10.1007/978-981-99-5410-0_14

$$\left(-\partial^2 + m^2\right)\phi_{\text{in}}(x) = 0 \tag{14.2}$$

$$\left[\phi_{\text{in}}(x), \frac{\partial}{\partial y^0}\phi_{\text{in}}(y)\right]\delta(x^0 - y^0) = i\delta(x - y) \tag{14.3}$$

$$[\phi_{\text{in}}(x), \phi_{\text{in}}(y)]\delta(x^0 - y^0) = 0, \quad \left[\frac{\partial}{\partial x^0}\phi_{\text{in}}(x), \frac{\partial}{\partial y^0}\phi_{\text{in}}(y)\right]\delta(x^0 - y^0) = 0 \tag{14.4}$$

This free field theory has a solution in terms of creation and annihilation operators

$$\phi_{\text{in}}(x) = \int \frac{d^3k}{\sqrt{(2\pi)^3 2E(\vec{k})}}\left[e^{ikx}a_{\text{in}}(\vec{k}) + e^{-ikx}a_{\text{in}}^\dagger(\vec{k})\right]$$

$$k^0 = E(\vec{k}) = \sqrt{\vec{k}^2 + m^2} \tag{14.5}$$

and the in-coming states are the usual multi-particle states of free field theory created by beginning with the vacuum state, defined by

$$a_{\text{in}}(\vec{k})|\mathcal{O}> = 0, \quad <\mathcal{O}|a_{\text{in}}^\dagger(\vec{k}) = 0, \quad <\mathcal{O}|\mathcal{O}> = 1 \tag{14.6}$$

and multi-particle states by operating with creation operators,[1]

$$|k_1, ..., k_n>_{\text{in}} = \frac{1}{\sqrt{n!}}a_{\text{in}}^\dagger(\vec{k}_1)...a_{\text{in}}^\dagger(\vec{k}_n)|\mathcal{O}> \tag{14.7}$$

To discuss transition probabilities, we ideally need unit normalized, rather than continuum normalized states. We would create such a state as the wave packet

$$|\zeta>_{\text{in}} = \int d^3k_1...d^3k_n\zeta(\vec{k}_1, ..., \vec{k}_n)|k_1, ..., k_n>_{\text{in}} \tag{14.8}$$

which is unit normalized

$$_{\text{in}}<\zeta|\zeta>_{\text{in}} = 1$$

when the wave packet is unit normalized,

$$\int d^3k_1...d^3k_n|\zeta(\vec{k}_1, ..., \vec{k}_n)|^2 = 1$$

---

[1] Note that much of the literature on this subject uses a different normalization of the states. In that normalization, one includes a factor of $\sqrt{2E(\vec{k})}$ with the operation of each creation operator so that single-particle states have the Lorentz covariant inner product $<\vec{k}|\vec{\ell}> = \delta(\vec{k} - \vec{\ell})2E(\vec{k})$. Or, often, the creation operators themselves are normalized with this factor included.

This in-particle state is prepared by the experimenter. It then participates in the scattering experiment. Eventually the particles emerge and fly away from the region where they interacted and they evolve to another quantum state which we could call $|\tilde{\zeta}>_{out}$. This outgoing state must be a quantum superposition of the possible incoming states. The precise nature of this superposition is contained in the S matrix.

The wave-packet states $|\zeta>_{in}$ are linear superpositions of the basis states $|k_1, k_2, ..., k_n>_{in}$. Time evolution is also a linear process in that if, in the process of the scattering experiment $|\zeta>_{in}$ evolves to the state $|\zeta>_{out}$ and a different incoming state $|\tilde{\zeta}>_{in}$ evolves to the state $|\tilde{\zeta}>_{out}$ then the linear superposition of the two incoming states

$$c_1|\zeta>_{in} + c_2|\tilde{\zeta}>_{in}$$

evolves to the same superposition of the outgoing states

$$c_1|\zeta>_{out} + c_2|\tilde{\zeta}>_{out}$$

For this reason, to study the transition, it is sufficient to consider the evolution of the individual basis vectors in the Fock space of the incoming states. By this reasoning, during the scattering process, $|k_1, ..., k_n>_{in}$ evolves to some other quantum state which we will call $|k_1, ..., k_n>_{out}$. The latter is labeled by the same momenta $\vec{k}_1, ..., \vec{k}_n$, as the in-state. However, it is not an in-state. If interactions occur, it is generally a superposition of any of the possible in-states (which are compatible with conservation laws such as conservation of energy and momentum),

$$
|k_1, ..., k_n>_{out}
$$
$$
= \sum_{m=0}^{\infty} \int d^3\ell_1 ... d^3\ell_m |\ell_1, ..., \ell_m>_{in} S^\dagger(\ell_1, ..., \ell_m; k_1, ..., k_n) \tag{14.9}
$$

where $|k_1, ..., k_n>_{in}$ is the state defined in equation (14.7). The coefficients in this superposition, the functions

$$
S^\dagger(\ell_1, ..., \ell_m; k_1, ..., k_n) = {}_{in}<\ell_1, ..., \ell_m|k_1, ..., k_n>_{out} \tag{14.10}
$$

are called the S matrix. (The dagger is there so that it matches conventional definitions.) We could also think of these coefficients as the matrix elements of an operator, $S^\dagger$, so that

$$
|k_1, ..., k_n>_{out} = S^\dagger |k_1, ..., k_n>_{in} \tag{14.11}
$$

and

$$
S^\dagger(\ell_1, ..., \ell_m; k_1, ..., k_n) = {}_{in}<\ell_1, ..., \ell_m|k_1, ..., k_n>_{out}
$$
$$
= {}_{in}<\ell_1, ..., \ell_m|S^\dagger|k_1, ..., k_n>_{in} \tag{14.12}
$$

Then, for the unit normalized stated defined in equation (14.8),

$$|\zeta>_{\text{out}} = S^{\dagger} |\zeta>_{\text{in}} \qquad (14.13)$$

Since the S matrix is the operator which implements time evolution, it should be a unitary operator,

$$S S^{\dagger} = \mathcal{I} \qquad (14.14)$$

where the right-hand-side is the unit operator in the space of quantum states.

We expect that the S matrix has the symmetries of the underlying theory. For example, since by Bose statistics, the states $|k_1, ..., k_n>_{\text{in}}$ are completely symmetric under permutations of the labels on the variables $k_1, ..., k_n$ we would expect that $S^{\dagger}(\ell_1, ..., \ell_m; k_1, ..., k_n)$ is a completely symmetric function of the incoming momenta $k_1, ..., k_n$ and also a completely symmetric function of the outgoing momenta $\ell_1, ..., \ell_m$. Translation invariance of the field theory and the conservation of energy and momentum which follows from it mean that the S matrix commutes with translations that are generated by the total momentum operator, that is

$$S e^{i P_{\mu} a^{\mu}} = e^{i P_{\mu} a^{\mu}} S \quad \text{or} \quad S = e^{i P_{\mu} a^{\mu}} S e^{-i P_{\mu} a^{\mu}} \qquad (14.15)$$

and that, $S^{\dagger}(\ell_1, ..., \ell_m; k_1, ..., k_n)$ in equation (14.9) can be non-zero only when the sums over the incoming and the outgoing four-momenta are equal,

$$\sum_i \ell_i^{\mu} = \sum_j k_j^{\mu} \qquad (14.16)$$

The S-matrix should also be Lorentz invariant. It should commute with the unitary operators $U(\Lambda)$ which implement Lorentz transformations of the states,

$$U(\Lambda) S = S U(\Lambda) \quad \text{or} \quad U(\Lambda) S U^{\dagger}(\Lambda) = S \qquad (14.17)$$

The constraints posed by Poincare invariance are quite useful in analyzing specific S matrix elements.

14.1.1 The T Matrix

It is often useful to separate the S matrix into a trivial part, which acts on the incoming states like the unit matrix in Hilbert space—which is what we would expect the S matrix to do if we turn off the interactions between particles, and the non-trivial part which accounts for how interactions drive transitions between different states. The first of these, the unit operator, we denote by $\mathcal{I}$. It maps every state in the Hilbert space onto itself. The second is another matrix, called the T matrix, which we denote by the symbol **T**. It contains all of the effects of interactions between particles. We thus write the S matrix as

$$S = \mathcal{I} + i\mathbf{T} \tag{14.18}$$

and we can take this expression as defining the T-matrix **T**. Use of the T matrix is convenient as it gives a direct expression for the transition amplitude, absent the amplitude for no scattering which is represented by the unit operator $\mathcal{I}$ in equation (14.18).

Unitarity of the S matrix, $SS^\dagger = \mathcal{I}$, implies that the T matrix obeys the non-linear identity

$$\mathbf{T}^\dagger \, \mathbf{T} \; = \; i(\mathbf{T} \, - \, \mathbf{T}^\dagger) \tag{14.19}$$

This identity, or matrix elements of this identity are called the optical theorem. When the optical theorem is applied to the T matrix, it contains the information that the S matrix is unitary.

14.2   The LSZ Formula

There is a beautiful formula, called the Lehman–Symanzik–Zimmerman or LSZ reduction formula, which relates the matrix elements of the S matrix to the n point functions of the quantum field theory. This formula tells us how to compute S matrix elements. We will present it here and give an outline of how to prove it in section 14.5.

To get the elements of the S matrix, one should take specific matrix elements in the Fock space of the in-fields, $\phi_{\text{in}}(x)$, of the following operator

$$S = \; : \; \exp\left(\int dx\, \phi_{\text{in}}(x)[-\partial^2 + m^2]\frac{\delta}{\delta J(x)}\right) \; : \; Z[J]\bigg|_{J=0} \tag{14.20}$$

$$S(k_1, ..., k_n; \ell_1, ..., \ell_m) = \; _{\text{in}} < k_1, \ldots, k_n | S | \ell_1, \ldots, \ell_m >_{\text{in}}$$

$$= \; _{\text{out}} < k_1, \ldots, k_n | S | \ell_1, \ldots, \ell_m >_{\text{in}} \tag{14.21}$$

(To be clear, $S(k_1, ..., k_n; \ell_1, ..., \ell_m)$ is the complex conjugate of the amplitude for finding the state $|\ell_1, \ldots, \ell_m >_{\text{in}}$ in the final state given that the initial state was $|k_1, \ldots, k_m >_{\text{in}}$.) For our purposes, the exponential of the operators in the above formula can be defined by its Taylor expansion,

$$e^{\int dx\, \phi_{\text{in}}(x)[-\partial^2 + m^2]\frac{\delta}{\delta J(x)}} \equiv \sum_{n=0}^{\infty} \frac{1}{n!}\left[\int dx\, \phi_{\text{in}}(x)[-\partial^2 + m^2]\frac{\delta}{\delta J(x)}\right]^n \tag{14.22}$$

The notation : : means "normal ordering" of whatever free field operators appear between the colons. When we normal order an operator, we put all of the annihilation operators to the right and all of the creation operators to the left. For example

$$: a^\dagger(\vec{k}_1)a(\vec{k}_2) :\ =\ a^\dagger(\vec{k}_1)a(\vec{k}_2), \quad : a(\vec{k}_1)a^\dagger(\vec{k}_2) :\ =\ a^\dagger(\vec{k}_2)a(\vec{k}_1)$$

The normal ordering implies that the S matrix in equation (14.20) can also be written as

$$S = e^{\int dx\, \phi_{\rm in}^{(-)}(x)[-\partial^2 + m^2]\frac{\delta}{\delta J(x)}}\ e^{\int dx\, \phi_{\rm in}^{(+)}(x)[-\partial^2 + m^2]\frac{\delta}{\delta J(x)}}\ Z[J]\Big|_{J=0} \qquad (14.23)$$

where $\phi_{\rm in}^{(+)}(x)$ and $\phi_{\rm in}^{(-)}(x)$ are the positive and negative frequency components of the free field $\phi_{\rm in}(x) = \phi_{\rm in}^{(+)}(x) + \phi_{\rm in}^{(-)}(x)$. Remember that the positive frequency components contain the annihilation operators and the negative frequency components contain the creation operators.

We should remark that we have chosen a particular phase for the S matrix in (14.20) and (14.23). Since, in the end, we are only interested in transition probabilities, which will always be linearly proportional to both S and $S^\dagger$, we could change the overall phase of S, $S \to e^{i\theta}S$, without changing any of the results that we would obtain for transition probabilities. There are definitions of S which differ from ours by (sometimes infinite) phases.

In the formulae for the S matrix in equations (14.20) or (14.23) we have assumed that the contributions to the generating functional $Z[J]$ have been properly renormalized and that functional derivatives of it by J yield n point correlation functions of the renormalized fields, so that these correlation functions are free of ultraviolet divergences. It is assumed that the 2 point function as well as all of the other correlation functions are normalized using the on-shell subtraction scheme. In particular this requires that the mass m^2 which appears in the formula is the position of the pole in the 2 point function. Recall that, in the on-shell subtraction scheme, the renormalization constants are fixed so that the pole and the residue at the pole of the 2 point function obey equation (14.1). Also, recall that the coupling constant λ which will govern corrections to free field theory behaviour is defined in the on-shell subtraction scheme as the limit of the irreducible 4 point function at vanishing momentum transfer,

$$\Gamma_I(p, p, -p, -p) = -i\lambda, \quad p^2 = -m^2$$

We can find any matrix element of the S-matrix by taking a matrix element of the formula in equation (14.20) or (14.23). The functional derivatives in equation (14.20) operate on the generating functional and they produce the n point functions of the interacting scalar field theory. The formula is used by expanding the exponential in powers of the exponent to the appropriate order, normal ordering the terms and computing the matrix elements of the terms.

To understand the expression for the S matrix better, we can study some of its simplest properties. For example, we anticipate that, since the vacuum of the quantum field theory is defined as that eigenstate of the Hamiltonian which has the smallest eigenvalue, aside from a possible overall phase, it is unchanged by time evolution. To explore this idea, let us operate the S matrix on the in-field vacuum state,

$$S|\mathcal{O}> = e^{\int dx \phi_{in}^{(-)}(x)[-\partial^2+m^2]\frac{\delta}{\delta J(x)}}|\mathcal{O}> Z[J]\Big|_{J=0} \qquad (14.24)$$

Here, we have used the normal ordered expression for the S matrix in equation (14.23) and the fact that $|\mathcal{O}>$ is the vacuum state of the in-fields so that

$$\phi_{in}^{(+)}(x)|\mathcal{O}> = 0$$

When it is Taylor expanded, the right-hand-side of equation (14.24) contains a series of terms, the generic one of which is[2]

$$\frac{1}{n!} \int \frac{d^3 p_1 \dots d^3 p_n}{(2\pi)^{\frac{3}{2}n}\sqrt{2E(p_1)\dots 2E(p_n)}} a^\dagger(p_1)\dots a^\dagger(p_n)\cdot$$

$$\cdot \lim_{p_1^\mu \to (E(p_1),\vec{p}_1)} \dots \lim_{p_n^\mu \to (E(p_n),\vec{p}_n)} (p_1^2+m^2)\dots(p_n^2+m^2)\cdot$$

$$\cdot \Gamma(p_1,\dots,p_n)(2\pi)^4\delta(p_1+\dots+p_n)$$

The right-hand-side of the above equation vanishes since the delta function requires that all of the energies sum to zero, $p_1^0+\dots+p_n^0 = 0$. But all of the energies have the same sign and they cannot sum to zero. Thus the contribution of all of the terms of the type written above, except the leading, trivial n=0 term vanish and we conclude that

$$S|\mathcal{O}> = |\mathcal{O}>$$

As we anticipated, this is something that one might have expected at the outset, the vacuum remains the vacuum.

Let us ask the question as to what happens to the state which has one particle. Again, our intuition would tell us that, a state with one particle should evolve like it would in free field theory, that is, in the absence of interactions, since there are no other particles for that single particle to interact with. In other words, a single

---

[2] The quantity $\lim_{p_1^\mu \to (E(p_1),\vec{p}_1)}(p_1^2+m^2)\dots\lim_{p_n^\mu \to (E(p_n),\vec{p}_n)}(p_n^2+m^2)\Gamma(p_1,\dots,p_n)$ has a finite limit, called the "on-shell amputated n point function".

particle state should remain a single particle state. Let us confirm this. Consider what we obtain when the S matrix acts on a one-particle state,

$$\mathcal{S}a^{\dagger}(k)|\mathcal{O}> = \tag{14.25}$$

$$e^{\int dx \phi_{in}^{(+)}(x)[-\partial^2+m^2]\frac{\delta}{\delta J(x)}} e^{\int dx \phi_{in}^{(-)}(x)[-\partial^2+m^2]\frac{\delta}{\delta J(x)}} Z[J]\Bigg|_{J=0} \quad : a^{\dagger}(k)|\mathcal{O}> \tag{14.26}$$

The second exponential in the above expression, when Taylor expanded, results in two terms which survive. One is the unit matrix, and the full term that this contribution produces has structure identical to what we found when we discussed stability of the vacuum state above. Analyzing it, we find its contribution equal to the unit operator acting on the one-particle state.

The other contribution contains an annihilation operator which removes the creation operator $a^{\dagger}(k)$ and replaces it with

$$\int d^3x \, \frac{e^{ikx}}{\sqrt{(2\pi)^3 2E(k)}}(-\partial^2+m^2)\frac{\delta}{\delta J(x)}$$

Then, a generic term in a Taylor expansion of the first exponential in equation (14.26) produces terms of the form

$$\frac{1}{n!}\int \frac{d^3p_1 \dots d^3p_n}{(2\pi)^{\frac{3}{2}n}\sqrt{2E(p_1)\dots 2E(p_n)}}a^{\dagger}(p_1)\dots a^{\dagger}(p_n)|\mathcal{O}>$$

$$\lim_{p_1^{\mu}\to(E(p_1),\vec{p}_1)} \dots \lim_{p_n^{\mu}\to(E(p_n),\vec{p}_n)} \lim_{k^{\mu}\to(E(k),\vec{k})} (p_1^2+m^2)\dots(p_n^2+m^2)(k^2+m^2)\cdot$$

$$\cdot\Gamma(p_1,\dots,p_n,-k)(2\pi)^4\delta(p_1+\dots+p_n-k)$$

Because of the four-momentum conserving Dirac delta function, the above equation can only be nonzero if the energies and momenta satisfy

$$E(k) = E(p_1) + \dots + E(p_n), \quad \vec{k} = \vec{p}_1 + \dots + \vec{p}_n$$

A solution of these constraints exists only when $n = 1$ where it is given by $\vec{p} = \vec{k}$. We conclude that

$$\mathcal{S}a^{\dagger}(k)|\mathcal{O}> = a^{\dagger}(k)|\mathcal{O}>$$

The one-particle state is stable.[3]

---

[3] Of course it is possible for a particle to be unstable. If there are more than one type of particle and one of them has a mass that is greater than the sum of two or more masses of some lighter particles, the kinematical constraint that we have noted here does have a solution and if the process is not suppressed by symmetries, the one-particle state of the heavy particle would not be stable. Indeed, though we shall not discuss it in detail here, the LSZ formalism can be used to compute the decay rate of the heavy particle.

As a third simple test of the formula for the S matrix, we expect that in the free field theory, $\lambda \to 0$, limit the S matrix should become trivial. Without interactions, there can be no scattering. To see how this is compatible with our formula for the S matrix, consider the following. We recall that, when $\lambda \to 0$ we know the generating functional exactly. It is given by the expression

$$Z[J] \;=\; e^{-\frac{1}{2}\int dydz J(y)\Delta(y,z)J(z)}$$

Plugging this generating functional into equation (14.20) leads to

$$S_{\lambda=0} =: \exp\left(\int dx\,\phi_{\text{in}}(x)[-\partial^2 + m^2]\frac{\delta}{\delta J(x)}\right) : e^{-\frac{1}{2}\int dydz J(y)\Delta(y,z)J(z)}\Bigg|_{J=0}$$

$$= \; : \exp\left\{-\frac{i}{2}\int dxdy\,\phi_{\text{in}}(x)(\overrightarrow{\partial}_x^2 - m^2)\Delta(x,y)(\overleftarrow{\partial}_y^2 + m^2)\phi_{\text{in}}(y)\right\} \; :$$

$$= \; : \exp\left\{-\frac{i}{2}\int dx\,\phi_{\text{in}}(x)(\partial^2 - m^2)\phi_{\text{in}}(x)\right\} \; : \; = \mathcal{I}$$

which is the unit operator in the Hilbert space of quantum states. We have confirmed that, if we turn off the interaction, the S matrix goes to the unit matrix in the Hilbert space.

Having demonstrated a few rather trivial facts about the S matrix, we are now prepared to study a situation where interactions are important. The prototypical example studies incoming states with two particles. To conserve energy and momentum, the outgoing state must then have at least two particles and, if the incoming energies are not sufficient to create more than two, the outgoing state will be a superposition of two-particle states. The transition between these is called elastic scattering and we will discuss it in more detail in the next section.

14.3   Elastic Two-Particle Scattering

If we turn on the coupling constant λ the S matrix becomes nontrivial in that the T-matrix will be nonzero. As an example of the use of the T matrix, we can use the equation which defines the S matrix (14.20) to find the expression for the T matrix for elastic two-particle scattering which is contained in the transition amplitude from the in-coming state with two particles $|k_1, k_2 >_{\text{in}}$ to another state with two particles, $|\ell_1, \ell_2 >_{\text{in}}$. We shall implement this computation by taking the appropriate matrix element of the S matrix as given in equation (14.20) and dropping the term which is the contribution of the unit matrix, so that we are computing the T matrix. We also observe that, when we expand the exponential in that formula, only even powers of the exponent can contribute. This is due to the $\phi \to -\phi$ symmetry. Moreover, it is easy to see that the second order term gives a vanishing contribution. We are then left with the fourth order term which gives us the expression

$$_{\text{in}} < k_1, k_2| \, i \, \mathbf{T} \, |\ell_1, \ell_2 >_{\text{in}}$$
$$= \frac{1}{4!} \int dx_1...dx_4 \; _{\text{in}} < k_1, k_2|\phi_{\text{in}}(x_1)...\phi_{\text{in}}(x_4)|\ell_1, \ell_2 >_{\text{in}} \cdot$$
$$\cdot (-\partial_1^2 + m^2)...(-\partial_4^2 + m^2) < \mathcal{O}|T\phi(x_1)...\phi(x_4)|\mathcal{O} >$$

Upon inserting the decomposition of the free fields into plane waves with creation and annihilation operators we obtain the expression

$$_{\text{in}} < k_1, k_2| \, i\mathbf{T} \, |\ell_1, \ell_2 >_{\text{in}} = \lim_{k_1^2, k_2^2, \ell_1^2, \ell_2^2 \to -m^2} \cdot$$
$$\cdot \frac{(k_1^2 + m^2)(k_2^2 + m^2)(\ell_1^2 + m^2)(\ell_2^2 + m^2)}{\sqrt{(2\pi)^{12}2E(\vec{k}_1)2E(\vec{k}_2)2E(\vec{\ell}_1)2E(\vec{\ell}_2)}} \Gamma(k_1, k_2, -\ell_1, -\ell_2) \cdot$$
$$\cdot (2\pi)^4 \delta(k_1 + k_2 - \ell_1 - \ell_2) \tag{14.27}$$

We remind the reader that $\Gamma(k_1, k_2, -\ell_1, -\ell_2)$ is the Fourier transform of the 4 point function with the overall momentum conserving delta function removed. It uses the convention where the momenta are all assumed to be ingoing. This gives the minus signs for the two momenta ℓ_1, ℓ_2 of the outgoing particles. These functions were defined in equations (13.45)–(13.47). We note that the limits of $k^2 \to -m^2$ of the factors $(k^2 + m^2)$ times the 4 point function $\Gamma(k_1, k_2, -\ell_1, -\ell_2)$ will turn out to be finite.

Equation (14.27) indicates that all of the information that can be obtained about elastic two-particle scattering is contained in the 4 point function. Let us review some of its properties. To begin, we observe that the 4 point function and the connected 4 point function are related to each other by the identity

$$\Gamma(y_1, y_2, y_3, y_4) = D(y_1, y_2)D(y_3, y_4) + D(y_1, y_3)D(y_2, y_4)$$
$$+ D(y_1, y_4)D(y_2, y_3) + \Gamma_C(y_1, y_2, y_3, y_4) \tag{14.28}$$

where the $D(x, y)$'s are 2 point functions. Moreover, the connected and the irreducible 4 point function are related by

$$\Gamma_C(k_1, k_2, k_3, k_4) = D(k_1)D(k_2)D(k_3)D(k_4)\Gamma_I(k_1, k_2, k_3, k_4) \tag{14.29}$$

Using the property of the 2 point function computed in the on-shell subtraction scheme,

$$\lim_{p^2 \to -m^2} (p^2 + m^2)D(p^2) = -i \tag{14.30}$$

we can see that the amplitude for elastic 2 particle scattering is given by the irreducible 4 point function with the four four-momenta put onto their mass shells,

$$_{in} < k_1, k_2| \, i\mathbf{T} \, |\ell_1, \ell_2 >_{in} =$$
$$\lim_{k_1^2, k_2^2, \ell_1^2, \ell_2^2 \to -m^2} \frac{\Gamma_I(k_1, k_2, -\ell_1, -\ell_2)}{\sqrt{(2\pi)^{12} 2E(\vec{k}_1) 2E(\vec{k}_2) 2E(\vec{\ell}_1) 2E(\vec{\ell}_2)}} (2\pi)^4 \delta(k_1 + k_2 - \ell_1 - \ell_2)$$

$$(14.31)$$

Thus the problem of computing the amplitude for two particle elastic scattering reduces to the problem of computing the irreducible 4 point function for the scalar field theory with the four momentum arguments each obeying the mass shell condition $k^2 + m^2 = 0$.

Now that we have the amplitude, we can ask the question as to what the transition probability between different states must be. For this question, we need to pay attention to the normalization of the ingoing and outgoing states. We generally should consider the transition amplitude between wave-packet states, however, there is a faster route to the version of the transition probability that is most relevant to experiment. We will simply "normalize" the plane-wave states that we are using, by observing that their continuum norm is $< k|k >= \delta(0) = \frac{1}{(2\pi)^3} V$ where V is the volume of space. We unit normalize the incoming and outgoing states by multiplying the amplitude by $\left[1/\sqrt{V/(2\pi)^2} \right]^4$. Then we square the amplitude and divide by the factor $1/VT$ to get the transition probability per unit volume per unit time as[4]

$$\frac{1}{VT} |_{in} < k_1, k_2| \, \mathbf{T} \, |\ell_1, \ell_2 >_{in} |^2 \left(\frac{(2\pi)^3}{V} \right)^4 \qquad (14.32)$$

---

[4] The trick of defining the square of the delta function as

$$\frac{1}{VT} [(2\pi)^4 \delta(k_1 + k_2 - \ell_1 - \ell_2)][(2\pi)^4 \delta(k_1 + k_2 - \ell_1 - \ell_2)]$$

$$= (2\pi)^4 \delta(k_1 + k_2 - \ell_1 - \ell_2) \cdot \frac{1}{VT} \lim_{k_1 + k_2 - \ell_1 - \ell_2 \to 0} (2\pi)^4 \delta(k_1 + k_2 - \ell_1 - \ell_2)$$

and we write the latter expression as

$$\frac{1}{VT} \lim_{k_1 + k_2 - \ell_1 - \ell_2 \to 0} (2\pi)^4 \delta(k_1 + k_2 - \ell_1 - \ell_2)$$

$$= \frac{1}{VT} \lim_{k_1 + k_2 - \ell_1 - \ell_2 \to 0} (2\pi)^4 \int dx \frac{e^{-i(k_1 + k_2 - \ell_1 - \ell_2)x}}{(2\pi)^4}$$

$$= \int dx \frac{1}{VT} = 1$$

VT the infinite volume of spacetime.

What we have now is the probability of a transition between specific unit normalized states of the incoming to outgoing particles. We should multiply it by the number of states of the outgoing particles in a small cell of momentum space, which is the factor $V \frac{d^3\ell}{(2\pi)^3}$ for each outgoing particle. This produces an overall factor of $V \frac{d^3\ell_1}{(2\pi)^3} V \frac{d^3\ell_2}{(2\pi)^3}$. This will give the transition probability from normalized initial single particle states to final states occupying the infinitesimal regions $d^3\ell_1$ and $d^3\ell_2$ of momentum space.

However, if a scattering experiment is performed by scattering beams of particles, or a beam of particles from a fixed target, it is more conventional to use the transition rate per unit of incident particle beam flux rather than the rate per incident particle that we have so far. In the rest frame of one of the particles, this flux is proportional the velocity of the other particle, and to the density of each of the particles. The state $|\vec{k}>$ has a single particle distributed uniformly over space and it therefore has density $1/V$. The relative velocity boosted to the frame where the incoming particles have momenta $\vec{k}_1$ and $\vec{k}_2$ is the relative relativistic velocity $|\vec{v}_1 - \vec{v}_2| = \frac{\sqrt{(k_{1\mu}k_2^\mu)^2 - M^4}}{E(\vec{k}_1)E(\vec{k}_2)}$. The net result is that we should divide by the factor

$$\frac{1}{V^2}|\vec{v}_1 - \vec{v}_2| = \frac{1}{V^2}\frac{\sqrt{(k_{1\mu}k_2^\mu)^2 - M^4}}{E(\vec{k}_1)E(\vec{k}_2)}$$

Putting all of the factors together we get the differential cross-section

$$d\sigma =$$

$$\frac{1}{VT}|_{\text{in}} < k_1, k_2| \mathbf{T} |\ell_1, \ell_2 >_{\text{in}}|^2 \frac{(2\pi)^{12}}{V^4} \frac{V^2 E(\vec{k}_1)E(\vec{k}_2)}{\sqrt{(k_{1\mu}k_2^\mu)^2 - M^4}} V\frac{d^3\ell_1}{(2\pi)^3} V\frac{d^3\ell_2}{(2\pi)^3}$$

$$(14.33)$$

The factors of V cancel and we are left with the expression for the differential cross-section

$$d\sigma =$$

$$\frac{|\Gamma_I(k_1, k_2, -\ell_1, -\ell_2)|^2}{4\sqrt{(k_{1\mu}k_2^\mu)^2 - M^4}} \frac{d^3\ell_1}{(2\pi)^3 2E(\vec{\ell}_1)} \frac{d^3\ell_2}{(2\pi)^3 2E(\vec{\ell}_2)}(2\pi)^4\delta(k_1 + k_2 - \ell_1 - \ell_2)$$

$$(14.34)$$

where we remember that the four-momenta in $|\Gamma(k_1, k_2, -\ell_1, -\ell_2)|^2$ are to be put on their mass shell. A partial or even total cross section is obtained by integrating equation (14.34) over part of or all of the outgoing momenta $\vec{\ell}_1$ and $\vec{\ell}_2$. We note that the cross-section that we have found in equation (14.34) is explicitly Lorentz invariant.

In previous sections we have done an explicit computation of the irreducible 4 point function to the first few orders in perturbation theory. The result, in the on-shell subtraction scheme, which is the one which should be used for n point functions that insert into the LSZ formula is quoted in equation (13.84). In the leading nonzero order, $\Gamma_I(k_1, k_2, -\ell_1, -\ell_2) = -i\lambda + \ldots$ is a constant and computing the cross section is a straightforward exercise in computing whatever integrals over $\vec{\ell}_1$ and $\vec{\ell}_2$ of equation (14.34) are needed.

14.4 Connected and Irreducible Generating Functionals

As we have already observed in our perturbative computations, there are several redundancies in the diagrammatic computation of an n point function. Some of the diagrams which would be used to compute an n point function $\Gamma(k_1, \ldots, k_n)$ simply contribute to the factors in a factorization of $\Gamma(k_1, \ldots, k_n)$ into lower-point correlation functions.

For example, the set of all Feynman diagrams which contribute to a 4 point function has some contributions which simply contribute to 2 point functions which connect pairs of external points in any of three different ways. Such contributions build up the first three terms on the right-hand-side of the equation

$$\Gamma(x_1, x_2, x_3, x_4) = D(x_1, x_2)D(x_3, x_4) + D(x_1, x_3)D(x_2, x_4)$$
$$+D(x_1, x_4)D(x_2, x_3) + \Gamma_C(x_1, x_2, x_3, x_4)$$

and they are redundant with computing the 2 point function, which must usually be done separately to the same order in order to fix the counterterms. The remainder, $\Gamma_C(x_1, x_2, x_3, x_4)$, which we call the connected 4 point function, is obtained by summing connected Feynman diagrams only. It has the advantage that, when it is studied in perturbation theory, only connected Feynman diagrams contribute to it and all diagrams which are not connected can be ignored.

In addition, the perturbative contribution to an n point function can contribute to Feynman diagrams which are reducible, in that they can be rendered disconnected by removing a single internal line. Such contributions can often be taken into account more efficiently by computing irreducible correlation functions and then combining them to find the reducible function. For example, consider the connected 4 point function, $\Gamma_C(x_1, x_2, x_3, x_4)$. We could write it as an irreducible function with 2 point functions attached as follows,[5]

$$\Gamma_C(x_1, x_2, x_3, x_4) =$$
$$\int dw_1 dw_2 dw_3 dw_4 D(x_1, w_1)D(x_2, w_2)D(x_3, w_3)D(x_4, w_4)\Gamma_I(w_1, w_2, w_3, w_4) \quad (14.35)$$

---

[5] The $\Gamma_I(w_1, w_2, w_3, w_4)$ which is defined by this formula might also be called the "amputated" 4 point function as it simply has the 2 point functions from its external lines removed. In this particular case, because of the $\phi \to -\phi$ symmetry, it coincides with the irreducible 4 point function. If that symmetry were absent, there would be additional terms on the right-hand side of this equation with pairs of irreducible three-point functions connected by D's.

Connected Feynman diagrams which contribute to this 4 point function come in two types. One type simply contributes to building up the D's which are attached to the four external points. All such contributions are reducible. We can also see that all reducible diagrams only contribute to the D's. For this we need to remember that, because of symmetry, the 3 point function vanishes.

Moreover, when we consider the contributions to the two point function D, they also come in two types, reducible and irreducible. If we consider the sum of all irreducible diagrams only and define this sum as the irreducible 2 point function $\Pi(x, y)$, then we shall show that the full 2 point function is indeed determined by this irreducible part, by the equation

$$D(x, y) = \Delta(x, y) + \int dw_1 dw_2 \Delta(x, w_1) \Pi(w_1, w_2) D(w_2, y) \qquad (14.36)$$

Thus, to compute the 4 point function we really only need to compute the irreducible 4 point function and the irreducible 2 point function to a given order in perturbation theory. From those, we could reassemble the full 2 point function, the connected 4 point function or the full 4 point functions.

14.4.1 Connected Correlation Functions and the Linked Cluster Theorem

In this section, we will present a definition of what we mean by the term "connected correlation function" in the general case. We will do this by finding a systematic relationship between correlation functions and connected correlation functions. Then we will prove the linked cluster theorem, the fact that the connected correlation functions obtain contributions from connected Feynman diagrams only and, conversely, that the sum of all connected Feynman diagrams is equal to the connected correlation function.

We can define connected correlation functions for any n point function. The definition states that, whenever the correlation functions are obtained from a generating functional by functional derivatives using a formula such as

$$\Gamma(x_1, \ldots, x_n) = \frac{1}{i} \frac{\delta}{\delta J(x_1)} \cdots \frac{1}{i} \frac{\delta}{\delta J(x_n)} Z[J]|_{J=0}$$

the connected correlation functions are also obtained by taking functional derivatives of a connected generating functional,

$$\Gamma_C(x_1, \ldots, x_n) = \frac{1}{i} \frac{\delta}{\delta J(x_1)} \cdots \frac{1}{i} \frac{\delta}{\delta J(x_n)} W[J]|_{J=0}$$

where the original and the connected generating functionals are related by

$$W[J] = \ln Z[J], \quad Z[J] = e^{W[J]}$$

This gives us a way to write a connected correlation function in terms of correlation functions or the inverse, a correlation function in terms of connected correlation functions. This definition is purely combinatorial and it does not depend on perturbation theory or the expansion of correlation functions in Feynman diagrams.

In the following, we will prove the linked cluster theorem. This theorem contains the additional statement that the generating functional $W[J]$ for connected correlation functions is computed in perturbation theory by summing connected Feynman diagrams only.

A simple consequence of the linked cluster theorem shows that the role of the functional integral in the denominator of the ratio of functional integrals which is the starting point of our perturbative computation of a correlation function is to cancel those contributions in the numerator which are called vacuum diagrams. In this section, we will find a simple proof of this theorem in the context of our self-interacting scalar quantum field theory. However, both the linked cluster theorem and this simple consequence of it are much more general than this specific example. Suitably modified for the field content, they apply to the perturbative expansion of any quantum field theory.

14.4.2 Connected Correlation Functions

The functional given by the logarithm of the generating functional

$$W[J] \equiv \ln Z[J]$$

is a generating functional for connected Feynman diagrams. A connected Feynman diagram is defined as a Feynman diagram which has the property that one can trace a path between any two vertices, internal or external in the entire diagram by continually following internal and external lines which are in the diagram. A diagram is said to be disconnected if it is not connected.

For example, in free field theory, that is, when we set the coupling constant λ to zero, we know the generating functional explicitly,

$$Z_{\lambda=0}[J] = e^{-\int dy dz \frac{1}{2} J(y) \Delta(y,z) J(z)}$$

If we then find the connected generating functional by taking the logarithm we find

$$W_{\lambda=0}[J] = -\int dy dz \frac{1}{2} J^{\mu}(y) \Delta_{\mu\nu}(y,z) J^{\nu}(z)$$

This formula tells us the rather obvious fact that, in free field theory, the only connected correlation functions are 2 point functions. Thus, the connected n point functions with $n > 2$ vanish in the free field theory and they can only be nonzero when there are interactions.

We shall now set out to prove that, when interactions are present, $W[J]$ generates correlation functions which are the summations of connected Feynman diagrams only. The generating functional for connected correlation functions is given by the functional expressions

$$e^{W[J]} = \frac{\int [d\phi] e^{iS_0[\phi]+iS_I[\phi]+i\int dx J(x)\phi(x)}}{\int [d\phi] e^{iS_0[\phi]+iS_I[\phi]}}$$

$$= \frac{e^{iS_I[\frac{1}{i}\frac{\delta}{\delta J}]}e^{-\int dydx \frac{1}{2}J(y)\Delta(y,z)J(z)}}{e^{iS_I[\frac{1}{i}\frac{\delta}{\delta J}]}e^{-\int dydx \frac{1}{2}J(y)\Delta(y,z)J(z)}}\bigg|_{J=0}$$

We take the derivative $\frac{d}{d\lambda}W[J]$ and using the expression for $W[J]$ from the second equation above, we get the expression

$$\frac{d}{d\lambda}W[J]\,e^{W[J]}$$
$$= \left\{ i\frac{d}{d\lambda}S_I\left[\frac{1}{i}\frac{\delta}{\delta J}\right] \right\} e^{W[J]} - e^{W[J]}\left[i\frac{d}{d\lambda}S_I\left[\frac{1}{i}\frac{\delta}{\delta J}\right] e^{W[J]} \right]_{J=0}$$

The first term on the right-hand-side of this equation comes from taking the derivative of the numerator in the equation above. The second term comes from the derivative of the denominator. We can rewrite the above expression as

$$\frac{d}{d\lambda}W[J] = i\frac{d}{d\lambda}S_I\left[\frac{1}{i}\frac{\delta}{\delta J} + \frac{1}{i}\frac{\delta W}{\delta J}\right] - \left[i\frac{d}{d\lambda}S_I\left[\frac{1}{i}\frac{\delta}{\delta J}\right] e^{W[J]} \right]_{J=0} \qquad (14.37)$$

where we have used the identity

$$e^{-W}\frac{\delta}{\delta J}e^{W} = \frac{\delta}{\delta J} + \frac{\delta W}{\delta J},$$

and the remaining derivatives $\frac{\delta}{\delta J}$ operate on whatever functionals appear to the right of them. For example

$$\left(\frac{\delta}{\delta J} + \frac{\delta W}{\delta J}\right)^4 = \left(\frac{\delta}{\delta J} + \frac{\delta W}{\delta J}\right)^3 \frac{\delta W}{\delta J} = \left(\frac{\delta}{\delta J} + \frac{\delta W}{\delta J}\right)^2\left(\frac{\delta^2 W}{\delta J^2} + \left(\frac{\delta W}{\delta J}\right)^2\right)$$

$$= \left(\frac{\delta}{\delta J} + \frac{\delta W}{\delta J}\right)\left(\frac{\delta^3 W}{\delta J^3} + 3\frac{\delta^2 W}{\delta J^2}\frac{\delta W}{\delta J} + \left(\frac{\delta W}{\delta J}\right)^3\right)$$

$$= \left(\frac{\delta^4 W}{\delta J^4} + 4\frac{\delta^3 W}{\delta J^3}\frac{\delta W}{\delta J} + 3\frac{\delta^2 W}{\delta J^2}\frac{\delta^2 W}{\delta J^2} + 6\frac{\delta^2 W}{\delta J^2}\left(\frac{\delta W}{\delta J}\right)^2 + \left(\frac{\delta W}{\delta J}\right)^4\right)$$

Now, with the above equation, we can make use of the explicit expression for the interaction terms, to write equation (14.37) as

$$
\frac{d}{d\lambda} W[J] = -\frac{i}{4!} \int dw \frac{d(\lambda Z_\lambda)}{d\lambda} \left\{ \frac{\delta^4 W}{\delta J(w)^4} + 4 \frac{\delta^3 W}{\delta J(w)^3} \frac{\delta W}{\delta J(w)} \right.
$$
$$
\left. + 3 \frac{\delta^2 W}{\delta J(w)^2} \frac{\delta^2 W}{\delta J(w)^2} + 6 \frac{\delta^2 W}{\delta J(w)^2} \left(\frac{\delta W}{\delta J(w)} \right)^2 + \left(\frac{\delta W}{\delta J(w)} \right)^4 \right\}
$$
$$
- \frac{i}{2} \int dw \frac{d(z_{m^2} m^2)}{d\lambda} \left(-\frac{\delta^2 W}{\delta J^2(w)} - \left(\frac{\delta W}{\delta J(w)} \right)^2 \right)
$$
$$
- \frac{i}{2} \int dw \frac{dz}{d\lambda} \left(-\partial^\mu \partial'_\mu \frac{\delta^2 W}{\delta J(w) \delta J(w')} + \partial_\mu \frac{\delta W}{\delta J(w)} \partial^\mu \frac{\delta W}{\delta J(w)} \right)_{w' \to w}
$$
$$
- \Big[J \to 0 \Big] \tag{14.38}
$$

Now, let us assume that $W[J]$ is given by a series in powers of the coupling constant λ

$$
W[J] = \sum_{k=0}^{\infty} \lambda^k W_k[J] \tag{14.39}
$$
$$
W_0[J] = -\frac{1}{2} \int dx\, dy\, J(x) \Delta(x, y) J(y) \tag{14.40}
$$

where we have recalled we know the free field theory limit of the generating functional $W_0[J]$.

Let us assume that we have computed the terms in this expansion up to and including the k'th order, $W_k[J]$. In renormalized perturbation theory, to get $W_k[J]$ we must also have computed the counterterms up to a certain order

$$
\lambda Z_\lambda = \lambda + \lambda^2 z_\lambda^{(1)} + \ldots + \lambda^k z_\lambda^{(k-1)}
$$

$$
m^2 z_{m^2} = m^2 \lambda z_{m^2}^{(1)} + \ldots + m^2 \lambda^k z_{m^2}^{(k)}
$$

$$
z = \lambda^2 z^{(2)} + \ldots + \lambda^k z^{(k)}
$$

Then, knowing all of the above, up to order k, we can plug them into the right-hand-side of equation (14.38) and we find that it determines $W_{k+1}[J]$. For example, since we know $W_0[J]$, we can set λ to zero in equation (14.38) and use our expression for

$W_0[J]$ and its functional derivatives to find

$$
\begin{aligned}
W_1[J] = &-\frac{i}{4!} \int dw \left[\int dy \Delta(w, y) J(y) \right]^4 \\
&- \frac{i}{2} \int dw \left[\int dy \Delta(w, y) J(y) \right] \left[\frac{1}{2} \Delta(w, w) \right] \left[\int dz \Delta(w, z) J(z) \right] \\
&- \frac{i}{2} \int dw \left[\int dy \Delta(w, y) J(y) \right] \left[-z^{(1)} \partial^2 + z_{m^2}^{(1)} m^2 \right] \left[\int dz \Delta(w, z) J(z) \right]
\end{aligned}
$$

$$(14.41)$$

What we obtain, $W_1[J]$, is a connected functional. To see this, we note that, in the first line of equation (14.41), one can continuously follow external or internal lines, the Δ's, between the vertex at w and any of the J's or between any of the J's. Taking four functional derivatives of this term would strip off the J's and would give us the leading order approximation to the connected 4 point function.

The second and third lines of equation (14.41) contain the leading corrections to the connected 2 point function together with the counterterms. This correction to the 2 point function agrees with the one which we found in equation (13.36).

We could continue this approach to perturbation theory by plugging $W_0[J]$ and $W_1[J]$ into the right-hand side of equation (14.38) and determining $W_2[J]$ and iterate this process to find, in principle any order $W_k[J]$.

Now, we can fashion an inductive proof that $W[J]$ is a connected functional of J. By this we mean that each contribution to $W[J]$ is a sum of terms of various degree in J and each term is connected in the sense that we have discussed. Let us assume that we have computed the generating functional up to some order, that is, the contributions $W_1[J], W_2[J], \ldots, W_k[J]$ and that we have found that they are all connected functionals of J.

When we plug these into the right-hand-side of equation (14.38) we find that functional derivatives of them appear there. It is easy to see that the functional derivative of a connected functional is itself connected. Then, in each term on the right-hand-side of equation (14.38), the functional derivatives are connected to each other at the point w which is then subsequently integrated over. What results is a connected functional. Thus, if $W_1[J], W_2[J], \ldots, W_k[J]$ are connected functionals then $W_{k+1}[J]$ must be a connected functional. In particular, this means that it is obtained by a summation of connected Feynman diagrams. We see this very explicitly in our simple example above where $W_0[J]$ and $W_1[J]$ are connected functionals and the individual terms in $W_1[J]$ correspond to sums of topologically identical Feynman diagrams. Then, by the hypothesis of mathematical induction, every perturbative order $W_k[J]$ is a connected functional and $W[J]$ is a connected functional to all orders in perturbation theory.

This completes the proof of the linked cluster theorem, that $W[J]$ is a connected functional and that the connected n point functions which are obtained by taking functional derivates of it are computed by summing only the connected Feynman diagrams.

14.4.3 Cancelation of Vacuum Diagrams

Now that we have established that $W[J]$ is a generating functional for connected correlation functions, it is a simple consequence that the correlation functions that are generated by $Z[J] = e^{W[J]}$ cannot have vacuum diagrams. Functional derivatives of $Z[J]$ are given by polynomials in functional derivatives of $W[J]$ and in such an expression, all of the parts of all of the terms are connected to some external line. Therefore, they can contain no vacuum diagrams.

From this, we conclude that the vacuum diagrams that are generated by perturbation theory must cancel. Moreover, the only contributions that corrections to the functional integral in the denominator of the generating functional can contain are those with vacuum diagrams, since such corrections cannot be connected to any external line. Thus, the absence of vacuum diagrams in correlation functions allows us to ignore them whenever they occur in our use of Wick's theorem and, simultaneously, to ignore perturbative contributions to the denominator. This is a great simplification of many computations.

14.4.4 Irreducible Correlation Function

We have found the generating functional, $Z[J]$ to be an important tool for encoding the data contained in the quantum field theory which we have been studying. Taking n functional derivatives of it by the source $J(x)$ yields the n point correlation functions. We have also found that the logarithm of the generating functional, $W[J] = \ln Z[J]$, is itself a generating functional for connected correlation functions. In this section, we will introduce a third generating functional, which we shall denote by $\Gamma[\bar{\phi}]$. It will be a generating functional for irreducible correlations functions, which we shall define shortly. We will then have three generating functionals

$$Z[J] = 1 + \sum_{n=1}^{\infty} \frac{i^n}{n!} \int dx_1 \dots dx_n \Gamma(x_1, \dots, x_n) J(x_1) \dots J(x_n) \qquad (14.42)$$

$$W[J] = \sum_{n=1}^{\infty} \frac{i^n}{n!} \int dx_1 \dots dx_n \Gamma_C(x_1, \dots, x_n) J(x_1) \dots J(x_n) \qquad (14.43)$$

$$\Gamma[\bar{\phi}] = \sum_{n=2}^{\infty} \frac{1}{n!} \int dx_1 \dots dx_n \Gamma_I(x_1, \dots, x_n) [\bar{\phi}(x_1) - \bar{\phi}_0] \dots [\bar{\phi}(x_n) - \bar{\phi}_0]$$

$$(14.44)$$

$$\frac{\delta}{\delta \bar{\phi}(x)} \Gamma[\bar{\phi}] \bigg|_{\bar{\phi} = \bar{\phi}_0} = 0 \qquad (14.45)$$

for the n point functions, connected n point functions and irreducible n point functions, respectively. When they are computed in perturbation theory, $\Gamma(x_1, \dots, x_n)$ is the sum of all Feynman diagrams (excluding vacuum graphs) with n external lines,

$\Gamma_C(x_1, \ldots, x_n)$ and $\Gamma_I(x_1, \ldots, x_n)$ is the sum of all irreducible Feynman diagrams with n external lines, more precisely, n external points where external lines would be attached. We call $W[J]$ a connected functional of J and $\Gamma[\bar{\phi}]$ an irreducible functional of $\bar{\phi}$. Notice that the expansion of $\Gamma[\bar{\phi}]$ contains a classical field $\phi_0(x)$. This is due to the fact that, after taking some functional derivatives of $\Gamma[\bar{\phi}]$ by $\bar{\phi}$ we must set J to zero to obtain the correlation function. That means that we must find the specific field $\bar{\phi}(x) = \bar{\phi}_0(x)$ where $\delta\Gamma[\bar{\phi}]/\delta\bar{\phi}(x) = 0$. We call this field $\bar{\phi}_0(x)$. It is the value of the 1 point function, $\bar{\phi}_0(x) = <\mathcal{O}|\phi(x)|\mathcal{O}>$. If the Z_2 $\phi \to -\phi$ symmetry is not spontaneously broken, all of the odd point functions, including the 1 point function $\bar{\phi}_0(x)$ vanish. We will assume that this is the case in the remainder of this chapter.

Let us consider a perturbative computation of the 2 point function, $D(x, y) \equiv \Gamma(x, y)$, of the real scalar field. Since the 1 point function vanishes, it is also equal to the connected 2 point function $D(x, y) = \Gamma_C(x, y)$. This function is given by the sum over all of the connected Feynman diagrams which have two external lines.

There is a very useful way to organize this sum. We begin by considering the sum over all irreducible Feynman diagrams with two "external vertices", that is, two points where a external lines would attach. An irreducible diagram is defined as a diagram which cannot be cut into two disconnected pieces by cutting a single internal line. We denote the result of this sum over all irreducible diagrams with two connections for external lines by $\Pi(x, y)$ and we call it the irreducible two point function.

Then, we can find the summation over all possible connected Feynman diagrams with two external lines, $D(x, y)$, by beginning with the sum of the irreducible diagrams $\Pi(x, y)$ and by restoring the reducible diagrams and the external lines. This is a simple process. If we consider the Fourier transforms $D(k)$, $\Pi(k)$ and the propagator $\Delta(k)$,[6] the result is

$$D(k) = \Delta(k) + \Delta(k)\Pi(k)\Delta(k) + \Delta(k)\Pi(k)\Delta(k)\Pi(k)\Delta(k) + \ldots \qquad (14.46)$$

$$= \frac{-i}{k^2 + m^2 - i\epsilon + i\Pi(k)} \qquad (14.47)$$

The series in equation (14.46) is a geometric series. It can therefore be easily summed with the result in equation (14.47). It is the appearance of $k^2 + m^2 - i\epsilon + i\Pi(k)$ in the denominator which inspires the term "self-energy" for the function $i\Pi(k)$.

It might not be clear to the reader why all of the terms on the right-hand-side of equation (14.46) indeed form a geometric series. For this to be so the numerical

---

[6] Here we remind the reader that we are assuming spacetime translation invariance which implies that $D(x, y)$, $\Pi(x, y)$ and $\Delta(x, y)$ are functions of $x - y$ and that their Fourier transforms with delta function removed, $D(k)$, $\Pi(k)$ and $\Delta(k)$ depend on only one momentum, k^μ, and in fact Lorentz invariance renders them functions of k^2 only. The present discussion is purely combinatorial and it would easily be generalized to non Lorentz or translation invariant states of the quantum field theory.

factors which multiply the diagrams must cooperate and we have not derived those factors here. Detailed arguments to this effect can be made, however, we will not present them. Instead, later in this section, we will use functional methods to show that $\Pi(k)$, defined by equation (14.47) is a sum over irreducible diagrams only. Then, a Taylor expansion of equation (14.47) reproduces the series in equation (14.46). We note that, in momentum space, in assembling a reducible diagram from its irreducible components, no further momentum integrals need to be done. It is simply algebra, which is comparatively easy.

An irreducible correlation function is defined as one whose perturbative computation involves only irreducible Feynman diagrams. There is also a purely combinatorial definition of irreducible correlation functions which gives them in terms of connected correlation functions. Irreducible correlation functions can be obtained from a generating functional. To find this generating functional, we begin by defining the classical field $\bar{\phi}(x)$ as the functional derivative

$$\bar{\phi}(x) \equiv \frac{1}{i} \frac{\delta}{\delta J(x)} W[J] \tag{14.48}$$

Note that we do not put the source $J(x)$ to zero in the right-hand-side of this equation. If we put $J \to 0$ we obtain the 1 point function, which we are assuming vanishes due to Z_2 symmetry.

Then let us consider the generating functional which is a Legendre transform of $W[J]$. With the definition of $\bar{\phi}(x)$ in equation (14.48), we consider the following functional

$$\Gamma[\bar{\phi}] = W[J] - i \int dx \, \bar{\phi}(x) J(x) \tag{14.49}$$

Here, as in generic Legendre transformations, we are supposed to solve equation (14.48) to find the source $J(x)$ as a functional of the classical field, $\bar{\phi}(x)$. Then we plug this into equation (14.49) to get $\Gamma[\bar{\phi}]$ as a functional of $\bar{\phi}(x)$.

The basic property of the Legendre transformation is that equation (14.48) is replaced by the equation

$$\frac{1}{i} \frac{\delta}{\delta\bar{\phi}(x)} \Gamma[\bar{\phi}] = -J(x) \tag{14.50}$$

which we can easily derive.[7] To set $J(x)$ to zero, we are supposed to adjust the classical field $\bar{\phi}(x)$ to an extremum of the functional $\Gamma[\bar{\phi}]$ (which we are assuming is

---

[7] To find this equation, consider

$$\frac{1}{i} \frac{\delta \Gamma[\bar{\phi}]}{\delta\bar{\phi}(x)} = \frac{1}{i} \frac{\delta}{\delta\bar{\phi}(x)} \left[W[J] - \int iJ\bar{\phi} \right]$$

$$= \int dy \frac{\delta J(y)}{\delta\bar{\phi}(x)} \frac{1}{i} \frac{\delta W[J]}{\delta J(y)} - \int dy \frac{\delta J(y)}{\delta\bar{\phi}(x)} \bar{\phi}(y) - J(x)$$

$$= -J(x)$$

at $\bar{\phi} = 0$). Then further functional derivatives yield irreducible correlation functions,

$$\Gamma_I(x_1, \ldots, x_n) = \frac{\delta}{\delta\bar{\phi}(x_1)} \cdots \frac{\delta}{\delta\bar{\phi}(x_n)} \Gamma[\bar{\phi}] \Big|_{\bar{\phi}=0} \qquad (14.51)$$

In the following, we will prove that, at the same time, $\Gamma_I(x_1, \ldots, x_n)$ is the sum over all of the connected irreducible Feynman diagrams which would contribute to the n point function.

To gain some intuition, let us study $\Gamma[\bar{\phi}]$ to a few orders in perturbation theory. In the previous section, we obtained $W[J]$ up to order λ in equations (14.40) and (14.41) which we can reorganize a little to get

$$
\begin{aligned}
W[J] = {}& \frac{i}{2} \int dz \Big[i \int dy\, \Delta(x, y) J(y) \Big] \Big[-\partial^2 + m^2 \Big] \Big[i \int dy\, \Delta(x, y) J(y) \Big] \\
& - \frac{i\lambda}{4!} \int dw \Big[i \int dy\, \Delta(w, y) J(y) \Big]^4 \\
& + \frac{i\lambda}{2} \int dw \Big[i \int dy\, \Delta(w, y) J(y) \Big] \times \\
& \quad \times \Big[-z^{(1)}\partial^2 + z^{(1)}_{m^2} m^2 + \frac{\Delta(w, w)}{2} \Big] \Big[i \int dz\, \Delta(w, z) J(z) \Big] \\
& + \ldots
\end{aligned}
\qquad (14.52)
$$

The ellipses here stand for contributions of order λ^2 and higher. From this equation we can find

$$
\begin{aligned}
\bar{\phi}(x) = {}& \frac{1}{i} \frac{\delta}{\delta J(x)} W[J] \\
= {}& i \int dy\, \Delta(x, y) J(y) - \frac{i\lambda}{3!} \int dw\, \Delta(x, w) \Big[i \int dy\, \Delta(w, y) J(y) \Big]^3 \\
& + i\lambda \int dw\, \Delta(x, w) \Big[-z^{(1)}\partial^2 + z^{(1)}_{m^2} m^2 + \frac{\Delta(w, w)}{2} \Big] \Big[i \int dy\, \Delta(w, y) J(y) \Big] \\
& + \ldots
\end{aligned}
\qquad (14.53)
$$

We are supposed to solve this equation for J as a functional of $\bar{\phi}$. We get

$$
\begin{aligned}
J(x) = {}& (-\partial^2 + m^2 - i\epsilon)\bar{\phi}(x) + \frac{\lambda}{3!}\bar{\phi}(x)^3 \\
& - \lambda \Big[-z^{(1)}\partial^2 + z^{(1)}_{m^2} m^2 + \frac{\Delta(x, x)}{2} \Big] \bar{\phi}(x) + \ldots
\end{aligned}
\qquad (14.54)
$$

where the ellipses stand for terms of order λ^2 and higher.

Then, using equation (14.49), we can form

$$\Gamma[\bar{\phi}] = i \int dx \left(-\frac{1}{2}\partial_\mu\bar{\phi}(x)\partial^\mu\bar{\phi}(x) - \frac{m^2 - i\epsilon}{2}\bar{\phi}(x)^2 - \frac{\lambda}{4!}\bar{\phi}(x)^4 \right.$$
$$\left. -\frac{\lambda}{2}\bar{\phi}(x)\left[-z^{(1)}\partial^2 + z^{(1)}_{m^2}m^2 + \frac{\Delta(x,x)}{2} \right]\bar{\phi}(x) \right) + \dots \qquad (14.55)$$

The right-hand-side of this equation is an expansion in λ to the zeroth and first order. Corrections to it, represented by the ellipses, are of order λ^2 and higher.

We see that the first line of $\Gamma[\bar{\phi}]$ in equation (14.55) is simply equal to i times the action $S[\phi]$, where ϕ is replaced by $\bar{\phi}$. The second line of $\Gamma[\bar{\phi}]$ in equation (14.55) contains what we could regard as the leading quantum corrections to the classical action. This important quantity is sometimes called the effective action.[8]

Now, we see that the equation

$$\frac{1}{i}\frac{\delta}{\delta\bar{\phi}(x)}\Gamma[\bar{\phi}] = 0$$

is indeed satisfied by setting $\bar{\phi}(x) = 0$. Then two and higher point irreducible correlation functions are obtained by taking functional derivatives of the functional $\Gamma[\bar{\phi}]$ in equation (14.55) and evaluating them at $\bar{\phi}(x) = 0$. Taking two functional derivatives of $\Gamma[\bar{\phi}]$ in equation (14.55) generates $\Gamma_I(x, y)$ which has the explicit form

$$\Gamma_I(x, y) = \frac{\delta}{\delta\bar{\phi}(x)}\frac{\delta}{\delta\bar{\phi}(y)}\Gamma[\bar{\phi}]\bigg|_{\bar{\phi}=0} \qquad (14.56)$$

$$= -i\left[(-\partial^2 + m^2) + \frac{\lambda}{2}\Delta(x,x) + \lambda(-z^{(1)}\partial^2 + z^{(1)}_{m^2}m^2) \right]\delta(x-y) + \dots \qquad (14.57)$$

where equation (14.56) is the definition of $\Gamma_I(x, y)$ and equation (14.57) is the expansion of $\Gamma_I(x, y)$ to the first order in λ obtained by taking the functional derivatives of the expression in equation (14.55). Now, we recognize that equation (14.57) is (-1) times the perturbative expansion of the inverse of the 2 point function,

$$\Delta^{-1}(x, y) = i(-\partial^2 + m^2)\delta(x - y)$$
$$\Pi(x, y) = -i\left[\frac{\lambda}{2}\Delta(x,x) + \lambda(-z^{(1)}\partial^2 + z^{(1)}_{m^2}m^2) \right]\delta(x-y) + \dots \qquad (14.58)$$

which indeed contains the $\Pi(x, y)$ which is the sum of irreducible Feynman diagrams. Note that this result matches our perturbative computation of the self-energy given in equation (13.38).

---

[8] Reader beware, there are several different things that are called effective actions.

Thus, in the context of our perturbative computation, we see that

$$\Gamma_I(x, y) = -\Delta^{-1}(x, y) + \Pi(x, y) = -D^{-1}(x, y) \qquad (14.59)$$

Of course, this is what we might have expected as it can be derived from our formal definition of the generating functional,

$$\frac{\partial}{\partial\bar{\phi}(x)} \frac{\delta}{\delta\bar{\phi}(y)} \Gamma[\bar{\phi}]\Big|_{\bar{\phi}=\bar{\phi}_0} = -i\frac{\partial J(y)}{\partial\bar{\phi}(x)}\Big|_{\bar{\phi}=\bar{\phi}_0} = -\left[\frac{1}{i}\frac{\partial\bar{\phi}(x)}{\partial J(y)}\right]^{-1}\Big|_{J=0}$$

$$= -\left[\frac{1}{i}\frac{\partial}{\partial J(x)}\frac{1}{i}\frac{\partial}{\partial J(y)}W[J]\Big|_{J=0}\right]^{-1} = -D^{-1}(x, y) \qquad (14.60)$$

where we have used the fact that

$$\int dw\, \frac{\partial\bar{\phi}(x)}{\partial J(w)}\frac{\partial J(w)}{\partial\bar{\phi}(y)} = \delta(x - y) \qquad (14.61)$$

so that we can regard the bilocal object $\frac{\partial\bar{\phi}(x)}{\partial J(y)}$ as the inverse of the bilocal object $\frac{\partial J(x)}{\partial\bar{\phi}(y)}$,

$$\frac{\partial J(x)}{\partial\bar{\phi}(y)} = \left[\frac{\partial\bar{\phi}(y)}{\partial J(x)}\right]^{-1} \qquad (14.62)$$

and $D^{-1}(x, y)$ is the bilocal object which has the property

$$\int dw\, D^{-1}(x, w)D(w, y) = \delta(x - y) \qquad (14.63)$$

The inverse of the 2 point function has the general form

$$D^{-1}(x, y) = \Delta^{-1}(x, y) - \Pi(x, y) \qquad (14.64)$$

where $\Pi(x, y)$ is the self-energy function. We can confirm from equations (14.57) and (14.58) that the first term in equation equation (14.58) is indeed $-\Delta^{-1}(x, y)$ and the subsequent terms indeed correspond to the sum of irreducible Feynman diagrams which contribute to the 2 point function at order λ. We note at this point that the discussion above contains a proof that the series in equation (14.46) is indeed a geometric series.

We can also use the perturbative computation of the generating functional in equation (14.55) to find the irreducible 4 point function by taking four functional derivatives. The result is

$$\Gamma_I(x_1, x_2, x_3, x_4) = \frac{\delta}{\delta\phi(x_1)} \frac{\delta}{\delta\phi(x_2)} \frac{\delta}{\delta\phi(x_3)} \frac{\delta}{\delta\phi(x_4)} \Gamma[\bar{\phi}] \Big|_{\bar{\phi}=\bar{\phi}_0}$$

$$= -i\lambda \int dy \delta(y-x_1)\delta(y-x_2)\delta(y-x_3)\delta(y-x_4) + \mathcal{O}(\lambda^2) \tag{14.65}$$

If we take the Fourier transform (and remove the overall momentum conserving delta function) we find

$$\Gamma_I(k_1, k_2, k_3, k_4) = -i\lambda + \mathcal{O}(\lambda^2) \tag{14.66}$$

This is precisely the first contribution to the irreducible 4 point function coming from the one and only irreducible Feynman diagram at first order in λ. This agrees with what we expect from the formal expression for the irreducible 4 point function which we obtain as

$$\Gamma_I(x_1, x_2, x_3, x_4) = \frac{\delta}{\delta\phi(x_1)} \frac{\delta}{\delta\phi(x_2)} \frac{\delta}{\delta\phi(x_3)} \frac{\delta}{\delta\phi(x_4)} \Gamma[\bar{\phi}] \Big|_{\bar{\phi}=0}$$

$$= \frac{\delta}{\delta\phi(x_1)} \frac{\delta}{\delta\phi(x_2)} \frac{\delta}{\delta\phi(x_3)} [-iJ(x_4)] \Big|_{J=0}$$

$$= \frac{\delta}{\delta\phi(x_1)} \frac{\delta}{\delta\phi(x_2)} \left[-i\frac{\delta J(x_4)}{\delta\phi(x_3)} \right] \Big|_{J=0} = \frac{\delta}{\delta\phi(x_1)} \frac{\delta}{\delta\phi(x_2)} \left[-\frac{1}{i}\frac{\delta\phi(x_3)}{\delta J(x_4)} \right]^{-1} \Big|_{J=0}$$

$$= -\int dy_1 \frac{\delta J(y_1)}{\delta\phi(x_1)} \frac{\delta}{\delta J(y_1)} \int dy_2 \frac{\delta J(y_2)}{\delta\phi(x_2)} \frac{\delta}{\delta J(y_1)} \left[\frac{1}{i}\frac{\delta}{\delta J(x_3)} \frac{1}{i}\frac{\delta}{\delta J(x_4)} W[J] \right]^{-1} \Big|_{J=0}$$

$$= \int dy_1 \dots dy_4 D^{-1}(x_1, y_1) D^{-1}(x_2, y_2) D^{-1}(x_3, y_3) D^{-1}(x_4, y_4) \Gamma_C[y_1, y_2, y_3, y_4] \tag{14.67}$$

where we obtain the last line by simply taking successive functional derivatives and assuming that the 1 and 3 point functions vanish. We see that the irreducible 4 point function is equal to the connected 4 point function with the external lines and all of the corrections to the external lines removed. The latter is also called the "amputated 4 point function".

When we imagine that we have a computation of this functional in perturbation theory. Each of the correlation functions $\Gamma_I(y_1, \dots, y_n)$ is expressed as a sum of Feynman diagrams. Our hypothesis is that all of the diagrams that contribute to them are irreducible.

In order to prove that $\Gamma_I(y_1, \dots, y_n)$ is irreducible, we need to find a way to remove one propagator from each of the Feynman diagrams that contribute to it in all possible ways and we must test each of the results that we obtain to see if they are then connected or disconnected. If they are all connected then $\Gamma_I(y_1, \dots, y_n)$

is indeed irreducible. If any of then are disconnected then $\Gamma_I(y_1, \ldots, y_n)$ is not irreducible.

A way to do this is to take a functional derivative of $\Gamma_I(y_1, \ldots, y_n)$ by the propagator $\Delta(x, y)$. This generates a sum of functions, each of which is a Feynman diagram contributing to $\Gamma_I(y_1, \ldots, y_n)$ with one propagator removed. The sum of these functions would be connected only if every function in the sum was itself connected. Thus we see that, if the derivative of $\Gamma_I(y_1, \ldots, y_n)$ by $\Delta(x, y)$ is connected, then $\Gamma_I(y_1, \ldots, y_n)$ is irreducible.

We must therefore find a way to take a functional derivative by $\Delta(x, y)$. To this end, recall that the first few, quadratic terms in the action have the form

$$S = -\frac{i}{2} \int dxdy \, \phi(x) \Delta^{-1}(x, y) \phi(y) + \ldots \tag{14.68}$$

$$\Delta^{-1}(x, y) = i(-\partial^2 + m^2)\delta(x - y) \tag{14.69}$$

We want to take a derivative of the generating functional by $\Delta(x, y)$. To do this, we introduce a new, infinitesimal bilocal source in to the action

$$S \rightarrow S + \delta S \tag{14.70}$$

$$\delta S = \frac{i}{2} \int dxdydw_1dw_2 \, \phi(x) \Delta^{-1}(x, w_1)\delta\Delta(w_1, w_2)\Delta^{-1}(w_2, y)\phi(y) \tag{14.71}$$

Then, by Taylor expanding the generating functional to first order in $\delta\Delta(w, z)$ we would obtain something proportional to the functional derivative of the generating functional by $\Delta(x, y)$. To see how this works, consider the functional integral representation of the generating functional for connected correlation functions in the theory described by the perturbed action (14.71),

$$W[J] = \ln \frac{\int d\phi e^{iS[\phi]+\frac{1}{2}\int \phi\Delta^{-1}\delta\Delta\Delta^{-1}\phi+i\int J\phi}}{\int d\phi e^{iS[\phi]+\frac{1}{2}\int \phi\Delta^{-1}\delta\Delta\Delta^{-1}\phi}} \tag{14.72}$$

and its first order variation

$$\delta W[J] = -\frac{\int d\phi e^{iS(\phi)+i\int J\phi}\frac{1}{2}\int \phi\Delta^{-1}\delta\Delta\Delta^{-1}\phi}{\int d\phi e^{iS[\phi]+i\int J\phi}} + \frac{\int d\phi e^{iS(\phi)}\frac{1}{2}\int \phi\Delta^{-1}\delta\Delta\Delta^{-1}\phi}{\int d\phi e^{iS[\phi]}}$$

$$+ i\int \delta J\bar{\phi} \tag{14.73}$$

The first line in the above expression comes from expanding the exponentials $e^{\frac{1}{2}\int \phi\Delta^{-1}\delta\Delta\Delta^{-1}\phi}$ in the numerator and denominator of equation (14.72) and the second line comes from the fact that, when we vary the propagator, the source which we want to keep fixed is $\bar{\phi}(x)$ rather than $J(x)$. If $\bar{\phi}(x)$ is to remain fixed, $J(x)$ must vary in a way that depends on $\delta\Delta$ but we shall not need the explicit expression for.

Now, we recall that $\Gamma[\bar\phi]$ is defined by the Legendre transformation and we compute

$$\delta\Gamma[\bar\phi] = \delta\left[W[J] - i\int J\bar\phi\right]$$

$$= -\frac{\int d\phi\, e^{iS(\phi)+i\int J\phi}\frac{1}{2}\int \phi\Delta^{-1}\delta\Delta\Delta^{-1}\phi}{\int d\phi\, e^{iS[\phi]+i\int J\phi}} + \frac{\int d\phi\, e^{iS(\phi)}\frac{1}{2}\int \phi\Delta^{-1}\delta\Delta\Delta^{-1}\phi}{\int d\phi\, e^{iS[\phi]}}$$

$$+ i\int \delta J\bar\phi - i\int \delta J\bar\phi \tag{14.74}$$

We see in the last line of the above equation that the terms with δJ cancel. What remains is

$$\delta\Gamma[\bar\phi] = -\frac{1}{2}\int dx\,dy\,dw_1\,dw_2\, \Delta^{-1}(x,w_1)\delta\Delta(w_1,w_2)\Delta^{-1}(w_2,y)\times$$

$$\times\left[\frac{\int d\phi\, e^{iS(\phi)+i\int J\phi}\phi(x)\phi(y)}{\int d\phi\, e^{iS[\phi]+i\int J\phi}} - (J\to 0)\right] \tag{14.75}$$

This equation contains

$$\frac{\int d\phi\, e^{iS[\phi]+i\int J\phi}\phi(x)\phi(y)}{\int d\phi\, e^{iS[\phi]+i\int J\phi}} = \frac{1}{i^2}\frac{\delta^2 W[J]}{\delta J(x)\delta J(y)} - \frac{1}{i}\frac{\delta W[J]}{\delta J(x)}\frac{1}{i}\frac{\delta W[J]}{\delta J(y)}$$

$$= \frac{1}{i^2}\frac{\delta^2 W[J]}{\delta J(x)\delta J(y)} - \bar\phi(x)\bar\phi(y)$$

With this observation, we can write equation (14.75) as

$$\delta\Gamma[\bar\phi] = \frac{1}{2}\int dx\,dy\,dw_1\,dw_2\,\bar\phi(x)\Delta^{-1}(x,w_1)\delta\Delta(w_1,w_2)\Delta^{-1}(w_2,y)\bar\phi(y)$$

$$+ \frac{1}{2}\int dx\,dy\,dw_1\,dw_2\, \Delta^{-1}(x,w_1)\delta\Delta(w_1,w_2)\Delta^{-1}(w_2,y)\times$$

$$\times\left[\frac{1}{i^2}\frac{\delta^2 W[J]}{\delta J(x)\delta J(y)} - (J\to 0)\right] \tag{14.76}$$

If we remember that the quadratic part of the generating functional contains the inverse of the 2 point function,

$$\Gamma[\bar\phi] = -\frac{1}{2}\int dx\,dy\,\bar\phi(x)D^{-1}(x,y)\bar\phi(y) + \text{ higher orders in } \bar\phi$$

$$= -\frac{1}{2}\int dx\,dy\,\bar\phi(x)\Delta^{-1}(x,y)\bar\phi(y) + \frac{1}{2}\int dx\,dy\,\bar\phi(x)\Pi(x,y)\bar\phi(y)$$

$$+ \text{ higher orders in } \bar\phi \tag{14.77}$$

we see that the first term on the right-hand-side of equation (14.76) is precisely what we would find by varying the first term on the second line of equation (14.77).

In the second term on the right-hand-side of equation (14.76),

$$\frac{1}{2} \int dx dy dw_1 dw_2 \Delta^{-1}(x, w_1) \delta \Delta(w_1, w_2) \Delta^{-1}(w_2, y) \times$$

$$\times \left[\frac{1}{i^2} \frac{\delta^2 W[J]}{\delta J(x) \delta J(y)} - (J \to 0) \right]$$

the last term does not depend on J and therefore it does not depend on $\bar{\phi}$ and it forms a constant, trivial part of the generating functional. The first term in the above equation is a connected functional of J.—since it contains functional derivatives of $W[J]$, which is a connected functional of J. Any number of functional derivatives of a connected functional is still a connected functional. Since then, in turn, J is a functional of $\bar{\phi}$ we only need to know that J is a connected functional of $\bar{\phi}$ and we will have established that this functional is a connected functional of $\bar{\phi}$. We will have demonstrated that removing one propagator from $\Gamma[\bar{\phi}]$, aside from the outlier first term in (14.76), results in a connected functional of $\bar{\phi}$. This means that it is an irreducible functional, which is what we set out to prove.

To show that J is a connected functional of $\bar{\phi}$ we need to examine equation (14.48) which we are supposed to use to determine $J(x)$ as a functional of $\bar{\phi}(x)$. It is a functional derivative of $W[J]$ and we have established that $W[J]$ is a connected functional. $\bar{\phi}$ is therefore a connected functional of J. We need the same statement about the inverse functional, that J is a connected functional of $\bar{\phi}$.

We will give a perturbative proof that $J(x)$ is a connected functional of $\bar{\phi}(x)$. Consider the equation (14.48) which defines the classical field and the fact that $W[J]$ is a generating functional for connected correlation functions, so that

$$\bar{\phi}(x) \equiv \frac{1}{i} \frac{\delta W[J]}{\delta J(x)}$$

$$\bar{\phi}(x) = \sum_{n=1}^{\infty} \frac{i^n}{n!} \int dy_1 \dots dy_n \Gamma_C(x, y_1, \dots, y_n) J(y_1) \dots J(y_n)$$

We can re-arrange this summation as

$$J(x) = \int dy \, \Gamma_C^{-1}(x, y) \bar{\phi}(y)$$

$$- \sum_{n=2}^{\infty} \frac{i^n}{n!} \int dy dy_1 \dots dy_n \Gamma_C^{-1}(x, y) \Gamma_C(y, y_1, \dots, y_n) J(y_1) \dots J(y_n)$$

$$(14.78)$$

Equation (14.78) can be solved iteratively for J as a function of $\bar{\phi}$. We can present the solution as an expansion in the number of $\bar{\phi}$'s.

$$
\begin{aligned}
J(x) = & \int dy\, \Gamma_C^{-1}(x, y)\bar{\phi}(y) \\
& + \frac{1}{2} \int dy dy_1 dy_2 dy_3 dz_1 dz_2 dz_3 \Gamma_C^{-1}(x, y)\Gamma_C(y, y_1, y_2, y_3) \times \\
& \times \Gamma_C^{-1}(y_1, z_1)\Gamma_C^{-1}(y_2, z_2)\Gamma_C^{-1}(y_2, z_3)\bar{\phi}(z_1)\bar{\phi}(z_2)\bar{\phi}(z_3) + \dots \quad (14.79)
\end{aligned}
$$

which is clearly a sum over connected functionals to the order that is displayed and it is also clear that once it is iterated to find all orders, it will be connected to all orders. Since a sum over connected functionals is itself connected, this establishes that J is a connected functional of $\bar{\phi}$, at least in the domain where the expansions that we have used above are valid. This establishes that the right-hand-side of equation (14.49) and (14.76) are connected functionals of $\bar{\phi}$. This in turn establishes that the appropriate parts of $\Gamma[\bar{\phi}]$ is an irreducible functional of $\bar{\phi}$.

Taking derivatives of the generating functional $\Gamma[\bar{\phi}]$ will, at any order, give us combinatorial formulae which relate the irreducible n point functions to the connected n point and lower order connected functions. Like the relationships between n point functions and connected n point functions, they are combinatorial and they define what we mean by a connected or irreducible correlation function in a way that is independent of perturbation theory. In addition, in perturbation theory the connected and irreducible n point functions are gotten by summing only connected or only connected and irreducible Feynman diagrams.

14.5   Derivation of the LSZ Formula

In our setup of a scattering thought experiment, we visualized the particles which participate in the scattering as being prepared in a quantum state where they are very far separated, so much so that they do not interact with each other. Then we modelled that state by matching it with a similar state in a free field theory that results in quantizing the scalar field but ignoring its interaction. We could say that, the initial conditions of the scattering experiment were such that

$$
\text{weak} \lim_{x^0 \to -\infty} \phi(x) = \phi_{\text{in}}(x) \quad (14.80)
$$

The notation "weak lim" is intended to denote the limit in the weak sense, where one should always take matrix elements of the operators on each side of the equation, in normalizable states, before taking the limit. The definition of the free field theory of the in-fields is given in equations (14.2)–(14.4) and solved in equations (14.5)–(14.7).

The same could be said of late times after the scattering process is completed. As the particles fly away from the scattering event, they eventually become separated by large distances, Their interactions are again negligible. The final state can thus also be modelled by free fields which we call the out-fields. They are defined by the weak limit

$$w \lim_{x^0 \to +\infty} \phi(x) = \phi_{\text{out}}(x) \tag{14.81}$$

They have identical structure to the in-fields including their definition and quantization which are identical to that of the in-fields in equations (14.2)–(14.7) with the subscript "in" replaced by "out". However, the basis states $a_{\text{in}}^\dagger(\vec{\ell}_1) \ldots a_{\text{in}}^\dagger(\vec{\ell}_m)|>$ and $a_{\text{out}}^\dagger(\vec{\ell}_1) \ldots a_{\text{out}}^\dagger(\vec{\ell}_m)|>$, although they are both obtained by quantizing identical free field theories, are not identical. They are related by a unitary transformation where the unitary operator which relates them is the S matrix,

$$a_{\text{out}}^\dagger(\vec{\ell}_1) \ldots a_{\text{out}}^\dagger(\vec{\ell}_m)|> = S^\dagger a_{\text{in}}^\dagger(\vec{\ell}_1) \ldots a_{\text{in}}^\dagger(\vec{\ell}_m)|> \tag{14.82}$$

We are interested in computing matrix elements of the S matrix which are the quantities

$$< \mathcal{O}|a_{\text{in}}(\vec{k}_1) \ldots a_{\text{in}}(\vec{k}_n)Sa_{\text{in}}^\dagger(\vec{\ell}_1) \ldots a_{\text{in}}^\dagger(\vec{\ell}_m)|>$$
$$=< \mathcal{O}|a_{\text{out}}(\vec{k}_1) \ldots a_{\text{out}}(\vec{k}_n)a_{\text{in}}^\dagger(\vec{\ell}_1) \ldots a_{\text{in}}^\dagger(\vec{\ell}_m)|> \tag{14.83}$$

The LSZ formula relates these matrix elements to time ordered correlation functions. Its derivation is outlined below.

For the following discussion, we recall the formulae which project creation and annihilation operators from a free scalar field,

$$a_{\text{in/out}}(\vec{k}) = i \int d^3x \frac{e^{iE(\vec{k})t - i\vec{k}\cdot\vec{x}}}{\sqrt{(2\pi)^3 2E(\vec{k})}} \overleftrightarrow{\frac{\partial}{\partial x^0}} \phi_{\text{in/out}}(x)$$

$$a_{\text{in/out}}^\dagger(\vec{k}) = -i \int d^3x \frac{e^{-iE(\vec{k})t + i\vec{k}\cdot\vec{x}}}{\sqrt{(2\pi)^3 2E(\vec{k})}} \overleftrightarrow{\frac{\partial}{\partial x^0}} \phi_{\text{in/out}}(x)$$

where $\overleftrightarrow{\frac{\partial}{\partial x^0}} \equiv \left(\overrightarrow{\frac{\partial}{\partial x^0}} - \overleftarrow{\frac{\partial}{\partial x^0}} \right)$ and we note that the right-hand-sides of these equations, the space integrals which project out the creation and annihilation operators, are independent of time. They can thus be evaluated at any time.

Then, we can develop a formula which relates an in-field and out-field annihilation operator, perhaps in the presence of other operators, which can be summarized as follows,

$$a_{\text{out}}(k)\, \mathcal{T} e^{i\int J\phi} - \mathcal{T} e^{i\int J\phi} a_{\text{in}}(k) = \tag{14.84}$$

$$i \int d^3x \, \frac{e^{iE(\vec{k})t - i\vec{k}\cdot\vec{x}}}{\sqrt{(2\pi)^3 2E(\vec{k})}} \overset{\leftrightarrow}{\frac{\partial}{\partial x^0}} \left[\phi_{\text{out}}(x)\, \mathcal{T} e^{i\int J\phi} - \mathcal{T} e^{i\int J\phi} \phi_{\text{in}}(x) \right] \tag{14.85}$$

$$= i \left[w \lim_{t\to\infty} - w \lim_{t\to-\infty} \right] \int d^3x \, \frac{e^{iE(\vec{k})t - i\vec{k}\cdot\vec{x}}}{\sqrt{(2\pi)^3 2E(\vec{k})}} \overset{\leftrightarrow}{\frac{\partial}{\partial x^0}} \mathcal{T}\phi(x) e^{i\int J\phi} \tag{14.86}$$

$$= \left[w \lim_{t\to\infty} - w \lim_{t\to-\infty} \right] \int d^3x \, \frac{e^{iE(\vec{k})t - i\vec{k}\cdot\vec{x}}}{\sqrt{(2\pi)^3 2E(\vec{k})}} \overset{\leftrightarrow}{\frac{\partial}{\partial x^0}} \frac{\delta}{\delta J(x)} \mathcal{T} e^{i\int J\phi} \tag{14.87}$$

$$= \int_{-\infty}^{\infty} dx^0 \frac{\partial}{\partial x^0} \int d^3x \, \frac{e^{iE(\vec{k})t - i\vec{k}\cdot\vec{x}}}{\sqrt{(2\pi)^3 2E(\vec{k})}} \overset{\leftrightarrow}{\frac{\partial}{\partial x^0}} \frac{\delta}{\delta J(x)} \mathcal{T} e^{i\int J\phi} \tag{14.88}$$

$$= \int dx \, \frac{e^{iE(\vec{k})t - i\vec{k}\cdot\vec{x}}}{\sqrt{(2\pi)^3 2E(\vec{k})}} \left(-\overset{\rightarrow}{\partial}^2 + m^2 \right) \frac{\delta}{\delta J(x)} \mathcal{T} e^{i\int J\phi} \tag{14.89}$$

In equation (14.85) we have projected the in-field and out-field annihilation operators out of the free fields $\phi_{\text{in/out}}(x)$ using the formula that is appropriate for a free scalar field. Then we noted that the right-hand-side of (14.86) does not depend on x^0, so x^0 could be set to an infinitely late or infinitely early time in the case of the out- or in-field. We also recognized that, at very late or early times, the in-field $\phi_{\text{out/in}}(x)$ can be replaced by the field operator $\phi(x)$ of the interacting field theory. We remind the reader that this is valid only for matrix elements of this formula in normalizable wave-packet states. It is in such states where the particles would indeed become far separated at late or at early times. We also note that we handled the issue of the ordering of the operators on the right-hand-side by making use of the time ordering which, for the large positive and negative time limits will place $\phi(x)$ in the correct position. In equation (14.88) we write the difference of infinite late and early time limits as the integral over all time of a derivative by time. Then, in equation (14.89) we refine the formula for the first term using the fact that the plane wave satisfies the wave equation so that

$$\left(\frac{\partial^2}{\partial x^0} \right)^2 e^{-ikx} = \vec{\nabla}^2 e^{-ikx}$$

We integrate the spatial derivatives in $\vec{\nabla}^2 e^{-ikx}$ by parts and we assume that the surface terms vanish.

A similar formula with creation operators can be derived by identical reasoning. In summary, we have the master formulae

$$a_{\text{out}}(k) \, T e^{i \int J \phi} - T e^{i \int J \phi} a_{\text{in}}(k) =$$

$$\int dx \frac{e^{i E(\vec{k})t - i \vec{k} \cdot \vec{x}}}{\sqrt{(2\pi)^3 2 E(\vec{k})}} \left(-\vec{\partial}^2 + m^2\right) \frac{\delta}{\delta J(x)} T e^{i \int J \phi} \qquad (14.90)$$

$$T e^{i \int J \phi} a_{\text{in}}^\dagger(k) - a_{\text{out}}^\dagger(k) \, T e^{i \int J \phi} =$$

$$\int dx \frac{e^{-i E(\vec{k})t + i \vec{k} \cdot \vec{x}}}{\sqrt{(2\pi)^3 2 E(\vec{k})}} \left(-\partial^2 + m^2\right) \frac{\delta}{\delta J(x)} T e^{i \int J \phi} \qquad (14.91)$$

We can use the formulae in equations (14.90) and (14.91) to study the S matrix element (if we assume that none of the k's are equal to any of the ℓ's)

$$< \mathcal{O}|a_{\text{in}}(\vec{k}_1) \dots a_{\text{in}}(\vec{k}_n) \, S \, a_{\text{in}}^\dagger(\vec{\ell}_1) \dots a_{\text{in}}^\dagger(\vec{\ell}_m)|\mathcal{O} >$$

$$= < \mathcal{O}|a_{\text{out}}(\vec{k}_1) \dots a_{\text{out}}(\vec{k}_n) \, a_{\text{in}}^\dagger(\vec{\ell}_1) \dots a_{\text{in}}^\dagger(\vec{\ell}_m)|\mathcal{O} > \qquad (14.92)$$

$$= \prod_{i=1}^{m} \int \frac{dx_i \, e^{i \ell_i x_i}}{\sqrt{(2\pi)^3 2 E(\vec{\ell}_i)}} (-\partial_{x_i}^2 + m^2) \frac{\delta}{\delta J(x_i)} \cdot$$

$$\cdot \prod_{j=1}^{n} \int \frac{dy_j \, e^{-i k_j y_j}}{\sqrt{(2\pi)^3 2 E(\vec{k}_j)}} (-\partial_{y_j}^2 + m^2) \frac{\delta}{\delta J(y_j)} Z[J] \Bigg|_{J=0} \qquad (14.93)$$

When some of the k's and ℓ's can be equal this formula is corrected by delta function terms. The latter are summarized by taking the exponential, so that the S matrix as an operator has the form

$$S = \; : e^{\int dx \phi_{\text{in}}(x)(-\partial^2 + m^2) \frac{\delta}{\delta J(x)}} : Z[J] \Big|_{J=0} \qquad (14.94)$$

This is the LSZ formula (14.20) quoted in Sect. 14.2. Taking any matrix elements of both sides of equation (14.94) in the in-field states reproduces the expression (14.93) for matrix elements of the S matrix. It gives us a direct relationship between the S matrix and time ordered correlation functions.

It is easy to derive a similar formula for photons and Dirac fields that would apply to quantum electrodynamics. We quote the general formula here, in the Feynman gauge, without derivation,

$$S = \lim_{\eta, \bar{\eta}, J \to 0}$$

$$: e^{\int dx \left[A_{\text{in}}^\mu(x)(-\partial^2) \frac{\vec{\delta}}{\delta J^\mu(x)} + \bar{\psi}_{\text{in}}(x)(\slashed{\partial} + m) \frac{\vec{\delta}}{i \delta \bar{\eta}(x)} \right]} Z e^{\int dx \frac{\overleftarrow{\delta}}{i \delta \eta(x)} (\overleftarrow{\slashed{\partial}} + m) \psi_{\text{in}}(x)} :$$

$$Z = < \mathcal{O}|T e^{i \int dy [\bar{\eta}(y) \psi(y) + \bar{\psi}(y) \eta(y) + A_\mu(y) J^\mu(y)]}|\mathcal{O} > \qquad (14.95)$$

Perturbative Quantum Electrodynamics

15

As we have already discussed in previous chapters, particularly in Sects. 10.5 and 12.5, the quantum field theory which is composed of the coupled Maxwell and Dirac theories is called quantum electrodynamics. In this chapter we will give an elementary introduction to the techniques of perturbation theory that can be used to analyze this important physical model. It can describe any system which has a massless vector field interacting with a massive Dirac field. However, in line with its most successful application, from here on, we will call the Dirac field the "electron" and the massless scalar field the "photon".

A generating functional for the time-ordered correlation functions of quantum electrodynamics is defined by the expression

$$Z[J, \eta, \bar{\eta}] = < \mathcal{O}|\mathcal{T}e^{i \int dx \left[A_\mu(x)J^\mu(x)+\bar{\eta}(x)\psi(x)+\bar{\psi}(x)\eta(x)\right]}|\mathcal{O} > \qquad (15.1)$$

where, as usual, the symbol $\mathcal{T}$ indicates that operators that are placed to the right of it should be ordered so that their time arguments are descending as read from left to right. This generating functional is given by the ratio of functional integrals

$$Z[J, \eta, \bar{\eta}] = \frac{\int [dAd\psi d\bar{\psi}]e^{i\mathbf{S}+i \int dx \left[A_\mu(x)J^\mu(x)+\bar{\eta}(x)\psi(x)+\bar{\psi}(x)\eta(x)\right]}}{\int [dAd\psi d\bar{\psi}]e^{i\mathbf{S}}} \qquad (15.2)$$

where the action and Lagrangian density are

$$\mathbf{S} = \int dx \, \mathcal{L}(x) \qquad (15.3)$$

© The Author(s), under exclusive license to Springer Nature Singapore Pte Ltd. 2023
G. W. Semenoff, *Quantum Field Theory*, Graduate Texts in Physics,
https://doi.org/10.1007/978-981-99-5410-0_15

$$\mathcal{L}(x) = -i\bar{\psi}(x)\left[\partial\!\!\!/ - ie A\!\!\!/(x) + m\right]\psi(x) - \frac{1}{4}F_{\mu\nu}(x)F^{\mu\nu}(x) - \frac{\xi}{2}(\partial_{\mu}A^{\mu}(x))^{2}$$

$$(15.4)$$

The equations above are identical to equations (12.71) and (12.72), except that we have dropped the subscript ξ, which indicated that gauge fixing had already been taken care of, from Z, $\mathbf{S}$ and $\mathcal{L}$. We will assume the relativistic gauge fixing which results in the term multiplied by ξ in the Lagrangian density (15.4).

The photon is a boson field so the integration variable $A_{\mu}(x)$ in the functional integral and the source $J^{\mu}(x)$ are ordinary commuting functions. The electron, on the other hand, is a fermion and the fields $\psi(x)$ and $\bar{\psi}(x)$ which are integrated in the functional integral as well as the sources $\eta(x)$ and $\bar{\eta}(x)$ are anti-commuting functions.

We will use this functional integral representation of the generating functional as the starting point of our study of quantum electrodynamics using perturbation theory. Our main goal will be to learn how to compute correlation functions in an expansion in the parameter e which appears in front of the only term in the Lagrangian density which is not quadratic in the fields. Being cubic, the term $-e\bar{\psi}(x)A\!\!\!/(x)\psi(x)$ represents an interaction between electrons and photons. It is also the term which prevents us from writing down an exact solution of this quantum field theory, as we have already done in previous chapters for the case where $e = 0$.

15.1   Counterterms

Before we proceed, there are some observations that are required in order to complete our formulation of the quantum field theory. First of all, in four spacetime dimensions, the quantum field theory with Lagrangian density (15.4) is a renormalizable quantum field theory. The Lagrangian density contains all of the marginal and relevant local operators that one could write down, given the field content and the constraints of symmetry. The requisite symmetries are translation invariance, Lorentz invariance and gauge invariance. Inclusion of all of the marginal operators is required for this be a renormalizable quantum field theory. We can also rescale the fields and redefine the parameters to obtain independent coefficients for all of the terms in the Lagrangian density. This rescaling would replace the Lagrangian density in equation (15.4) by

$$\mathcal{L}(x) = -iZ_{2}\bar{\psi}(x)\partial\!\!\!/\psi(x) - eZ_{1}\bar{\psi}(x)A\!\!\!/(x)\psi(x) - imZ_{m}\bar{\psi}(x)\psi(x)$$

$$-\frac{Z_{3}}{4}F_{\mu\nu}(x)F^{\mu\nu}(x) - \frac{\xi Z_{\xi}}{2}(\partial_{\mu}A^{\mu}(x))^{2}$$

$$(15.5)$$

where the parameters e and m will be taken to be the some physical parameters related to the electric charge and mass of the electron. The precise identification of these parameters with physically observable parameters will depend on the renormalization scheme, and we will discuss this in more detail once we have introduced renormalization. There is, for example, a renormalization scheme where they are simply the electric charge and the mass of the electron, which is represented by the Dirac field. The five adjustable parameters of the theory are then Z_1, Z_2, Z_3, Z_m, Z_ξ. They are called renormalization constants. They will be adjusted in order to model the behaviour of the physical system that we are describing.

Once the parameters are fixed by computing a few predictions for physical behaviour of photons and electrons and fitting them to experimental data, every other computation we would subsequently do is a prediction about the behaviour of photons and electrons. These predictions either match observed physical phenomena and the model is correct or they do not and the model is wrong and it does not describe the physical system. It is an absolutely remarkable fact that, for modelling the behaviour of electrons and photons, this model, to as high a degree of precision as can currently be computed, and for every process which it is expected to describe, agrees with what is measured in the laboratory.

There are circumstances where some of the counterterms are related by symmetry. With a gauge invariant regularization and an on-shell subtraction scheme, $Z_1 = Z_2$. In fact, the ultraviolet divergent parts of Z_1 and Z_2 are always equal. This is a result of the Ward–Takahashi identities that are discussed in Sect. 15.13. Since the exact equality of the finite parts of these two renormalization constants is scheme dependent we will not assume that they are equal at the outset.

As well as being ultraviolet divergent, some quantum amplitudes that we compute are also infrared divergent. This is a divergence which comes from the small momentum regime in Feynman integrals which can have a singularity due to the fact that the photon is massless. Such an infrared divergence will appear when we evaluate the vertex function and when we put the electron four-momenta in that function on their mass shells. We will also discuss a way to deal with this divergence. A way to regularize the infrared divergent which preserves Poincare symmetry is to introduce a small mass for the photon. This is done by adding the term

$$-\frac{\kappa^2}{2} Z_\kappa A_\mu(x) A^\mu(x)$$

to the Lagrangian density. This is a relevant operator. It doesn't upset renormalizability. In fact, if we do not require gauge invariance, it is the one term that we would add to the Lagrangian density so that it contains all of the relevant and marginal Poincare and phase invariant operators. We have included a renormalization constant Z_κ which would generally be needed to renormalize the theory after this term is added.

The experimental upper bound for the mass of the photon is very small and electrodynamics does behave as though the photon is massless, so eventually one would like to take a limit where κ^2 is very small or zero.

In Sect. 15.13 we shall examine an infinite set of identities for correlation functions called the Ward–Takahashi identities. Those identities are a consequence of the phase symmetry of the Dirac theory and they pose restrictions on how the electrons which are described by the Dirac theory can couple to photons. There we will use these identities to argue that the gauge fixing term and the photon mass term do not renormalize if we regularize the theory in a way that is consistent with Poincare symmetry and with gauge invariance. This implies that, at the outset, we can set

$$Z_\xi = 1 = Z_\kappa$$

From now on, we will assume that these renormalization constants are indeed equal to one. We will eventually confirm that the leading orders of perturbation theory are consistent with this choice.

Beyond simple modelling, the renormalization constants Z_1, Z_2, Z_3, Z_m play another important role. Perturbative computations of correlation functions will encounter infinities coming from the large energy and momentum regime in Feynman integrals. These infinities are called ultraviolet divergences. In this renormalizable quantum field theory, the renormalization constants can be tuned, order by order in perturbation theory, so that the ultraviolet infinities are canceled and so that certain renormalization conditions are met.

The constants Z_1, Z_2, Z_3, Z_m are each given by power series in the coupling constant e,

$$Z_1 = 1 + z_1 \qquad\qquad z_1 = \sum_{k=2}^{\infty} z_1^{(k)} \quad z_1^{(k)} \sim e^k \tag{15.6}$$

$$Z_2 = 1 + z_2 \qquad\qquad z_2 = \sum_{k=2}^{\infty} z_2^{(k)} \quad z_2^{(k)} \sim e^k \tag{15.7}$$

$$Z_m = 1 + z_m \qquad\qquad z_m = \sum_{k=2}^{\infty} z_m^{(k)} \quad z_m^{(k)} \sim e^k \tag{15.8}$$

$$Z_3 = 1 + z_3 \qquad\qquad z_3 = \sum_{k=2}^{\infty} z_3^{(k)} \quad z_3^{(k)} \sim e^k \tag{15.9}$$

The expansion of these counterterms will turn out to contain only even powers of e.

The gauge-fixed action and Lagrangian density now appear as

$$S = S_0 + S_{\text{int}} \tag{15.10}$$

$$S_0 = \int dx \, \mathcal{L}_0(x) \tag{15.11}$$

$$S_{\text{int}} = \int dx \, \mathcal{L}_{\text{int}}(x) = \sum_{n=1}^{\infty} S_{\text{int}}^{(n)} = \sum_{n=1}^{\infty} \int dx \, \mathcal{L}_{\text{int}}^{(n)}(x) \tag{15.12}$$

$$\mathcal{L}_0(x) = -i\bar{\psi}[\slashed{\partial} + m]\psi - \frac{1}{4} F_{\mu\nu} F^{\mu\nu} - \frac{\kappa^2}{2} A_\mu A^\mu - \frac{\xi}{2}(\partial_\mu A^\mu)^2 \tag{15.13}$$

$$\mathcal{L}_{\text{int}}^{(1)}(x) = -e\bar{\psi}A\!\!\!/\psi \tag{15.14}$$

$$\mathcal{L}_{\text{int}}^{(n>1)}(x) = -z_1^{(n-1)}e\bar{\psi}A\!\!\!/\psi - iz_2^{(n)}\bar{\psi}\partial\!\!\!/\psi - iz_m^{(n)}m\bar{\psi}\psi$$
$$- \frac{z_3^{(n)}}{4}F_{\mu\nu}F^{\mu\nu} \tag{15.15}$$

where we have included the terms with z_1, z_2, z_3, z_m in the interactions. These terms in the interaction are called counterterms.

15.2   The Generating Functional in Perturbation Theory

Due to the presence of the cubic coupling term in the exponent in the integrand, it is not known how to take the functional integral of interacting quantum electrodynamics exactly. However, for quantum electrodynamics, the parameter in front of the cubic term is small and perturbation theory is a viable and important tool for studying this functional integral. To implement perturbation theory, we consider the Taylor expansion of the numerator and the denominator of equation (12.71) in powers of the interactions, which will be proportional to powers of e. This involves expanding the exponential of the action so that the order e^k term is

$$e^{i\mathbf{S}} = e^{i\mathbf{S}_0}e^{i\mathbf{S}_{\text{int}}} = e^{i\mathbf{S}_0}\sum_{k=0}^{\infty}\sum_{n_1,n_2\ldots=0}^{\infty}\frac{\delta_{k,n_1+n_2+\ldots}}{n_1!n_2!\ldots}\left[i\mathbf{S}_{\text{int}}^{(1)}\right]^{n_1}\left[i\mathbf{S}_{\text{int}}^{(2)}\right]^{n_2}\ldots$$
$$= e^{i\mathbf{S}_0}\sum_{k=0}^{\infty}\sum_{n_1,n_2\ldots=0}^{\infty}\frac{\delta_{k,n_1+n_2+\ldots}}{n_1!n_2!\ldots}\left[i\int dx\mathcal{L}_{\text{int}}^{(1)}(x)\right]^{n_1}\left[i\int dy\mathcal{L}_{\text{int}}^{(2)}(y)\right]^{n_2}\ldots \tag{15.16}$$

where $\mathbf{S}_0$ is the free-field theory action and $\mathbf{S}_{\text{int}}^{(n)}$ is the term in the interaction action which is of order e^n. The $\mathcal{L}_{\text{int}}^{(n)}(x)$ are the order e^n components of the interaction Lagrangian density given in equations (15.12), (15.14) and (15.15). This expansion can either be plugged into the functional integral representation of the generating functional (15.2), into both the numerator and the denominator integrals, or applied directly to the computation of correlation functions by plugging it into the functional integral representation of the individual correlation functions,

$$\Gamma_{\mu_1\ldots\mu_n a_1\ldots a_k, b_1\ldots, b_k}(x_1, \ldots, x_n, y_1, \ldots, y_k, z_1, \ldots, k_k) \equiv$$
$$< \mathcal{O}|TA_{\mu_1}(x_1)\ldots A_{\mu_n}(x_n)\psi_{a_1}(y_1)\ldots\psi_{a_k}(y_k)\bar{\psi}_{b_1}(z_1)\ldots\bar{\psi}_{b_\ell}(z_\ell)|\mathcal{O} >$$
$$= \frac{\int[dAd\psi d\bar{\psi}]e^{i\mathbf{S}}A_{\mu_1}(x_1)\ldots A_{\mu_n}(x_n)\psi_{a_1}(y_1)\ldots\psi_{a_k}(y_k)\bar{\psi}_{b_1}(z_1)\ldots\bar{\psi}_{b_\ell}(z_\ell)}{\int[dAd\psi d\bar{\psi}]e^{i\mathbf{S}}} \tag{15.17}$$

When the expansion in equation (15.16) is plugged into the numerator and denominator in equation (15.17) what remains, for each integral, is a series of Gaussian integrals with monomials of the fields and their derivatives in the integrand. These integrals can be done explicitly and we have already done them in previous chapters. Wick's theorem, which we will review shortly, summarizes the results of these integrations.

An alternative way of writing the generating functional which is sometimes valuable is to use the explicit forms of the generating functionals for the non-interacting fields and correct it for interactions. The expression is

$$Z[J, \eta, \bar\eta] = \dfrac{e^{iS_{\text{int}}\left[\frac{1}{i}\frac{\delta}{\delta J}\cdot\frac{1}{i}\frac{\partial}{\partial\eta},\,-\frac{1}{i}\frac{\delta}{\delta\eta}\right]}e^{-\int dxdy\left[\frac{1}{2}J^\mu(x)\Delta_{\mu\nu}(x,y)J^\nu(y)+\bar\eta(x)g(x,y)\eta(y)\right]}}{e^{iS_{\text{int}}\left[\frac{1}{i}\frac{\delta}{\delta J}\cdot\frac{1}{i}\frac{\partial}{\partial\eta},\,-\frac{1}{i}\frac{\delta}{\delta\eta}\right]}e^{-\int dxdy\left[\frac{1}{2}J^\mu(x)\Delta_{\mu\nu}(x,y)J^\nu(y)+\bar\eta(x)g(x,y)\eta(y)\right]}}\Bigg|_{J\eta\bar\eta=0}$$
(15.18)

$$g_{ab}(y, z) = \int \frac{dp}{(2\pi)^4}e^{ip(y-z)}\frac{i\slashed{p}-m}{p^2+m^2-i\epsilon}$$
(15.19)

$$\Delta_{\mu\nu}(x, y) = -i\int \frac{dk}{(2\pi)^4}e^{ik(x-y)}\left(\frac{\eta_{\mu\nu}-k_\mu k_\nu/k^2}{k^2+\kappa^2-i\epsilon}+\frac{k_\mu k_\nu/k^2}{\xi k^2+\kappa^2-i\epsilon}\right)$$
(15.20)

where $g_{ab}(y, z)$ and $\Delta_{\mu\nu}(x, y)$ are the propagators for the Dirac and photon fields, respectively. We have included the photon mass κ in the photon propagator. Its role there is to regulate infrared divergences and the massless photon propagator is gotten by putting $\kappa \to 0$. The quantity $iS_{\text{int}}\left[\frac{1}{i}\frac{\delta}{\delta J}, \frac{1}{i}\frac{\partial}{\partial\eta}, -\frac{1}{i}\frac{\delta}{\delta\eta}\right]$ in equation (15.18) is gotten by taking the full interaction action from equations (15.12), (15.14) and (15.15) and making the substitutions $A_\mu(x) \to \frac{1}{i}\frac{\delta}{\delta J^\mu(x)}$, $\psi(x) \to \frac{1}{i}\frac{\delta}{\delta\bar\eta(x)}$ and $\bar\psi(x) \to -\frac{1}{i}\frac{\delta}{\delta\eta(x)}$ for the fields, wherever they occur. Then, in formula (15.18) one can expand the exponential, as in equation (15.16) and the contributions to perturbation theory are gotten by taking functional derivatives.

15.2.1 Wick's Theorem for Photons and Electrons

In order to evaluate the Gaussian functional integrals with additional monomials of the fields located in the integrands, we can use formulae which we derived in the previous chapters for dealing with similar integrals in the study of free photons and non-interacting Dirac fields. These formulae are summarized in the following version of Wick's theorem. The master formula is

$$\frac{\int [d\,A d\psi d\bar\psi] e^{iS_0} A_{\mu_1}(x_1)\dots A_{\mu_n}(x_n)\psi_{a_1}(y_1)\dots\psi_{a_m}(y_m)\bar\psi_{b_1}(z_1)\dots\bar\psi_{b_\ell}(z_\ell)}{\int [d\,A d\psi d\bar\psi] e^{iS_0}}$$

$$= \sum_{\text{pairings}} (-1)^{\text{deg}} \prod_{\text{photon pairs}} \Delta_{\mu_i \mu_j}(x_i, x_j) \prod_{\psi\bar\psi \text{ pairs}} g_{a_j b_k}(y_j, z_k) \qquad (15.21)$$

where "deg" is the number of exchanges of positions that is needed to change the order of the electron integration variables ψ and $\bar\psi$ from the ordering in the integrand to the ordering in the pairing. In the pairings, photons are only paired with photons and ψ's are always paired with $\bar\psi$'s. Each photon field is labeled by its position coordinate x_i and its vector index μ_i. Each electron field has a position and Dirac index y_j, a_j or z_k, b_k. These become indices on the propagators $\Delta_{\mu_i \mu_j}(x_i, x_j)$ and $g_{a_j b_k}(y_j, z_k)$ which are associated with each pair. Clearly, the pairings can only be implemented if there are an even number of A_μ's and the same number of ψ's as $\bar\psi$'s in the integrand, that is, if n is even and if $m = \ell$. If the number of A_μ's is odd or if the numbers of ψ's and $\bar\psi$'s are not equal, the contribution vanishes. Equation (15.21) is called Wick's theorem.

As an example of an application of the Wick's theorem (15.21), consider the functional integral

$$\frac{\int [d\,A d\psi d\bar\psi] e^{iS_0}\psi_{a_1}(y_1)\psi_{a_2}(y_2)\bar\psi_{b_1}(z_1)\bar\psi_{b_2}(z_2)}{\int [d\,A d\psi d\bar\psi] e^{iS_0}}$$

$$= -g_{a_1 b_1}(y_1, z_1) g_{a_2 b_2}(y_2, z_2) + g_{a_1 b_2}(y_1, z_2) g_{a_2 b_1}(y_2, z_1)$$

There are two pairings, corresponding to the two terms on the right-hand side. Each term contains a product over pairs of the electron propagators for each pair. The first term has a minus sign because the number of interchanges of neighbouring electron fields to make the ordering in the pairing is odd for the first term, whereas it is even for the second term.

As another example, consider

$$\frac{\int [d\,A d\psi d\bar\psi] e^{iS_0}\bar\psi(w_1)\slashed{A}(w_1)\psi(w_2)\,\bar\psi(w_2)\slashed{A}(w_2)\psi(w_2)}{\int [d\,A d\psi d\bar\psi] e^{iS_0}}$$

$$= -g_{ab}(w_1, w_2)\gamma^\mu_{bc} g_{cd}(w_2, w_1)\gamma^\nu_{da}\Delta_{\mu\nu}(w_2, w_1)$$

$$= -\text{Tr}[g(w_1, w_2)\gamma^\mu g(w_2, w_1)\gamma^\nu]\Delta_{\mu\nu}(w_2, w_1)$$

Here, there is only one pairing and the repeated indices in the second line are assumed to be summed over. The minus sign is due to the fact that the ordering in the pairing differs from the ordering of the electron fields in the integrand by an odd number of interchanges. In the third line the sums over Dirac field indices are presented as a product of Dirac matrices and electron propagators which occurs in the form of a trace over the resulting matrix.

15.3 Feynman Diagrams

The perturbative expansion in quantum field theory has a useful diagrammatic representation where the diagrams are called Feynman diagrams. These diagrams were a great advance in the formulation of field theory in that they make lower order perturbation theory quite easy and intuitive.

In a Feynman diagram, the propagator $g_{ab}(y, z)$ is represented by a oriented solid line, as depicted in Fig. 15.1. The photon propagator $\Delta_{\mu\nu}(x, y)$ is represented by an un-oriented wiggly line as depicted in Fig. 15.2. The interaction vertex is an intersection of a photon line and an electron line as depicted in Fig. 15.3. The vertex has a Dirac gamma matrix whose indices are matched with the photon and electron propagators which are eventually attached to it. The counterterm vertices are depicted in Fig. 15.4.

We can associate a Feynman diagram with any of the pairings that is produced by the application of Wick's theorem. Consider, as an example, the leading contribution to the 3-point function

$$\Gamma_{\mu ab}(x, y, z) \equiv \ <\mathcal{O}|T A_\mu(x)\psi_a(y)\bar{\psi}_b(z)|\mathcal{O}> \tag{15.22}$$

Let us first study this 3-point function by doing an explicit expansion of the functional integral and use Wick's theorem. Then, we will find the same result by drawing Feynman diagrams.

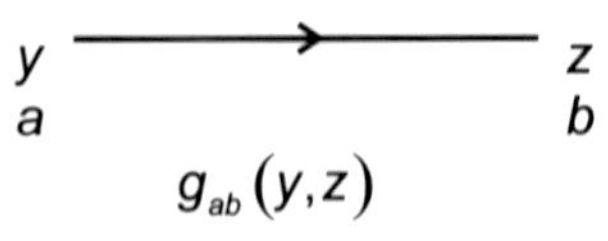

Fig. 15.1 The propagator for the Dirac field is represented by an oriented solid line where the orientation is from a, y to b, z and it corresponds to the analytic expression $g_{ab}(y, z) = \int \frac{dp}{(2\pi)^4} e^{ip(y-z)} \frac{i\not{p}-m}{p^2+m^2-i\epsilon}$

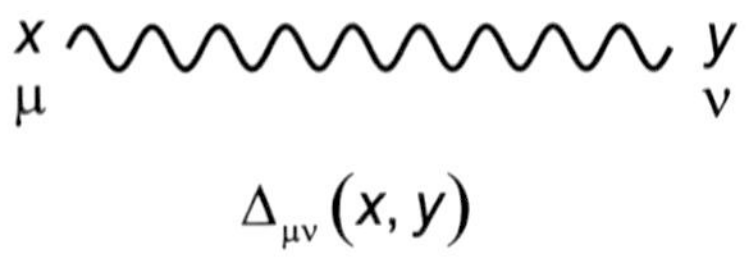

Fig. 15.2 The propagator for the photon is represented by an un-oriented wiggly line and it corresponds to the analytic expression $\Delta_{\mu\nu}(x, y) = -i \int \frac{dk}{(2\pi)^4} e^{ik(x-y)} \left(\frac{\eta_{\mu\nu}-k_\mu k_\nu/k^2}{k^2+\kappa^2-i\epsilon} + \frac{k_\mu k_\nu/k^2}{\xi k^2+\kappa^2-i\epsilon} \right)$. We have presented the propagator with the photon mass κ which will be used to regulate infrared divergences

Fig. 15.3 The interaction vertex absorbs and emits an electron line and it absorbs a photon line at a spacetime point w. It is represented by $-ie\gamma^\lambda_{cd}$

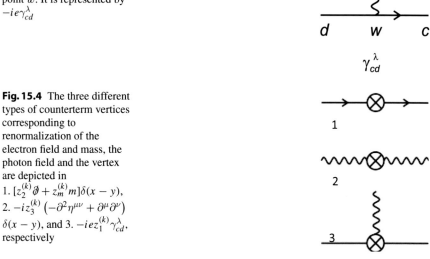

Fig. 15.4 The three different types of counterterm vertices corresponding to renormalization of the electron field and mass, the photon field and the vertex are depicted in
1. $[z_2^{(k)}\slashed{\partial} + z_m^{(k)}m]\delta(x-y)$,
2. $-iz_3^{(k)}\left(-\partial^2\eta^{\mu\nu} + \partial^\mu\partial^\nu\right)$ $\delta(x-y)$, and 3. $-iez_1^{(k)}\gamma^\lambda_{cd}$, respectively

To implement the first approach, we begin with the functional integral representation,

$$\Gamma_{\mu ab}(x,y,z) = \frac{\int [dAd\psi d\bar\psi]e^{iS}A_\mu(x)\psi_a(y)\bar\psi_b(z)}{\int [dAd\psi d\bar\psi]e^{iS}} \tag{15.23}$$

and expand the exponential of the action as in equation (15.16). The result, expanded up to the leading order in e, is

$$\Gamma_{\mu ab}(x,y,z) = \frac{\int [dAd\psi d\bar\psi]e^{iS_0}A_\mu(x)\psi_a(y)\bar\psi_b(z)}{\int [dAd\psi d\bar\psi]e^{iS_0}}$$

$$+\frac{\int [dAd\psi d\bar\psi]e^{iS_0}[iS_{\text{int}}^{(1)}]A_\mu(x)\psi_a(y)\bar\psi_b(z)}{\int [dAd\psi d\bar\psi]e^{iS_0}}$$

$$-\frac{\int [dAd\psi d\bar\psi]e^{iS_0}A_\mu(x)\psi_a(y)\bar\psi_b(z)}{\int [dAd\psi d\bar\psi]e^{iS_0}}\frac{\int [dAd\psi d\bar\psi]e^{iS_0}[iS_{\text{int}}^{(1)}]}{\int [dAd\psi d\bar\psi]e^{iS_0}} + \dots$$

We can write the above formula more explicitly as

$$\Gamma_{\mu ab}(x, y, z) = \frac{\int [dA d\psi d\bar{\psi}] e^{iS_0} A_\mu(x)\psi_a(y)\bar{\psi}_b(z)}{\int [dA d\psi d\bar{\psi}] e^{iS_0}} \tag{15.24}$$

$$-ie\int dw \frac{\int [dA d\psi d\bar{\psi}] e^{iS_0} \bar{\psi}(w) \not{A}(w)\psi(w) \, A_\mu(x)\psi_a(y)\bar{\psi}_b(z)}{\int [dA d\psi d\bar{\psi}] e^{iS_0}} \tag{15.25}$$

$$+ ie\int dw \frac{\int [dA d\psi d\bar{\psi}] e^{iS_0} A_\mu(x)\psi_a(y)\bar{\psi}_b(z)}{\int [dA d\psi d\bar{\psi}] e^{iS_0}} \times$$

$$\times \frac{\int [dA d\psi d\bar{\psi}] e^{iS_0} \bar{\psi}(w) \not{A}(w)\psi(w)}{\int [dA d\psi d\bar{\psi}] e^{iS_0}} + \dots \tag{15.26}$$

In the above formula, the line labeled (15.24) is the zero'th order contribution. The line labeled (15.25) comes from inserting one power of the interaction Lagrangian density into the functional integral in the numerator and the line labeled (15.26) arises from inserting the leading order interaction vertex into the functional integral in the denominator. The ellipses denote higher orders in e. Note that, because each of the counterterms are at least of order e^2, they do not contribute at order e.

The second step is to use Wick's theorem to evaluate the Gaussian functional integrals. The first, zero'th order term, line (15.24), vanishes, as does the third term, line (15.26), with the correction to the denominator. The only contributions which survive are in the line labeled (15.25) and when we use Wick's theorem there are two pairings of the fields in the integrand. The result is the two contributions given in lines (15.27) and (15.28) of the equation below

$$\Gamma_{\mu ab}(x, y, z) = -ie\int dw \, g_{ac}(y, w)\gamma^\nu_{cd} g_{db}(w, z) \, \Delta_{\nu\mu}(w, x) \tag{15.27}$$

$$+ ieg_{ab}(y, z)\int dw \, \gamma^\nu_{cd} g_{dc}(w, w) \, \Delta_{\nu\mu}(w, x) + \dots \tag{15.28}$$

Let us now try to get this result by drawing Feynman diagrams. We begin studying Feynman diagrams which have one external photon line and one ingoing and one outgoing electron line. The first diagrams which can be drawn have a single internal vertex and they are therefore of order e. (The counterterm vertices are all of too high an order to contribute here.) This starting point is depicted in Fig. 15.5. There, the

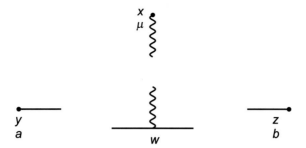

Fig. 15.5 We begin to draw the Feynman diagrams contributing to $\Gamma_{\mu ab}(x, y, z)$ by putting the external points, x, μ for the photon, y, a for the incoming electron and z, b for the outgoing electron at the edge of the diagram and the vertex in the centre. We will then proceed to draw Feynman diagrams by connecting the points with the appropriate lines in all possible ways

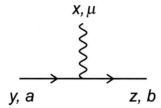

Fig. 15.6 When we begin with Fig. 15.5 and we connect the incoming and outgoing electron lines to the vertex, the result is the Feynman diagram that is depicted here

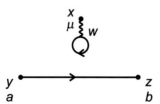

Fig. 15.7 When we begin with Fig. 15.5 and we connect the incoming and outgoing lines so that the line leaving and entering the vertex must connect with itself, the result is the Feynman diagram that is depicted here. This diagram has a sub-diagram with a closed electron loop emitting a single photon line. The contribution of such a diagram vanishes as a result of symmetry, a fact that will be encapsulated in Furry's theorem which we will study shortly. Furry's theorem allows us to ignore all such diagrams

three external fields, the photon, the ingoing and outgoing electron are drawn at the edges of the diagram. The vertex is drawn in the centre.

We connect the external lines and the vertex in all possible ways and we obtain the Feynman diagrams in Figs. 15.6 and 15.7.

We then write a mathematical expression for the result by associating the lines of the diagrams with the propagators, $g_{ab}(y, z)$ and $\Delta_{\mu\nu}(x, y)$ and the vertex with γ^{μ}_{ab}. The result is the two analytic expressions in equations (15.27) corresponding to Fig. 15.6 and (15.28) corresponding to Fig. 15.7.

Note that the contributions in equations (15.27) and (15.28) differ by a minus sign. In the application of Wick's theorem, this difference came from the number of interchanges of electron operators which are needed to go from the ordering of the fields in the integrand of the functional integral to the ordering of fields in the pairing which should always be arranged as $(\psi\bar{\psi})(\psi\bar{\psi})$.... The two Feynman diagrams which correspond to equations (15.27) and (15.28), those in Figs. 15.6 and 15.7 respectively, differ in that the one in Fig. 15.7 has a closed electron loop and the diagram in Fig. 15.6 does not have a closed electron loop. It will turn out that, in a Feynman diagram, a minus sign is always associated with each closed electron loop.

We can also immediately observe that since, due to translation invariance and Lorentz invariance, we would expect that the quantity in equation (15.28), $\int dw \; \gamma^{\nu}_{cd} g_{dc}(w, w) = 0$. This would imply that the contribution in (15.28), equation vanishes, which is indeed so. This is a result of a more general fact, known as Furry's theorem, which we shall discuss in detail in Sect. 15.12 and which applies to Feynman diagrams. The theorem states that any Feynman diagram which contains a closed Dirac field loop and where the loop emits an odd number of photon lines vanishes. With this fact in hand, we would routinely drop the contribution from the diagrams with such closed loops. The contribution in Fig. 15.7 should therefore be dropped.

The final result is then the Feynman diagram given in Fig. 15.6,

$$\Gamma_{\mu ab}(x, y, z) = -ie \int dw \; g_{ac}(y, w) \gamma^{\nu}_{cd} g_{db}(w, z) \, \Delta_{\nu\mu}(w, x) + \ldots \qquad (15.29)$$

which is our result for the computation of this particular correlation function to order e. The ellipses denote corrections of order e^3 and higher.

This 3 point function, at this order, is connected. In fact, it is possible to show that it is connected at all orders in perturbation theory. It can be made irreducible by removing the propagators which connect the central point to the external legs. The result is

$$\Gamma_{I \; \mu ab}(x, y, z) = -ie\gamma_{\mu cd}\delta(x - z)\delta(y - x) + \ldots \qquad (15.30)$$

which, in momentum space is given by

$$\Gamma_{I \; \mu ab}(k, p, -p') = -ie\gamma_{\mu cd} + \ldots \qquad (15.31)$$

where we have extracted the overall momentum conserving delta function $(2\pi)^4\delta(k + p - p')$ – see equation (15.39) below.

15.4   Feynman Rules

The procedure for drawing the Feynman diagrams follows a systematic procedure called the Feynman rules. We can list these rules as follows:

1. We place the external vertices at the boundary of the Feynman diagram. We place the desired number of interaction vertices and counterterm vertices in the centre of the diagram. The external electron lines for ψ's are labeled with incoming arrows and the external lines with $\bar{\psi}$'s have outgoing arrows. The incoming and outgoing photon lines are not oriented.

2. Then we draw all of the distinct topologically inequivalent Feynman diagrams which have the requisite number of external lines of each kind. We do this by using the appropriate lines to connect the external points to vertices and the vertices to other vertices in all possible ways that are compatible with the orientation of the electron lines.

3. We can drop all vacuum diagrams (all diagrams which have a sub-diagram which is not connected to any external lines) and all diagrams which have a closed electron loop emitting an odd number of photon lines. The first of these facts is a result of the linked cluster theorem, which we have discussed in detail for the case of scalar fields in Sect. 14.4.3. It has an easy generalization to quantum electrodynamics which we adopt here. The second of these facts is due to Furry's theorem which is discussed in Sect. 15.12.

4. We can then construct the mathematical expression corresponding to the Feynman diagrams. We associate propagators to the internal and external lines in the diagram, ensuring that the coordinates and the Dirac indices on the electron propagators and the coordinates and the Lorentz indices on the photon propagators are correctly matched with each other, with the γ_{ab}^{μ}'s, or with the counterterm contributions that they are connected to.

5. We then integrate over the positions of each of the internal vertices.

6. The symmetry factors which were required in the scalar field theory that we studied in the previous chapter are all unity in quantum electrodynamics. The overall factor is therefore quite simple. It is $(-ie)^k(-1)^{N_\ell}$ for a diagram with k vertices and/or counterterms so that the total order in e adds to k and N_ℓ electron loops. If there are counterterm vertices, some of the factors $-ie$ will be replaced by the appropriate factors for the counterterms.

15.5   The Electron 2 Point Function

As another example, where we can apply the Feynman rules which we have enumerated in Sect. 15.4 above, let us consider the computation of the 2 point function of the electron,

$$S_{ab}(y, z) \equiv < \mathcal{O}|T\psi_a(y)\bar{\psi}_b(z)|\mathcal{O} > \tag{15.32}$$

$$= \frac{\int [dAd\psi d\bar{\psi}]e^{iS}\psi_a(y)\bar{\psi}_b(z)}{\int [dAd\psi d\bar{\psi}]e^{iS}} \tag{15.33}$$

$$= g_{ab}(x, y) + \mathcal{O}(e^2) \tag{15.34}$$

We will compute the order e^2 contributions.

There will be two sets of diagrams, one with interaction vertices and the other with counterterms. Since the leading order counterterms are of order e^2 and each vertex introduces a factor of e, there will be no contributions at order e^2 which have both counterterms and vertices.

Let us begin with the diagrams with interaction vertices. We begin by writing the external points y, z, together with their Dirac indices a, b at the edges of the diagram. Then, since the computation is of the order e^2, we draw two vertices in the interior of the graph. We label the vertices by their coordinates w_1, w_2. This starting point is depicted in Fig. 15.8.

Then, secondly, we consider the diagram with counterterms. This starting point with the counterterms is depicted in Fig. 15.9.

Then, with each of these starting points, we connect the lines in all of the possible ways that are compatible with their orientation. The results are the diagrams depicted in figure 15.10.

Then, we implement the linked cluster theorem and drop vacuum diagrams. We also implement Furry's theorem and drop diagrams which have an electron loop emitting an odd number of photons. These two theorems allow us to drop the diagrams labeled $b)$, $c)$, $d)$, $f)$ in Fig. 15.10. That leaves only two Feynman diagrams, the ones that are labeled $a)$ and $e)$ in Fig. 15.10.

Once we have the Feynman diagrams, we can find the explicit mathematical formulae for the corrections by associating the lines with photon and electron propagators, the vertices with Dirac matrices and the appropriate counterterms to get the explicit mathematical expression for the corrections that are represented by the two remaining diagrams,

$$S_{ab}(y, z) = g_{ab}(y, z)$$
$$- e^2 \int dw_1 dw_2 \left[g(y, w_1) \gamma^\mu g(w_1, w_2) \gamma^\nu g(w_2, z) \right]_{ab} \Delta_{\mu\nu}(w_1, w_2)$$

$$(15.35)$$

Fig. 15.8 We begin to draw the order e^2 Feynman diagrams with two interaction vertices by putting the external points, y and z at the edge of the graph and the two interaction vertices in the centre

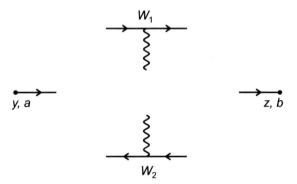

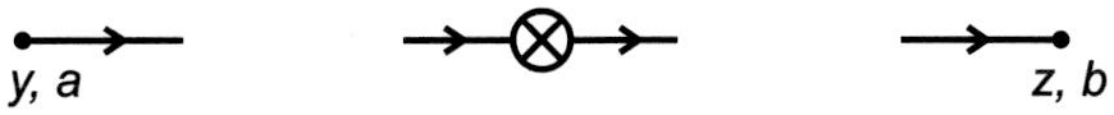

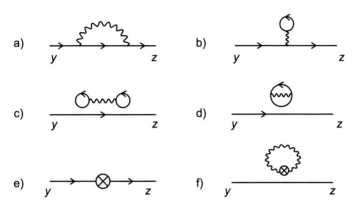

Fig. 15.9 We begin to draw the order e^2 Feynman diagrams with counterterms by putting the external points, y and z at the edge of the graph and the order e^2 counterterm in the centre. We choose only those counterterms which can be joined to the other lines in the diagram, in this case, to the two external lines

Fig. 15.10 The Feynman diagrams which are obtained by connecting the lines in Figs. 15.8 and 15.9 in all possible ways. Furry's theorem allows us to drop the diagram **b** and the linked cluster theorem instructs us to drop diagrams **c**, **d** and **f**. What remains is diagrams **a** and **e**, the one-loop correction to the electron 2 point function and the counterterms

$$+ e^2 \int dw \left[g(y, w)(z_2^{(2)} \not{\partial} + m z_m^{(2)}) g(w, z) \right]_{ab}$$
$$+ \dots \tag{15.36}$$

where the ellipses denote contributions of orders e^4 and higher.

15.6 Feynman Rules in Momentum Space

The result for the electron 2 point function quoted in equations (15.35) and (15.36) have convolutions of propagators which are functions only of the differences of the two coordinates which occur in each propagator. This leads to the fact that these equations have a simpler form in momentum space. If we recall the definitions of the Fourier transforms of the 2 point functions and propagators,

$$D_{\mu\nu}(x, y) = \int \frac{dk}{(2\pi)^4} e^{ik(x-y)} D_{\mu\nu}(k), \quad \Delta_{\mu\nu}(x, y) = \int \frac{dk}{(2\pi)^4} e^{ik(x-y)} \Delta_{\mu\nu}(k)$$

$$S_{ab}(x, y) = \int \frac{dp}{(2\pi)^4} e^{ip(x-y)} S_{ab}(p), \quad g_{ab}(x, y) = \int \frac{dp}{(2\pi)^4} e^{ip(x-y)} g_{ab}(p)$$

equations (15.35) and (15.36) become

$$S_{ab}(p) = g_{ab}(p)$$

$$- e^2 \int \frac{dk}{(2\pi)^4} \left[g(p)\gamma^\mu g(p-k)\gamma^\nu g(p) \right]_{ab} \Delta_{\mu\nu}(k) \qquad (15.37)$$

$$+ e^2 \left[g(p)(iz_2^{(2)} \not{p} + mz_m^{(2)})g(p) \right]_{ab} \qquad (15.38)$$

$$+ \dots$$

where, we can see, there is only one remaining four-dimensional integral. If we did that integral, we would have the explicit result.

In order to find this momentum space expression for corrections to an n point function, it is normally more convenient to draw Feynman diagrams and associate them with Feynman integrals directly in momentum space. We recall the notation for a generic correlation function that was defined in equation (15.17). We define the Fourier transform of the correlation function as

$$\Gamma_{\mu_1\dots\mu_n a_1\dots a_k b_1\dots b_k}(x_1\dots, x_n, y_1, \dots, y_k, z_1, \dots, z_k) \equiv$$

$$< \mathcal{O}|T A_{\mu_1}(x_1)\dots A_{\mu_n}(x_n)\psi_{a_1}(y_1)\dots\psi_{a_k}(y_k)\bar\psi_{b_1}(z_1)\dots\bar\psi_{b_k}(z_k)|\mathcal{O} >$$

$$= \int \frac{dk_1}{(2\pi)^4} \cdots \frac{dk_n}{(2\pi)^4} \frac{dp_1}{(2\pi)^4} \cdots \frac{dp_k}{(2\pi)^4} \frac{dq_1}{(2\pi)^4} \cdots \frac{dq_n}{(2\pi)^4} \cdot$$

$$\cdot \Gamma_{\mu_1\dots\mu_n a_1\dots a_k b_1\dots b_k}(k_1\dots, k_n, p_1, \dots, p_k, q_1, \dots, q_k) \cdot$$

$$\cdot (2\pi)^4\delta(\sum_{i=1}^{n} k_i + \sum_{i=1}^{k} p_i + \sum_{i=1}^{k} q_i)e^{i\sum p_i x_i + i\sum p_j y_j + i\sum q_j z_i} \qquad (15.39)$$

We similarly denote the Fourier transforms of connected correlation functions and irreducible correlation functions as

$$\Gamma_{C\mu_1\dots\mu_n a_1\dots a_k b_1\dots b_k}(k_1\dots, k_n, p_1, \dots, p_k, q_1, \dots, q_k)$$

and

$$\Gamma_{I\mu_1\dots\mu_n a_1\dots a_k b_1\dots b_k}(k_1\dots, k_n, p_1, \dots, p_k, q_1, \dots, q_k)$$

respectively. Connected correlation functions are defined as those correlation functions which composed only of connected Feynman diagrams. They are generated by the logarithm of the generating functional for general correlation functions, that is, by taking functional derivatives of

$$W[J, \eta, \bar\eta] = \ln Z[J, \eta, \bar\eta]$$

Irreducible correlation functions are correlation functions which are composed of irreducible Feynman diagrams only. A Feynman diagram is irreducible if it is a connected Feynman diagram and if it cannot be rendered disconnected by cutting one internal line. Irreducible correlation functions are generated by a Legendre transformation of $W[J, \eta, \bar{\eta}]$.

Note that in equation (15.39) we have extracted the momentum conserving delta function from the Fourier transform. This momentum conservation is a consequence of translation invariance.

We have reserved a special notation for 2 point functions and irreducible 2 point functions,

$$D_{\mu\nu}(k) = \Gamma_{\mu\nu}(k, -k) \tag{15.40}$$

$$\Pi_{\mu\nu}(k) = \Gamma_{I\mu\nu}(k, -k) \tag{15.41}$$

$$S_{ab}(p) = \Gamma_{ab}(p, -p) \tag{15.42}$$

$$\Sigma_{ab}(p) = \Gamma_{Iab}(p, -p) \tag{15.43}$$

They are related to each other by the formulae

$$D_{\mu\nu}(k) = \Delta_{\mu\nu}(k) + \Delta_{\mu\rho}(k)\Pi^{\rho\sigma}(k)\Delta_{\sigma\nu}(k)$$
$$\quad + \Delta_{\mu\rho}(k)\Pi^{\rho\kappa}(p)\Delta_{\kappa\sigma}(k)\Pi^{\sigma\xi}(k)\Delta_{\xi\nu}(k) + \dots$$

$$D^{-1}(k) = \Delta^{-1}(k) - \Pi(k) \tag{15.44}$$

$$D(k) = [\Delta^{-1}(k) - \Pi(k)]^{-1} \tag{15.45}$$

$$S(p) = g(p) + g(p)\Sigma(p)g(p) + g(p)\Sigma(p)g(p)\Sigma(p)g(p)$$
$$\quad + g(p)\Sigma(p)g(p)\Sigma(p)g(p)\Sigma(p) + \dots$$

$$S^{-1}(p) = g^{-1}(p) - \Sigma(p) \tag{15.46}$$

$$S(p) = -[i\not{p} + m + \Sigma(p)]^{-1} \tag{15.47}$$

Then we can reformulate the Feynman rules in momentum space as follows:

1. To compute

$$\Gamma_{\mu_1\dots\mu_n a_1\dots a_k b_1\dots b_k}(k_1 \dots, k_n, p_1, \dots, p_k, q_1, \dots, q_k)$$

or

$$\Gamma_{C\mu_1\dots\mu_n a_1\dots a_k b_1\dots b_k}(k_1 \dots, k_n, p_1, \dots, p_k, q_1, \dots, q_k)$$

or

$$\Gamma_{I\mu_1\dots\mu_n a_1\dots a_k b_1\dots b_k}(k_1 \dots, k_n, p_1, \dots, p_k, q_1, \dots, q_k)$$

we draw all of the Feynman diagrams or all of the connected Feynman diagrams or all of the connected and irreducible Feynman diagrams, respectively, with n external photon lines, k external inwardly oriented electron lines and k external outwardly oriented electron lines. The orientations and momentum assignments of the external lines of a general multi-point function is illustrated in Fig. 15.11. As well, we include a number of vertices and counterterm vertices in the interior of the diagram.

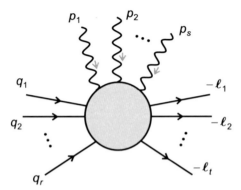

Fig. 15.11 The orientation of the momentum assignments to external lines is depicted. We will take all of the external momenta as incoming. The fact that they must sum to zero is a result of the conservation of momentum. The momentum flow in an electron line is along its orientation. Thus, the outgoing electron lines carry negative momenta. Photon lines must also be assigned a direction of momentum flow. For external photons lines, as depicted here, the momentum is taken as flowing inward

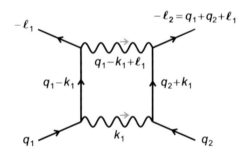

Fig. 15.12 An example of a Feynman diagram with momentum assignments to external and internal lines. The total momentum is conserved at each of the four vertices. The total momentum of the diagram is also conserved, $q_1 + q_2 + \ell_1 + \ell_2 = 0$. This leaves one momentum arbitrary, k_1^μ. The analytic expression for the diagram must be integrated over this loop momentum $\int \frac{dk_1}{(2\pi)^4}$

2. We drop all vacuum diagrams.
3. We drop all diagrams which have a closed electron loop emitting an odd number of photons.
4. We give each of the photon lines, both internal and external, an orientation. External photon lines are always oriented as if they are directed into the diagram. The orientation will determine the sign of the momentum that is flowing through the line. The electron lines are oriented and, for electron lines which are internal to a Feynman diagram, we take the direction of momentum flow in the as being the same as that of the orientation of the line. We also do this for all external lines. However, we must remember that, for outgoing oriented external electron lines labeled by incoming momentum q, the corresponding propagator would be $g(-q)$. An example of a Feynman diagram with the appropriate momentum assignments to the internal lines is given in Fig. 15.12.

5. We label the external photon lines by the momenta and Lorentz indices (k_1, μ_1), $\ldots, (k_n, \mu_n)$ for photons, the inwardly oriented electron lines by momenta and Dirac indices $(p_1, a_1), \ldots, (p_k, a_k)$ and the outwardly oriented electron lines by the momenta and Dirac indices $(-q_1, b_1) \ldots, (-q_k, b_k)$.

6. We assign momenta to all of the internal lines of the diagram in such a way that it is consistent with the conservation of momenta at each of the internal vertices and with the values of the momenta of the external lines. The number of momenta that are left undetermined by these constraints must be equal to the number of closed loops in the diagram.

7. To each internal photon line we ascribe a factor of $\Delta_{\mu\nu}(\ell)$ where ℓ is the momentum of the line and μ and ν are the Lorentz indices associated with the endpoints of the line. Similarly, to each electron line, we ascribe a propagator $g_{ab}(\ell)$ where ℓ is the momentum of that line and a and b are the Dirac indices of its endpoints.

8. To each vertex, we ascribe a factor of $-ie\gamma^\mu_{ba}$ if it is an interaction vertex, $-iez_1^{(k)}\gamma^\mu_{ba}$ if it is an interaction vertex counterterm, $-iz_3^{(k)}[\eta_{\mu\nu} - k_\mu k_\nu] - i\xi z_\xi^{(k)} k_\mu k_\nu$ if it is a photon counterterm and $[z_2^{(k)} i \not{p} + m z_m^{(k)}]_{ab}$ if it is a electron counterterm.

9. We then integrate $\int \prod_i \frac{d\ell_i}{(2\pi)^4} \ldots$ for each of the independent momenta of the internal lines. There is an independent four-momentum for each closed loop in the diagram.

10. There is a additional overall factor of (-1) for each closed electron loop in the diagram.

As an example. we can consider the irreducible 2 point function for the electron. which we have already found in equations (15.35) and (15.36).

The connected irreducible Feynman diagrams which would contribute to $\Sigma(k)$ at order e^2 are diagrams $a)$ and $e)$ displayed in Fig. 15.10 if in those diagrams we removed the external electron lines. Diagram $a)$ is the only legal irreducible diagram with two interaction vertices. Diagram $e)$ is the contribution of the electron counterterms. The result is

$$\Sigma_{ab}(p) = -e^2 \int \frac{d\ell}{(2\pi)^4} [\gamma^\mu g(p+\ell)\gamma^\nu]_{ab} \Delta_{\mu\nu}(\ell) + \left[z_2^{(k)} i \not{p} + m z_m^{(k)} \right]_{ab} + \ldots$$

(15.48)

where the ellipses represent contributions of order e^4 and higher.

15.7 The Photon 2 Point Function

The photon 2 point function is defined by

$$D_{\mu\nu}(x, y) \equiv \Gamma_{\mu\nu}(x, y) = <\mathcal{O}|\mathcal{T} A_\mu(x) A_\nu(y)|\mathcal{O}>$$

(15.49)

or, in momentum space

$$D_{\mu\nu}(k) \equiv \Gamma_{\mu\nu}(k, -k) \tag{15.50}$$

The irreducible 2 point function is denoted by

$$\Pi_{\mu\nu}(x, y) \equiv \Gamma_{I\mu\nu}(x, y) \tag{15.51}$$
$$\Pi_{\mu\nu}(k) \equiv \Gamma_{I\mu\nu}(k, -k) \tag{15.52}$$

It can be shown, using the definition of the generating functional for irreducible correlation functions, that the photon 2 point function can be reconstructed from the irreducible 2 point function using the equation

$$D_{\mu\nu}^{-1}(k) = \Delta_{\mu\nu}^{-1}(k) - \Pi_{\mu\nu}(k) \tag{15.53}$$

where $D_{\mu\nu}^{-1}(k)$ is the matrix inverse of $D_{\mu\nu}(k)$ in that $D_{\mu\nu}^{-1}(k)D^{\nu\lambda}(k) = \delta_\mu^\lambda$. The propagator $\Delta_{\mu\nu}(k)$ and its inverse are given by

$$\Delta_{\mu\nu}(k) = \frac{-i}{k^2 + \kappa^2 - i\epsilon}\left[\eta_{\mu\nu} - \frac{k_\mu k_\nu}{k^2}\right] - \frac{i}{\xi k^2 + \kappa^2 - i\epsilon}\frac{k_\mu k_\nu}{k^2} \tag{15.54}$$
$$\Delta_{\mu\nu}^{-1}(k) = i(k^2 + m^2 - i\epsilon)[\eta_{\mu\nu} - k_\mu k_\nu/k^2] + i(\xi k^2 + m^2 - i\epsilon)k_\mu k_\nu/k^2 \tag{15.55}$$

In the absence of interaction, that is, in the limit $e^2 \to 0$, the 2 point function of the photon field in quantum electrodynamics is given by the propagator, $D_{\mu\nu}(x, y) \to \Delta_{\mu\nu}(x, y)$. When the interactions are turned on, this 2 point function is modified by quantum effects and becomes the full photon 2 point function $D_{\mu\nu}(x, y)$. In

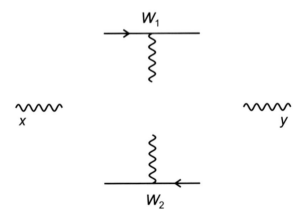

Fig. 15.13 We begin finding the Feynman diagrams which correct the 2 point function of the photon by drawing two external photon lines and two vertices. There will be another possibility where the two vertices are replaced by an order two counterterm. We will not draw the counterterm diagrams, but of course we will take them into account in the final result

Fig. 15.14 The Feynman diagram which contributes to the 2 point correlation function of the photon field at order e^2. There will also be a diagram with a counterterm which we have not displayed

Fig. 15.15 The Feynman diagram which contributes to the 2 point correlation function of the electron field at order e^2 in momentum space

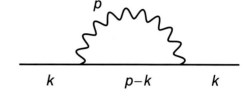

Fig. 15.16 The Feynman diagram which contributes to the 2 point correlation function of the photon field at order e^2 in momentum space

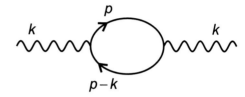

particular, it becomes a function of e^2 which we assume has an expansion in powers of e^2 so that

$$D_{\mu\nu}(k) = \Delta_{\mu\nu}(k) + \mathcal{O}(e^2)$$

In this section, we will compute the order e^2 correction. We will use the technique of writing the Feynman diagrams that contribute at that order, converting the diagrams to Feynman integrals and performing the integrations. Along the way we will encounter and deal with ultraviolet divergences. As we have done for other functions, it is useful to begin with the irreducible 2 point function

The irreducible 2 point function $\Pi_{\mu\nu}(k)$ is sometimes called the "vacuum polarization" or the "photon self-energy". It is given by the summation of connected irreducible Feynman diagrams.

The Feynman diagrams which contribute to the vacuum polarization will have two points where external photon lines attach and two internal vertices. The starting point for drawing these diagrams is depicted in Fig. 15.13. Discarding the vacuum diagrams and those which can be discarded using Furry's theorem we immediately arrive at the diagram in Fig. 15.14. The diagram has one closed loop, and therefore an extra overall minus sign. Explicitly, the photon irreducible 2 point function is

$$\Pi^{\mu\nu}(x, y) = e^2 \text{Tr}\left[\gamma^\mu g(x, y) \gamma^\nu g(y, x)\right] + i z_3^{(2)} (\eta^{\mu\nu} \partial^2 - \partial^\mu \partial^\nu) \delta(x - y) + \dots \tag{15.56}$$

Note that we have not included a counterterm $z_\kappa^{(2)}$ for the photon mass. This is because we anticipate that we shall not need it. We will see this explicitly as we proceed.

In momentum space, the leading contribution to the vacuum polarization is gotten by taking the Fourier transform of equation (15.56). It can also be gotten by applying the Feynman rules in momentum space. The result is

$$\Pi^{\mu\nu}(k) = e^2 \int \frac{d^4\ell}{(2\pi)^4} \mathrm{Tr}[g(\ell)\gamma^\mu g(\ell - k)\gamma^\nu] - iz_3^{(2)} \left[k^2\eta^{\mu\nu} - k^\mu k^\nu\right] \quad (15.57)$$

It is this four-dimensional momentum space volume integral which must now be taken in order to find the explicit form of this contribution to $\Pi^{\mu\nu}(k)$ (see Figs. 15.13, 15.14, 15.15 and 15.16).

The Feynman integral on the right-hand-side of the above equation (15.57) is divergent. The divergence comes from the large energy and momentum regime. When ℓ is large the integrand is of order $1/\ell^2$ and integration over the four dimensional volume of momentum space diverges. In order to deal with this divergence, we must first regularize the integral. For this purpose, we will use dimensional regularization where we replace the dimension 4 of spacetime and of momentum space by a parameter 2ω to get

$$\begin{aligned}\Pi^{\mu\nu}(k) =& e^2\mu^{4-2\omega} \int \frac{d^{2\omega}\ell}{(2\pi)^{2\omega}} \mathrm{Tr}[g(\ell)\gamma^\mu g(\ell - k)\gamma^\nu] \\ & - iz_3^{(2)} \left[k^2\eta^{\mu\nu} - k^\mu k^\nu\right]\end{aligned} \quad (15.58)$$

We have included the factor $\mu^{4-2\omega}$ which is introduced so that e^2 can remain dimensionless. As usual, with dimensional regularization we will assume that the dimension 2ω is small enough that the integral converges.

Then, we perform a Wick rotation and, with the explicit expressions for the electron propagators, we have

$$\Pi^{\mu\nu}(k) = ie^2\mu^{4-2\omega} \int \frac{d^{2\omega}\ell}{(2\pi)^{2\omega}} \frac{\mathrm{Tr}[(i\ell\!\!\!/ - m)\gamma^\mu(i\ell\!\!\!/ - i k\!\!\!/ - m)\gamma^\nu]}{[\ell^2 + m^2][(\ell - k)^2 + m^2]} \quad (15.59)$$

$$-iz_3^{(2)} \left[k^2\delta^{\mu\nu} - k^\mu k^\nu\right] \quad (15.60)$$

The overall factor of i came from $d^{2\omega}\ell \to id^{2\omega}\ell$. In the Wick rotation, as well as replacing ℓ^0 and $\ell^0 - k^0$ by $i\ell^0$ and $i(\ell^0 - k^2)$, respectively, we have also re-defined the components of the tensor $\Pi^{00} \to -\Pi^{00}$, $\Pi^{0b} \to i\Pi^{0b}$ and $\Pi^{a0} \to i\Pi^{a0}$. This is seen in the counterterm contribution where $\eta_{\mu\nu}$ has been replaced by the Euclidean metric $\delta_{\mu\nu}$. The Dirac matrices are also replaced by $(\gamma_E^0, \vec{\gamma}_E) = (i\gamma^0, \vec{\gamma})$. so that all of the 2ω matrices are now Hermitian and they obey the algebra

$$\{\gamma_E^\mu, \gamma_E^\nu\} = 2\delta^{\mu\nu} \quad (15.61)$$

and, in equation (15.59), we define

$$\ell \equiv \sum_{\mu=0}^{2\omega-1} \gamma_E^\mu \ell_\mu$$

We will define the dimension of the Dirac matrices in 2ω dimensions as being $4 \cdot f(\omega)$ where $f(2) = 1$ and

$$4 \cdot f(\omega) = 4 - 4 \cdot f'(2)(2 - \omega) + \dots \tag{15.62}$$

We do not have a way of knowing more about the function $f(\omega)$. As we will see, beyond knowing $f(2) = 1$, we will not need it.

In order to take the trace over the Euclidean Dirac matrices, we do need the following formulae, which can be demonstrated as consequences of the matrix algebra (15.61) and the dimension of the matrices (15.62),

$$\mathrm{Tr}[\mathcal{I}] = 4 \cdot f(\omega)$$
$$\mathrm{Tr}[\gamma_E^\mu] = 0, \quad \text{in fact } \mathrm{Tr}[\gamma_E^{\mu_1} \gamma_E^{\mu_2} \dots \gamma_E^{\mu_{2k+1}}] = 0$$
$$\mathrm{Tr}[\gamma_E^\mu \gamma_E^\nu] = 4 \cdot f(\omega)\delta^{\mu\nu}$$
$$\mathrm{Tr}[\gamma_E^\mu \gamma_E^\nu \gamma_E^\rho \gamma_E^\sigma] = 4 \cdot f(\omega) \left(\delta^{\mu\nu} \delta^{\rho\sigma} - \delta^{\mu\rho} \delta^{\nu\sigma} + \delta^{\mu\sigma} \delta^{\rho\nu} \right)$$

With these formulae, we find that the numerator in the integrand in equation (15.59) becomes

$$\mathrm{Tr}[(i\ell - m)\gamma^\mu (i\ell - i\slashed{k} - m)\gamma^\nu] =$$

$$4f(\omega) \left[-(\ell - k)^\mu \ell^\nu - \ell^\mu (\ell - k)^\nu + \delta^{\mu\nu} \ell \cdot (\ell - k) + \delta^{\mu\nu} m^2 \right] \tag{15.63}$$

Now, in order to simplify the integrand in equation (15.59), we will use the Feynman parameter formula (see Sect. 13.8.2)

$$\frac{1}{[(\ell - k)^2 + m^2][\ell^2 + m^2]} = \int_0^1 d\alpha \frac{1}{\{\alpha[(\ell - k)^2 + m^2] + (1 - \alpha)[\ell^2 + m^2]\}^2}$$

$$= \int_0^1 d\alpha \frac{1}{[(\ell - \alpha k)^2 + \alpha(1 - \alpha)k^2 + m^2]^2}$$

Plugging this back into our integral and changing the integration variable $\ell \to \ell + \alpha k$ yields

$$\Pi^{\mu\nu}(k) = i4f(\omega)e^2\mu^{4-2\omega} \cdot$$

$$\cdot \int_0^1 d\alpha \int \frac{d^{2\omega}\ell}{(2\pi)^{2\omega}} \frac{2\alpha(1-\alpha)k^\mu k^\nu + \delta^{\mu\nu}\left[\frac{\omega-1}{\omega}\ell^2 - \alpha(1-\alpha)k^2 + m^2\right]}{[\ell^2 + m^2 + \alpha(1-\alpha)k^2]^2}$$

$$- iz_3^{(2)}\left[k^2\delta^{\mu\nu} - k^\mu k^\nu\right] \tag{15.64}$$

where, we have used the symmetry of the integration to drop all terms in the numerator which are odd in k and to replace $2\ell^\mu \ell^\nu$ where it appears in the numerator of the integrand by $\frac{1}{\omega}\delta^{\mu\nu}\ell^2$.

Now, we rewrite the numerator in the integrand as

$$2\alpha(1-\alpha)k^\mu k^\nu + \delta^{\mu\nu}\left[\frac{\omega-1}{\omega}\ell^2 - \alpha(1-\alpha)k^2 + m^2\right]$$

$$= \frac{\omega-1}{\omega}\delta^{\mu\nu}\left[\ell^2 + \alpha(1-\alpha)k^2 + m^2\right]$$

$$+ 2\alpha(1-\alpha)k^\mu k^\nu + \frac{1}{\omega}\left[(1-2\omega)\alpha(1-\alpha)k^2 + m^2\right]\delta^{\mu\nu}$$

Then the integrals over the loop momenta are elementary and we can use the dimensional regularization integral formula (13.119) (see Sect. 13.8.3) to get

$$\Pi^{\mu\nu}(k) = 4if(\omega)e^2\mu^{4-2\omega} \times$$

$$\times \left\{ \frac{\Gamma[2-\omega]}{(4\pi)^\omega} \int_0^1 d\alpha \frac{2\alpha(1-\alpha)k^\mu k^\nu + \frac{1}{\omega}\left[(1-2\omega)\alpha(1-\alpha)k^2 + m^2\right]\delta^{\mu\nu}}{[m^2 + \alpha(1-\alpha)k^2]^{2-\omega}} \right.$$

$$\left. + \frac{\Gamma[1-\omega]}{(4\pi)^\omega} \int_0^1 d\alpha \frac{\frac{\omega-1}{\omega}\delta^{\mu\nu}}{[m^2 + \alpha(1-\alpha)k^2]^{1-\omega}} \right\} \tag{15.65}$$

$$- iz_3^{(2)}\left[k^2\delta^{\mu\nu} - k^\mu k^\nu\right] \tag{15.66}$$

Now, we can use the property of the Euler gamma function, $(1-\omega)\Gamma[1-\omega] = \Gamma[2-\omega]$, to get

$$\Pi^{\mu\nu}(k) =$$

$$- i\left[k^2\delta^{\mu\nu} - k^\mu k^\nu\right]\frac{8f(\omega)e^2\mu^{4-2\omega}\Gamma[2-\omega]}{(4\pi)^\omega} \int_0^1 d\alpha \frac{\alpha(1-\alpha)}{[m^2 + \alpha(1-\alpha)k^2]^{2-\omega}}$$

$$- iz_3^{(2)}\left[k^2\delta^{\mu\nu} - k^\mu k^\nu\right] \tag{15.67}$$

This is the result for the dimensionally regularized self-energy of the photon. From the appearance of the tensor $[k^2\delta^{\mu\nu} - k^\mu k^\nu]$ from the Feynman integrals which gave us the first line, we now confirm that the photon remains massless. Generation of a photon mass would have given an additional tensor structure with $\delta^{\mu\nu}$ or $k^\mu k^\nu$ in a different combination. This is a result of our use of dimensional regularization which is consistent with gauge invariance, as it is gauge invariance which suppresses the photon mass. This justifies our ignoring the photon mass counterterm from the beginning.

The contribution in equation (15.67) is ultraviolet divergent. This divergence resides in the singularity of the Euler gamma function $\Gamma[2 - \omega]$ as we put the dimension to four, that is, when we put $\omega \to 2$, $\Gamma[2 - \omega] \sim \frac{1}{2-\omega}$.

We must study equation (15.67) for the photon self-energy in the vicinity of four spacetime dimensions. The asymptotic expansion of the Euler gamma function is discussed in Sect. 13.8.3. Near four dimensions it is

$$\Gamma[2 - \omega] = \frac{1}{2 - \omega} - \gamma + \ldots$$

We use this expansion to examine the vacuum polarization in the vicinity of four dimensions,

$$\Pi^{\mu\nu}(k) = -i\left[k^2\delta^{\mu\nu} - k^\mu k^\nu\right]\left\{z_3^{(2)} + \frac{e^2}{24\pi^2}\frac{1}{2 - \omega}\right.$$

$$\left. - \frac{e^2}{2\pi^2}\int_0^1 d\alpha\, \alpha(1 - \alpha)\ln\left[\frac{m^2 + \alpha(1 - \alpha)k^2}{4\pi e^{-\gamma + f'(2)}\mu^2}\right]\right\} \tag{15.68}$$

where we have dropped all contributions in the asymptotic expansion about $\omega = 2$ which vanish when $\omega \to 2$. The divergence of the Feynman integral is manifest in the singularity of the right-hand-side of the above equation as $\omega \to 2$.

We can see that the counterterm $z_3^{(2)}$ must be determined so that it cancels the singular term $\frac{e^2}{24\pi^2}\frac{1}{2-\omega}$ in equation (15.68). This determines the singular part of the counterterm. The finite part of the counterterm depends on the subtraction scheme.

In the previous chapter we introduced three subtraction schemes, the on-shell subtraction and the MS and $\overline{MS}$ schemes. The on-shell subtraction scheme requires that the coefficient of the projection operator $[k^2\delta^{\mu\nu} - k^\mu k^\nu]$ in equation (15.68) vanishes when $k^2 \to 0$. This determines the counterterm which appears in that formula.

The result for the counterterm and the vacuum polarization tensor are

on − shell subtraction:

$$z_3^{(2)} = -\frac{e^2}{24\pi^2}\frac{1}{2-\omega} + \frac{e^2}{12\pi^2}\ln\left[\frac{m^2}{4\pi e^{-\gamma+f'(2)}\mu^2}\right] \tag{15.69}$$

$$\Pi^{\mu\nu}(k) = i\left[k^2\delta^{\mu\nu} - k^\mu k^\nu\right]\left\{\frac{e^2}{2\pi^2}\int_0^1 d\alpha\,\alpha(1-\alpha)\ln\left[1+\alpha(1-\alpha)\frac{k^2}{m^2}\right]\right\} \tag{15.70}$$

Alternatively, in the $\overline{MS}$ subtraction scheme, the counterterm is determined so that it just cancels the ultraviolet divergence as well as some of the constants that always accompany dimensional regularization. The counterterm and the vacuum polarization tensor in the $\overline{MS}$ scheme are

$\overline{MS}$:

$$z_3^{(2)} = -\frac{e^2}{24\pi^2}\frac{1}{2-\omega} - \frac{e^2}{12\pi^2}\ln\left[4\pi e^{-\gamma+f'(2)}\right] \tag{15.71}$$

$$\Pi^{\mu\nu}(k) = i\left[k^2\delta^{\mu\nu} - k^\mu k^\nu\right]\left\{\frac{e^2}{2\pi^2}\int_0^1 d\alpha\,\alpha(1-\alpha)\ln\left[\frac{m^2+\alpha(1-\alpha)k^2}{\mu^2}\right]\right\} \tag{15.72}$$

Note that, in both subtraction schemes, the vacuum polarization no longer depends on the constant $f'(2)$. In both cases, this constant has been absorbed into a finite correction of the counterterm $z_3^{(2)}$. In the on-shell subtraction scheme the vacuum polarization is independent of the regularization scale, μ. It depends only on the constants e^2 and m. On the other hand, in the $\overline{MS}$ scheme, the vacuum polarization retains a μ-dependence.

Let us proceed with the vacuum polarization tensor as determined by the on-shell subtraction scheme. Our next step is to do an inverse Wick rotation and go back to Minkowski space. The result is

$$\Pi^{\mu\nu}(k) =$$

$$i\left[(k^2-i\epsilon)\eta^{\mu\nu} - k^\mu k^\nu\right]\left\{\frac{e^2}{2\pi^2}\int_0^1 d\alpha\,\alpha(1-\alpha)\ln\left[1+\alpha(1-\alpha)\frac{k^2-i\epsilon}{m^2}\right]\right\} \tag{15.73}$$

The net effect of the inverse Wick rotation is to replace $\delta^{\mu\nu}$ by $\eta^{\mu\nu}$ in the projection operator and k^2 by $k^2 - i\epsilon$. The latter is needed since k^2 can now be zero or negative and we need to define the logarithm in the integrand in the region where its argument is negative, that is, where $k^2 < -4m^2$. The logarithm can be regarded as a function of a complex variable, k^2, which has a branch singularity on the negative real axis, in the interval $k^2 \in (-\infty, -4m^2)$.

We can use our result in equation (15.73) and equation (15.53) to reconstruct the photon 2 point function as

$$D_{\mu\nu}(k) =$$

$$\frac{-i \left(\eta_{\mu\nu} - \frac{k_\mu k_\nu}{k^2 - i\epsilon} \right)}{(k^2 - i\epsilon) \left[1 - \frac{e^2}{2\pi^2} \int\limits_0^1 d\alpha \, \alpha(1 - \alpha) \, \ln\left[1 + \alpha(1 - \alpha)\frac{k^2 - i\epsilon}{m^2} \right] \right]}$$

$$- \frac{i}{\xi} \frac{k_\mu k_\nu}{(k^2 - i\epsilon)^2} \tag{15.74}$$

Although the last integral over the Feynman parameter α that remains to be done in (15.74) is elementary, we find it sufficient for our purposes to leave it in integral form. Although the vacuum polarization has been incorporated into the inverse of the photon 2 point function in equation (15.74) we have a right to trust this formula only up to order e^2 in an expansion of it in this parameter.

We might ask how accurate our result is. If we regard everything after and including the integral over α as being of order one, we see that the quantum correction to the photon 2 point function is controlled by the factor $e^2/2\pi^2 \sim 0.005$. The corrections are indeed at a one percent level. Further corrections, if they are similarly suppressed, would be at the 0.01 percent level. That means that the computation that we have done here already gets the photon 2 point function to four digit accuracy.

15.8   Quantum Corrections of the Coulomb Potential

In the last section, we computed the leading perturbative correction to the 2 point function of the photon. We are now prepared to extract some of the physical implications of the result of that section. For example, the Coulomb interaction between electric charges will be modified by the quantum effects in quantum electrodynamics. In order to analyze how this comes about, we can ask the question as to how the energy of the vacuum state of the quantum field theory is modified if we introduce a time-independent charge distribution. We will assume that the charge is sufficiently small that we can analyze the problem in perturbation theory. Then our task will be to find the energy of the lowest energy state of the quantum field theory in the presence of the charge distribution compared to the energy of the vacuum state in the absence of the charge distribution and to the leading orders in perturbation theory.

Let us assume that the charge distribution is given by the density $\rho(\vec{x})$. In classical electrodynamics, this charge distribution would have an energy due to its Coulomb self-interaction. This classical energy is given by the integral

$$\delta E_{\text{classical}} = \frac{1}{2} \int d^3x \int d^3y \, \rho(\vec{x}) \frac{1}{4\pi |\vec{x} - \vec{y}|} \rho(\vec{y}) \tag{15.75}$$

In the following, we will re-derive this formula in the context of quantum electrodynamics where it will represent a certain limit of the full quantum mechanical result. Then we will find corrections to this formula that come from quantum effects.

We couple a charge distribution to quantum electrodynamics by adding a term to the Hamiltonian which takes the interaction energy of the classical charge with the quantum field theory into account. We will denote the Hamiltonian of quantum electrodynamics by H and the additional Hamiltonian $\tilde{H}$ so that

$$H \to H + \tilde{H}, \quad \tilde{H} = \int d^3x \, A_0(x)\rho(\vec{x})$$

where $\rho(\vec{x})$ is the charge distribution and $A_0(x)$ is the time component of the photon field.

Then, we take $\rho(\vec{x})$ to be so small that quantum mechanical perturbation theory is valid. The shift of the vacuum energy to the first order in perturbation theory is the expectation value of the perturbation to the Hamiltonian in the vacuum state of the unperturbed theory, that is,

$$\delta E^{(1)} = <\mathcal{O}|\tilde{H}|\mathcal{O}>$$

It is easy to see that this correction is zero. We only need to recall that the 1 point function of the photon, $<\mathcal{O}|A_0(x)|\mathcal{O}>$, vanishes in quantum electrodynamics. We thus conclude that

$$\delta E^{(1)} = 0 \tag{15.76}$$

The shift of the vacuum energy to the second order of perturbation theory is given by the expression

$$\delta E^{(2)} = - <\mathcal{O}|\tilde{H}\frac{1}{H - E_0}\tilde{H}|\mathcal{O}>$$

where E_0 is the vacuum energy, $H|\mathcal{O}> = E_0|\mathcal{O}>$. We can rewrite the right-hand-side of the above equation as

$$\delta E^{(2)} = -i \int_0^\infty dt \, <\mathcal{O}|\tilde{H}e^{-it(H - E_0 - i\epsilon)}\tilde{H}|\mathcal{O}>$$

$$= -i \int_0^\infty dt \, <\mathcal{O}|e^{iHt}\tilde{H}e^{-itH - \epsilon t}\tilde{H}|\mathcal{O}>$$

where we have used the fact that $< \mathcal{O}|e^{itE_0} =< \mathcal{O}|e^{iHt}$. We now define $\tilde{H}(t) \equiv e^{iHt}\tilde{H}e^{-iHt}$ and the above expression becomes

$$\delta E^{(2)} = -i \int_0^\infty dt < \mathcal{O}|\tilde{H}(t)\tilde{H}(0)|\mathcal{O} >$$

$$= -i \int_0^\infty dt \int d^3x \int d^3y \rho(\vec{x})\rho(\vec{y}) < \mathcal{O}|A_0(\vec{x},t)A_0(\vec{y},0)|\mathcal{O} >$$

$$= -\frac{i}{2} \int_{-\infty}^\infty dt \int d^3x \int d^3y \rho(\vec{x})\rho(\vec{y}) < \mathcal{O}|T A_0(\vec{x},t)A_0(\vec{y},0)|\mathcal{O} >$$

The result is

$$\delta E^{(2)} = -\frac{i}{2} \int dx \int d^3y \rho(\vec{x}) D_{00}(\vec{x},x^0;\vec{y},0)\rho(\vec{y}) \tag{15.77}$$

which is similar to the classical expression given in equation (15.75) with the Coulomb potential $\frac{1}{4\pi|\vec{x}-\vec{y}|}$ replaced by the time integral of the 2 point function $-i \int_{-\infty}^\infty dt D_{00}(\vec{x},x^0;\vec{y},0)$. The corrected Coulomb potential must be

$$V_0(|\vec{x}-\vec{y}|) = -i \int_{-\infty}^\infty dt D_{00}(\vec{x},t;\vec{y},0)$$

$$= -i \int \frac{dk}{(2\pi)^4} \int_{-\infty}^\infty dt e^{i\vec{k}\cdot(\vec{x}-\vec{y})-ik^0t} D_{00}(k^0,\vec{k})$$

$$= -i \int \frac{d^3k}{(2\pi)^3} e^{i\vec{k}\cdot(\vec{x}-\vec{y})} D_{00}(0,\vec{k})$$

To understand this expression better, we could expand the right-hand-side in perturbation theory, in an expansion in e^2 where we remember that the 2 point function has the expansion

$$-i D_{00}(k_0 = 0,\vec{k}) = -i \Delta_{00}(k_0,\vec{k}) + \mathcal{O}(e^2) = \frac{1}{\vec{k}^2} + \mathcal{O}(e^2)$$

By plugging this expression into equation (15.77) we see that

$$V_0(|\vec{x}-\vec{y}|) = \int \frac{d^3k}{(2\pi)^3} \frac{e^{i\vec{k}\cdot(\vec{x}-\vec{y})}}{\vec{k}^2} + \mathcal{O}(e^2) = \frac{1}{4\pi|\vec{x}-\vec{y}|} + \mathcal{O}(e^2) \tag{15.78}$$

The leading order term indeed reproduces the classical Coulomb potential of equation (15.75).

To find the corrections to the classical Coulomb potential, we must examine the corrections to the vacuum polarization in equation (15.74) which gives us

$$-i D_{00}(k^0 = 0, \vec{k}) = \frac{1}{\vec{k}^2 \left[1 - \frac{e^2}{2\pi^2} \int_0^1 d\alpha \, \alpha (1 - \alpha) \, \ln \left[1 + \alpha (1 - \alpha) \frac{\vec{k}^2}{m^2} \right] \right]} \tag{15.79}$$

Here, we have used the on-shell subtraction scheme. In that scheme, m is the physical mass of the electron and e is its electric charge.

If we are interested in the the effects of this potential at distance scales which are much larger than the Compton wave-length of the electron, or at wave-numbers much smaller than the electron mass, we can use the Taylor expansion of the 2 point function in powers of $\frac{\vec{k}^2}{m^2}$. To get this expansion, we first consider an expansion of the terms in the denominator where, when $\vec{k}^2 \ll m^2$,

$$1 - \frac{e^2}{2\pi^2} \int_0^1 d\alpha \, \alpha (1 - \alpha) \, \ln \left[1 + \alpha (1 - \alpha) \frac{\vec{k}^2 - i\epsilon}{m^2} \right] \tag{15.80}$$

$$= 1 - \frac{e^2}{2\pi^2} \left(\frac{1}{30} \frac{\vec{k}^2}{m^2} - \frac{1}{280} \frac{(\vec{k}^2)^2}{m^4} + \frac{1}{1890} \frac{(\vec{k}^2)^3}{m^6} + \dots \right) \tag{15.81}$$

Then the expansion in equation (15.80) in a series in e^2 and $\frac{\vec{k}^2}{m^2}$ is

$$-i D_{00}(k^0 = 0, \vec{k}) = \frac{1}{\vec{k}^2} + \frac{e^2}{60\pi^2} \frac{1}{m^2} + \dots$$

or, in the Fourier transform

$$V(|\vec{r}|) = \frac{1}{4\pi |\vec{r}|} + \frac{e^2}{60\pi^2 m^2} \delta^3(\vec{r}) + \dots + \dots$$

The leading correction to the Coulomb potential is a contact interaction which is called the Uehling term. As a perturbation, it affects only the s-wave atomic orbitals and it accounts for about $-50\,\text{MHz}$ of the $1000\,\text{MHz}$ Lamb shift. The measurement of the Lamb shift and the detailed understanding of the other contributions to it then give us a stringent check on the validity of this term. Of course, the theory passes this test.

The Coulomb energy of the charge distribution with its first correction becomes

$$\delta E = \frac{1}{2} \int d^3x \int d^3y \rho(\vec{x}) \frac{1}{4\pi |\vec{x} - \vec{y}|} \rho(\vec{y}) + \frac{e^2}{60\pi^2 m^2} \int d^3x \rho(\vec{x})^2 + \dots \tag{15.82}$$

Another interesting limit is the high-energy or large wave-number limit. In equation (15.79) we see that, if we begin with small values $\vec{k}^2$, we find that the potential is an increasing function of $\vec{k}^2$ which means that the strength of the Coulomb interaction gets stronger as we increase the magnitude of the wave-number. When the wavenumber is much greater than the electron mass, $\vec{k}^2 >> m^2$, where using equation (15.79) gives us

$$-i\,D_{00}(\vec{k}^2) = \frac{1}{\vec{k}^2\left[1 - \frac{e^2}{12\pi^2}\ln\frac{\vec{k}^2}{m^2} + \dots\right]}$$
(15.83)

In this high energy regime, the combination $\frac{e^2}{12\pi^2}\ln\frac{\vec{k}^2}{m^2}$ can still be small for an interesting range of $|\vec{k}|$, however, eventually it becomes large and perturbation theory is no longer valid. We can fix this to some extent by using the renormalization group to re-sum the offending logarithms. For this, it is best to go to the $\overline{MS}$ subtraction scheme where the effective Coulomb potential would be replaced

$$-i\,D_{00}(\vec{k}^2) = \frac{1}{\vec{k}^2\left[1 - \frac{e^2}{2\pi^2}\int\limits_0^1 d\alpha\,\alpha(1-\alpha)\ln\left[\frac{m^2+\alpha(1-\alpha)k^2}{\mu^2}\right] + \dots\right]}$$
(15.84)

Now we can recall our discussion of the renormalization group analysis of the behaviour of correlation functions under a change of scale of the momentum variables. The solution of the renormalization group equation would have the form

$$-i\,D_{00}(s^2\vec{k}^2) =$$

$$\frac{Z_3(s)/Z_3(1)}{s^2\vec{k}^2\left[1 - \frac{e^2(s)}{2\pi^2}\int\limits_0^1 d\alpha\,\alpha(1-\alpha)\ln\left[\frac{m^2(s)/s^2+\alpha(1-\alpha)k^2}{\mu^2}\right] + \dots\right]}$$
(15.85)

where the running variables $e(s)$, $m(s)$ and $Z_3(s)$ are solutions of renormalization group equations. Now the large k behaviour can be deduced from the large s behaviour which is controlled by the flows of $e(s)$, $m(s)$ and $Z_3(s)$. Of most interest to us is the coupling constant flow where the magnitude of $e^2(s)$ governs the accuracy of the above result. The beta function of quantum electrodynamics (which is scheme independent at this order) is

$$\beta(e) = \frac{e^3}{12\pi^2} + \dots$$
(15.86)

The solution of the flow equation is

$$e^2(s) = \frac{e^2(1)}{1 - \frac{e^2(1)}{12\pi^2}\ln s^2}$$
(15.87)

If we plug this result into equation (15.85) we can see that we have indeed re-summed the dangerous logarithms. We can now explore the small or large $|\vec{k}|$ regime by making s small or large. We are interested in the large s regime. We see that the effective coupling is indeed an increasing function of s, which means that perturbation theory should fail when s gets large enough. This will happen only for exponentially large s. Together with the fact that $e^2(1)$ is small, this means that the above result should be valid up to a phenomenally large values of $|\vec{k}|$. One implication is that this running of the coupling constant with scale should be visible to experiments. It is indeed observed in high energy scattering experiments.

It would seem that the running coupling in equation (15.87) grows large and then diverges at a large value of s. This behaviour is called a "Landau pole" or "Moscow zero". However, the regime where it the coupling grows large is precisely the regime where the perturbation theory which was used to get this result is not valid. Whether quantum electrodynamics has a Landau pole or some other asymptotic behaviour depends on whether the beta function remains positive as e is taken large and if so, whether the large e limit of the beta function grows faster or slower than a linear function of e when e is large. (See the discussion in the latter parts of Sect. 13.7.)

15.9   The Electron 2 Point Function

We showed in a previous section that the leading correction to the self-energy of the electron is given by the expression which we found in equation (15.48) and which we recopy here

$$\Sigma_{ab}(p) = -e^2 \int \frac{d\ell}{(2\pi)^4} [\gamma^\mu g(p+\ell)\gamma^\nu]_{ab} \Delta_{\mu\nu}(\ell) + \left[z_2^{(k)} i \not{p} + m z_m^{(k)} \right]_{ab} + \dots$$

Like the vacuum polarization, this expression is ultraviolet divergent and a regularization is required on order to define it. We will use dimensional regularization. It will turn out that it has a singular behaviour due to an infrared divergence. In order to regularize infrared singularities, we also keep the photon mass. We will us the Feynman gauge where the gauge fixing parameter is set to $\xi = 1$ so that the photon propagator becomes

$$\Delta_{\mu\nu}(\ell) = \frac{-i\eta_{\mu\nu}}{\ell^2 + \kappa^2 - i\epsilon}$$

Then, a series of computations similar to those that we outlined for the vacuum polarization leave us with an expression for the electron self-energy

$$\Sigma(k) = i\!\!\not{k} z_2^{(2)} + m z_m^{(2)} +$$

$$i\!\!\not{k}\frac{e^2}{(4\pi)^2}\left\{ \frac{1}{2-\omega} - \int_0^1 d\alpha 2(1-\alpha)\ln\frac{\alpha(1-\alpha)k^2 + \alpha m^2 + (1-\alpha)\kappa^2 - i\epsilon}{4\pi e^{-1-\gamma}\mu^2} \right\}$$

$$+ m\frac{e^2}{(4\pi)^2}\left\{ \frac{4}{2-\omega} - 4\int_0^1 d\alpha\ln\frac{\alpha(1-\alpha)k^2 + \alpha m^2 + (1-\alpha)\kappa^2 - i\epsilon}{4\pi e^{-\frac{1}{2}-\gamma}\mu^2} \right\}$$

$$\text{(15.88)}$$

$$\Sigma(k) = i\!\!\not{k}\,\Sigma_1(k) + m\,\Sigma_2(k) \tag{15.89}$$

$$\Sigma_1(k) = z_2^{(2)} + \frac{e^2}{(4\pi)^2}\frac{1}{2-\omega}$$

$$- \frac{e^2}{(4\pi)^2}\int_0^1 d\alpha 2(1-\alpha)\ln\frac{\alpha(1-\alpha)k^2 + \alpha m^2 + (1-\alpha)\kappa^2 - i\epsilon}{4\pi e^{-1-\gamma}\mu^2}$$

$$\text{(15.90)}$$

$$\Sigma_2(k) = z_m^{(2)} + \frac{e^2}{(4\pi)^2}\frac{4}{2-\omega}$$

$$- \frac{e^2}{(4\pi)^2}4\int_0^1 d\alpha\ln\frac{\alpha(1-\alpha)k^2 + \alpha m^2 + (1-\alpha)\kappa^2 - i\epsilon}{4\pi e^{-\frac{1}{2}-\gamma}\mu^2} \tag{15.91}$$

Now, we can see that both of the counterterms, $z_2^{(2)}$ and $z_m^{(2)}$ are needed in order to cancel the ultraviolet divergences which are manifest as the poles in equations (15.90) and (15.91) at $\omega \to 2$. Of course, the finite parts of these counterterms depend on the subtraction scheme.

Let us apply the on-shell subtraction scheme. In that scheme, we need to guarantee that the pole in the electron 2 point function $S(k)$ remains at $-k^2 = m^2$ and the residue of that pole remains as it was in $g_{ab}(k)$, the electron propagator. The requirement is thus

$$S(k)_{k^2 \to -m^2} = \frac{i\!\!\not{k} - m}{k^2 + m^2 - i\epsilon} + (\text{non-singular}) \tag{15.92}$$

Using this condition, the counterterms are

on−shell subtraction scheme:

$$z_2^{(2)} = -\frac{e^2}{(4\pi)^2}\frac{1}{2-\omega} - \frac{e^2}{(4\pi)^2}\int_0^1 d\alpha\, 2(1-\alpha)\ln\frac{\alpha^2 m^2 + (1-\alpha)\kappa^2}{4\pi e^{-\gamma-1}\mu^2}$$

$$-\frac{e^2}{(4\pi)^2}\int_0^1 d\alpha\frac{4\alpha(1-\alpha)(1+\alpha)}{\alpha^2 + (1-\alpha)\kappa^2/m^2} \tag{15.93}$$

$$z_m^{(2)} = -4\frac{e^2}{(4\pi)^2}\frac{1}{2-\omega} - 4\frac{e^2}{(4\pi)^2}\int_0^1 d\alpha\ln\frac{\alpha^2 m^2 + (1-\alpha)\kappa^2}{4\pi e^{-\frac{1}{2}-\gamma}\mu^2}$$

$$-\frac{e^2}{(4\pi)^2}\int_0^1 d\alpha\frac{4\alpha(1-\alpha)(1+\alpha)}{\alpha^2 + (1-\alpha)\kappa^2/m^2} \tag{15.94}$$

Notice that, as well as having the $\frac{1}{2-\omega}$ ultraviolet singularity, the counterterms also have an infrared singularity coming from the last term in each expression. When this term is integrated it is singular in the limit of small infrared cutoff, κ. It behaves as $\sim \ln\kappa$ then $\kappa \to 0$. This is in spite of the fact that the expression that we have found for $\Sigma(k)$ is not singular at $\kappa \to 0$. However it is not analytic there since its derivatives by $-k^2$ diverges. We need its derivative at $k^2 = -m^2$ in order to get the correct residue of the pole in the 2 point function.

The renormalized irreducible 2 point function is

on−shell subtraction scheme :

$$\Sigma(k) = -i\slashed{k}\frac{e^2}{(4\pi)^2}\int_0^1 d\alpha 2(1-\alpha)\ln\frac{\alpha(1-\alpha)k^2 - i\epsilon + \alpha m^2 + (1-\alpha)\kappa^2}{\alpha^2 m^2 + (1-\alpha)\kappa^2}$$

$$- m\frac{e^2}{(4\pi)^2}4\int_0^1 d\alpha\ln\frac{\alpha(1-\alpha)k^2 - i\epsilon + \alpha m^2 + (1-\alpha)\kappa^2}{\alpha^2 m^2 + (1-\alpha)\kappa^2}$$

$$+ (i\slashed{k} + m)\left[\frac{e^2}{(4\pi)^2}\int_0^1 d\alpha\frac{4\alpha(1-\alpha)(1+\alpha)}{\alpha^2 + (1-\alpha)\kappa^2/m^2}\right] \tag{15.95}$$

Notice that we have now guaranteed that (15.92) holds. However the irreducible 2 point function still contains the last term in equation (15.95) which is logarithmically infrared divergent at the infrared cutoff κ is sent to zero. It is proportional to $(i\slashed{k} + m)$ which vanishes when $-k^2 \to m^2$ and we operate it on any spinor $\psi(k)$ obeying the Dirac equation $(i\slashed{k} + m)\psi(k) = 0$ and $\bar{\psi}(k)(i\slashed{k} + m) = 0$.

For the electron, the on-shell renormalization scheme chooses the counterterms in such a way that the 2 point function has the property

$$\bar{\psi}(k)\left[\lim_{k^2 \to -m^2} (i\not{k} + m)S(k)\right] = -1$$

$$\left[\lim_{k^2 \to -m^2} (i\not{k} + m)S(k)\right]\psi(k) = -1$$

This normalization is very convenient when we use the LSZ formula since the weak asymptotic condition – the large time limits where the interacting field becomes a free field – has the form

$$\text{weak} \lim_{x^0 \to \pm\infty} \psi(x) = \psi_{\text{in/out}}(x)$$

without an additional constant multiplying $\psi_{\text{in/out}}(x)$.

15.10   Radiative Correction of the Vertex

Let us consider the irreducible Feynman diagrams which contribute to the 3 point function $\Gamma^\rho_{Iab}(p' - p, p, -p')$. We have already computed this 3 point function to the lowest order where it is nonzero, order e, in equation (15.30). At order e^3 there are contributions from only two Feynman diagrams, one with a counterterm and one with an internal loop. The diagram with the internal loop is displayed in Fig. 15.17. The contributions of that diagram and the counterterm diagram are given by

$$\Gamma^\rho_{Iab}(p' - p, p, -p') = -i[e\mu^{2-\omega}]\gamma^\rho_{ab} - iz_1^{(2)}[e\mu^{2-\omega}]\gamma^\rho_{ab} + i[e\mu^{2-\omega}]e^2\mu^{2(2-\omega)} \times$$

$$\times \int \frac{d^{2\omega}k}{(2\pi)^{2\omega}} \frac{[\gamma^\mu\left(i\not{p} - i\not{k} - m\right)\gamma^\rho\left(i\not{p'} - i\not{k} - m\right)\gamma^\nu]_{ab}}{[(p-k)^2 + m^2 - i\epsilon][(p'-k)^2 + m^2 - i\epsilon]} \frac{-i\eta_{\mu\nu}}{[k^2 + \kappa^2 - i\epsilon]}$$

$$+ \ldots \tag{15.96}$$

where we have used the Feynman gauge, $\xi = 1$. The ellipses denote further corrections, of order e^5 and higher. The constant $z_1^{(2)}$ is the contribution of the vertex counterterm.

We have recognized the fact that the integral over the loop momentum k^μ will be ultraviolet divergent when the spacetime dimension is four and we have introduced dimensional regularization in order to define the integral. As is the usual practice, we have replaced e by $e\mu^{2-\omega}$ so that the parameter e remains dimensionless, at the expense of introducing the arbitrary dimensionful parameter μ. Note that we have done this systematically everywhere that e appears.

We have also recognized the fact that the integral will be infrared divergent when we put the electron momenta p and p' on-shell. In order to define it in that case, we have retained the photon mass κ which provides an infrared cutoff.

Fig. 15.17 The Feynman
diagram which contributes to
the irreducible 3 point
correlation function
corresponding to the
irreducible vertex function at
order e^3

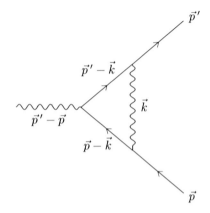

Before we proceed with doing some of the integrals in equation (15.96), let
us pause for a moment and try to confirm that the vertex function in the integral
form given in that equation obeys the Ward–Takahashi identity. The identity that
we are interested in is equation (15.176) which is derived in Sect. 15.13. The Ward–
Takahashi identities are relationships that correlation functions must satisfy independently of perturbation theory. They are consequences of symmetries of the theory. In
this case it is the phase symmetry which results in conservation of electric charge and
the conservation of the associated current operator $\mathbf{J}^\mu(x) = -e\bar\psi(x)\gamma^\mu\psi(x)$. What
the survival of this symmetry needs in the quantum field theory is that the details of
our computation, particularly the ultraviolet regularization are compatible with the
symmetry. Indeed we expect that dimensional regularization must be. Also, the introduction of a small photon mass, which is needed to regularize infrared divergences
does not violate the phase symmetry.

We can check to see that our perturbation theory calculation respects the Ward–
Takahashi identity (15.176). For this we begin with equation (15.96) and we consider

$$(p'-p)_\rho \Gamma^\rho_{lab}(p'-p,p,-p') = -i(1+z_1^{(2)})[e\mu^{2-\omega}](p'-p)_{ab}+$$

$$[e\mu^{2-\omega}]^3 \int \frac{d^{2\omega}k}{(2\pi)^{2\omega}} \frac{[\gamma^\mu(i\not p - i\not k - m)(\not p' - \not p)(i\not p' - i\not k - m)\gamma_\mu]_{ab}}{[(p-k)^2 + m^2 - i\epsilon][(p'-k)^2 + m^2 - i\epsilon][k^2 + \kappa^2 - i\epsilon]}$$

$$+\dots \tag{15.97}$$

Then we can use the identity

$$-i\not p' + i\not p = [-i\not p' - m] - [-i\not p - m] = g^{-1}(p') - g^{-1}(p)$$

and

$$-i\not p' + i\not p = [-i\not p' + i\not k - m] - [-i\not p + i\not k - m] = g^{-1}(p'-k) - g^{-1}(p-k)$$

to rewrite the right-hand-side of equation (15.97) as the difference of two quantities,

$$(p' - p)_\rho \Gamma^\rho_{lab}(p' - p, p, -p') = (1 + z_1^{(2)} - z_2^{(2)})[e\mu^{2-\omega}]\left[S^{-1}(p') - S^{-1}(p) \right]$$
(15.98)

where

$$S^{-1}(p) = -i\not{p}(1 + z_2^{(2)}) - (1 + z_m^{(2)})m$$
$$- i[e\mu^{2-\omega}]^2 \int \frac{d^{2\omega}k}{(2\pi)^{2\omega}} \frac{\gamma^\mu \left(i\not{p} - i\not{k} - m \right) \gamma_\mu}{[(p-k)^2 + m^2 - i\epsilon][k^2 + \kappa^2 - i\epsilon]}$$
$$+ \dots$$
(15.99)

We recall that the irreducible 2 point function of the electron, $\Sigma(p)$, is included in the inverse of the full 2 point function, $S^{-1}(p)$, according to the formula

$$S^{-1}(p) = -i\not{p} - m - \Sigma(p)$$

and we see that order e^2 contribution in equation (15.99) matches the computation of the irreducible electron 2-point function which we found in equation (15.48). This confirms that equation (15.99) is indeed an expansion of $S^{-1}(p)$ to order e^2 and then, moreover, that equation (15.98) is indeed the expansion of the Ward–Takahashi identity (15.176) to order e^3.

Note that the dependence of $S^{-1}(p)$ on the counterterm $z_m^{(2)}$ cancels in the difference in equation (15.98) but the counterterm $z_2^{(2)}$ remains there. Since both sides of equation (15.98) contain only renormalized correlation functions, the entire equation must be free of ultraviolet divergences. This tells us that, in the limit $\omega \to 2$, at least the singular parts, $\sim \frac{1}{2-\omega}$, of $z_1^{(2)}$ and $z_2^{(2)}$ must be identical. Whether the finite parts are also identical depends on the renormalization scheme.

Now, let us return to equation (15.96) and attempt to perform some of the integrations. We will begin with the momentum integral. For the momentum integral, we begin by doing a Wick rotation which results in an additional factor of i. After that the Dirac matrices are Euclidean, as are the inner products of vectors and $\eta^{\mu\nu}$ is replaced by $\delta^{\mu\nu}$.

To simplify the loop integral we need to use the Feynman parameter formula (see Sect. 13.8.2) which, for three denominators, has the form

$$\frac{1}{ABC} = 2 \int_0^1 d\alpha \, d\beta \, d\gamma \frac{\delta(1 - \alpha - \beta - \gamma)}{[\alpha A + \beta B + \gamma C]^3}$$
(15.100)

We use this formula and then we translate the variable k^μ in order to simplify the denominator in the integrand. The result is

$$
\Gamma^\rho_{1ab}(p'-p, p, -p') =
$$
$$
- i[e\mu^{2-\omega}]\gamma^\rho_{ab} - iz^{(2)}_1[e\mu^{2-\omega}]\gamma^\rho_{ab} + i[e\mu^{2-\omega}]e^2\mu^{2(2-\omega)} \times
$$
$$
\times \int_0^1 d\alpha d\beta d\gamma \delta(1 - \alpha - \beta - \gamma) \int \frac{d^{2\omega}k}{(2\pi)^{2\omega}} \frac{-[\gamma^\mu \slashed{k} \gamma^\rho \slashed{k} \gamma_\mu]_{ab} + N^\rho_{ab}}{[k^2 + D]^3}
$$
$$
+ \dots \tag{15.101}
$$

$$
D = (\alpha + \beta)m^2 + \alpha p^2 + \beta p'^2 - (\alpha p + \beta p')^2 + \gamma \kappa^2 \tag{15.102}
$$

$$
N^\rho = \gamma^\mu \left(i(1-\alpha)\slashed{p} - i\beta \slashed{p}' - m\right) \gamma^\rho \left(i(1-\beta)\slashed{p}' - i\alpha \slashed{p} - m\right) \gamma_\mu \tag{15.103}
$$

The numerator in the integrand can be simplified by realizing that, inside the Euclidean rotationally symmetric integral, we can make the replacement

$$
k_\mu k_\nu \;\rightarrow\; \frac{1}{2\omega}\delta_{\mu\nu}k^2
$$

and hence

$$
\gamma^\mu \slashed{k} \gamma^\rho \slashed{k} \gamma_\mu = \frac{1}{2\omega}k^2 \gamma^\mu \gamma^\nu \gamma^\rho \gamma_\nu \gamma_\mu = 2\frac{(\omega - 1)^2}{\omega}k^2 \gamma^\rho
$$

where we have used the formula

$$
\gamma^\mu \slashed{p} \gamma_\mu = (2 - 2\omega)\slashed{p}
$$

Then, we can use the dimensional regularization integration formulae (see Sect. 13.8.3) to do the integral over k to get

$$
\int \frac{d^{2\omega}k}{(2\pi)^{2\omega}} \frac{-2\frac{(\omega-1)^2}{\omega}\gamma^\rho k^2 + N^\rho}{(k^2 + D)^3} = \frac{\Gamma[2-\omega]}{(4\pi)^\omega D^{2-\omega}} \left[-(\omega - 1)^2 \gamma^\rho + \frac{2-\omega}{2}\frac{N^\rho}{D}\right] \tag{15.104}
$$

and equation (15.101) becomes

$$
\Gamma^\rho_{1ab}(p'-p, p, -p') = -i[e\mu^{2-\omega}]\gamma^\rho_{ab} - iz^{(2)}_1[e\mu^{2-\omega}]\gamma^\rho_{ab} + 2i[e\mu^{2-\omega}]e^2\mu^{2(2-\omega)} \times
$$
$$
\times \int_0^1 d\alpha d\beta d\gamma \delta(1 - \alpha - \beta - \gamma) \frac{\Gamma[2-\omega]}{(4\pi)^\omega D^{2-\omega}} \left[-(\omega - 1)^2 \gamma^\rho + \frac{2-\omega}{2}\frac{N^\rho}{D}\right]
$$
$$
+ \dots \tag{15.105}
$$

Now, we take an asymptotic expansion near $\omega \sim 2$ to get

$$\Gamma^\rho_{Iab}(p' - p, p, -p') = -ie\gamma^\rho_{ab} - iz^{(2)}_1 e\gamma^\rho_{ab} - ie\frac{e^2}{16\pi^2}\frac{1}{2-\omega}\gamma^\rho_{ab}$$

$$- ie\frac{e^2}{8\pi^2}\int_0^1 d\alpha d\beta d\gamma \delta(1 - \alpha - \beta - \gamma)\left[\gamma^\rho \ln\frac{D}{4\pi e^{-\gamma-2}\mu^2} - \frac{1}{2}\frac{N^\rho}{D}\right] + \cdots$$

$$(15.106)$$

$$D = (\alpha + \beta)m^2 + \alpha p^2 + \beta p'^2 - (\alpha p + \beta p')^2 + \gamma\kappa^2 \tag{15.107}$$

$$N^\rho = \gamma^\mu\left(i(1 - \alpha)\not{p} - i\beta\not{p}' - m\right)\gamma^\rho\left(i(1 - \beta)\not{p}' - i\alpha\not{p} - m\right)\gamma_\mu \tag{15.108}$$

The ultraviolet divergence has now been identified as the pole term $\sim \frac{1}{2-\omega}$ and it must be canceled by the counterterm $z^{(2)}_1$. It is also clear that, upon comparing the divergent part of equation (15.106) and the expression for $z^{(2)}_2$ given in equation (15.93), the divergent parts of $z^{(2)}_1$ and $z^{(2)}_2$ are indeed equal, which is what we expected from our discussion of the Ward–Takahashi identity.

We can also identify an infrared divergence in the last integral in equation (15.106). If we put $\kappa^2 \to 0$ and if we put the electrons on their mass shells, $p^2 \to -m^2$, $p'^2 \to -m^2$,

$$D \to (\alpha + \beta)^2 m^2 + \alpha\beta q^2$$

and the Feynman parameter integral in equation (15.106) is logarithmically divergent, the singularity coming from the $\alpha \sim 0$, $\beta \sim 0$, $\gamma \sim 1$ regime of the integration. Unlike an ultraviolet divergence which is canceled by renormalization, an infrared divergence must be dealt with by some physical reasoning. We will return to this important issue after we study the use of equation (15.106) in computing a scattering amplitude.

15.10.1   Electromagnetic form Factors

So far in this section, we have studied the general irreducible 3 point function $\Gamma^\rho_{Iab}(q, p, -p')$, which we call the vertex function. In particular, we have established that it is renormalizable in that the ultraviolet singularity is canceled by the appropriate choice of $z^{(2)}_1$. Many of the applications of this vertex function use it with the electron momenta put on-shell, $-p^2 = m^2 = -p'^2$, and take matrix elements of it between Dirac spinors which represent the wave-functions of physical states of the electron. It is thus worthwhile to study the quantity

$$\bar{\psi}_a(p)\Gamma^\rho_{Iab}(q, p, -p')\psi'_b(p') \tag{15.109}$$

where we put $-p^2 = -p'^2 = m^2$ and the spinors obey the Dirac equation

$$(i\not{p}' + m)\psi'(p') = 0, \quad \bar{\psi}(p)(i\not{p} + m) = 0 \qquad (15.110)$$

and, in all cases in the following,

$$q^\mu = (p' - p)^\mu$$

The photon momentum, q remains off-shell. In fact it must be so, since energy and momentum conservation of the vertex cannot be satisfied when all three of its external momenta are on-shell, unless $q = 0$. Here, we are assuming that m is the physical mass of the electron and that the mass counterterm $z_m^{(2)}$ and the electron wavefunction renormalization counterterm $z_2^{(2)}$ are determined using the on-shell renormalization scheme. Although, by the discussion above, we know its ultraviolet divergent part, we will reserve determining the vertex counterterm $z_1^{(2)}$ until later.

This matrix element of the vertex function (15.109) would be used in the LSZ formula for a scattering amplitude where, for example, a sub-process in the scattering event had an electron interacting with other electrically charged matter through the exchange of a single off-shell photon. It can also describe an electron scattering from a source of electric or magnetic fields. Let us focus on the scenario where the electron is scattering from such a source which produces a classical electromagnetic field, which we denote by the four-vector potential $A_\rho^{(\mathrm{cl})}(x)$ with Fourier transform

$$A_\rho^{(\mathrm{cl})}(x) = \int dq e^{iqx} A_\rho^{(\mathrm{cl})}(q) \qquad (15.111)$$

To linear order in this field, the quantum amplitude for such a process would be

$$< p|i\mathbf{T}|p' > = \bar{\psi}_a(p)\Gamma_{Iab}^\rho(q, p, -p')\psi'_b(p')A_\rho^{(\mathrm{cl})}(q) \qquad (15.112)$$

where $\Gamma_{Iab}^\rho(q, p, -p')$ is the irreducible vertex function and $q = p' - p$.

As well as the Dirac equations for the spinors (15.110), the on-shell subtraction scheme guarantees that

$$\bar{\psi}(p)S^{-1}(p) = 0, \quad S^{-1}(p')\psi'(p') = 0$$

where $S(p)$ is the full 2 point function for the electron. These equations, together with equation (15.98), which follows from the Ward–Takahashi identity, implies the identity

$$(p' - p)_\rho \bar{\psi}_a(p)\Gamma_{Iab}^\rho(p' - p, p, -p')\psi'_b(p') = 0 \qquad (15.113)$$

This in turn implies that the **T**-matrix element in equation (15.112) is gauge invariant, that is, it is invariant under the replacement

$$A_\rho^{(\mathrm{cl})}(x) \rightarrow A_\rho^{(\mathrm{cl})}(x) + \partial_\rho\chi(x) \text{ or } A_\rho^{(\mathrm{cl})}(q) \rightarrow A_\rho^{(\mathrm{cl})}(q) + iq_\rho\chi(q) \qquad (15.114)$$

An interesting fact about the matrix element that we are discussing is that, after symmetry considerations and equation (15.113) are taken into account, its content can be expressed in terms of two "form factors", the two functions $F_1(q^2)$ and $F_2(q^2)$ which each depend only on the photon momentum in the combination q^2. They appear as the formula

$$\bar{\psi}_a(p)\Gamma^\rho_{Iab}(p'-p,p,-p')\psi'_b(p') =$$
$$- ieF_1(q^2)\bar{\psi}(p)\gamma^\rho\psi'(p') + i\frac{e}{m}F_2(q^2)q_\mu\bar{\psi}(p)\Sigma^{\rho\mu}\psi'(p') \qquad (15.115)$$

Here, $\Sigma^{\mu\nu} = -\frac{i}{4}[\gamma^\mu,\gamma^\nu]$ is the spin tensor for the electron. The functions $F_1(q^2)$ and $F_2(q^2)$ are called the electromagnetic form factors and they characterize the interaction of the electron with electric and magnetic fields.

To show that the right-hand-side of equation (15.115) indeed has the form that is given there, we observe that there are three possible forms which are compatible with Lorentz invariance, parity and current conservation (15.113). They are

$$\sim \bar{\psi}(p)\gamma^\rho\psi'(p'), \quad \sim (p'-p)_\mu\bar{\psi}(p)\Sigma^{\rho\mu}\psi'(p'), \quad \sim (p'+p)^\rho\bar{\psi}(p)\psi'(p')$$

each of which could be multiplied by a function of $q^2 = (p'-p)^2$. Then we can reduce this to two forms using the Gordon identity,

$$(p'+p)^\rho\bar{\psi}(p)\psi'(p') = 2im\bar{\psi}(p)\gamma^\rho\psi'(p') - 2i(p'-p)_\mu\bar{\psi}(p)\Sigma^{\rho\mu}\psi'(p')$$
$$(15.116)$$

which holds for any two Dirac spinors $\psi(p)$, $\psi'(p')$, all that is needed is that they satisfy the Dirac equation in momentum space, that is, equation (15.110). The Gordon identity can be easily be confirmed by using those Dirac equations for the spinors and some elementary algebra.

The formula (15.115) is interesting in that it is universal. Once we determine the form factors, $F_1(q^2)$ and $F_2(q^2)$, this decomposition holds for whatever the electron states are, that is, whatever spinors $\psi'(p')$ and $\bar{\psi}(p)$ we use it with.

We now have a characterization of how an electron interacts with an electromagnetic field. Our expression for the transition amplitude is explcitly Lorentz covariant and gauge invariant. It is useful to understand the interaction of the electron with electric and magnetic fields separately. Of course, Lorentz transformations mix electric and magnetic fields and any distinction between them must be reference frame dependent. To distinguish them, we must choose a reference frame. This can be done in the non-relativistic limit where the spatial components of the momenta $\vec{p}$ and $\vec{p}\,'$ are small compared to the electron mass,

$$|\vec{p}|, |\vec{p}\,'| << m, \quad p^0 = m + \frac{\vec{p}^{\,2}}{2m} + \dots, \quad p'^0 = m + \frac{\vec{p}^{\,\prime\,2}}{2m} + \dots$$

$$\vec{q} = \vec{p}\,' - \vec{p}, \quad q^0 = \frac{\vec{p}^{\,\prime 2} - \vec{p}^{\,2}}{2m} + \dots$$

To proceed, we must understand the non-relativistic limit of the matrix elements that are taken with Dirac spinors in equation (15.115). For this purpose, we recall our explicit solution of the Dirac equation

$$\psi_{++}(\vec{p}) = \frac{1}{\sqrt{2}} \begin{bmatrix} i\sqrt{1 - \frac{|\vec{p}|}{p^0}} u_+ \\ \sqrt{1 + \frac{|\vec{p}|}{p^0}} u_+ \end{bmatrix}, \quad \psi_{+-}(\vec{p}) = \frac{1}{\sqrt{2}} \begin{bmatrix} i\sqrt{1 + \frac{|\vec{p}|}{p^0}} u_- \\ \sqrt{1 - \frac{|\vec{p}|}{p^0}} u_- \end{bmatrix}$$

which were found using an explicit representation of the Dirac matrices,

$$\gamma^a = \begin{bmatrix} 0 & \sigma^a \\ \sigma^a & 0 \end{bmatrix}, \quad \gamma^0 = \begin{bmatrix} 0 & 1 \\ -1 & 0 \end{bmatrix}$$

Here the first "+" subscript on $\psi_{++}(p)$ and $\psi_{+-}(p)$ denotes a positive energy solution and we are assuming that it is the positive energy solutions that describe electrons (as opposed to positrons). The second subscript denotes the helicity states where the 2 component spinors $u_\pm(p)$ are helicity eigenvectors,

$$\vec{\sigma} \cdot \vec{p}\, u_\pm(p) = \pm |\vec{p}|\, u_\pm(p)$$

In the limit where the momentum is small we can re-instate the helicity matrix to write the positive energy solution of the Dirac equation as

$$\psi(\vec{p}) \approx \frac{1}{\sqrt{2}} \begin{bmatrix} i[1 - \frac{\vec{\sigma} \cdot \vec{p}}{2m}]u \\ [1 + \frac{\vec{\sigma} \cdot \vec{p}}{2m}]u \end{bmatrix}$$

Here, $u = \begin{bmatrix} u_\uparrow \\ u_\downarrow \end{bmatrix}$ is the normalized two-component wave-function of the non-relativistic electron. The components, $u_\uparrow$ and $u_\downarrow$ are the amplitudes for its spin to be aligned along the z-axis. This is an approximate solution of the Dirac equation which is accurate when $|\vec{p}|/m \ll 1$ for any choice of the 2 component object u.

We can then find the non-relativistic limit of the matrix elements that we need,

$$\bar{\psi}(p)\psi'(p') = -iu^*u' + \dots \tag{15.117}$$

$$\bar{\psi}(p)\gamma^0\psi'(p') = -u^*u' + \dots \tag{15.118}$$

$$\bar{\psi}(p)\gamma^a\psi'(p') = -i\frac{(p+p')^a}{m}u^*u' + \frac{(p-p')^b}{m}u^*\sigma^c u' + \dots \tag{15.119}$$

$$\bar{\psi}(p)\Sigma^{0a}\psi'(p') = i\frac{(p-p')^a}{m}u^*u' - \epsilon^{abc}\frac{(p'+p)^c}{m}u^*\sigma^b u' + \dots \tag{15.120}$$

$$\bar{\psi}(p)\Sigma^{bc}\psi'(p') = -i\epsilon^{bca}u^*\sigma^c u' + \dots \tag{15.121}$$

where the ellipses denote contributions that are suppressed by higher powers of $|\vec{p}|/m$ and $|\vec{p}\,'|/m$ than those terms which are displayed. With these, we conclude that, in this nonrelativistic limit

$$
\bar{\psi}_a(p)\Gamma^\rho_{lab}(q, p, -p')\psi'_b(p')A^{(cl)}_\rho(q) = ieF_1(q^2)A^{(cl)}_0(q)u^*u'
$$
$$
+ \frac{e}{m}\left[F_1(q^2) + F_2(q^2)\right]\vec{B}^{(cl)}(q)\cdot u^*\frac{\vec{\sigma}}{2}u' + \dots \tag{15.122}
$$

where $\vec{B}^{(cl)}(q) = i\vec{q} \times \vec{A}^{(cl)}(q)$ is the Fourier transform of the magnetic field and we must remember that $q = p' - p$. Now we can see that the form factor $F_1(q^2)$ governs the coupling of the electron to the electric field and $\frac{e}{m}\left[F_1(q^2) + F_2(q^2)\right]$ governs the coupling of the electron to the magnetic field.

Moreover, the leading order, e^0, in perturbation theory, we have

$$
F_1(q^2) = 1 + \mathcal{O}(e^2), \quad F_2(q^2) = 0 + \mathcal{O}(e^2)
$$

and our computation of the radiative correction to the vertex in principle gives us the corrections to these form factors at order e^2.

After a significant amount of algebra we can use equations (15.106), (15.115) and (15.116) to determine the form factors in terms of Feynman parameter integrals. The algebra for simplifying $\bar{\psi}N^\rho\psi'$ is summarized in Sect. 15.10.2.1 and the result is

$$
\bar{\psi}(p)N^\rho\psi'(p') =
$$
$$
\left[2q^2(1 - \alpha)(1 - \beta) + 2m^2[2 - 2(\alpha + \beta) - (\alpha + \beta)^2]\right]\bar{\psi}(p)\gamma^\rho\psi'(p')
$$
$$
- 4m[\alpha + \beta - (\alpha + \beta)^2]q_\nu\bar{\psi}(p)\Sigma^{\rho\nu}\psi'(p') \tag{15.123}
$$

We can use this in conjunction with equation (15.106) to get

$$
F_1(q^2) = 1 + z^{(2)}_1 + \frac{e^2}{16\pi^2}\frac{1}{2 - \omega} +
$$
$$
\frac{e^2}{8\pi^2}\int_0^1 d\alpha d\beta d\gamma \delta(1 - \alpha - \beta - \gamma)\left\{-\ln\frac{m^2(1 - \gamma)^2 + q^2\alpha\beta + \kappa^2\gamma}{4\pi e^{-\gamma - 2}\mu^2}\right.
$$
$$
\left. + \frac{m^2(2 - 2(\alpha + \beta) - (\alpha + \beta)^2) + q^2(1 - \alpha)(1 - \beta)}{m^2(1 - \gamma)^2 + q^2\alpha\beta + \kappa^2\gamma}\right\} + \dots \tag{15.124}
$$
$$
F_2(q^2) = \frac{e^2}{8\pi^2}\int_0^1 d\alpha d\beta d\gamma \delta(1 - \alpha - \beta - \gamma)\frac{2m^2\gamma(1 - \gamma)}{m^2(1 - \gamma)^2 + q^2\alpha\beta + \kappa^2\gamma} + \dots
$$
$$
\tag{15.125}
$$

The remaining Feynman parameter integrals for $F_2(q^2)$ is straightforward. It has no ultraviolet divergence and it is finite in the limit where $\kappa^2 \to 0$ and therefore it is also free of infrared divergences. We will do the integral explicitly in Sect. 15.10.2 below where we discuss its application to computing the anomalous magnetic moment of the electron.

On the other hand, $F_1(q^2)$ contains the counterterm and it also has ultraviolet and infrared divergent contributions. We will begin studying it by fixing the counterterm. We will do this by requiring that $F_1(0) = 1$. This is in line with the on-shell subtraction scheme that we have been using and one could regard this renormalization condition as fixing the value of the parameter e as the charge of the electron as it is seen in the low energy limit of scattering of the electron from a static point charge. This renormalization condition determines the counterterm as

$$
z_1^{(2)} = -\frac{e^2}{16\pi^2}\frac{1}{2-\omega} +
$$
$$
\frac{e^2}{8\pi^2}\int_0^1 d\alpha\, d\beta\, d\gamma\, \delta(1-\alpha-\beta-\gamma)\left\{\ln\frac{m^2(1-\gamma)^2+\kappa^2\gamma}{4\pi e^{-\gamma-2}\mu^2}\right.
$$
$$
\left. +\frac{m^2(2-2(\alpha+\beta)-2(\alpha+\beta)^2)}{m^2(1-\gamma)^2+\kappa^2\gamma}\right\}+\dots \qquad (15.126)
$$

and the renormalized form factor in the on shell renormalization scheme is thus

on shell renormalization:

$$
F_1(q^2) = 1 - \frac{e^2}{8\pi^2}\int_0^1 d\alpha\, d\beta\, d\gamma\, \delta(1-\alpha-\beta-\gamma)\times
$$
$$
\times\left\{\ln\frac{m^2(1-\gamma)^2+q^2\alpha\beta+\kappa^2\gamma}{m^2(1-\gamma)^2+\kappa^2\gamma}\right.
$$
$$
+\frac{m^2(2-2(\alpha+\beta)-2(\alpha+\beta)^2)+q^2(1-\alpha)(1-\beta)}{m^2(1-\gamma)^2+q^2\alpha\beta+\kappa^2\gamma}
$$
$$
\left.-\frac{m^2(2-2(\alpha+\beta)-2(\alpha+\beta)^2)}{m^2(1-\gamma)^2+\kappa^2\gamma}\right\}+\dots \qquad (15.127)
$$

Now the remaining expression for $F_1(q^2)$ is free of ultraviolet divergences and the remaining integrations that are presented in equation (15.127) should have a finite result. However, the last two terms in that formula will still become divergent in the limit where we put the photon mass to zero, $\kappa^2 \to 0$. This is the infrared divergence. The offending terms are

$$
-\frac{e^2}{8\pi^2}\int_0^1 d\alpha\, d\beta\, d\gamma\, \delta(1-\alpha-\beta-\gamma)\left\{\frac{2m^2+q^2}{m^2(1-\gamma)^2+q^2\alpha\beta+\kappa^2\gamma}-(q^2\to 0)\right\}
$$

To study this integral, we shall begin by re-organizing the Feynman parameters. We insert the identity

$$1 = \int_0^1 \delta(\rho - \alpha - \beta)$$

into the integrand and we subsequently rescale the integration variables

$$\alpha \to \rho\alpha, \ \beta \to \rho\beta$$

to rewrite the integral as

$$-\frac{e^2}{8\pi^2} \int_0^1 d\alpha\, d\beta\, d\gamma\, d\rho\, \delta(1 - \rho - \gamma)\delta(1 - \alpha - \beta) \times$$

$$\times \left\{ \frac{(2m^2 + q^2)\rho}{m^2\rho^2 + q^2\rho^2\alpha\beta + \kappa^2(1 - \rho)} - (q^2 \to 0) \right\}$$

$$= -\frac{e^2}{8\pi^2} \int_0^1 d\alpha\, d\rho \left\{ \frac{(2m^2 + q^2)\rho}{m^2\rho^2 + q^2\rho^2\alpha(1 - \alpha) + \kappa^2(1 - \rho)} - (q^2 \to 0) \right\}$$

We can now do the integral over ρ. The result is

$$-\frac{e^2}{16\pi^2} \int_0^1 d\alpha \frac{2m^2 + q^2}{m^2 + q^2\alpha(1 - \alpha)} \left\{ \ln \frac{m^2 + q^2\alpha(1 - \alpha)}{\kappa^2} - (q^2 \to 0) \right\}$$

We can now present the form factor, in the limit $\kappa \to 0$ as

on shell renormalization:

$$F_1(q^2) = 1 - \frac{e^2}{8\pi^2} \int_0^1 d\alpha\, d\beta\, d\gamma\, \delta(1 - \alpha - \beta - \gamma) \times$$

$$\times \left\{ \ln \frac{m^2(1 - \gamma)^2 + q^2\alpha\beta}{m^2(1 - \gamma)^2} + \frac{2(1 + (\alpha + \beta))}{(1 - \gamma)} + \right.$$

$$\left. + \frac{m^2(-2(\alpha + \beta) - 2(\alpha + \beta)^2) + q^2(\alpha\beta - \alpha - \beta)}{m^2(1 - \gamma)^2 + q^2\alpha\beta} \right\}$$

$$- \frac{e^2}{16\pi^2} \int_0^1 d\alpha \frac{2m^2 + q^2}{m^2 + q^2\alpha(1 - \alpha)} \ln \frac{m^2 + q^2\alpha(1 - \alpha)}{m^2}$$

$$+ \frac{e^2}{8\pi^2} \ln \frac{\kappa}{m} \int_0^1 d\alpha \left[\frac{2m^2 + q^2}{m^2 + q^2\alpha(1 - \alpha)} - 2 \right]$$

$$+ \dots \tag{15.128}$$

The ellipses denote corrections that are at least of order e^4. In the last line, we have separated the part which diverges as $\sim \ln \kappa$ when κ is set to zero. This is the infrared divergent part of the form factor. It indeed contributes to the transition probability which is the quantity

$$\mathbf{P}(p \to p') = |<p|i\mathbf{T}|p'>|^2$$

$$= \left| F_1(q^2) \left[-ie\bar{\psi}(p) A^{(cl)}(q)\psi'(p') \right] + F_2(q^2) \left[-i\frac{e}{m}\bar{\psi}(p) \Sigma^{\rho\mu} F^{(cl)}_{\rho\mu}(q)\psi'(p') \right] \right|^2$$

$$(15.129)$$

as an infrared divergent term which appears at order e^4 as

$$\mathbf{P}(p \to p') = \ldots +$$

$$+ \frac{e^2}{4\pi^2} \ln \frac{\kappa}{m} \int_0^1 d\alpha \left[\frac{2m^2 + q^2}{m^2 + q^2\alpha(1-\alpha)} - 2 \right] \left| -ie\bar{\psi}(p) A^{(cl)}(q)\psi'(p') \right|^2$$

$$+ \mathcal{O}(e^6) \tag{15.130}$$

Corrections to the infrared divergent part are at least of order e^6. The remaining integral in the infrared divergent part is elementary, however, for our purposed here, the simple integral form which is displayed above will be good enough for us. As for the remaining Feynman parameter integrals that are needed to get $F_1(q^2)$, they have been done and the results can be found in published literature.[1]

Although at this point we have to admit that the full implications of the infrared divergent part are not completely understood, in the following sections we will see that this infrared divergence actually cancels with similar terms in other contributions if we ask the appropriate questions about physical processes. We shall put off further discussion of this form factor until that point.

15.10.2   Anomalous Magnetic Moment

It turns out that, to the order in perturbation theory that we have computed, our formula for $F_2(q^2)$ in equation (15.125) which we recopy here:

$$F_2(q^2) = \frac{e^2}{8\pi^2} \int_0^1 d\alpha d\beta d\gamma \delta(1 - \alpha - \beta - \gamma) \frac{2m^2\gamma(1-\gamma)}{m^2(1-\gamma)^2 + q^2\alpha\beta + \kappa^2\gamma} + \ldots$$

is free of ultraviolet divergences and it does not depend on the counterterms. It is therefore insensitive to the subtraction scheme that is used for renormalization.

---

[1] R.P.Feynman, Physical Review **76**, 769 (1949); R.Karplus and N.M.Kroll, Physical Review **77**, 536 (1950).

Moreover, to this order in perturbation theory, $F_2(q^2)$ does not have an infrared divergence. For this reason we can set the infrared regulator $\kappa^2 \to 0$.

It is straightforward to do the remaining Feynman parameter integrals to get $F_2(q^2)$ explicitly. We begin with the above equation and we introduce another Feynman parameter by inserting the identity

$$1 = \int\limits_0^1 d\rho\,\delta(\rho - \alpha - \beta)$$

into the integral. Then we scale some of the integration variables as $\alpha, \beta \to \rho\alpha, \rho\beta$ and use the equation $\delta(\rho - \rho\alpha - \rho\beta) = \frac{1}{\rho}\delta(1 - \alpha - \beta)$ to get

$$F_2(q^2) = \frac{e^2}{8\pi^2}\int\limits_0^1 d\alpha\, d\beta\, d\gamma\, \rho\, d\rho\,\delta(1 - \alpha - \beta)\delta(1 - \rho - \gamma)\frac{2m^2\gamma(1 - \gamma)}{m^2(1 - \gamma)^2 + \rho^2 q^2\alpha\beta} + \cdots$$

We then use the delta function to integrate over ρ and we get

$$F_2(q^2) = \frac{e^2}{8\pi^2}\int\limits_0^1 d\alpha\, d\beta\, d\gamma\,\delta(1 - \alpha - \beta)\frac{2m^2\gamma(1 - \gamma)^2}{m^2(1 - \gamma)^2 + (1 - \gamma)^2 q^2\alpha\beta} + \cdots$$

We can then cancel the factors of $(1 - \gamma)$, integrate over γ and use the delta function to integrate over β. We obtain

$$F_2(q^2) = \frac{e^2}{16\pi^2}\int\limits_0^1 d\alpha\,\frac{2m^2}{m^2 + q^2\alpha(1 - \alpha)} + \cdots$$

where what remains is an elementary integral over α which yields

$$F_2(q^2) = \frac{e^2}{16\pi^2}\frac{8m^2}{\sqrt{4m^2q^2 + q^4}}\,\text{arctanh}\,\frac{q^2}{\sqrt{4m^2q^2 + q^4}} \qquad (15.131)$$

This expression has an expansion in small q^2 as

$$F_2(q^2) = \frac{e^2}{16\pi^2}\left[-2 + \frac{q^2}{3m^2} - \frac{1}{15}\frac{q^4}{m^4} + \cdots\right] \qquad (15.132)$$

The interactions of an electron with an electromagnetic field are characterized by the form factors as they enter equation (15.122). The small momentum transfer limit of the electric interaction in equation (15.122) is fixed by our definition of the electric charge. We have chosen the renormalization constants so that $F_1(0) = 1$. This defines the parameter e as the charge of the electron.

On the other hand, the interaction with a magnetic field is given by the second term in equation (15.122). In the limit where the magnetic field is very slowly varying, its coefficient in that term is called its magnetic moment. If the magnetic field is constant we use the $q^2 \to 0$ limit of the form factors and we obtain the magnetic moment of the electron as

$$\mu = \frac{e}{2m}g = \frac{e}{m}\left[F_1(0) + F_2(0)\right] = \frac{e}{m}\left[1 + F_2(0)\right] \qquad (15.133)$$

Here g is called the Landé g-factor. For a free spin $J = \frac{1}{2}$ Dirac particle, $g = 2$. This agrees with the leading order of our result (15.133) when F_2 is ignored. When we include F_2 we obtain a radiative correction to this classical value. Using equation (15.132), we get that

$$\frac{g}{2} = 1 + F_2(0) = 1 + \frac{e^2}{8\pi^2} + \ldots \qquad (15.134)$$

where the ellipses denote corrections which are at least of order e^4. If we plug in the value of the electric charge in natural units we get

$$\frac{g}{2} = 1.001\,161\ldots \qquad (15.135)$$

where higher order corrections are expected (and confirmed by calculation) to affect this result only in the last (sixth) decimal place. This precise prediction turns out to have very good agreement with experiment. In fact, the correction to the Landé g-factor is one of the easier of the quantum electrodynamics computations. For the electron it has been computed up to order e^{10}

$$\frac{g}{2} = 1.001\,159\,652\,181\,643(764)$$

where the numbers in brackets are the uncertainty. This prediction agrees with the experimentally measured value to more than ten significant figures, making the magnetic moment of the electron the most accurately verified prediction of a physical theory.

15.10.2.1 Appendix: Simplification of N^ρ

Let us consider the expression for the matrix element $\bar{\psi}(p)N^\rho\psi'(p')$,

$$\bar{\psi}(p)N^\rho\psi'(p') =$$
$$\bar{\psi}(p)\gamma^\mu\left(i(1-\alpha)\not{p} - i\beta\not{p}' - m\right)\gamma^\rho\left(i(1-\beta)\not{p}' - i\alpha\not{p} - m\right)\gamma_\mu\psi'(p') \quad (15.136)$$

We can use the Dirac equation $(i\not{p}' + m)\psi'(p') = 0$ and $\bar{\psi}(p)(i\not{p} + m) = 0$ and anti-commutator of Dirac matrices $\{\gamma^\mu, \gamma^\nu\} = 2\delta^{\mu\nu}$ to derive the following two equations:

$$\bar{\psi}(p)\gamma^\mu\left(i(1-\alpha)\not{p} - i\beta\not{p}' - m\right) \to 2ip^\mu\bar{\psi}(p) + \bar{\psi}(p)\gamma^\mu\left(-i\alpha\not{p} - i\beta\not{p}'\right)$$

and

$$\left(i(1-\beta)\slashed{p}' - i\alpha\slashed{p} - m\right)\gamma_\mu\psi'(p') \rightarrow 2i\,p'_\mu\psi'(p') + \left(-i\beta\slashed{p}' - i\alpha\slashed{p}\right)\gamma_\mu\psi'(p')$$

whence the matrix element becomes a sum of three simplified terms,

$$\bar{\psi}(p)N^\rho\psi'(p') = -4p\cdot p'\,\bar{\psi}(p)\gamma^\rho\psi'(p')$$
$$+ 2i\left[\bar{\psi}(p)\slashed{p}'\left(-i\beta\slashed{p}' - i\alpha\slashed{p}\right)\gamma^\rho\bar{\psi}'(p') + \bar{\psi}(p)\gamma^\rho\left(-i\beta\slashed{p}' - i\alpha\slashed{p}\right)\slashed{p}\bar{\psi}'(p')\right]$$
$$+ \bar{\psi}(p)\gamma^\mu\left(-i\beta\slashed{p}' - i\alpha\slashed{p}\right)\gamma^\rho\left(-i\beta\slashed{p}' - i\alpha\slashed{p}\right)\gamma_\mu\psi'(p') \tag{15.137}$$

in which we can simplify the second line by remembering that $\slashed{p}\slashed{p} = p^2 = -m^2$ and $\slashed{p}'\slashed{p}' = p'^2 = -m^2$ to get

$$\bar{\psi}(p)N^\rho\psi'(p') = \left[-4p\cdot p' - 2m^2(\alpha+\beta)\right]\bar{\psi}(p)\gamma^\rho\bar{\psi}'(p')$$
$$+ 2\bar{\psi}(p)(\alpha\slashed{p}'\slashed{p}\gamma^\rho + \beta\gamma^\rho\slashed{p}'\slashed{p})\bar{\psi}'(p')$$
$$+ \bar{\psi}(p)\gamma^\mu\left(-i\beta\slashed{p}' - i\alpha\slashed{p}\right)\gamma^\rho\left(-i\beta\slashed{p}' - i\alpha\slashed{p}\right)\gamma_\mu\psi'(p') \tag{15.138}$$

Then, applying the anti-commutator of Dirac matrices and the Dirac equation to simplify the second line leads to

$$\bar{\psi}(p)N^\rho\psi'(p') = \left[-4p\cdot p' - 2m^2(\alpha+\beta) + 4p\cdot p'(\alpha+\beta)\right]\bar{\psi}(p)\gamma^\rho\bar{\psi}'(p')$$
$$- 2im\bar{\psi}(p)(\alpha\slashed{p}'\gamma^\rho + \beta\gamma^\rho\slashed{p})\bar{\psi}'(p')$$
$$+ \bar{\psi}(p)\gamma^\mu\left(-i\beta\slashed{p}' - i\alpha\slashed{p}\right)\gamma^\rho\left(-i\beta\slashed{p}' - i\alpha\slashed{p}\right)\gamma_\mu\psi'(p') \tag{15.139}$$

and we can then further simplify the second line to get

$$\bar{\psi}(p)N^\rho\psi'(p') = \left[-4p\cdot p' - 4m^2(\alpha+\beta) + 4p\cdot p'(\alpha+\beta)\right]\bar{\psi}(p)\gamma^\rho\bar{\psi}'(p')$$
$$- 2im(\alpha+\beta)(p+p')^\rho\bar{\psi}(p)\bar{\psi}'(p')$$
$$+ \bar{\psi}(p)\gamma^\mu\left(-i\beta\slashed{p}' - i\alpha\slashed{p}\right)\gamma^\rho\left(-i\beta\slashed{p}' - i\alpha\slashed{p}\right)\gamma_\mu\psi'(p') \tag{15.140}$$

In the first line we have used $-2p\cdot p' = q^2 + 2m^2$ and in the second line we have symmetrized over α and β. This anticipates that, whenever we use this expression, it will be inside an integration where the rest of the integrand and the integration measure are symmetric under interchange of α and β.

Now, to simplify the third line, we observe that

$$\bar{\psi}(p)\gamma^\mu\left(-i\beta\slashed{p}' - i\alpha\slashed{p}\right)\gamma^\rho\left(-i\beta\slashed{p}' - i\alpha\slashed{p}\right)\gamma_\mu\psi'(p')$$
$$= -2i(\alpha p + \beta p')^\rho\bar{\psi}(p)\gamma^\mu\left(-i\beta\slashed{p}' - i\alpha\slashed{p}\right)\gamma_\mu\psi'(p') \tag{15.141}$$

$$+ (\alpha p + \beta p')^2 \bar{\psi}(p) \gamma^\mu \gamma^\rho \gamma_\mu \psi'(p')$$

$$= 4im(\alpha p + \beta p')^\rho (\alpha + \beta) \bar{\psi}(p) \psi'(p') - 2(\alpha p + \beta p')^2 \bar{\psi}(p) \gamma^\rho \psi'(p')$$

$$= 2im(p + p')^\rho (\alpha + \beta)^2 \bar{\psi}(p) \psi'(p') - 2(\alpha p + \beta p')^2 \bar{\psi}(p) \gamma^\rho \psi'(p') \quad (15.142)$$

$$= 2im(p + p')^\rho (\alpha + \beta)^2 \bar{\psi}(p) \psi'(p')$$

$$+ [2m^2(\alpha + \beta)^2 + 2\alpha\beta q^2] \bar{\psi}(p) \gamma^\rho \psi'(p') \quad (15.143)$$

where, in (15.142), we symmetrized over α and β. Now, we can plug the result (15.141) into (15.140) to get

$$\bar{\psi}(p) N^\rho \psi'(p') =$$

$$\left[2q^2(1 - \alpha)(1 - \beta) + 2m^2[2 - 4(\alpha + \beta) + (\alpha + \beta)^2] \right] \bar{\psi}(p) \gamma^\rho \bar{\psi}'(p')$$

$$- 2im(\alpha + \beta - (\alpha + \beta)^2)(p + p')^\rho \bar{\psi}(p) \bar{\psi}'(p') \quad (15.144)$$

We can use the Gordon identity (15.116) to see that

$$-2im(p + p')^\rho \bar{\psi}(p) \psi'(p') = 4m^2 \bar{\psi}(p) \gamma^\rho \psi'(p') - 4mq_\nu \bar{\psi}(p) \Sigma^{\rho\nu} \psi'(p') \quad (15.145)$$

and, using this expression in the last line of (15.144) we get

$$\bar{\psi}(p) N^\rho \psi'(p') =$$

$$\left[2q^2(1 - \alpha)(1 - \beta) + 2m^2[2 - 2(\alpha + \beta) - (\alpha + \beta)^2] \right] \bar{\psi}(p) \gamma^\rho \bar{\psi}'(p')$$

$$- 4m[\alpha + \beta - (\alpha + \beta)^2] q_\nu \bar{\psi}(p) \Sigma^{\rho\nu} \psi'(p') \quad (15.146)$$

which is our final result.

15.11 Photon Production, the Soft Photon Theorem

In the last section we considered the process where an incoming electron with momentum p is deflected by a classical field $A_\mu^{(\text{cl})}(x)$ and emerges with momentum p'. We computed quantum the amplitude, $< p|i\mathbf{T}|p' >$, for this process to occur, to the order e^3 in perturbation theory in the electric charge e. In this section we will consider a competitor to that process. Whenever an electron is scattered by a classical electromagnetic field, as well as the process which we have studied, there is another very similar process where the deflection of the electron induces the emission of one or more photons. To the order in e that we have been considering, the competitor is the emission of a single photon. In the following we will compute the amplitude $< p|i\mathbf{T}|p', k >$ that, as well as deflection of the electron, a photon with wave-number k is created.

Of course, the final states of the two process are different. The first process has only the outgoing electron whereas the second has the outgoing electron and an outgoing photon. However, observationally, they could be for all intents and purposes identical if the photon that is emitted in the second process has such a long wave-length or small frequency that it is undetectable by any physical apparatus.

We call such a long wavelength, small frequency photon a "soft photon" and soft photon emission is indeed a physical effect which we must learn how to take into account.

Let us consider such a process in the leading order of perturbation theory. The amplitude for the scattering of an electron from a classical electromagnetic field together with the production of a single photon is given by the expression

$$< p|i\mathbf{T}|p'.k >=$$

$$(-ie)^2\bar{\psi}(p)\gamma^\mu \frac{\epsilon^s_\mu(k)}{\sqrt{(2\pi)^3 2k^0}} \frac{i\not{p} - i\not{k} - m}{(p-k)^2 + m^2 - i\epsilon} A^{(cl)}(p'+k-p)\psi'(p')$$

$$+ (-ie)^2\bar{\psi}(p)A^{(cl)}(p'+k-p)\frac{i\not{p}' + i\not{k} - m}{(p'+k)^2 + m^2 - i\epsilon}\gamma^\mu \frac{\epsilon^s_\mu(k)}{\sqrt{(2\pi)^3 2k^0}}\psi'(p')$$

$$+ \ldots \tag{15.147}$$

where the ellipses denote corrections of order e^4 and higher, k_μ is the wave-number and $\epsilon^s_\mu(k)$ is the polarization vector of the photon which is emitted. The single photon wave-function is $\frac{\epsilon^s_\mu(k)}{\sqrt{(2\pi)^3 2k^0}}$ and the electron wave-functions are $\psi(p)$ and $\psi'(p')$. In the first term on the right-hand-side of equation (15.147) the soft photon is emitted by the incoming electron before it scatters from the classical field. In the second term the soft photon is emitted by the outgoing electron.

Here, the final state photon as well as the incoming and outgoing electrons are on-shell, so that $p^2 = -m^2 = p'^2$ and $k^2 = -\kappa^2$ and $k^0 = \sqrt{\vec{k}^2 + \kappa^2}$. We have retained the photon mass κ since we anticipate that the probability for photon production will have an infrared divergence. We are, however, always interested in the limit where $\kappa << m$. Also, we will only consider photons in a narrow range of energies where they are soft, that is where they have wave-vectors $|\vec{k}| < E_{\text{res}}$ with E_{res} a parameter which we call the "detector resolution". We will consider the limit where the photon mass is much smaller than the detector resolution and the detector resolution is much smaller than other quantities with the dimension of momentum, mass or inverse length, like the electron mass,

$$\kappa << E_{\text{res}} << m \tag{15.148}$$

We can use the mass shell conditions as well as the Dirac equations for the spinors

$$(i\not{p}' + m)\psi(p') = 0, \quad \bar{\psi}(p)(i\not{p} + m) = 0$$

to simplify equation (15.147),

$$
< p|i\mathbf{T}|p', k >=
$$
$$
\frac{\epsilon_\mu^s(k)}{\sqrt{(2\pi)^3 2k^0}}\left[\frac{2ep^\mu}{-2p\cdot k - \kappa^2} + \frac{2ep'^\mu}{2p'\cdot k - \kappa^2}\right]\left[-ie\bar{\psi}(p)\slashed{A}^{(cl)}(p'-p)\psi(p')\right]
$$

$$(15.149)$$

Here, we have dropped a sub-leading term of order k from each of the numerators. We have assumed that $A_\rho^{(cl)}(p'+k-p)$ is a sufficiently smooth function of momenta and k is sufficiently small that $A_\rho^{(cl)}(p'+k-p)$ can be replaced by $A_\rho^{(cl)}(p'-p)$. We note that the denominators $-2p\cdot k - \kappa^2$ and $2p'\cdot k - \kappa^2$ cannot be zero when the momenta are on-shell and when $\kappa < 2m$. We have therefore dropped the $i\epsilon$'s.

Notice the universal form of the result in equation (15.149). The right-hand-side is equal to the first factor, which accounts for soft photon emission, times the matrix element of the vertex function where no photon has been emitted. The soft photon factor does not depend on the helicity or any other details of the state of the electron beyond its momentum. This formula is called the "soft photon theorem".

The probability of soft photo-emission is gotten by squaring the amplitude for soft photo-emission and then summing over the polarizations and integrating over the wave-numbers of the soft photon final states. We obtain

$$
\int\limits_{\text{soft } k} \mathbf{P}(p \to p' + k) = \int\limits^{E_{\text{res}}} d^3k| < p|i\mathbf{T}|p', k > |^2
$$

$$
= \left|-ie\bar{\psi}(p)\gamma^\rho\psi(p')A_\rho^{(cl)}(p'-p)\right|^2 \int\limits^{E_{\text{res}}} \frac{d^3k}{(2\pi)^3 2\sqrt{\vec{k}^2 + \kappa^2}}\sum_s \epsilon_\mu^s(k)\epsilon_\nu^{*s}(k) \times
$$

$$
\times \left[\frac{2ep^\mu}{-2p\cdot k - \kappa^2} + \frac{2ep'^\mu}{2p'\cdot k - \kappa^2}\right]\left[\frac{2ep^\nu}{-2p\cdot k - \kappa^2} + \frac{2ep'^\nu}{2p'\cdot k - \kappa^2}\right] \quad (15.150)
$$

The sum over polarizations for a massive photon is

$$
\sum_s \epsilon_\mu^s(k)\epsilon_\nu^{*s}(k) = \eta_{\mu\nu} + \frac{k_\mu k_\nu}{\kappa^2} \quad (15.151)
$$

and equation (15.150) becomes

$$
\int\limits_{\text{soft } k} \mathbf{P}(p \to p' + k) = \left|-ie\bar{\psi}(p)\gamma^\rho\psi(p')A_\rho^{(cl)}(p'-p)\right|^2 \times
$$

$$
\times \frac{e^2}{(2\pi)^3}\int\limits^{E_{\text{res}}} \frac{d^3k}{2\sqrt{\vec{k}^2 + \kappa^2}}\left[\frac{-8p\cdot p'}{[2p\cdot k + \kappa^2][2p'\cdot k - \kappa^2]}\right.
$$

$$
\left. + \frac{4p\cdot p}{[2p\cdot k + \kappa^2][2p\cdot k - \kappa^2]} + \frac{4p'\cdot p'}{[2p'\cdot k + \kappa^2][2p'\cdot k - \kappa^2]}\right] \quad (15.152)
$$

To simplify the integrals, we can use Feynman parameters so that, for example, the first term on the right-hand side becomes

$$\frac{e^2}{(2\pi)^3} \int^{E_{\text{res}}} \frac{d^3k}{2\sqrt{\vec{k}^2 + \kappa^2}} \frac{-8p \cdot p'}{[2p \cdot k + \kappa^2][2p' \cdot k - \kappa^2]}$$

$$= \frac{e^2}{(2\pi)^3} \int_0^1 d\alpha \int^{E_{\text{res}}} \frac{d^3k}{2\sqrt{\vec{k}^2 + \kappa^2}} \frac{-8p \cdot p'}{[\alpha[2p \cdot k + \kappa^2] + (1 - \alpha)[2p' \cdot k - \kappa^2]]^2}$$

$$= \frac{e^2}{(2\pi)^3} \int_0^1 d\alpha \int^{E_{\text{res}}} \frac{d^3k}{2\sqrt{\vec{k}^2 + \kappa^2}} \frac{-2p \cdot p'}{[[\alpha p + (1 - \alpha)p'] \cdot k + (\alpha - 1/2)\kappa^2]^2}$$

Now, when we take the limit $\kappa \to 0$, $E_{\text{res}} \to 0$ $E_{\text{res}}/\kappa \to \infty$ the integration over the magnitude of $\vec{k}$ yields

$$= \frac{e^2}{4\pi^2} \ln \frac{E_{\text{res}}}{\kappa} \int_0^1 d\alpha \int_{-1}^1 \frac{d \cos \theta (-p \cdot p')}{[[\alpha p + (1 - \alpha)p']^0 - |\alpha p + (1 - \alpha)p'| \cos \theta]^2}$$

$$= -\frac{e^2}{4\pi^2} \ln \frac{E_{\text{res}}}{\kappa} \int_0^1 d\alpha \frac{2m^2 + q^2}{m^2 + \alpha(1 - \alpha)q^2}$$

We have found that the integral depends only on q^2. This is due to Lorentz invariance and we could have deduced this at the start. The remaining terms are thus easy to include, they are just the same integral evaluated at $q^2 = 0$. The net result for the transition probability is

$$\int_{\text{soft } k} \mathbf{P}(p \to p' + k) = \left| -ie\bar{\psi}(p)\gamma^\rho\psi(p')A_\rho^{(\text{cl})}(p' - p) \right|^2 \times$$

$$\times \frac{e^2}{4\pi^2} \ln \frac{E_{\text{res}}}{\kappa} \int_0^1 d\alpha \left[\frac{2m^2 + q^2}{m^2 + \alpha(1 - \alpha)q^2} - 2 \right] \qquad (15.153)$$

This transition probability is infrared divergent in that it is $\sim \ln \kappa$ in the limit as $\kappa \to 0$. It should be compared with the expression that we found for a different process in equation (15.130). We see that, when we add the two together, the result contains the term

$$\mathbf{P}(p \to p') + \int\limits_{\text{soft } k} \mathbf{P}(p \to p' + k) = \ldots +$$

$$+ \frac{e^2}{4\pi^2} \ln \frac{\kappa}{m} \int\limits_0^1 d\alpha \left[\frac{2m^2 + q^2}{m^2 + q^2\alpha(1-\alpha)} - 2 \right] \left| -ie\bar{\psi}(p)\mathcal{A}^{(\text{cl})}(q)\psi'(p') \right|^2$$

$$\frac{e^2}{4\pi^2} \ln \frac{E_{\text{res}}}{\kappa} \int\limits_0^1 d\alpha \left[\frac{2m^2 + q^2}{m^2 + \alpha(1-\alpha)q^2} - 2 \right] \left| -ie\bar{\psi}(p)\mathcal{A}^{(\text{cl})}(q)\psi'(p') \right|^2$$

$$+ \mathcal{O}(e^6) \tag{15.154}$$

We see that the logarithm of the photon mass, κ, cancels in the sum so that the result is

$$\mathbf{P}(p \to p') + \int\limits_{\text{soft } k} \mathbf{P}(p \to p' + k) = \ldots +$$

$$+ \frac{e^2}{4\pi^2} \ln \frac{E_{\text{res}}}{m} \int\limits_0^1 d\alpha \left[\frac{2m^2 + q^2}{m^2 + q^2\alpha(1-\alpha)} - 2 \right] \left| -ie\bar{\psi}(p)\mathcal{A}^{(\text{cl})}(q)\psi'(p') \right|^2$$

$$+ \mathcal{O}(e^6) \tag{15.155}$$

It is now finite in the limit as κ goes to zero. What we obtain is the same result that we would have had if we ignored the soft photon emission contribution and used E_{res} as an infrared cutoff when we computed the form factor $F_1(q^2)$. It leaves one extra finite parameter, E_{res}.

It is indeed possible to argue that infrared divergences cancel in this way in all higher orders of perturbation theory. The details are beyond the spope of what we can discuss in this introductory monograph. Also, there are simple questions that one can ask about transition amplitudes where the infrared divergences do not cancel. One simple example occurs when, in the above discussion, the incoming electron is in a superposition of momentum states, $\frac{1}{\sqrt{2}}(|p_1 > + |p_2 >)$. If, in this case, by generalizing the arguments of this and the previous section, it is possible to see that infrared divergences do not cancel from the interference terms in the probability, those which go like

$$< p_1 |i\mathbf{T}| p' > < p_2 |i\mathbf{T}| p' >^* + < p_2 |i\mathbf{T}| p' > < p_1 |i\mathbf{T}| p' >^* +$$

$$+ \int\limits_{\text{soft} k} \left[< p_1 |i\mathbf{T}| p'k > < p_2 |i\mathbf{T}| p'k >^* + < p_2 |i\mathbf{T}| p'k > < p_1 |i\mathbf{T}| p'k >^* \right]$$

The implications of this non-cancellation are not completely understood, though the most conservative view simply regards this as a limitation of the application of the

S matrix formalism to a theory with long ranged interactions, where it is not strictly valid.

15.12 Furry's Theorem

Furry's theorem states that an electron loop emitting an odd number of photons must vanish. It is useful since it allows us to eliminate some possibilities for Feynman diagrams. It applies to an electron loop if it is the entire Feynman diagram or if is a sub-diagram of a more complicated Feynman diagram. The vanishing of the tadpole, which is an electron loop emitting one photon line is a simple example.

To prove this theorem, we consider a single closed electron loop emitting n photons, which contributes the Feynman integral

$$\int \frac{dp}{(2\pi)^4} \text{Tr}\left[g(p)\gamma^{\lambda_1} g(p+q_1)\gamma^{\lambda_2} g(p+q_1+q_2)\ldots g(p+q_1+\ldots+q_{n-1})\gamma^{\lambda_n} \right]$$

(15.156)

(with an additional factor of $(-ie)^n$ which we omit for brevity). This Feynman integral could be the contribution of a sub-diagram to a more complicated Feynman integral describing a high order perturbative contribution to some correlation function.

To prove the theorem, we will need to use a discrete symmetry of the full theory called charge conjugation invariance. However, we shall not need the full machinery that implements this symmetry. It suffices to note that there exists a unitary matrix $\mathcal{C}$ with the following properties

$$\mathcal{C}^\dagger \mathcal{C} = 1 = \mathcal{C}\mathcal{C}^\dagger, \quad \mathcal{C}^\dagger \gamma^\mu \mathcal{C} = (\gamma^\mu)^* = (\gamma^\mu)^{\dagger t} = (\gamma^0 \gamma^\mu \gamma^0)^t$$

The superscript t stands for transpose. Here, we have remembered that $\gamma^{0\dagger} = -\gamma^0$ and $\gamma^{a\dagger} = \gamma^a$. These properties do not depend on the specific choice of the Dirac matrices, only on the fact that they satisfy the correct anti-commutator algebra. However, the specific form of $\mathcal{C}$ depends on the choice of Dirac matrices. With the choice that we made in our previous chapters, γ^2 is purely imaginary, whereas the other matrices are purely real. In that case, we could choose $\mathcal{C} = \gamma^2 \gamma^5$ where $\gamma^5 = i\gamma^0 \gamma^1 \gamma^2 \gamma^3$ is the chirality matrix with the properties $\gamma^5 = \gamma^{5\dagger}$ and $\gamma^5 \gamma^\mu \gamma^5 = -\gamma^\mu$. We can easily see that, in that case,

$$\mathcal{C}^\dagger \gamma^\mu \mathcal{C} = (\gamma^2 \gamma^5)^\dagger \gamma^\mu (\gamma^2 \gamma^5) = \gamma^5 \gamma^2 \gamma^\mu \gamma^2 \gamma^5 = -\gamma^5 \gamma^{\mu*} \gamma^5 = \gamma^{\mu*} = (\gamma^0 \gamma^\mu \gamma^0)^t$$

which is what is needed.

If we conjugate the momentum space propagator of the electron using $\mathcal{C}$, we find that

$$\mathcal{C}^\dagger g(k)\mathcal{C} = -(\gamma^0 g(-k)\gamma^0)^t$$

Then, inserting $1 = CC^\dagger$ between each matrix product in the trace in the integrand in equation (15.156) yields

$$
\int \frac{dp}{(2\pi)^4} \mathrm{Tr}\left[g(p)\gamma^{\lambda_1} g(p+q_1)\gamma^{\lambda_2} g(p+q_1+q_2)\dots g(p+q_1+\dots+q_{n-1})\gamma^{\lambda_n} \right]
$$
$$
= \int \frac{dp}{(2\pi)^4} \mathrm{Tr}\left[CC^\dagger g(p) CC^\dagger \gamma^{\lambda_1} C \dots C^\dagger g(p+q_1+\dots+q_{n-1}) CC^\dagger \gamma^{\lambda_n} CC^\dagger \right]
$$
$$
= \int \frac{dp}{(2\pi)^4} \mathrm{Tr}\left[(-\gamma^0 g(-p)\gamma^0)^t (\gamma^0 \gamma^{\lambda_1} \gamma^0)^t \dots \right.
$$
$$
\left. \dots (-\gamma^0 g(-p-q_1-\dots-q_{n-1})\gamma^0)^t (\gamma^0 \gamma^{\lambda_n} \gamma^0)^t \right]
$$
$$
= (-1)^n \int \frac{dp}{(2\pi)^4} \mathrm{Tr}\left[(\gamma^0 \gamma^{\lambda_n} \gamma^0)(\gamma^0 g(-p-q_1-\dots-q_{n-1})\gamma^0) \dots \right.
$$
$$
\left. \dots (\gamma^0 \gamma^{\lambda_1} \gamma^0)(\gamma^0 g(-p)\gamma^0) \right]
$$

where we have observed that the product of the transposes of matrices is equal to the transpose of the product in the reverse order and that the trace of the transpose of a matrix is equal to the trace of the matrix. We have also taken out the factor of $(-1)^n$ which are the minus signs in front of each of the g's. Now we observe that $\mathrm{Tr}[\gamma^0 \dots \gamma^0] = \mathrm{Tr}[\dots \gamma^0 \gamma^0]$, that $\gamma^0 \gamma^0 = -1$ and there there are $2n$ such products, so no overall sign change. The above expression becomes

$$
(-1)^n \int \frac{dp}{(2\pi)^4} \mathrm{Tr}\left[\gamma^{\lambda_n} g(-p-q_1-\dots-q_{n-1}) \dots \gamma^{\lambda_1} g(-p) \right]
$$

Now, changing the integration variable, $p \to -p$ recovers the expression

$$
(-1)^n \int \frac{dp}{(2\pi)^4} \mathrm{Tr}\left[\gamma^{\lambda_n} g(p-q_1-\dots-q_{n-1}) \dots \gamma^{\lambda_1} g(p) \right]
$$

This final expression is the original loop but with reversed orientation times a minus sign when n is odd and a plus sign when n is even. In applying the Feynman rules to the computation of any particular correlation function, when the electron loop is present, both the original forward and this reverse orientations of the loop will occur in the application of Wick's theorem which leads to the Feynman diagrams.[2] The contributions of the two orientations must therefore cancel when n is odd. This completes the proof of Furry's theorem.

The utility of Furry's theorem is that when we are doing a perturbative computation of a correlation function, we can drop all closed electron loops which contain an odd number of photon vertices.

---

[2] In fact, all permutations of the set $\{(\lambda_1, q_1), \dots, (\lambda_n, q_n)\}$ will occur. This imposes Bose symmetry on the photons that are emanating from the loop subgraph.

15.13   The Ward–Takahashi Identities

The Ward–Takahashi identities are a manifestation in the gauge fixed quantum field theory of the original gauge symmetry of electrodynamics and of the phase symmetry of the electron theory which survives even after the gauge fixing is done. The symmetry is manifest in some identities that the correlation functions must obey. These are called the Ward–Takahashi identities.

The Ward–Takahashi identities are exact statements about correlation functions. Therefore they must hold order by order in perturbation theory. They can be very useful in simplifying some perturbative computations, or they can be used as a check on the consistency of such computations. They of course depend on the survival of the symmetries in question in the quantized field theory. In quantum electrodynamics, there is a gauge and phase invariant regularization available, dimensional regularization (amongst others), and we employ such an invariant regularization, we expect that the Ward–Takahashi identities are valid.

To find the Ward–Takahashi identities for quantum electrodynamics, we begin with the gauge-fixed action written as,

$$S = \int dx \left\{ -\frac{Z_3}{4} F_{\mu\nu}(x) F^{\mu\nu}(x) - \frac{\xi}{2} Z_\xi (\partial_\mu A^\mu(x))^2 - \frac{\kappa^2 Z_\kappa}{2} A_\mu(x) A^\mu(x) \right.$$

$$\tag{15.157}$$

$$\left. -i Z_2 \bar{\psi}(x) \partial\!\!\!/ \psi(x) - im Z_m \bar{\psi}(x)\psi(x) - Z_1 e \bar{\psi}(x) A\!\!\!/(x)\psi(x) \right\} \tag{15.158}$$

Under an infinitesimal gauge transformation where

$$\delta A_\mu(x) = \frac{Z_2}{Z_1} \partial_\mu \chi(x), \quad \delta\psi(x) = ie\chi(x)\psi(x), \quad \delta\bar{\psi}(x) = -ie\chi(x)\bar{\psi}(x) \tag{15.159}$$

the action plus the source terms varies as

$$\delta \left\{ S + \int dx \left[A_\mu(x) J^\mu(x) + \bar{\eta}(x)\psi(x) + \bar{\psi}(x)\eta(x) \right] \right\} =$$
$$\int dx \chi(x) \left[-\frac{Z_2}{Z_1} (\xi Z_\xi \partial^2 - \kappa^2 Z_\kappa) \partial_\nu A^\nu(x) - \frac{Z_2}{Z_1} \partial_\nu J^\nu(x) \right.$$
$$\left. + ie\bar{\eta}(x)\psi(x) - ie\bar{\psi}(x)\eta(x) \right]$$

The measure in the functional integral should be invariant under this change of variables. In a practical sense, for perturbation theory, what we need is that we have a Lorentz invariant cutoff and renormalization scheme. The dimensional regularization that we have used in our explicit computations is one such scheme. Let us assume that this is so. Then, upon doing this infinitesimal gauge transformation to the integration variables in the functional integral, and noting that the result of the functional integral cannot depend on the function $\chi(x)$ that appears in the change of variables, we obtain an identity that the generating functional must obey

$$\left[(\xi Z_\xi \partial^2 - \kappa^2 Z_\kappa)\partial_\mu \frac{1}{i}\frac{\delta}{\delta J^\mu(x)} + \partial_\nu J^\nu(x)\right. \tag{15.160}$$

$$\left. - e\frac{Z_1}{Z_2}\left(\bar{\eta}(x)\frac{\delta}{\delta\bar{\eta}(x)} - \eta(x)\frac{\delta}{\delta\eta(x)}\right)\right]Z[A, \eta, \bar{\eta}] = 0 \tag{15.161}$$

Equation (15.161) is more useful as an identity for the generating functional for connected correlation functions, which we obtain by dividing it by $Z[A, \eta, \bar{\eta}]$ to get

$$0 = \partial_\nu J^\nu(x) + \left[(-\xi Z_\xi \partial^2 + \kappa^2 Z_\kappa)\partial_\mu \frac{1}{i}\frac{\delta}{\delta J^\mu(x)}\right.$$

$$\left. + e\frac{Z_1}{Z_2}\left(\bar{\eta}(x)\frac{\delta}{\delta\bar{\eta}(x)} - \eta(x)\frac{\delta}{\delta\eta(x)}\right)\right]W[A, \eta, \bar{\eta}] \tag{15.162}$$

$$W[A, \eta, \bar{\eta}] = \ln Z[A, \eta, \bar{\eta}]$$

This is our master formula for the Ward–Takahashi identity for quantum electrodynamics. Taking functional derivatives of it yields an infinite hierarchy of Ward–Takahashi identities for the connected correlation functions.

The Ward–Takahashi identities have some interesting implications. For example, upon taking n functional derivatives of equation (15.162) by J^μ and then setting J, η and $\bar{\eta}$ to zero, we find,

$$-\partial^2 \frac{\partial}{\partial x^\nu} < \mathcal{O}|T A^\nu(x)A_{\lambda_1}(y_1)\ldots A_{\lambda_n}(y_n)|\mathcal{O} >_c= 0, \quad n \ge 2$$

which, together with symmetry of the time ordered product, tells us that the connected correlation function of $r \ge 4$ photons must obey

$$k_1^{\mu_1}\Gamma_{C\mu_1\ldots\mu_r}(k_1, k_2, \ldots, k_r) = 0, \quad r \ge 3 \tag{15.163}$$

Another interesting expression can be obtained by taking one functional derivative, $\frac{1}{i}\frac{\delta}{\delta J^\lambda(y)}$, of equation (15.162) and putting the sources to zero. In this way we obtain an identity for the photon 2 point function,

$$(\xi Z_\xi \partial^2 - \kappa^2 Z_\kappa)\partial_\nu < \mathcal{O}|T A^\nu(x)A_\lambda(y)|\mathcal{O} >= i\partial_\lambda\delta(x - y) \tag{15.164}$$

which, upon taking the Fourier transforms is

$$-i(\xi Z_\xi k^2 + \kappa^2 Z_\kappa)k_\nu D^\nu{}_\lambda(k) = k_\lambda \tag{15.165}$$

Multiplying from the right by the inverse of D yields

$$k_\lambda[D^{-1}(k)]^{\lambda\nu}(k) = -i(\xi Z_\xi k^2 + \kappa^2 Z_\kappa)k^\nu \tag{15.166}$$

Epilogue

16

We have come to the end of our brief introduction to quantum field theory. What I hope that I have presented here is a simple and logical gateway to the subject and a beginners introduction to the language as well as the techniques of the subject. I have made no attempt to be complete or to give credit in the way of citations of original research papers. I have also for the most part resisted introducing material which would not have fit into the one semester course on introductory quantum field theory from whose notes this monograph originated. This has left many interesting and immediately accessible directions wide open and my best hope is that the student who has read through this introduction is now well prepared to take on a deeper and more comprehensive study of the subject.

© The Author(s), under exclusive license to Springer Nature Singapore Pte Ltd. 2023
G. W. Semenoff, *Quantum Field Theory*, Graduate Texts in Physics,
https://doi.org/10.1007/978-981-99-5410-0_16

Index

A
Action, 50, 51
Analyticity, 123
Angular momentum, 82
Annihilation operator, 16
Anomalous magnetic moment, 384
Anyons, 8

B
Bare coupling, 264
Bare field, 264
Bare mass, 262, 264
Bogoliubov transformation, 44
Bose gas, 35
Bose gas, pressure, 45
Bose gas, quasiparticle, 44
Bose–Einstein condensate, 35
Bose–Einstein statistics, 8

C
Canonical momenta, 55
Charge conjugation, 209
Charge conservation, 58
Charge density, 58
Chemical potential, 33
Conformal algebra, 134
Conformal transformations, 134
Connected correlation function, 274–320
Conservation, charge, 58
Conservation law, 57
Contact interaction, 9
Continuity equation, 57
Coordinate transformations, 93
Coordinates, 92

Correlation function, connected, 274–320
Correlation function, irreducible, 274–325
Correlation functions, 111
Coulomb interaction, 9
Counterterm, 262, 263, 340
CPT, 210
Creation operator, 16
Critical velocity, 40
Cross section, 318
Current density, 58
Cutoff, 261

D
Debye function, 145
Debye theory of solids, 143
Degenerate Bose gas, 35
Degenerate Fermi gas, 28
Density, 32
Density operator, 17
Differential cross section, 318
Dimensional regularization, 281, 305
Dimensional transmutation, 297
Dimension of operator, 264
Dirac equation, 157
Dirac equation, Lorentz invariance, 165
Dirac equation, solution of, 163
Dirac Hamiltonian, 159
Dirac matrices, 159

E
Einstein summation convention, 93
Elastic scattering, 317
Electrodynamics, 207
Electromagnetic form factors, 377

© The Editor(s) (if applicable) and The Author(s), under exclusive license to Springer
Nature Singapore Pte Ltd. 2023
T. Shimobaba and T. Ito (eds.), *Hardware Acceleration of Computational Holography*,
https://doi.org/10.1007/978-981-99-1938-3

Emergent symmetry, 137
Empty vacuum, 16, 25
Energy, free, 33
Energy, internal, 32
Energy-momentum tensor, 66–68
Equal-time anti-commutations relations, 21
Equal-time commutation relations, 21
Equivalence relation, 195
Euclidean field theory, 286
Euler-Lagrange equations, 52, 55, 107
Euler's gamma function, 301

F
Faddeev-Popov substitution, 234
Fermi-Dirac statistics, 8
Fermi energy, 29
Fermi gas, 28
Fermi gas, pressure, 35
Fermi momentum, 29
Fermi surface, 29
Fermi wave-number, 29
Feynman diagrams, 270, 277
Feynman gauge, 236
Feynman parameters, 282, 303, 375
Feynman rules, 276–277
Feynman slash, 160
Fine structure constant, 182
Form factor, 386
Free energy, 33
Free energy, degenerate Bose gas, 43
Free energy, grand canonical, 34
Fried-Yennie gauge, 236
Functional derivative, 212
Functional integral, 215, 252
Functional integral, Gaussian, 216
Functional methods for fermions, 236–242
Furry's theorem, 349–353, 359, 393

G
Galilean symmetry, 38, 68–73
Galilei algebra, 83
Gauge fixing, relativistic, 185, 235
Gauge invariance, 185, 253
Gaussian functional integral, 216
General coordinate transformations , 93
Generating functional, 218, 250, 255, 319
Generating functional, equation for, 220
Goldstone boson, 47
Gordon identity, 379, 388
Grand canonical free energy, 34
Graphene, 147
Ground state, 28

H
Hamiltonian density, 57
Heisenberg picture, 20
Helicity, 162
Hermitian conjugate, 15
Holes, 30

I
Identical particles, 7
Improved stress tensor, 76, 133
Indistinguishable particles, 7
Infrared divergence, 374, 382, 392
Internal energy, 32
Irreducible correlation function, 274–325

K
Killing equation, 98–99
Killing vector, 98–99

L
Lagrangian density, 50
Lamb shift, 368
Landau criterion, 38
Landau gauge, 236
Landau pole, 298, 370
Linked cluster theorem, 272, 320, 351
Local operator, 264
Lorentz transformation, 99
LSZ formula, 311

M
Mandelstam variables, 290
Many particle Schrödinger equation, 7
Massive photon, 203
Mass tadpole, 285
Maxwell's equations, 182–184
Metric tensor, 96
Minkowski space, 98
MSbar scheme, 289–291
MS scheme, 289

N
Natural units, 102
Negative normed states, 190
Noether's theorem, 57, 61–62, 109, 171
Noether's theorem, alternative proof, 172
N-point function, 111
Null states, 194–195
Number operator, 19

O
One loop approximation, 277
On-shell subtraction, 288–289

Operator ordering, 50
Orthochronous Lorentz transformation, 101

P
Parity, 209
Particle number, 63
Particles and holes, 30
Permutations, 7
Phase symmetry, 59, 63
Phonon, 138
Photon, 191–198, 230
Photon mass, 203
Photon production, 388
Physical state condition, 190–191
Poincare algebra, 111
Poincare transformation , 99
Poisson bracket, 55
Pressure, 35
Primary operator, 135
Propagator, 114
Proper time, 97

Q
Quantum electrodynamics, 207–209, 253, 339

R
Reeh–Schlieder theorem, 126
Regularization, dimensional, 281
Relativistic field, 103
Relativistic quantum field theory, 92
Renormalization group, 292
Renormalization group alpha, beta, gamma functions, 294
Renormalization group equation, 294
Renormalized coupling, 264
Renormalized field, 264
Renormalized mass, 264
Rotation symmetry, 82

S
Scalar field, 95
Scalar field propagator, 114
Scalar field theory, 105
Scale invariance, 73–76
Scheme, subtraction, 288
Schrödinger algebra, 83
Schrödigner equation, 6, 18
Schwinger parameters, 303
Second quantization, 5

Second quantized Hamiltonian, 18
Self-energy, 326, 330, 363, 370
Slater determinant, 13
S matrix, 307
Soft photon theorem, 388
Space-time symmetry, 98
Special Schrödinger symmetry, 79–80
Spectral theorem, 120–122
Spin, 13
Spin dependent interaction, 14
Spontaneous symmetry breaking, 46
Stationary state, 11
Stress tensor, 66–68, 109, 132, 174, 201–202
Stress tensor, improved, 76, 133
Subtraction scheme, 288
Subtraction scheme, MS, 288–289
Subtraction scheme, MSbar, 288–289
Subtraction scheme, on-shell, 288–289
Summation convention, 93
Superfluid, 38
Superfluid, critical velocity of, 40
Symmetry, 59
Symmetry breaking, 46
Symmetry, emergent, 137
Symmetry of Minkowski space, 99–102

T
Tadpole, mass, 285
Tensor field, 96
Time-ordered function, 112
Time reversal, 209
T matrix, 310
Translation invariance, 60
Tree approximation, 277

U
Uehling term, 368

V
Vacuum, 16, 29, 164–165, 189
Vacuum diagrams, 325
Vacuum expectation value, 40
Vector field, 95
Vertex correction, 373

W
Ward–Takahashi identities, 341, 374, 395
Wick rotation, 281, 284
Wick's theorem, 222
Wightman function, 111

Printed in the United States
by Baker & Taylor Publisher Services